T0140102

Lecture Notes on Data Engineering and Communications Technologies

193

Series Editor

Fatos Xhafa, *Technical University of Catalonia, Barcelona, Spain*

The aim of the book series is to present cutting edge engineering approaches to data technologies and communications. It will publish latest advances on the engineering task of building and deploying distributed, scalable and reliable data infrastructures and communication systems.

The series will have a prominent applied focus on data technologies and communications with aim to promote the bridging from fundamental research on data science and networking to data engineering and communications that lead to industry products, business knowledge and standardisation.

Indexed by SCOPUS, INSPEC, EI Compendex.

All books published in the series are submitted for consideration in Web of Science.

Leonard Barolli
Editor

Advances in Internet, Data & Web Technologies

The 12th International Conference on Emerging Internet, Data & Web Technologies (EIDWT-2024)

 Springer

Editor
Leonard Barolli
Faculty of Information Engineering
Fukuoka Institute of Technology
Fukuoka, Japan

ISSN 2367-4512   ISSN 2367-4520 (electronic)
Lecture Notes on Data Engineering and Communications Technologies
ISBN 978-3-031-53554-3   ISBN 978-3-031-53555-0 (eBook)
https://doi.org/10.1007/978-3-031-53555-0

This Springer imprint is published by the registered company Springer Nature Switzerland AG
The registered company address is: Gewerbestrasse 11, 6330 Cham, Switzerland

Paper in this product is recyclable.

Welcome Message of EIDWT-2024 International Conference Organizers

Welcome to the 12th International Conference on Emerging Internet, Data and Web Technologies (EIDWT-2024), which will be held at the University of Naples "Federico II," Naples, Italy, from February 21 to February 23, 2024.

The EIDWT is dedicated to the dissemination of original contributions that are related to the theories, practices, and concepts of emerging Internet and data technologies yet most importantly their applicability in business and academia toward a collective intelligence approach.

In EIDWT-2024 will be discussed topics related to data modeling, data visualization, data mining, databases, knowledge discovery, semantics, cloud computing, IoT, social networks, security issues, and other web implementations toward a collaborative and collective intelligence approach leading to advancements of virtual organizations and their user communities. This is because web implementations will store and continuously produce a vast amount of data, which if combined and analyzed through a collective intelligence manner will make a difference in the organizational settings and their user communities. Thus, the scope of EIDWT-2024 includes methods and practices which bring various emerging Internet and data technologies together to capture, integrate, analyze, mine, annotate, and visualize data in a meaningful and collaborative manner. Finally, EIDWT-2024 aims to provide a forum for original discussion and prompt future directions in the area.

An international conference requires the support and help of many people. A lot of people have helped and worked hard for a successful EIDWT-2024 technical program and conference proceedings. First, we would like to thank all authors for submitting their papers. We are indebted to Program Area chairs, Program Committee members, and reviewers who carried out the most difficult work of carefully evaluating the submitted papers. We would like to give our special thanks to the Honorary Chair of EIDWT-2024 Prof. Makoto Takizawa, Hosei University, Japan, for his guidance and support. We would like to express our appreciation to our keynote speakers for accepting our invitation and delivering very interesting keynotes at the conference.

EIDWT-2024 Organizing Committee

Honorary Chair

Makoto Takizawa   Hosei University, Japan

General Co-chairs

Flora Amato	University of Naples "Federico II," Italy
Juggapong Natwichai	Chiang Mai University, Thailand
Tomoya Enokido	Rissho University, Japan

Program Co-chairs

Francesco Moscato	University of Salerno, Italy
Tomoyuki Ishida	Fukuoka Institute of Technology, Japan
Admir Barolli	"Aleksander Moisiu" University of Durres, Albania

International Advisory Committee

Arjan Durresi	IUPUI, USA
Beniamino Di Martino	University of Campania "Luigi Vanvitelli," Italy
Wenny Rahayu	La Trobe University, Australia
Fang-Yie Leu	Tunghai University, Taiwan
Yoshihiro Okada	Kyushu University, Japan

Publicity Co-chairs

Salvatore Venticinque	University of Campania "Luigi Vanvitelli," Italy
Vincenzo Moscato	University of Naples "Federico II," Italy
Keita Matsuo	Fukuoka Institute of Technology, Japan
Pruet Boonma	Chiang Mai University, Thailand

International Liaison Co-chairs

Walter Balzano	University of Naples "Federico II," Italy
David Taniar	Monash University, Australia
Marek Ogiela	AGH University of Science and Technology, Poland
Tetsuya Oda	Okayama University of Science, Japan

Local Organizing Committee Co-chairs

Antonino Ferraro	University of Naples "Federico II," Italy
Giovanni Cozzolino	University of Naples "Federico II," Italy
Mattia Fonisto	University of Naples "Federico II," Italy
Silvia Stranieri	University of Naples "Federico II," Italy

Web Administrators

Phudit Ampririt	Fukuoka Institute of Technology, Japan
Shunya Higashi	Fukuoka Institute of Technology, Japan
Ermioni Qafzezi	Fukuoka Institute of Technology, Japan

Finance Chair

Makoto Ikeda	Fukuoka Institute of Technology, Japan

Steering Committee Chair

Leonard Barolli	Fukuoka Institute of Technology, Japan

PC Members

Akimitsu Kanzaki	Shimane University, Japan
Alba Amato	University of Campania "Luigi Vanvitelli," Italy
Alberto Scionti	LINKS, Italy
Antonella Di Stefano	University of Catania, Italy
Bhed Bista	Iwate Prefectural University, Japan

Carmen de Maio	University of Salerno, Italy
Dana Petcu	West University of Timisoara, Romania
Danda B. Rawat	Howard University, USA
Elinda Mece	Polytechnic University of Tirana, Albania
Eric Pardede	La Trobe University, Australia
Evjola Spaho	Polytechnic University of Tirana, Albania
Fabrizio Marozzo	University of Calabria, Italy
Fabrizio Messina	University of Catania, Italy
Farookh Hussain	University of Technology Sydney, Australia
Francesco Orciuoli	University of Salerno, Italy
Francesco Palmieri	University of Salerno, Italy
Gen Kitagata	Tohoku University, Japan
Giovanni Masala	Plymouth University, UK
Giovanni Morana	C3DNA, USA
Giuseppe Caragnano	LINKS, Italy
Giuseppe Fenza	University of Salerno, Italy
Harold Castro	Universidad de Los Andes, Bogotá, Colombia
Hiroaki Kikuchi	Meiji University, Japan
Hiroaki Yamamoto	Shinshu University, Japan
Isaac Woungang	Toronto Metropolitan University, Canada
Jiahong Wang	Iwate Prefectural University, Japan
Jindan Zhang	Xianyang Vocational Technical College, China
Jugappong Natwichai	Chiang Mai University, Thailand
Kenzi Watanabe	Hiroshima University, Japan
Kiyoshi Ueda	Nihon University, Japan
Lidia Fotia	Università Mediterranea di Reggio Calabria (DIIES), Italy
Lucian Prodan	Polytechnic University Timisoara, Romania
Luca Davoli	University of Parma, Italy
Mauro Marcelo Mattos	FURB Universidade Regional de Blumenau, Brazil
Minoru Uehara	Toyo University, Japan
Naohiro Hayashibara	Kyoto Sangyo University, Japan
Nobukazu Iguchi	Kindai University, Japan
Nobuo Funabiki	Okayama University, Japan
Olivier Terzo	LINKS, Italy
Omar Hussain	UNSW Canberra, Australia
Raffaele Pizzolante	University of Salerno, Italy
Sajal Mukhopadhyay	National Institute of Technology, Durgapur, India
Santi Caballe	Open University of Catalonia, Spain
Shigetomo Kimura	University of Tsukuba, Japan
Shinji Sakamoto	Kanazawa Institute of Technology, Japan

Sotirios Kontogiannis	University of Ioannina, Greece
Teodor Florin Fortis	West University of Timisoara, Romania
Tomoki Yoshihisa	Osaka University, Japan
Toshihiro Yamauchi	Okayama University, Japan
Xu An Wang	Engineering University of CAPF, China
Yusuke Gotoh	Okayama University, Japan

EIDWT-2024 Reviewers

Amato Flora	Kikuchi Hiroaki
Amato Alba	Kohana Masaki
Barolli Admir	Leu Fang-Yie
Barolli Leonard	Leung Carson
Bista Bhed	Matsuo Keita
Chellappan Sriram	Ogiela Lidia
Chen Hsing-Chung	Ogiela Marek
Cui Baojiang	Okada Yoshihiro
Di Martino Beniamino	Oliveira Luciana
Enokido Tomoya	Pardede Eric
Esposito Antonio	Paruchuri Vamsi Krishna
Fachrunnisa Olivia	Rahayu Wenny
Fun Li Kin	Spaho Evjola
Funabiki Nobuo	Taniar David
Gotoh Yusuke	Uehara Minoru
Iio Jun	Yoshihisa Tomoki
Ikeda Makoto	Venticinque Salvatore
Ishida Tomoyuki	Wang Xu An
Kamada Masaru	Woungang Isaac
Kayem Anne	Xhafa Fatos

EIDWT-2024 Keynote Talks

Empowering Education: Leveraging AI and Ethical Advancements in Online Learning

Santi Caballé

Open University of Catalonia, Barcelona, Spain

Abstract. This keynote discusses the transformative potential of Artificial Intelligence (AI) in online education by introducing groundbreaking AI-powered educational methods tailored for online learning, underpinned by a commitment to ethical principles. These methods encompass intelligent conversational agents, adaptive recommendations, and insightful learning analytics, all geared toward enhancing the online educational experience. The objectives revolve around improving teaching quality, promoting effective online engagement, assessing skill acquisition, and critically examining the ethical implications of AI in online education. We will explore current and future approaches of how AI can empower online learners with essential competencies while maintaining ethical integrity, paving the way for a more inclusive and ethically informed digital education landscape.

Validation and Verification of Smart Contracts in IoT Applications

Francesco Moscato

University of Salerno, Salerno, Italy

Abstract. Industry 4.0 is becoming more and more interested in blockchain and smart contract technology. With the use of smart contracts, new interactions are now possible, allowing complete automation of agreed-upon operations by contractors. This technique has a lot of potential in the Internet of Things (IoT) space since it enables complete automation by generating events for the software agents participating in a smart contract execution. Smart contracts must, however, abide by local, national, and international regulations as well as participant accountability. A smart contract legal compliance must be confirmed in order to determine its soundness. We discuss a mechanism for validating and verifying smart contracts adherence to the law in IoT environments in order to offer a formal model for validating law compliance of smart contracts and to determine potential responsibilities of failures. This talk will present a systematic approach to smart contracts validation and verification, which is based on model transformation techniques. The talk focuses on the use of a Multi-Agent System (MAS) model and a workflow language that can define and analyze agents' interactions in IoT applications.

Contents

Few Shot NER on Augmented Unstructured Text from Cardiology Records

Antonino Ferraro[✉], Antonio Galli, Valerio La Gatta, Mario Minocchi,
Vincenzo Moscato, and Marco Postiglione

Department of Electrical Engineering and Information Technology (DIETI),
University of Naples "Federico II", Via Claudio 21, Naples, Italy
{antonino.ferraro,antonio.galli,valerio.lagatta,mario.minocchi,
vincenzo.moscato,marco.postiglione}@unina.it

Abstract. The principal challenge encountered in the realm of Named-Entity Recognition lies in the acquisition of high-caliber annotated data. In certain languages and specialized domains, the availability of substantial datasets suitable for training models via traditional machine learning methodologies can prove to be a formidable obstacle [10]. In an effort to address this issue, we have explored a Policy-based Active Learning approach aimed at meticulously selecting the most advantageous instances generated through a Data Augmentation procedure [3,6]. This endeavor was undertaken within the context of a few-shot scenario in the biomedical field. Our study has revealed the superiority of this strategy in comparison to active learning techniques relying on fixed metrics or random instance selection, guaranteeing the privacy of patients from whose medical records the source data were obtained and used. However, it is imperative to note that this approach entails heightened computational demands and necessitates a longer execution duration [7].

Keywords: NER · Data Augmentation · Active Learning · Machine learning

1 Introduction

To obtain useful results for Named Entity Recognition (NER) tasks, a significant amount of labeled data is essential. However, acquiring such data, especially in specialized domains like biomedicine, can be challenging [1]. The "Few Shot" scenario, where conventional machine learning techniques lack sufficient data, poses additional challenges. Various methods, including Data Augmentation, have been devised to address this issue, creating synthetic data from existing sets. Different Data Augmentation techniques, such as Mention Replacement and Random Swap, aim to improve the quality of augmented data [4,9].

In Few Shot learning, some approaches enable cross-domain or cross-language knowledge transfer, building a knowledge base in data-rich domains or languages

L. Barolli (Ed.): EIDWT 2024, LNDECT 193, pp. 1–12, 2024.
https://doi.org/10.1007/978-3-031-53555-0_1

and transferring it to those with limited data [2,8]. However, determining the value of generated instances for model learning still requires human annotation.

In 2017, [5] introduced a Reinforcement Learning-based Active Learning algorithm, utilizing an intelligent agent as a substitute for the human oracle. This "Policy-based" approach was extended to cross-language scenarios, allowing instance selection from unlabeled datasets in different languages. Building on this, the paper aims to establish a framework for defining and refining a policy to select augmented data from source datasets with limited instances, enhancing NER model performance. The key aspects include explaining Active Learning principles, redesigning them with a Policy-based approach, implementing a Deep Q-Network for constructing the selection policy, and applying Data Augmentation.

2   Background

This section covers foundational concepts essential for understanding the paper. It begins with an explanation of Data Augmentation's application in Named Entity Recognition (NER), introduces Reinforcement Learning, and concludes with an in-depth explanation of Active Learning.

2.1   Named-Entity Recognition (NER)

NER in Natural Language Processing identifies entities in sentences. The labeled dataset, denoted as a group with elements N and H_i, uses the IOB2 tagging format ($\mathcal{Y} = \{O, B\text{-}, I\text{-}\}$) where O: No entity mentioned, B-: First token of an entity and I-: Subsequent tokens of the entity.

2.2   Data Augmentation

Data Augmentation generates new data by applying transformations. In NER, augmentation includes Word Replacement, Random Swap, Mention Replacement, and Generative Models [4].

2.3   Reinforcement Learning in Healthcare

Reinforcement Learning automates decision-making processes [15]. In healthcare NER, RL aids in recognizing entities from a limited corpus, addressing challenges in annotated medical texts. This paper uses RL to enhance existing tagged data.

3   Active Learning

Active Learning, a machine learning technique, selects a subset from a non-labelled dataset annotated by a human oracle (see Fig. 1). Pool-based active learning involves training a model with a small set L, querying new instances U from the unlabelled set, annotating them, and adding to L until a criterion is met [14].

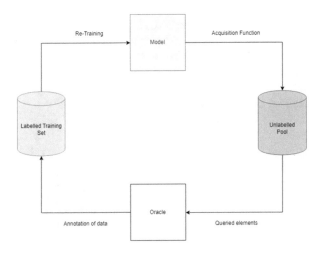

Fig. 1. Active learning cycle.

3.1 Policy-Based Active Learning

Active Learning involves selecting a subset from an unlabeled dataset for annotation, guided by acquisition functions [14]. In pool-based active learning, the
iterative process includes training a model (L) and querying new instances (U)
until predefined criteria are met.

Acquisition functions:

- **Uncertainty Sampling** [13]: Selects uncertain data points using metrics
  like least confidence, smallest margin, entropy, or maximum normalized log-
  probability.
- **Diversity Sampling** [11]: Maximizes diversity among instances based on
  their probability distribution.
- **Contrastive Learning** [12]: Uses a K-Nearest Neighbour algorithm to select
  instances based on Kullback-Leibler divergence.

Policy-Based Active Learning. [5] is a dynamic approach, transforming
data selection into a reinforcement learning problem (see Fig. 2). An intelligent agent evaluates instances, deciding to include them in the training set or
not, earning rewards. This process is a Markov Decision Process defined by
$[S, A, Pr(s_{i+1}|s_i, a), R(s, a)]$.

4 Methodology

This section details the developed framework, explaining the policy-based active
learning process and the data augmentation system.

Reinterpretation of Active Learning Cycle. The environment involves an already labeled augmented dataset. The cycle includes training a model, generating augmented data, and, in each step, evaluating a single instance by a reinforcement learning agent. The agent decides whether to add it to a new training set, initially empty. A budget limits insertions. The oracle is absent as augmented instances already have labels.

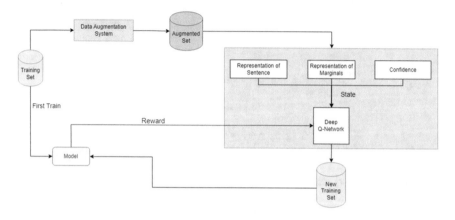

Fig. 2. Policy-based active learning.

Data Augmentation System. The PAL Framework uses a data augmentation system employing the Mention Replacement technique. Figure 3 illustrates the process, generating the dataset by extracting embeddings for each entity in the training set.

4.1   PAL Framework

State Representation. The state consists of sentence representation, probability representation, and confidence, formed by embedding, convolutional neural network, and Viterbi's algorithm.

Deep Q-Network. A Deep Q-Network is used for policy learning. It takes the state vector (sr, pr, C) as input, with an intermediate layer implementing the Q-Function. Actions (add or not add instance) are represented as binary values. Rewards are calculated based on accuracy improvement.

Training Process. A tuple (s_i, a_i, r_i, s_{i+1}) representing a state transition is saved in a memory replay. After a set number of actions, a batch of transitions is sampled, and the loss function is minimized to update the Q-Network.

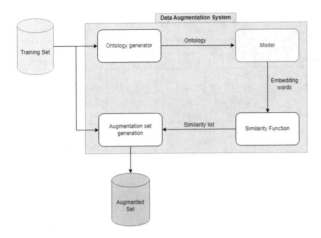

Fig. 3. Data augmentation system.

5   Experimental Protocol

This section outlines the experimental protocol, detailing the dataset, evaluation metrics, baseline, and results.

5.1   Dataset, Metrics and Baseline

The dataset is sourced from the Department of Advanced Biomedical Sciences at the University of Naples Federico II, Italy obtained from the Campania Salute (CS) network. The dataset, collected in May 2023, comprises medical event records of 57,147 individuals since 1980. Annotated by approximately 30 annotators in Biomedical Informatics, the dataset follows Uzuner et al.'s guidelines [16] for Named Entity Recognition (NER).

Mentions were annotated for three categories:

- *Medical Problems*: observations regarding a patient's body or mind that are considered abnormal or indicative of a sickness.
- *Treatments*: terminology used to describe actions taken, interventions performed, and medications administered to a patient.
- *Tests*: phrases used to describe actions conducted on a patient or a sample of body fluid or tissue to identify, exclude, or gain more insights about a medical problem.

Evaluation metrics include: *Accuracy (Acc)*, *Precision*, *Recall* and F_1 *Score*. The main focus is evaluating a Policy-based Active Learning approach on an augmented dataset. Benchmark methods include Random Selection and Uncertainty Selection. In all scenarios, the model undergoes pre-training with a consistent few-shot dataset derived from the original dataset.

Table 1. Augmentation results.

Entity Type	Few Shot Size	Final Length					Mean	Std. Dev.
		SEED 100	SEED 200	SEED 300	SEED 400	SEED 500		
MedicalProblem	10	40	15	25	25	25	26	8
	25	75	50	70	70	50	63	10.77
	50	150	115	140	140	140	137	11,66
	100	295	245	265	265	295	273	19,39
Treatment	10	20	35	35	30	30	30	5,48
	25	55	85	70	85	65	72	11,66
	50	120	155	140	155	130	140	13,78
	100	255	295	240	275	245	256	25,38
Test	10	35	40	35	25	45	36	6,63
	25	70	95	80	60	70	75	11,83
	50	135	150	180	140	135	148	16,91
	100	265	290	310	310	290	293	16,61

5.2   Results

This subsection presents outcomes from framework executions, organized by few-shot size. It highlights Policy-based Active Learning's (PAL) superior performance over Uncertainty Learning and Random Selection, particularly with limited instances. All methods operated on Augmented datasets from Data Augmentation, evaluated using the test set.

Few Shot: Size 100. The evaluation results are shown in Table 1. Starting with 100 Few Shots, PAL's performance is analyzed for MedicalProblem entity type using augmented datasets from Seeds 100 to 500 (see Fig. 4). PAL generally outperforms Uncertainty Learning and Random Selection, except at 50 instances, indicating the impact of dataset noise. Median values provide robustness against extreme values (see Fig. 5).

For Treatments entity type, PAL excels for 10 to 75 instances, although there are fluctuations in Seed 100 and Seed 300. Test entity type results show PAL's comparable performance, except in Seed 100 where PAL consistently outperforms the others (see Fig. 6, Fig. 7, Fig. 8, Fig. 9).

Few Shot: Size 50. Examining datasets with 50 instances, PAL shows varied performance across seeds. It outperforms other methods in Seeds 300 and 400, while struggling in Seeds 100 and 200 due to noisy instances. In Test entity type, PAL consistently outperforms Random Selection, with slight fluctuations at 25 instances:

Few Shot: Size 25. In this subsection, augmented datasets with 25 instances are analyzed. The Mean and Median of F1 score for Few Shots size 25 are shown in Table 2. PAL consistently outperforms other methods across seeds for MedicalProblem entity type (see Fig. 10). For Treatment entity type, PAL's F1 scores tend to be higher than others, reaching a peak at 0.4469 (see Fig. 11). In Test entity type, PAL achieves the highest score of 0.7644 in Seed 300 (see Fig. 12).

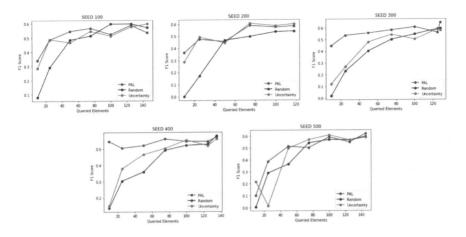

Fig. 4. Few Shot - Size 100: MedicalProblem F1 Score on test set

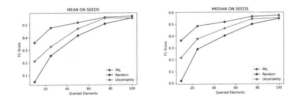

Fig. 5. MedicalProblem - Comparison of Mean and Median of F1 Score on test set (Size 100)

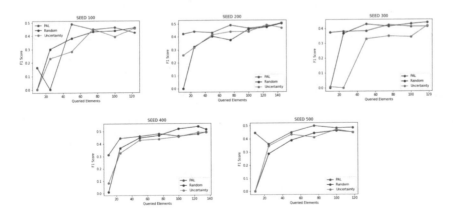

Fig. 6. Few Shot - Size 100: Treatment F1 Score on test set

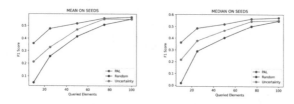

Fig. 7. Treatment - Comparison of Mean and Median of F1 Score on test set (Size 100)

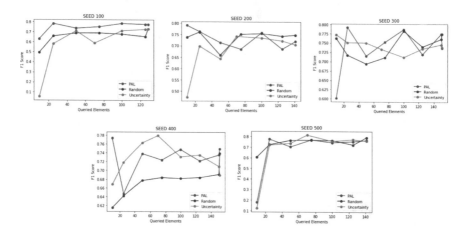

Fig. 8. Few Shot - Size 100: Test F1 Score on test set

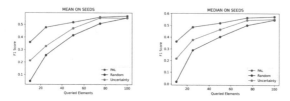

Fig. 9. Test - Comparison of Mean and Median of F1 Score on test set (Size 100)

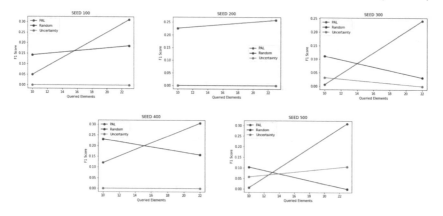

Fig. 10. Few Shot - Size 25: NCBI disease F1 Score on test set

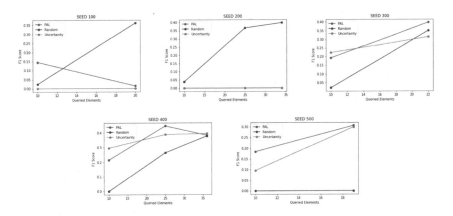

Fig. 11. Few Shot - Size 25: Treatment F1 Score on test set

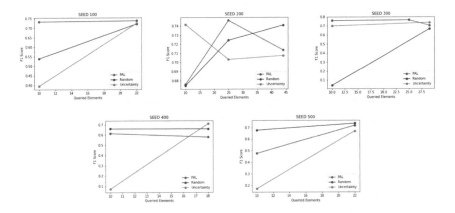

Fig. 12. Few Shot - Size 25: Test F1 Score on test set

Table 2. Few Shots Size 25 - Mean and Median of F1 Score.

Entity Type	Type	Mean	Median
MedicalProblem	PAL	0.1830	0.2338
	Random	0.0963	0.1072
	Uncertainty	0.0195	0
Treatment	PAL	0.2093	0.1886
	Random	0.1834	0.1513
	Uncertainty	0.1677	0.1599
Test	PAL	0.6933	0.7159
	Random	0.6219	0.6744
	Uncertainty	0.5753	0.7033

Few Shot: Size 10. Due to the low number of instances, scores for datasets with 10 instances are presented in Table 3, Table 4, Table 5 and Table 6. PAL demonstrates superior performance for MedicalProblem and Test entity types, while Random Selection is ineffective, scoring below 0.1. PAL achieves the best F1 score of 0.7329 for Test entity type.

Table 3. Few Shot - Size 10: Medical-Problem.

Type	Seed	Queried Elements	F1 Score
PAL	100	10	0.2581
	100	13	0.3401
	200	4	0.0041
	300	8	0.1589
	400	8	0.3190
	500	8	0.2568
Random	100	10	0.0832
	100	13	0.2790
	200	4	0
	300	8	0.0117
	400	8	0
	500	8	0.114
Uncertainty	100	10	0.3009
	100	13	0.3167
	200	4	0.0113
	300	8	0.2778
	400	8	0.3352
	500	8	0.3032

Table 4. Few Shot - Size 10: Treatment.

Type	Seed	Queried Elements	F1 Score
PAL	100	5	0.2553
	200	10	0.0027
	300	10	0.2286
	400	8	0.3179
	500	8	0.3263
Random	100	5	0.0152
	200	10	0.0089
	300	10	0
	400	8	0.0065
	500	8	0
Uncertainty	100	5	0
	200	10	0
	300	10	0.2438
	400	8	0.2139
	500	8	0.1763

Table 5. Few Shot - Size 10: Test.

Type	Seed	Queried Elements	F1 Score
PAL	100	10	0.7095
	200	10	0.6655
	200	13	0.6795
	300	10	0.6599
	400	8	0.5539
	500	10	0.6289
	500	19	0.7329
Random	100	10	0.5644
	200	10	0.6464
	200	13	0.6537
	300	10	0.2411
	400	8	0.6467
	500	10	0.7246
	500	19	0.6840
Uncertainty	100	10	0.7210
	200	10	0.7028
	200	13	0.0040
	300	10	0
	400	8	0.5694
	500	10	0.6002
	500	19	0.6627

Table 6. Few Shots Size 10 - Mean and Median of F1 Score.

Entity Type	Type	Mean	Median
MedicalProblem	PAL	0.2229	0.2574
	Random	0.0569	0.0117
	Uncertainty	0.2235	0.2009
Treatment	PAL	0.2310	0.2553
	Random	0.0061	0.0065
	Uncertainty	0.1268	0.1763
Test	PAL	0.6614	0.6655
	Random	0.5944	0.6467
	Uncertainty	0.4658	0.6002

6   Conclusions and Future Work

In this study, we developed a Progressive Active Learning (PAL) framework. This framework excels in the precise selection of instances from an augmented dataset, enhancing the model's performance while ensuring complete data anonymization. Empirical findings confirm PAL's effectiveness in scenarios with limited available instances, surpassing the performance of benchmarked methods. However, due to constraints in the computational platform used, further improvements in scores were not possible, even when PAL demonstrated superior performance. Augmenting the number of episodes could have mitigated this limitation.

One notable drawback of our approach is the extended execution time. In cases involving augmented datasets derived from a small initial number of instances (each containing 100 data points), the execution time stretched to around 5 h. This prolonged duration is mainly due to the need for model retraining after each instance insertion. To overcome this time constraint, future research could explore algorithmic modifications, such as substituting immediate reward assignment with a cumulative reward mechanism, building upon our framework.

References

1. Barolli, L., Ferraro, A.: A prediction approach in health domain combining encoding strategies and neural networks. In: Barolli, L. (ed.) 3PGCIC 2022. LNNS, vol. 571, pp. 129–136. Springer, Cham (2022). https://doi.org/10.1007/978-3-031-19945-5_12

2. Chen, S., Aguilar, G., Neves, L., Solorio, T.: Data augmentation for cross-domain named entity recognition (2021). https://doi.org/10.48550/ARXIV.2109.01758, https://arxiv.org/abs/2109.01758

3. Cubuk, E.D., Zoph, B., Mane, D., Vasudevan, V., Le, Q.V.: Autoaugment: learning augmentation policies from data. arXiv preprint arXiv:1805.09501 (2018)

4. Dai, X., Adel, H.: An analysis of simple data augmentation for named entity recognition (2020). https://doi.org/10.48550/ARXIV.2010.11683, https://arxiv.org/abs/2010.11683

5. Fang, M., Li, Y., Cohn, T.: Learning how to active learn: a deep reinforcement learning approach (2017). https://doi.org/10.48550/ARXIV.1708.02383, https://arxiv.org/abs/1708.02383

6. Ferraro, A., et al.: HEMR: hypergraph embeddings for music recommendation (2023)

7. Ferraro, A., et al.: Unsupervised anomaly detection in predictive maintenance using sound data (2023)

8. Ferraro, A., Galli, A., La Gatta, V., Postiglione, M.: A deep learning pipeline for network anomaly detection based on autoencoders. In: 2022 IEEE International Conference on Metrology for Extended Reality, Artificial Intelligence and Neural Engineering (MetroXRAINE), pp. 260–264. IEEE (2022)

9. Ferraro, A., Galli, A., La Gatta, V., Postiglione, M.: Benchmarking open source and paid services for speech to text: an analysis of quality and input variety. Front. Big Data 6 (2023). https://doi.org/10.3389/fdata.2023.1210559. https://www.frontiersin.org/articles/10.3389/fdata.2023.1210559

10. Houssein, E.H., Mohamed, R.E., Ali, A.A.: Machine learning techniques for biomedical natural language processing: a comprehensive review. IEEE Access 9, 140628–140653 (2021)

11. Krishna, K., Agal, A.: Diversity Sampling in Machine Learning

12. Margatina, K., Vernikos, G., Barrault, L., Aletras, N.: Active learning by acquiring contrastive examples (2021). https://doi.org/10.48550/ARXIV.2109.03764, https://arxiv.org/abs/2109.03764

13. Nguyen, V.L., Shaker, M., Hllermeier, E.: How to measure uncertainty in uncertainty sampling for active learning. Mach. Learn. 111, 89–122 (2022)

14. Settles, B.: Active learning literature survey. Computer Sciences Technical report 1648, University of Wisconsin–Madison (2009)

15. Sutton, R., Barto, A.: Reinforcement learning: an introduction. IEEE Trans. Neural Networks 9(5), 1054 (1998). https://doi.org/10.1109/TNN.1998.712192

16. Uzuner, Ö., South, B.R., Shen, S., DuVall, S.L.: 2010 i2b2/VA challenge on concepts, assertions, and relations in clinical text. J. Am. Medical Informatics Assoc. 18(5), 552–556 (2011). https://doi.org/10.1136/amiajnl-2011-000203

The Influence of Financial Literacy and Financial Technology on SMEs' Sustainability

Winarsih[✉], Desy Noviyanti, and Mutoharoh

Department of Accounting, Faculty of Economics, UNISSULA, Semarang, Indonesia
winarsih@unissula.ac.id

Abstract. SMEs' sustainability requires SME actors to develop businesses as a means of expanding employment opportunities and equalizing the economy of small people. Current developments in Small and Medium Enterprises are expected to use technology to increase efficiency and competitiveness in business. This research reveals the main problem, namely, how to improve the SMEs' sustainability with Financial Literacy and Financial technology so that they can compete in the digital era. The use of technology will provide significant changes to the ability of SMEs in business sustainability. This paper presents several methods and techniques for implementing models that support SMEs' sustainability, in this case, namely financial literacy and financial technology as independent variables. The data used is primary data with quantitative methods, and the respondents were 130 SME actors in Indonesia, especially in Semarang City, Central Java province.

Keywords: financial literacy · financial technology · SMEs' sustainability

1 Introduction

Economic development is one indicator to determine the success of a country, which is seen based on increasing production of industrial goods, increasing schools, and increasing production of capital goods or the service sector. SMEs are one sector that has an influence on economic development in various countries, especially in Indonesia. The public sector is believed to drive a country's economy. According to a survey conducted by the financial service authority, which states that the contribution of SMEs in Indonesia is significant for the national economy, contributing sixty per cent of the gross domestic product and absorbing ninety-seven per cent of the national workforce (OJK Press Release: sp-38 /DKNS/OJK/5/107). In general, SMEs often experience developmental delays. It is caused by various unresolved conventional problems (closed-loop problems), such as challenges to human resource capacity. SMEs need help competing with large businesses due to the challenges of ownership, financing, marketing and other challenges related to managing large corporate businesses. It can be seen from the absence of a concept of sustainable innovation and the inconsistency of the company's main business operations, as well as the development of the long-term performance of SMEs in the creative industry, which usually needs to be directed.

L. Barolli (Ed.): EIDWT 2024, LNDECT 193, pp. 13–19, 2024.
https://doi.org/10.1007/978-3-031-53555-0_2

Currently, the business sector has experienced rapid development, growth, and expansion, where many sectors, including the financial sector, are now starting to be digitalized [1]. One of them is the Small and Medium Enterprises sector, which is usually called SMEs. SMEs face several challenges as more and more people are affected by technological advances, which are reflected in the way they invest their money [2]. The need for careful and thorough financial literacy and personal financial management has become critical in balancing the rapid growth of technology and the ability to make sound investment decisions. It means that financial literacy is crucial for the survival of every human being. If every citizen is financially literate and able to make the right economic decisions, then it is hyperbole to say that the country can be considered prosperous. A financially literate population contributes to a healthier economy, and they are more likely to engage in productive financial activities such as investing in education, setting up a business or participating in the formal financial system that spurs economic growth and creates jobs [3].

SME financial literacy is the manager's ability to record financial reports, manage debt and prepare budgets. Recording financial reports requires SME managers to record business activities, business income and costs, business profitability, and other performance-related topics. SMEs are often unable to make their financial reports, and most commercial banks are reluctant to provide capital credit. Financial reporting is crucial for SMEs to determine the success of their business. Small and medium-sized businesses can fund their working capital and investments in two ways. According to debt management literature, there are two main sources of funding. Firstly, SMEs can take advantage of savings; secondly, they can contract obligations to third parties. When a business decides to borrow, the percentage of private capital to external capital/debt decreases. Budget planning skills can be used to plan future business activities. Financial literacy influences a person's way of thinking about financial conditions and how they make decisions and manage business owners' strategic finances. The ability of business owners to manage their finances is important for business performance and continuity. Business owners must acquire financial knowledge to improve their company's performance. It empowers SMEs to develop their businesses.

The whole world is experiencing rapid technological developments todays, as supported by financial technology. It has penetrated many industries, including the financial sector. Financial technology refers to technology that supports financial services. Many applications advance in the financial services sector, such as payment instruments, loan instruments, etc. They are increasingly popular in the digital era as a result of developments in financial technology. Meanwhile, modern society needs results for things to work quickly and easily, without restrictions and rules, and more people are taking advantage of financial technology. It shows that comfort, security, suitability of transactions, and ease of transactions are elements that encourage SME actors to utilize financial technology, accompanied by things related to supporting factors, such as ease of recording, ease of transaction processing, and sales growth.

Financial technology has various functions that are developing rapidly. Currently, financial technology is related to electronic money, virtual accounts, aggregators, lending, crowdfunding, and other online financial transactions. Financial technology makes transactions possible without a bank account. Although financial technology itself is not a bank, it is regulated by Bank Indonesia to protect consumers and the public. Financial technology companies must register with Bank Indonesia or the Financial Services Authority. Bank Indonesia states that financial technology can replace banks. Financial technology can 1) create markets for business actors, 2) become a means of payment, settlement and clearing, 3) facilitate more efficient investments, 4) mitigate the risks of conventional payment systems, and 5) help parties save, borrow, and participate in equity.

Therefore, there is a form of consistency in the state of the business, where sustainability is the ongoing stage of the business. It includes development, growth, and strategies for maintaining business continuity and developing the business, all of which focus on the existence and sustainability of the business. SMEs' sustainability is crucial for the future of the economy of both the country and the business sector itself because it shows the ability of an SME to realize business goals and add long-term value to the business or people who invest in the business. To advance and sustain the business, it is essential to increase the insight of SME actors in financial knowledge so that the business can continue to develop and increase business sustainability. The success of a business in carrying out innovation, realizing the welfare of employees and customers, as well as generating returns on business equity. It will show how the company can develop and be able to innovate sustainably.

Therefore, to develop the achievements and sustainability of SMEs for the long term, it is necessary to form a strategic business, namely with financial literacy and financial technology. This collaboration will renew innovations in existing products or processes in accordance with developments in the current era because innovation in business is the soul of a company to be able to continue to develop.

2 Financial Literacy

The importance of financial literacy is not only used by professionals in the fields of investment and banking but also by everyone who has the responsibility to manage finances for daily life. Financial literacy is not only limited to understanding knowledge, skills and confidence in a product, service or institution that is available, but behaviour and attitudes also influence increasing financial literacy to achieve survival. Previous research defines financial well-being as a state in which a person has control over daily finances can absorb financial shocks, can meet financial goals, and has the financial freedom to make choices that allow one to enjoy life [4].

Most of the literature on financial literacy focuses on three main areas: conceptual definition of financial literacy, measurement of financial literacy elements, and financial education [5]. Research conducted by [6] concluded that financial literacy had a positive effect on the sustainability of SMEs by 28.9%. Meanwhile, [7] examined the influence of financial literacy on the sustainability of SMEs, resulted in the conclusion that financial literacy had a significant positive effect on the performance and sustainability of SMEs. Financial knowledge has an influence; namely, the higher the knowledge obtained regarding financial knowledge, the longer the period for the sustainability of SMEs. One of the basic knowledge areas of finance is being able to read and make decisions through financial reports (cash flow reports, profit and loss, balance sheets, etc.)

for both short and long-term periods. If a company has financial insight, its management and accountability can be properly accounted for, like a large company. In this research, financial literacy is measured by indicator [8], namely, the ability to create a financial surplus periodically, the ability to make calculations about the use of funds owned, and the ability to analyse financial performance.

Table 1. Variable Description of Financial Literacy (X1).

No.	Item	SD (1)		D (2)		N (3)		A (4)		SA (5)		Total	
		F	%	F	%	F	%	F	%	F	%	F	
1	X1	0	0%	0	0%	23	18%	61	46%	46	35%	**130**	**100%**
2	X2	0	0%	0	0%	22	18%	59	45%	48	37%	**130**	**100%**
3	X3	0	0%	0	0%	32	25%	54	41%	44	34%	**130**	**100%**
4	X4	0	0%	0	0%	23	18%	61	47%	46	35%	**130**	**100%**
5	X5	0	0%	0	0%	34	26%	52	40%	44	44%	**130**	**100%**
6	X6	0	0%	0	0%	23	18%	61	47%	46	35%	**130**	**100%**
7	X7	0	0%	0	0%	33	26%	53	40%	44	34%	**130**	**100%**

Source: Processed data, 2021

Based on Table 1, the frequency distribution value of the Financial Literacy variable is highest for items X1, X4 and X6, with 61 respondents strongly agreeing. The information above shows that respondents agree that Financial Literacy can help in the SMEs' sustainability of Semarang.

The research results show that Financial Literacy has a positive influence on the SMEs' sustainability. It indicates that the better the implementation of Financial Literacy, the greater the sustainability of SMEs. It states that financial knowledge has an influence; namely, the higher the knowledge gained about financial knowledge, the longer the period for the sustainability of SMEs.

One of the basic knowledge areas of finance is being able to read and make decisions through financial reports for both the short and long term. If a company has financial insight, its management and accountability can be properly accounted for, like a large company. According to stakeholder theory, the link between financial literacy and the sustainability of SMEs is improving the quality of financial reports seen from the existing human resources in the company or existing stakeholders. Improving financial quality is carried out by providing sufficient financial knowledge for business actors and employees who are responsible for the company's finances. Therefore, it will increase the sustainability of the company's voice by improving the welfare of all existing stakeholders. The research is supported by research results [6] and [7], which state that Financial Literacy has a significant positive influence on the Sustainability of SMEs.

3 Financial Technology

Based on the research results, Financial Technology has a positive influence on the SMEs' sustainability. It indicates that the implementation of Financial Technology is getting better, so it will increase the SMEs' sustainability. Furthermore, the application of financial technology influences because the more we apply financial technology in business, the longer the sustainability period of SMEs will be.

If a company knows financial technology, sales will increase because, currently, financial technology is a technology that is very developed in a company or SME. Technology has also made financial services more affordable, easier and cheaper. These three elements are sometimes difficult for SMEs to obtain when applying for a loan from a bank. Thus, financial technology can be used as a solution to develop SMEs in the future. Based on the Technology Acceptance Model, attitudes towards using technology have a strong influence on interest in using it. This is because when technology provides benefits, people will remain interested in using technology. Therefore, the public will be aware of the benefits of financial technology in transaction and production activities carried out for the sustainability of SMEs.

In this case, financial technology for SMEs is expected to help improve their business. For example, SME loans to access financing. "financial technology" includes P2P loans, which make it possible to access SMEs that do not meet the qualifications for loans through banks. Thus, financial technology can create an additional supply of capital, such as People's Business Credit. Financial technology helps SMEs obtain efficiency and convenience related to finance. Financial Technology provides many solutions, especially for the sustainability of SMEs. Financial technology measurements in this research use indicators in research [9], namely, capital services, payment transaction services, and customer service.

Based on the results of the questionnaire regarding the Financial Technology variables, their description is shown in Table 2.

Table 2. Variable Description of Financial Technology (X2)

No.	Item	SD (1)		D (2)		N (3)		A (4)		SA (5)		Total	
		F	%	F	%	F	%	F	%	F	%	F	
1	X1	0	0%	0	0%	29	22%	63	49%	38	29%	130	100%
2	X2	0	0%	0	0%	30	23%	60	46%	40	31%	130	100%
3	X3	0	0%	0	0%	27	21%	63	48%	40	31%	130	100%
4	X4	0	0%	0	0%	21	16%	70	54%	39	30%	130	100%

Source: Processed data, 2021

The frequency distribution value of the Financial Technology variable is highest in item X4, with 70 respondents strongly agreeing. The information above shows that respondents agree that Financial Technology can help in the SMEs' sustainability of Semarang.

4   SMEs Sustainability

SMEs in Indonesia face critical new challenges that must be addressed. The strategies implemented include prioritizing the business by focusing on the core business, changing services from offline to online, and pivoting or developing a business model while remaining focused on its business goals The SME structure requires SME actors to work by following changing trends in society, evaluating behaviour, consumer satisfaction, after-sales guarantees, and building branding connected to online platforms.

Sustainability is a business condition in which several ways exist to develop, maintain resources, and meet the company's needs. The source of this method is through one's own experience with other individuals, and focusing on the economic conditions currently existing in the business continuity is a form of consistency in business conditions. It includes development, growth, ways to maintain business continuity and develop business [6]. The indicators used to measure sustainability, according to [10], consist of capital, human resources, production, and marketing. By understanding several aspects, business actors can maintain their business. Capital is the most important thing for business success. Capital can be obtained through loans or independently. Apart from capital, expertise and human resources are also needed so that the business can run well and then realize the desired goals. Production activities create new utility value from services or goods required to meet customer needs. Marketing also has a vital role in business continuity, and by having appropriate and suitable marketing techniques, a business can proliferate.

Based on the results of the questionnaire regarding the SMEs' Sustainability variable, a frequency distribution was obtained and shown in Table 3.

Table 3. Variable Description of SMEs' Sustainability (Y).

No.	Item	SD (1)		D (2)		N (3)		A (4)		SA (5)		Total	
		F	%	F	%	F	%	F	%	F	%	F	
1	Y1	0	0%	0	0%	23	18%	53	48%	54	34%	130	100%
2	Y2	0	0%	9	7%	32	25%	56	43%	33	25%	130	100%
3	Y3	0	0%	5	4%	27	21%	40	31%	58	45%	130	100%
4	Y4	0	0%	0	0%	23	18%	63	48%	44	34%	130	100%
5	Y5	0	0%	5	4%	28	21%	40	31%	57	44%	130	100%
6	Y6	0	0%	0	0%	23	18%	64	49%	43	33%	130	100%
7	X7	0	0%	5	4%	26	22%	42	30%	57	44%	130	100%
8	Y8	0	0%	0	0%	22	17%	66	50%	44	33%	130	100%
9	Y9	0	0%	0	0%	23	18%	63	48%	44	34%	130	100%

Source: Processed data, 2021

Based on Table 3, the frequency distribution value of the SME Sustainability variable is highest in item Y8, with 66 respondents strongly agreeing. From the information above, it shows that respondents agree that Sustainability for SMEs can help in the sustainability of SMEs in Semarang.

5 Conclusions

Financial literacy and financial technology have a positive influence on the SMEs' sustainability in Semarang. It makes it better to implement Financial Literacy (financial knowledge) regarding financial reports as a basis for decision-making, thereby minimizing errors in the management and accountability of SMEs and helping maintain business consistency. The more Financial Technology is implemented well. It can make it easier to prepare financial reports, make services more accessible and faster to consumers so that they have greater effectiveness in running their business and increase the quality of customer service speed so that SMEs will experience sustainability in business.

The theoretical implications of financial literacy as a basis for decision-makers in preparing financial reports are very influential on the sustainability of SMEs. SME players can make financial reports in line with existing regulations by studying the guidelines for preparing reports, and financial technology can make it easier to make financial reports, making services more accessible and faster to consumers. It can be one way for businesses to survive, especially SMEs in Indonesia and Semarang.

References

1. Utami, N., Sitanggang, M.L., Sitanggang, M.L.: The analysis of financial literacy and its impact on investment decisions: a study on generation Z In Jakarta. Inovbiz J Inov Bisnis. **9**(1), 33 (2021)
2. Copur, Z., Gutter, M.S.: Economic, sociological, and psychological factors of the saving behavior: Turkey case. J. Fam. Econ. Issues **40**(2), 305–322 (2019). https://doi.org/10.1007/s10834-018-09606-y
3. Pangestu, S., Karnadi, E.B.: The effects of financial literacy and materialism on the savings decision of generation Z Indonesians. Cogent. Bus. Manag. **7**(1), 1743618 (2020). https://doi.org/10.1080/23311975.2020.1743618. Foroudi P, editor
4. Michael Collins, J., Urban, C.: Measuring financial well-being over the lifecourse. Eur. J. Financ. **26**(4–5), 341–59 (2020). https://doi.org/10.1080/1351847X.2019.1682631
5. Kaiser, T., Lusardi, A., Menkhoff, L., Urban, C.: Financial education affects financial knowledge and downstream behaviors. J. Financ. Econ. **145**(2), 255–272 (2022)
6. Widayanti, R., Damayanti, R., Marwanti, F.: Pengaruh financial literacy Terhadap Keberlangsungan Usaha (business sustainability) Pada Umkm Desa Jatisari. J Ilm Manaj Bisnis. **18**(2), 153 (2017)
7. Idawati, I.A.A., Pratama, I.G.S.: Pengaruh Literasi Keuangan Terhadap Kinerja dan Keberlangsungan UMKM di Kota Denpasar. Warmadewa Manag Bus J. **2**(1), 1–9 (2020)
8. Puspitaningtyas, Z.: Manfaat Literasi Keuangan Bagi Business Sustainability. Semin Nas Kewirausahaan dan Inov Bisnis **VII**, 254–62 (2017). (ISSN: 2089-1040)
9. Muzdalifa, I, Rahma, I.A., Novalia, B.G.: Peran Fintech Dalam Meningkatkan Keuangan Inklusif Pada UMKM Di Indonesia (Pendekatan Keuangan Syariah). J Masharif al-Syariah J Ekon dan Perbank Syariah **3**(1) (2018)
10. Christoper, S.W.H., Kristianti, I.: Hubungan E-Commerce Dan Literasi Keuangan Terhadap Kelangsungan Usaha Di Boyolali. J Akunt ISSN 2303: 356 (2020)

The Effect of Digital and Human Capital Transformation on SME Performance with Accounting Information Systems as an Intervening Variable

Winarsih[✉], Kholida, and Chrisna Suhendi

Department of Accounting, Faculty of Economics, Universitas Islam Sultan Agung, Semarang, Indonesia
winarsih@unissula.ac.id

Abstract. SMEs have a big role as drivers of the country's economy. SMEs can optimize performance in their operations to ease achieving their goals. Several factors influence the performance of a business entity, including digital transformation and human capital. Through Digital Transformation, SMEs can adapt to existing developments to increase speed, scope, and efficiency in sales. It is supported by the existence of an accounting information system to facilitate the performance of SMEs in making decisions for SME owners for the sustainability of their business in the future. This research explains the influence of digital transformation and human capital on the performance of SMEs with accounting information systems as an intervening variable in Indonesia, especially the city of Semarang, Central Java province. This research is quantitative research using primary data collected through distributing questionnaires to SMEs. The number of respondents in this research was 125 SMEs.

Keywords: SME performance · digital transformation · human capital · accounting information systems

1 Introduction

The digital economy, as a "new engine' for the economy's evolution, has created new sectors and empowered traditional industries. It is also an essential way for countries to improve the quality of their economic development. Digital transformation allows companies to achieve greater flexibility and efficiency, optimize production processes, generate value propositions for innovation ecosystems, and respond to market needs promptly [1]. In addition, the digital transformation process is critical to maintain market competitiveness and remain at the forefront of technological innovation [2]. It needs to use digital transformation to increase production efficiency and innovation for small and medium enterprises (SMEs). Existing SMEs contribute significantly to the government. One of them is the creation of new jobs for the community. It can expand job

L. Barolli (Ed.): EIDWT 2024, LNDECT 193, pp. 20–30, 2024.
https://doi.org/10.1007/978-3-031-53555-0_3

opportunities in labor absorption, thereby reducing the unemployment rate in Indonesia. SMEs today are embracing digital transformation at an exponential pace to spur development by implementing new business models and digital technologies to improve customer experience. In [3], the authors believe that digitalization in the SME sector provides a sustainable flow, as well as presenting new opportunities to make organizational processes more competitive. Digital transformation provides an effective means for SMEs to successfully explore opportunities in foreign markets, including developing and developed countries [4]. Previous research shows that multinational enterprises (MNEs) are undergoing significant changes in their business strategies and structures to increase global integration [5]. Digital transformation will develop if it has qualified human resources. Human capital is collective intellectual capital in the form of competence, knowledge, and skills a person possesses. Employees are a valuable asset for all SME business actors. Meanwhile, human capital is one of the intangible assets. Research conducted by authors in [6] regarding Human Capital found that SME performance shows positive results. It is also supported by research conducted [7] and [8], showing positive results. Human capital, in practice, positions humans as helping decision makers to focus on human development to improve product quality and performance in SMEs, not as machines.

In addition, accounting information has a crucial role for business actors. Through accounting information, SME actors can evaluate all activities, programs, and processes in their business entity. Accounting information will provide results from a financial perspective on the performance implemented by SMEs. One of the things that supports the SMEs' performance in the current revolution is the existence of a digital-based accounting information system. To track incoming and outgoing funds, SMEs need an accounting information system (AIS) that has been carefully established and maintained, where AIS enables SMEs to achieve their goals. Therefore, an accounting information system is an IT-based solution that helps control the economic and financial operations of an organization. However, significant technological advances have enabled businesses to use this option strategically. In addition, SME accounting information systems are considered the primary source of information for making decisions because these systems provide various information and financial reports to beneficiaries needed for planning, organizing, and making decisions [9]. Likewise, the quality of the output provided by the AIS, the system characteristics, is the focus of service and information quality. As a result, proper provider system quality is necessary for an organization's profitability and success, as it is associated with increased profitability, customer loyalty, and competitive advantage. In addition, accounting information systems must be reliable in their operation and provide accurate and reliable information to their users in a timely and relevant manner. To protect the integrity of an SME's information and resources, acceptable internal controls must be implemented, and to do this, SMEs must prioritize such systems and consider system and employee-related issues when managing their information systems [10–12].

2 Digital Transformation

Digital transformation (DT) is completely changing businesses worldwide, especially for SMEs. Established companies from every industry, in all shapes and forms, are being pushed to change their business models, explicitly leveraging digital technologies. It aims to remain competitive in their respective markets. Several studies have explored how companies need to adapt and improve their internal capabilities to be successful in their DT attempts [13]. Digital transformation (DT) represents a paradigm shift affecting industries and companies worldwide [14]. It is even believed to have a positive influence on the development of society [15]. Therefore, it has attracted the interest of many researchers and practitioners in the last decades who seek to understand its nature and characteristics in the use of the SME field.

According to [16], factors drive Digital Transformation, namely perceived benefits, suitability, and costs that influence the adoption of e-commerce technology. One form of digital transformation that is currently widespread is the large number of e-commerce. This e-commerce is a virtual market as a forum that facilitates sellers and buyers. There are several types of e-commerce websites available, including, Business to Business (B2B), Business to Customer (B2C), Customer to Customer (C2C), Customer to Business (C2B), Business to Administration (B2A), Online To Offline (O2O). Research has been carried out previously regarding the relationship between Digital Transformation and SME performance, such as by [17]. They found that Digital Transformation affects SME performance. Digital Transformation is very influential in improving the performance of SMEs because, through Digital Transformation, SMEs can access a broader market, which will provide great opportunities to improve performance. It can be seen from the summary statistics in Table 1.

Table 1. Digital Transformation Statistics Summary

ITEM	N	M	Me	Mo	Min	Max	SD	CATEGORY
X1.1	125	3.58	4	4	1	5	0.960	HIGH
X1.2	125	3.78	4	4	2	5	0.851	HIGH
X1.3	125	3.59	4	4	1	5	0.934	HIGH
X1.4	125	3.59	4	4	1	5	0.899	HIGH
X1.5	125	3.71	4	4	1	5	0.949	HIGH
X1.6	125	4.08	4	4	2	5	0.725	HIGH
Total Mean		3.722						HIGH

Source: Processed primary data, 2021

Table 1 shows the results of descriptive statistics on the digital transformation variable. The average value of respondents' answers is 3,722, which means that the value in this variable is in the high category. This digital transformation has a minimum value of 1, which means strongly disagree and a maximum value of 5, which means strongly agree. Furthermore, the standard deviation value for the digital transformation variable in this study has a lower value than the mean. It indicates that the level of data diversity in this variable is low. Thus, digital transformation provides new opportunities for SMEs, namely increasing speed, scope, and efficiency in sales.

3 Human Capital

Modern economic growth theories indeed emphasize the importance of innovation and human capital, but these theories rarely look at the interaction between these two endogenous growth engines. The importance of intangible assets, especially human resources, for modern companies is increasing. Human capital possessing these skills intuitively captures vital areas of expertise in the modern workforce. It includes aspects such as business development (most common skills: business strategy, marketing strategy, business development) and Software Engineering. Human capital is knowledge, expertise, abilities, and skills that make humans or employees the company's capital or assets. Human capital, according to [6], is defined as humans with all the knowledge, abilities, skills, ideas, and innovations that are intangible assets in achieving company goals. Human ability has five components, including individual capability, individual motivation, leadership, organizational climate, and workgroup effectiveness.

Previous research regarding the relationship between Human Capital and SME performance, such as research conducted by [6], states that Human Capital influences SME performance. Resource-based View theory is a theory that describes that companies can increase their competitive advantage by developing resources so that they can direct the company to survive in the long term. Furthermore, human capital is very influential in improving performance because, through Human Capital, SMEs can increase worker productivity, produce professional services, and produce the best solutions for the company so that the performance achieved will be more optimal.

Table 2. Summary of Human Capital Statistics

ITEM	N	M	Me	Mo	Min	Max	SD	CATEGORY
X2.1	125	3.94	4	4	2	5	0.859	HIGH
X2.2	125	4.18	4	4	2	5	0.723	HIGH
X2.3	125	3.95	4	4	2	5	0.831	HIGH
X2.4	125	4.09	4	4	2	5	0.719	HIGH
X2.5	125	4.38	4	4	2	5	0.632	VERY HIGH
X2.6	125	4.07	4	4	2	5	0.720	HIGH
X2.7	125	4.23	4	4	2	5	0.709	VERY HIGH
X2.8	125	4.37	4	4	2	5	0.616	VERY HIGH
Total Mean		4.151						HIGH

Source: Processed primary data, 2021

Table 2 shows the results of descriptive statistics on the human capital variable that the average value of respondents' answers is 4,151. It implies that the value in this variable is in the high category. The research results of human capital having a positive effect on SME performance were rejected. Human Capital is humans with all the knowledge, abilities, skills, ideas, and innovations that are intangible assets in achieving company goals. Human capital does not affect SME performance, meaning that most SMEs in the city of Semarang need to understand and apply the concept of human capital well. It is reflected through the experience indicators contained in the variable questionnaire showing low values. In other words, SMEs are considered to have limited resource capabilities in terms of attracting consumer interest both in terms of satisfaction and in terms of communication. If it is related to most businesses that have only been running for 3–5 years, it can illustrate that their experience still needs to improve. It indicates that human capital has not been able to influence the performance of SMEs Summary of Human Capital Statistics.

4   Accounting Information System

An Accounting Information System is a network of all procedures, forms, notes, and tools to process financial data into a report form. It will be used by management in controlling its business activities and as a decision-making tool by management [18]. The accounting information system contains all activities regarding the condition of an SME, which can be used as a reference in making appropriate decisions to produce optimal performance. Previous research regarding the relationship between Accounting Information Systems and SME performance, such as research conducted by [10], states that Accounting Information Systems affect SME performance. The Accounting Information System is an intervening variable that connects digital transformation with SME performance and human capital on SME performance, which is the variable that influences research results.

Table 3. Summary of Accounting Information System Statistics

ITEM	N	M	Me	Mo	Min	Max	SD	CATEGORY
Z1	125	3.70	4	4	1	5	0.942	HIGH
Z2	125	3.81	4	4	2	5	0.913	HIGH
Z3	125	3.46	4	4	1	5	0.894	HIGH
Z4	125	4.09	4	4	2	5	0.696	HIGH
Z5	125	4.21	4	4	3	5	0.676	VERY HIGH
Z6	125	4.14	4	4	2	5	0.700	HIGH
Z7	125	4.06	4	4	2	5	0.770	HIGH
Z8	125	4.12	4	4	2	5	0.799	HIGH
Z9	125	3.86	4	4	2	5	0.931	HIGH

(*continued*)

Table 3. (*continued*)

ITEM	N	M	Me	Mo	Min	Max	SD	CATEGORY
Z10	125	3.99	4	4	2	5	0.866	HIGH
Z11	125	3.95	4	4	2	5	0.888	HIGH
Z12	125	3.67	4	4	2	5	0.850	HIGH
Total Mean		3.922						HIGH

Source: Processed primary data, 2021

Table 3 shows the results of descriptive statistics on accounting information system variables. The average value of respondents' answers is 3,922, which means that the value in this variable is in the high category.

4.1 The Effect of Digital Transformation on SME Performance

The relation of digital transformation, accounting information system and SME perormance is shown in Fig. 1. The results of the Sobel test using the Sobel test Calculator, which can be accessed from www.danielsoper.com are shown in Fig. 2.

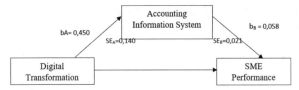

Fig. 1. Testing mediation effect of accounting information systems on effect of digital transformation on SME performance.

From the results of the Sobel test in Fig. 2, the Sobel test value is 2.094 > 1.96, and the two-tailed probability value was 0.036 < 5% significance level or 0.05. In conclusion, there is an influence of digital transformation on SME performance through the accounting information system.

The results that indicate digital transformation mediation positively affects SME performance through the Accounting Information System is **accepted**. Digital transformation is a change related to the application of digital technology in all aspects of society. Digital Transformation has an important role in the sustainability of SMEs; namely, through technological developments, SMEs can access a wider market. Additionally, digital transformation influences SME performance, and accounting information systems also have a role in improving SME performance. The greater the technology adoption in operations, the greater the performance obtained. An accounting information system allows SME operational activities to be controlled, which will later be used as material for decision-making.

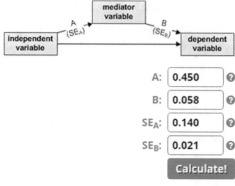

A: 0.450   ❓

B: 0.058   ❓

SE_A: 0.140   ❓

SE_B: 0.021   ❓

Calculate!

Sobel test statistic: 2.09480223
One-tailed probability: 0.01809429
Two-tailed probability: 0.03618858

Fig. 2. Calculation of the Sobel test on the First Mediation Effect

4.2   The Influence of Human Capital on SME Performance

The relation of human capital, accounting information system and SME perormance is shown in Fig. 3. The results of the Sobel test using the Sobel test Calculator, which can be accessed from www.danielsoper.com are shown in Fig. 4.

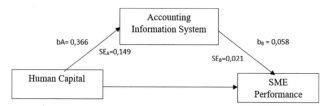

Fig. 3. Testing the Mediation Effect of Accounting Information Systems on the Influence of Human Capital on SME Performance

From the results of the Sobel test, a Sobel test value of 1,835 < 1.96 and a two-tailed probability value of 0.066 > 5% or 0.05 significance level were obtained. Accordingly, human capital has no influence on SME performance through the accounting information system.

The research results of human capital positively affecting SME performance through the Accounting Information System were rejected. It is depicted through indicators for people in the variable questionnaire showing low values. It means many SMEs still consider employees only as workers who are limited to helping in the business service process rather than developing or evaluating their business. Regarding the number of employees, most SMEs only have between 3 and 6 employees, which illustrates that there is still a need for more human resources. In other words, the development and

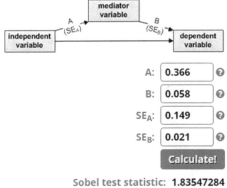

Fig. 4. Sobel test calculation on the second mediation effect

evaluation process, which is part of the accounting information system process, needs to be supported by human capital. Thus, the accounting information system does not influence mediating the relationship between human capital and SME performance.

5 SME Performance

These challenges are more important for SMEs because they usually have limited resources. The main challenges are trade and financial restrictions that complicate trade relations with other countries [19]. Previous research has mainly concentrated on post-economic crisis conditions, such as the global financial crisis, which appear to impact SMEs differently [20]. According to [10], company performance consists of financial, business, and organizational performance, where financial performance is at the center of organizational effectiveness. Performance can be interpreted as a person's ability to achieve predetermined goals and high success in carrying out a task. There are two performance measurement methods, namely, based on financial and non-financial performance. Financial performance is a formal effort to evaluate a company's efficiency and effectiveness in generating certain profits and cash positions [21]. The company's financial growth and development prospects can be seen by measuring financial performance. A company is said to be successful if the company achieves a certain predetermined performance. Non-financial performance includes customer, internal business, and innovation [22]. According to [10], the financial report reflects a company's performance. Financial reports are a source of information for users of financial information from both internal and external parties.

Table 4. Summary of SME Performance Statistics

ITEM	N	M	Me	Mo	Min	Max	SD	CATEGORY
Y1	125	3.82	4	4	2	5	0.862	HIGH
Y2	125	3.7	4	4	1	5	0.927	HIGH
Y3	125	3.58	4	4	1	5	0.960	HIGH
Y4	125	3.78	4	4	2	5	0.851	HIGH
Y5	125	3.58	4	4	1	5	0.943	HIGH
Y6	125	3.58	4	4	1	5	0.909	HIGH
Y7	125	3.69	4	4	1	5	0.971	HIGH
Y8	125	4.08	4	4	2	5	0.725	HIGH
Y9	125	3.72	4	4	2	5	0.799	HIGH
Y10	125	4.03	4	4	2	5	0.718	HIGH
Total Mean		3.756						HIGH

Source: Processed primary data, 2021

Table 4 shows the results of descriptive statistics on the SME Performance variable. The average value of respondents' answers is 3,756. It shows that the value in this variable is in the high category. Based on the results, it has a positive effect on the performance of SMEs, which is influenced by digital transformation accounting information systems, where digital transformation provides new opportunities for SMEs, namely providing increased speed, scope and efficiency in sales. Therefore, through this digital transformation, SMEs can improve their performance. The accounting information system contains all activities regarding the condition of an SME, which can be used as a basis for making appropriate decisions to produce optimal performance. However, Human Capital does not influence SME performance. This is because most SMEs in Semarang City need to properly understand and apply the concept of human capital. This can be seen from the experience indicators contained in the variable questionnaire showing low values. In other words, SMEs are considered to have limited resource capabilities in terms of attracting consumer interest both in terms of satisfaction and in terms of communication. If it is related to most businesses that have been running for less than five years, it can illustrate that their experience still needs to be improved. In conclusion, human capital has not been able to influence the performance of SMEs.

6   Conclusions

After researching factors that can influence the performance of SMEs in Indonesia, especially in Semarang City, Central Java, it is known that the implementation of digital transformation has gone quite well. It can be seen from the results of the research above. Most SMEs in Semarang City have taken advantage of digital transformation to improve their business performance, for instance, SMEs that promote their business via social media or join various types of marketplaces that currently exist. However, even though many SMEs have taken advantage of this digitalization era, they are expected to continue to enrich their knowledge to be aware of developments over time.

Nevertheless, it is found that the factor that does not influence the performance of SMEs is human capital. This is because there is still a need for understanding regarding the application of human capital by SMEs in their businesses. Considering that SMEs are one of the drivers of the nation's economy, it is expected that the government, through the Department of Cooperatives and SMEs, will provide education regarding understanding human capital to SMEs so that SMEs can apply human capital more widely. Therefore, Digital Transformation and Human Capital have an important role in the sustainability of SMEs; namely, through technological developments, SMEs can access a wider market. Digital Transformation influences improving SME performance, and accounting information systems also have a role. The greater the technology adoption in operations, the greater the performance obtained. Human Capital also has a role in increasing worker productivity, producing professional services and producing the best solutions for the company to achieve more optimal performance. Good performance will produce good accounting information so that SMEs can make the right decisions.

References

1. Queiroz, M.M., Fosso Wamba, S., Machado, M.C., Telles, R.: Smart production systems drivers for business process management improvement. Bus. Process. Manag. J. **26**(5), 1075–1092 (2020). https://doi.org/10.1108/BPMJ-03-2019-0134
2. Rangan, S., Sengul, M.: Information technology and transnational integration: theory and evidence on the evolution of the modern multinational enterprise. J. Int. Bus. Stud. **40**(9), 1496–514 (2009). https://doi.org/10.1057/jibs.2009.55
3. Board, I.A., Editors, A., Board, E.R.: Int. J. Softw. Sci. Comput. Intell. **12**(2) (2020)
4. Bertello, A., De Bernardi, P., Santoro, G., Quaglia, R.: Unveiling the microfoundations of multiplex boundary work for collaborative innovation. J. Bus. Res. **139**, 1424–1434 (2022). https://www.sciencedirect.com/science/article/pii/S0148296321007694
5. Pisani, N., Eelke, J.G.: Does it pay to be a multinational ? A large-sample, cross-national replication assessing the multinationality – performance relationship. Strategic Manage. J. **41**, 152–172 (2019)
6. Syarifah, I., Mawardi, M.K., Iqbal, M.: Pengaruh modal manusia terhadap orientasi pasar dan kinerja UMKM. J Ekon dan Bisnis. **23**(1), 69–96 (2020)
7. Syahsudarmi, S.: Pengaruh Aspek Keuangan Dan Modal Manusia Terhadap Kinerja Usaha Mikro Kecil dan Menengah (UMKM) (Studi Kasus: UMKM di Wilayah Kota Pekanbaru) **9**(1), 66–74 (2018)
8. Zuliyati, Z., Delima, Z.M.: Intellectual Capital dan Kinerja UMKM. Semin Nas dan Call Pap 2017 Strateg Pengemb Sumber Daya Mns melalui Publ J Ilm dalam Menyikapi Permenristekdikti No. 20 tahun 2017 (20), 280–290 (2017)
9. Almatarneh, Z., Akram, B., Jarah, F., Amin, M., Jarrah, A.L.: Uncertain Supply Chain Management The role of management accounting in the development of supply chain performance in logistics manufacturing companies **10**, 13–18 (2022)
10. Nurhikmah Esti Prastika DEP: Pengaruh Sistem Informasi Akuntansi Terhadap Kinerja Perusahaan Pada Usaha Mikro Kecil dan Menengah (UMKM) di Kota Pekalongan. J Ekon dan Bisnis (2019)
11. Nengsy, H.: Pengaruh Sistem Informasi Akuntansi dan Penggunaan Teknologi Informasi Akuntansi Terhadap Kinerja Manajerial Pada Perbankan di Tembilahan. J Akunt dan Keuang **7**, 1–17 (2018)

12. Wahyuni, T., Marsdenia, M., Soenarto, I.: Analisis Pengaruh Penerapan Sistem Informasi Akuntansi Terhadap Pengukuran Kinerja UMKM di Wilayah Depok. J Vokasi Indones **4**(2) (2018)

13. Verhoef, P.C., et al.: Digital transformation: a multidisciplinary reflection and research agenda. J. Bus. Res. **122**, 889–901 (2021). https://www.sciencedirect.com/science/article/pii/S01482 96319305478

14. Broekhuizen, T.L.J., Broekhuis, M., Gijsenberg, M.J., Wieringa, J.E.: Introduction to the special issue – digital business models: a multi-disciplinary and multi-stakeholder perspective. J. Bus. Res. **122**, 847–852 (2021). https://www.sciencedirect.com/science/article/pii/S01482 96320302307

15. Popkova, E.G., De Bernardi, P., Tyurina, Y.G., Sergi, B.S.: A theory of digital technology advancement to address the grand challenges of sustainable development. Technol. Soc. **68**, 101831 (2022). https://www.sciencedirect.com/science/article/pii/S0160791X21003067

16. Hanum, A.N., Sinarasri, A.: Analisis faktor-faktor yang mempengaruhi adopsi e commerce dan pengaruhnya terhadap kinerja umkm (studi kasus umkm di wilayah kota semarang). Maksimum Media Akunt. **1**(1), 1–15 (2017)

17. Nurlinda, F.V.: Determinan Adopsi E-Commerce Dan Dampaknya Pada Kinerja Usaha Mikro Kecil Menengah (Umkm). J Ris Akunt dan Keuang **7**(3), 445–464 (2019)

18. Marina, Ana D kawan kawan. Sistem Informasi Akuntansi Teori dan Pratikal. UMSurabaya Publishing, Surabaya (2017). 212 p.

19. Saeed Iranmanesh, M.I., Norallah Salehi, A., Seyyed Abdolmajid Jalaee, B.: Using fuzzy logic method to investigate the effect of economic sanctions on business cycles in the Islamic Republic of Iran. Appl. Comput. Intell. Soft. Comput. **2021**, 8833474. https://doi.org/10.1155/2021/8833474. Keshavarz-Ghorabaee M, editor

20. Cowling, M., Liu, W., Ledger, A., Zhang, N.: What really happens to small and medium-sized enterprises in a global economic recession? UK evidence on sales and job dynamics. Int. Small. Bus. J. **33**, 488–513 (2015). https://api.semanticscholar.org/CorpusID:53393782

21. Hery: Analisis Kinerja Manajemen. Gramedia Widiasarana Indonesia (2015). 216 p.

22. Ulum, I.: Intellectual Capital: Model Pengukuran, Framework Pengungkapan& Kinerja Organisasi. UMMPress (2017). 321 p.

OpenVSLAM-Based Development Framework for Web-Based VR Tours Using 360VR Videos and Its Extensions

Yoshihiro Okada[1][✉], Wei Shi[2], and Kosuke Kaneko[3]

[1] NOE, Data-Driven Innovation Initiative, Kyushu University, Fukuoka, Japan
okada@inf.kyushu-u.ac.jp
[2] Research Institute for Information Technology, Kyushu University, Fukuoka, Japan
[3] AiRIMaQ, Kyushu University, Fukuoka, Japan

Abstract. This paper proposes an extended development framework for web-based VR tours using 360VR videos based on OpenVSLAM. The authors have already proposed the development framework for web-based VR tours as navigation contents of 360VR videos and as walkthrough contents of 360VR images. However, it is not easy to prepare 360VR images of different locations for the walkthrough contents if the number of those images is too many. So, the authors have also proposed new development framework for web-based VR tours as both navigation contents and walkthrough contents using 360VR videos, not using 360VR images. For the walkthrough contents, we use 360VR videos because the required 360VR images can be extracted as keyframes of the same scenes from the 360VR videos. Furthermore, when there are many scenes, i.e. many 360VR images in one VR tour, it will become very convenient that the user can go to his/her desired scenes directly by the keyword search. Also, when 360VR videos created as educational contents include audio narrations, it will become also very convenient if the multi-language subtitles are available from the audio narrations. So, the authors extended their development framework for web-based 360VR tours to support the keyword search and multi-language subtitles using Google Cloud Vision AI and free AssemblyAI services.

Keywords: VR tours · 360VR · Development framework · Educational materials · OpenVSLAM

1 Introduction

This paper treats activities of NOE (the division of Next generation Open Education promotion). In April of last year, our Kyushu University established Data Driven Innovation Initiative consisting of several divisions. One of the divisions is our NOE. The missions of NOE are to provide educational materials using the latest ICT like XR (VR, AR, MR) and to support teachers for creating such educational materials. In this paper, we treat development frameworks for web-based VR tours of 360VR videos/images as the activities of NOE. Recently, many types of VR goggles have been commercially

L. Barolli (Ed.): EIDWT 2024, LNDECT 193, pp. 31–42, 2024.
https://doi.org/10.1007/978-3-031-53555-0_4

released and many entertainment applications for them have been created. However, there have been few educational XR contents so far because the creation of XR contents is tedious work that needs much time and human resources. Our NOE is considering enhancing educational efficiency using XR technology in the near future. On the other hand, many types of scanning devices have been researched and developed so far, e.g., lidar cameras, 360VR cameras and so on. NOE possesses these scanning devices to be used for developing educational XR contents. Using these scanning devices, we can obtain 3D data like Point Cloud Data (PCD), 3D model data, 360VR videos/images. To use such 3D data for educational XR contents, we have already proposed the development framework for web-based VR tours that support VR goggles and developed many web-based VR tour contents [12, 13].

Although we have proposed the development framework for navigation contents of 360VR videos and for walkthrough contents of 360VR images, it is not so easy for us to prepare 360VR images. For example, in the VR tours introduced in this paper, we prepared over 300 locations' 360VR images by spending much time and human resources. To reduce these costs, we have also proposed new OpenVSLAM-based development framework for web-based VR tours as both navigation and walkthrough contents of 360VR videos [14]. The development framework for navigation VR tours of 360VR videos need their moving path information and for walkthrough VR tours of 360VR images need their location information those were created manually so far. For generating them automatically, we decided to employ OpenVSLAM (Open Visual SLAM: Simultaneous Localization and Mapping) [16].

Furthermore, when there are many scenes, i.e. many 360VR images in one VR tour, it will become very convenient that the user can go to his/her desired scenes directly by the keyword search. Also, when 360VR videos created as educational contents include audio narrations, it will become also very convenient that multi-language subtitles are available. So, we extended our previous development framework for 360VR tours to support the keyword search and multi-language subtitles using Google Cloud Vision AI and free AssemblyAI services. In this paper, we explain these extensions besides introducing our OpenVSLAM-based development framework for 360VR tours.

The remainder of this paper is organized as follows: next Sect. 2 describes related work. We briefly explain the previously proposed framework [12, 13] and OpenVSLAM-based framework [14] in Sect. 3. In Sect. 4, we introduce the extended functionalities those are the keyword search and the multi-language subtitles. Finally, we conclude the paper and discuss about future work in Sect. 5.

2   Related Work

There have been many toolkit systems proposed for developing interactive 3D applications including educational materials. Indeed, we have proposed IntelligentBox [6] as one of them and we use it as a research system of our laboratory.

We have already developed many desktop applications including educational materials [3–5, 11] practically using IntelligentBox. We have also proposed the web-version of IntelligentBox [7] and there have been its web-based 3D applications. However, these web-based 3D applications cannot support native VR goggles such as Meta Quest 2. We need to use other toolkit systems for educational XR contents that support VR goggles. Unity is one of the most popular game engines in the world (https://unity.com/). We have developed a couple of serious games for education purposes using Unity [15]. Although Unity is very powerful and sophisticated tool, it requires programming knowledge and skills of its operations for developers. Therefore, the use of Unity is not easy for common end-users like teachers. We have developed several web-based 3D educational materials using our development frameworks [1, 2, 8–10, 17] based on Three.js, one of the popular WebGL-based 3D graphics JavaScript library systems (https://threejs.org/) and our development frameworks uses it.

For 360VR videos/images and PCD, there are several commercial services for creating interactive web VR contents using them. The service of RICOH (https://www.theta360.biz/) does support 360VR images but not 360VR videos nor PCD. As one of the remarkable points, this service provides a web-based editing function that enables even end-users to create their own web-based 360VR contents although our frameworks do not provide such a function. The service of Matterport (https://matterport.com/) does support 360VR images and PCD but not 360VR videos. As one of the remarkable points, this service provides a powerful tool based on AI technology called Matterport capture application that automatically generates a location map from 360VR images and PCD. Unfortunately, our previous framework does not provide such a tool so that we proposed the use of OpenVSLAM to generate moving paths and location information in our previous framework [14].

3   Previous Work

As previous work, this section introduces the development framework for web-based VR tours of 360VR images/videos and OpenVSLAM based development framework for web-based VR tours using 360VR videos.

3.1   360VR Cameras and Their Dedicated Software

When developing web-based VR tours, we have to take 360VR videos/images. Our NOE possesses Insta360 Pro shown in the left part of Fig. 1 that has six fish-eye lenses and takes high quality 360VR videos/images of up to 8K resolution. Also, our NOE possesses Insta360 ONE X2 shown in the middle part of the figure that has two fish-eye lenses and takes 360VR videos/images of up to 5.7K resolution. This portable camera is possible to walk around wearing a helmet with it as shown in the right part of the figure. For OpenVSLAM-based development framework, we use this portable 360VR camera. After taking two fish-eye lenses video/image shown in the left part of Fig. 2, we can convert them into one equirectangular video/image shown in the right part of the figure using the dedicated software, i.e., Insta360 STITCHER. By mapping an obtained equirectangular image/video onto the inside surface of a sphere object in a 3D space using the texture mapping technique, 360VR video/image contents become available.

Fig. 1. Insta360 Pro, Insta360 ONE X2, and a helmet with 360VR camera on its top.

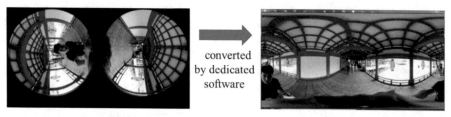

Fig. 2. Converting fish-eye images into equirectangular images using a dedicated software.

3.2 Development Framework for VR Tours of 360VR Images/Videos

Our development framework for VR tours of 360VR videos/images consists of two individual systems for walkthrough contents of 360VR images and navigation contents of 360VR videos. Figure 3 shows all JavaScript and HTML files of the two systems.

walk_map.html	JSM { Several JS files from Three.js library }
video_map.html	Assets
JS { walk_map.js,	Images { 360VR images, optional images }
walk_map_controls.js,	Videos { 360VR videos }
video_map.js,	Models { optional 3D model files }
video_map_controls.js	Movies { optional movie files }
}	Sounds { optional sound files }

Fig. 3. All files of two systems.

Each of the two systems includes of one HTML file like *_map.html, one main JavaScript file like *_map.js and one subsidiary JavaScript file like *_map_controls.js. *_map.js works with several functions derived from Three.js library. We do not need to change these programs when creating new contents. We need to prepare required media files, i.e., 360VR video/image files those are stored in Assets holder, and only to modify *_map_controls.js appropriately to read required media files into each of the two main JavaScript programs.

Actually, we created VR tours of the university library building for its PR those are navigation contents of 360VR videos and walkthrough contents of 360VR images. We introduce them in the followings.

3.2.1 Walkthrough Content of VR Tour of 360VR Images

There is the map of 3F, the library building shown in the left part and one screen shot of the walkthrough content in the right part of Fig. 4. You can see the same map in the left upper part of the screen shot. Over 100 small orange dots are displayed in the map that indicate the locations where 360VR images were taken. You can change the 360VR image into the next one taken at the location you want to go by clicking the corresponding orange dot. Also, thin gray cylinders are displayed in the screen shot that indicate the locations where 360VR images were taken. You can also change the 360VR image into the next one taken at the location by clicking the corresponding gray cylinder similarly to clicking one of the orange dots.

Fig. 4. The map of 3F, the library building (left) and one screen shot of the walkthrough content of 360VR images (right).

3.2.2 Navigation Contents of 360VR Videos

There is the map of 3F, the library building shown in the left part and one screen shot of the navigation content in the right part of Fig. 5. You can see the same map in the left upper part of the screen shot. There is a cyan color closed polyline that means the moving path that along with the person wearing a helmet with a potable 360VR camera walked for taking the 360VR video. You can go to the same location of the moving path by clicking on it. That means the current position moves to the corresponding position in the 360VR video. You can see the control panel in the middle lower part of the screen shot for controlling the 360VR video, i.e., play, pause, backward, forward, etc.

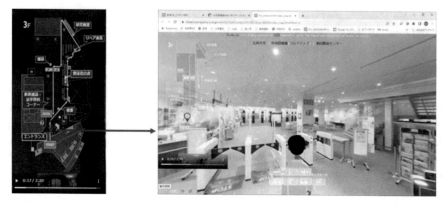

Fig. 5. The map of 3F, the library building (left) and one screen shot of the navigation content using 360VR video (right).

3.3 OpenVSLAM Based Development Framework for Web-Based VR Tours Using Portable 360VR Camera

For the walkthrough contents, when a floor is very large, we have to prepare many 360VR images and their location information. Such a task is very tedious and time-consuming. Therefore, we have proposed new development framework for web-based VR tours of both navigation and walkthrough contents using only 360VR videos. The main problem is the task to create map information that includes strict locations where 360VR images were taken and strict moving path where 360VR videos were taken. To create such map information from a video, we have to use any tools like SLAM (Simultaneous Localization and Mapping). We decided to employ OpenVSLAM (Open Visual SLAM), one of the Visual SLAM algorithms because it supports 360VR videos and its community version is downloadable from https://github.com/OpenVSLAM-Community/openvslam.git. In the following, we describe how OpenVSLAM works and show new 360VR tours of both navigation mode and walkthrough mode of 360VR videos realized by the new OpenVSLAM-based development framework.

3.3.1 OpenVSLAM

Figure 6 shows OpenVSLAM execution windows about the 360VR video of 3F, the library building. The right part is one frame of the 360VR video and the left part shows a moving path simultaneously generated by OpenVSLAM. There are many feature points called Landmarks and the moving path (purple polyline) with key frame locations (green triangles).

If you use the following command with the parameter '–eval-log-dir eval' when executing OpenVSLAM tool, you can obtain key frame location data as one text file called 'keyframe_trajectory.txt' in the 'eval' directory. Its contents are several lines each of those includes time_stamp (system time), x-, y- and z-coordinates, and quaternion values as rotation.

Fig. 6. OpenVSLAM execution windows for 360VR video of 3F, Library Building.

3.3.2 360VR Tours of Navigation Mode and Walkthrough Mode Using 360VR Videos

We made new OpenVSLAM-based development framework by a little modifying the previous development framework dedicated for the navigation contents of 360 VR videos. We do not need to use the previous development framework dedicated for the walkthrough contents of 360VR images due to the reason explained above. New OpenVSLAM-based development framework became very simple as shown in Fig. 7. There are three main files, video_map_v2.html, video_map_v2.js and video_map_controls_v2.js. One of the assets, 'Keyframes' directory was added that holds keyframe_trajectory.csv files.

walk_map_v2.html
video_map_v2.html
JS { video_map_v2.js,
    video_map_controls_v2.js }
JSM {
    Several JS files from Three.js library
    }
	Assets
	Images { optional images }
	Videos { 360VR videos }
	Keyframes { keyframe_trajectory.csv files }
	Models { optional 3D model files }
	Movies { optional movie files }
	Sounds { optional sound files }

Fig. 7. All files of new OpenVSLAM-based development framework.

Additional functionality is to display small orange dots those locations extracted from keyframe_trajectory.csv, that was generated from keyframe_trajectory.txt, instead of using the manually created map as shown in Fig. 8. The left and right parts of the figure show the navigation mode and walkthrough mode, respectively. For the walkthrough mode, the same html and JavaScript files of the navigation mode are used by adding the URL parameter 'WT = 1'. Thin gray cylinders are displayed by extracting the location information from keyframe_trajectory.csv file. For the walkthrough mode, it does not need to display a character object, i.e., a student model. As you see, the student model does not appear besides the control panel for the video in the walkthrough mode in the right part of Fig. 8.

Fig. 8. 360VR tours of navigation mode (left) and walkthrough mode (right) using 360VR video of 3F, the library building.

4   Extensions of Development Framework

Furthermore, we extended the development framework introduced in the previous section to support the keyword search and the multi-language subtitles. This section explains them. We tried to obtain keywords and multi-language subtitles about the 360VR videos of RIC (RadioIsotope Center, Kyushu University) because these videos contain voice narrations.

4.1   Keyword Search

The followings show the process to prepare keywords of each frame of the 360VR videos for the keyword search.

1) From the 360VR video, keyframe_trajectory.csv (time-code, x, y, z, quaternion) was generated by OpenVSLAM.
2) From the 360VR video, key frame images were generated by ffmpeg.exe and keyframe_trajectory.csv.
3) From the key frame images, keywords of each image were generated by Google Cloud Vison AI.

Once we obtained the keyframe information as 'keyframe_trajectory.csv' file by converting 'keyframe_trajectory.txt' file output from OpenVSLAM, we can extract key frames from the 360VR video based on time-stamps of the keyframe information. There are the five-stories and three-stories buildings of RIC. We extracted the corresponding key frame images from each of the eight 360VR videos using ffmpeg.exe by the command below.

> ffmpeg -ss time-code -i 360VR.mp4 -r 1 -vframes 1 -f image2 output.png

The parameter "-ss time-code (sec.)" specifies a key frame to be extracted. After that, we extracted keywords from each key frame image using Google Cloud Vision AI. Before that, we tried to execute another image OCR tool called 'Tessearact-OCR' (https://github.com/tesseract-ocr/tessdoc) because this tool is free software. Fortunately, it works for standard images containing recognized characters. However, our images are 360VR images, i.e., equirectangular images, for these images, 'Tessearact-OCR' does not work correctly at all. Therefore, we decided to use Google Cloud Vision AI although it is not free. For example, when we specified the image file of the left part of Fig. 9 by the following command of 'Tessearact-OCR', it returned the file contains some strange words.

> C:\Program Files\Tesseract-OCR\Tesseract.exe 18.png 18.txt -l jpn

Instead, when we specified the same image file in Google Cloud Vision AI, it returned the file contains many recognized words as shown in the right part of Fig. 9. These words can be used for the keyword search.

．．．．．

| RI仕分室画子7-11 「おねが
いサーベイメーターの電源は
必ずOFFELT FEL1.(水電池が非
常に価です。)プ総合セン
ターしませんアイソトープに
つき出し持出禁止サーベイ
メータ保管庫

．．．．．

Fig. 9. 360VR image used for OCR (left) and recognized keywords (right).

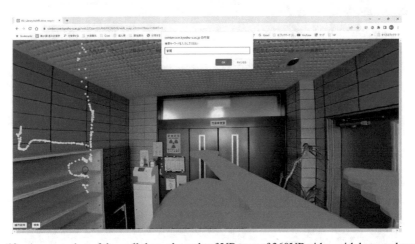

Fig. 10. A screen shot of the walkthrough mode of VR tour of 360VR video with keyword search button.

We modified our development framework furthermore a little for supporting the keyword search. Figure 10 shows a screen shot of the walkthrough mode of VR tour of 360 video with the keyword search button. By pushing the button, the dialog for entering a keyword will appear as shown in the middle upper part of the screen shot.

After entering a keyword, its corresponding scenes are searched and their corresponding dots become red color as shown in the left lower part of the screen shot. By clicking one of the red dots, you can go to the corresponding 360VR image.

4.2 Multi-language Subtitles

The followings show the process to prepare text transcription files for the multi-language subtitles.

1) From the 360VR video, *.wav file as the audio part of the 360VR video was generated by XMedia Recode.exe.
2) From *.wav file, Japanese transcript (*.vtt) was generated by AssemblyAI service.
3) From Japanese transcript, multi-language transcripts can be generated by any translation service.

Fig. 11. 360VR video (left) and its Japanese transcript in VTT format (right).

Fig. 12. A screen shot of the navigation mode of 360VR video with subtitles.

As shown in Fig. 11, firstly we used XMedia Recode.exe, one of the free software, to generate an audio file from the 360VR video that contains voice narration. Then, we obtained *.wav file as the audio part. Secondly, we used AssemblyAI service to generate Japanese transcript file in VTT format. The obtained Japanese transcript contains almost the same as the voice narration in the 360VR video. Finally, we obtain multi-language transcripts by any translation service such as Google translation service. Figure 12 shows a screen shot of the navigation mode of RIC VR tour of 360VR video. In the middle upper part of the figure, you can see Japanese transcript.

5 Conclusions

Many types of scanning devices e.g., lidar scanners, 360VR cameras and so on have been researched and developed. Using these devices, we can obtain 3D data like 3D models, Point Cloud Data (PCD) and 360VR videos/images. For creating educational XR contents using these kinds of data, we have already proposed the development framework for web-based VR tours of 360VR videos/images. However, it is still tedious work to prepare 360VR images and their location information if the number of them is too many. Therefore, we also proposed OpenVSLAM-based development framework. By using OpenVSLAM, we can obtain the location information and moving path information from 360VR videos. Furthermore, we extended the OpenVSLAM-based development framework to support the keyword search and multi-language subtitles, and we introduced their simple examples in this paper.

As future work, we will make more and more VR tour contents practically using the proposed development framework to justify the usefulness of it. Also, we will try to implement an editing function similar to the web-based editing function of the service of Ricoh theta.

Acknowledgement. This research was partially supported by JSPS KAKENHI Grant Number JP22H03705 and VISION EXPO Project of Kyushu University.

References

1. Ma, C., et al.: E-learning material development framework supporting VR/AR based on linked data for IoT security education. In: Proceedings of 6th International Conference on Emerging Internet, Data & Web Technologies (EIDWT 2018), pp. 479–491 (2018). ISBN 978-3-319-75928-9
2. Hirayama, D., et al.: Web-based interactive 3D educational material development framework and its authoring functionalities. In: Proceedings of the 22nd International Conference on Network-Based Information Systems (NBiS-2019), pp. 258–269 (2019). ISBN 978-3-030-29028-3
3. Kosuki, Y., Okada, Y.: 3D visual component-based development system for medical training systems supporting haptic devices and their collaborative environments. In: Proceedings of 6th International Conference on Complex, Intelligent, and Software Intensive Systems, CISIS 2012, pp. 687–692 (2012). https://doi.org/10.1109/CISIS.2012.131
4. Nomi, M., Okada, Y.: Improvement of dental treatment training system using a haptic device. In: Barolli, L., Takizawa, M., Enokido, T., Chen, H.-C., Matsuo, K. (eds.) BWCCA 2020. LNNS, vol. 159, pp. 143–153. Springer, Cham (2021). https://doi.org/10.1007/978-3-030-61108-8_14

5. Nomi, M., Okada, Y.: Dental treatment training system using haptic device and its user evaluations. In: Barolli, L. (eds.) CISIS 2022. LNNS, vol. 497, pp. 569–580. Springer, Cham (2022). https://doi.org/10.1007/978-3-031-08812-4_55

6. Okada, Y., Tanaka, Y.: IntelligentBox: a constructive visual software development system for interactive 3D graphic applications. In: Proceedings of Computer Animation 1995, pp. 114–125. IEEE CS Press (1995)

7. Okada, Y.: Web version of IntelligentBox (WebIB) and its integration with Webble world. In: Arnold, O., Spickermann, W., Spyratos, N., Tanaka, Y. (eds.) WWS 2013. CCIS, vol. 372, pp. 11–20. Springer, Heidelberg (2013).https://doi.org/10.1007/978-3-642-38836-1_2

8. Okada, Y., et al.: Framework for development of web-based interactive 3D educational contents. In: 10th International Technology, Education and Development Conference, pp. 2656–2663 (2016)

9. Okada, Y., et al.: Interactive educational contents development framework based on linked open data technology. In: 9th annual International Conference of Education, Research and Innovation, pp. 5066–5075 (2016)

10. Okada, Y., et al.: Interactive educational contents development framework and its extension for web-based VR/AR applications. In: Proceedings of the GameOn 2017, Eurosis, pp. 75–79 (2017). ISBN 978-90-77381-99-1

11. Okada, Y., Ogata, T., et al.: Component-based approach for prototyping of Tai Chi-based physical therapy game and its performance evaluations. Comput. Entertain. **14**(1) (2017). https://doi.org/10.1145/2735383

12. Okada, Y., Kaneko, K., Shi, W.: Development framework using 360VR cameras and lidar scanners for web-based XR educational materials supporting VR goggles. In: Advances in Internet, Data & Web Technologies, EIDWT 2023, pp. 401–412, February 2023

13. Okada, Y., Kaneko, K., Shi, W.: Development framework for web-based VR tours and its examples. In: 27th International Conference Information Visualisation (IV), pp. 420–425, July 2023

14. Okada, Y., Kaneko, K., Shi, W.: Development framework based on OpenVSLAM for web-based VR tours using 360VR videos. In: International Conference on WWW/Internet 2023, pp. 109–116, October 2023

15. Sugimura, R., et al.: Mobile game for learning bacteriology. In: Proceedings of IADIS 10th International Conference on Mobile Learning, pp. 285–289 (2014)

16. Sumikura, S., Shibuya, M., Sakurada, K.: OpenVSLAM: a versatile visual SLAM framework. In: MM 2019: Proceedings of the 27th ACM International Conference on Multimedia, pp. 2292–2295 (2019)

17. Yamamura, H., et al.: A development framework for RP-type serious games in a 3D virtual environment. In: VENOA 2020 in Proceedings of the 14th International Conference on Complex, Intelligent, and Software Intensive Systems (CISIS 2020), pp. 166–176 (2020)

Implementation of a Fuzzy-Based System for Assessment of Logical Trust Considering Reliability as a New Parameter

Shunya Higashi[1(✉)], Phudit Ampririt[1], Ermioni Qafzezi[2], Makoto Ikeda[2], Keita Matsuo[2], and Leonard Barolli[2]

[1] Graduate School of Engineering, Fukuoka Institute of Technology, 3-30-1 Wajiro-Higashi, Higashi-Ku, Fukuoka 811-0295, Japan
{mgm23108,bd21201}@bene.fit.ac.jp
[2] Department of Information and Communication Engineering, Fukuoka Institute of Technology, 3-30-1 Wajiro-Higashi, Higashi-Ku, Fukuoka 811-0295, Japan
qafzezi@bene.fit.ac.jp, makoto.ikd@acm.org, {kt-matsuo,barolli}@fit.ac.jp

Abstract. The human-to-human and human-to-devices communications are becoming increasingly intricate and unpredictable, which complicate the decision-making in different scenarios. So, many researcher are developing methods for assessing the trustworthiness of these kinds of communications. The Logical Trust (LT) is one crucial concept in trust computing that refer to the level of trust or confidence that an individual or system has in order to protect data, systems, or communications. In this paper, we introduce a Fuzzy-based system for evaluating LT, considering four parameters: Belief (Be), Experience (Ep), Rationality (Ra) and Reliability (Re), which is a new parameter. We evaluate the proposed system by computer simulations. We investigate the effect of each input parameter on the performance of the implemented system. The simulation results show the LT parameter increases when Be, Ep, Ra and Re are increasing. When Be and Ep values are 0.9 for all Ra and Re values, the LT values are more than 0.5, indicating that the people or devices are trustworthy.

1 Introduction

With the rapid growth of digital technologies, exchanging knowledge in the digital world has raised important questions about privacy and security. Issues related to data privacy, online surveillance, and cybersecurity have become central concerns in the social domain. Also, the interaction between humans and devices has become more intricate and problematic. Therefore, security algorithms or trust computing are required to guarantee reliability in the digital and social worlds.

The trust is the foundational element of decision-making and can be employed in various situations. Therefore, by contemplating these trust-related attributes within the digital world, the digital operators can distinguish individuals and determine their connections. To guarantee the reliability of a connection, it is essential to consider the acceptable level of risk involved in building the trust. For instance, when two individuals communicate with each other, they want a reliable medium to ensure the security

L. Barolli (Ed.): EIDWT 2024, LNDECT 193, pp. 43–52, 2024.
https://doi.org/10.1007/978-3-031-53555-0_5

of information and privacy, thereby fostering human-to-thing trust. Furthermore, they should have mutual trust before initiating any interaction [1, 2].

In 5th Generation (5G) and beyond networks, the concept of trust should be considered in order to quantify the system trustworthiness and maintain high levels of security and computer performance. Furthermore, trust in cloud, fog, and edge computing plays a crucial role in mitigating transaction risks and preventing data, privacy, and other information leaks [3–6].

The concept of trust is applied in a wide range of networks and systems. For instance, in Peer-to-Peer (P2P) networks, users are directly connected, enabling file transfers. The user assess and quantify the capabilities and experiences [7, 8].

The trust modeling in the digital world may not immediately apply to society. Social trends continue to hold significance in an increasingly information-driven digital society. Although more precise trust models could be implemented by incorporating social theories, it remains a challenging and complex issue. This complexity arises from the fact that humans express trust in qualitative, intricate and nuanced ways. For quantifying and evaluating the trust are needed many parameters, making this process NP-Hard. Consequently, a variety of algorithms, including Fuzzy Logic (FL), Hierarchical Bayesian Models, Game Theory, and others, become indispensable [9, 10].

In this paper, we introduce a Fuzzy-based system for evaluation of Logical Trust (LT). This system considers four parameters: Belief (Be), Experience (Ep), Rationality (Ra) and Reliability (Re), which is new parameter. To assess the performance of our proposed system, we conducted computer simulations. The simulation results show the LT parameter increases when Be, Ep, Ra and Re are increasing. Moreover, when Be and Ep reach 0.9 for all Ra and Re values, all LT values are more than 0.5, which indicate that a person or device is trustworthy.

The paper is organized into six sections. Following the introduction, Sect. 2 gives an overview of trust computing. Section 3 provides an introduction to Fuzzy Logic (FL). Section 4 presents the proposed Fuzzy-based system and its components. In Sect. 5 are discussed the simulation results. Finally, in Sect. 6, we conclude the paper and give future work.

2   Trust Computing

In recent years, the proliferation of devices such as smartphones, sensors, and IoT devices, along with the exponential growth of data, has facilitated a wide range of communications. The convergence of the physical and digital world is driven by the collaborative efforts of numerous technologies. Such knowledge sharing within the digital domain is poised to have a profound impact on social theory. Consequently, ensuring secure and highly reliable communication in digital space becomes imperative.

The trust computing represents a concept for bolstering the security and dependability of computing systems and digital transactions. Its primary aim is to establish a bedrock of trust in both the hardware and software components of a computer system or network. Trust computing involves a variety of technologies and practices to realize these objectives.

The trust computing can be implemented by involving the interactions of people, objects and digital entities. However, quantifying trust is a difficult and complex task because the perceptual mechanisms of physical space cannot be seamlessly applied to the digital context. Also, the convergence of physical and digital spaces introduces vulnerabilities that could potentially undermine authenticated access control and security mechanisms. Moreover, the presence of malicious entities within the digital network can lead to many issues [11]. Thus, trust modeling serves as a valuable model for evaluating and making decisions to tackle real-world challenges [12, 13].

The core concept of trust considers the relationship between the truster and trustee. The individual seeking a service is typically referred to as the trustor, entrusting their confidence in the trustee, often a business entity, to deliver the service. Depending on the context, this relationship may not always be strictly one-on-one, but it can extend to one person to a group or even a group to another group [14].

Trust assessment can be divided into individual trust and relational trust as shown in Fig. 1 [15].

- **Individual Trust Assessment** evaluates the degree of trustworthiness or credibility attributed to an individual, typically within a particular context or relationship by taking into account the trustee personal attributes. This process considers different factors, behaviors, and attributes that play a key role in determining whether others have their reliance, confidence, or trust in an individual. The ability to trust someone can be gauged by their emotional sense, including sentiments and inclinations.
- **Relational Trust Assessment** assesses the trust value by taking into account the interaction between the trustor and trustee. It considers interpersonal relationships of groups, organizations, or institutions and the primary focus lies in comprehending the dynamics of trust between individuals or entities, enabling the identification of both the robust aspects and vulnerabilities in trust within these relationships.

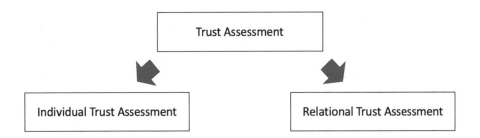

Fig. 1. Trust assessment process.

3   FL Overview

The FL systems can deal with nonlinear mapping of input data vectors to scalar outputs. These systems can process linguistic and numerical data distinctly, considering truth, falsity, or intermediate truth values [16, 17]. Moreover, FL is closely related with human cognition and natural language processing. It can process nonlinear functions of any level of complexity [18].

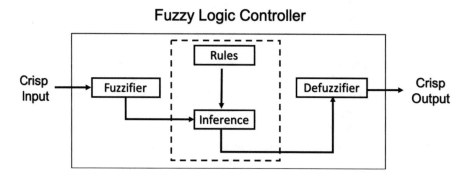

Fig. 2. FLC structure.

The main component of a Fuzzy-based system is the Fuzzy Logic Controller (FLC) shown in Fig. 2. The FLC structure has four elements: fuzzifier, inference engine, Fuzzy Rule Base (FRB), and defuzzifier. To generate fuzzy input, the fuzzifier employs linguistic variables to map fuzzy sets and crisp values. The inference engine utilizes FRB and fuzzified input values to deduce fuzzy outputs. Three commonly used fuzzy inference techniques are Sugeno, Mamdani, and Tsukamoto fuzzy inferences. The fuzzy rules can be formulated based on expert knowledge or derived from numerical data. Finally, defuzzifier transforms fuzzy outputs into precise control outputs [18, 19].

4   Proposed Fuzzy-Based System

In this paper, we consider FL to implement the Fuzzy-based System for Assessment of Logical Trust (FSALT). The structure of FSALT is shown in Fig. 3. For the implementation of the system four input parameters are taken into consideration: Belief (Be), Experience (Ep), Rationality (Ra) and Reliability (Re), which is a new parameter. The output parameter is Logical Trust (LT).

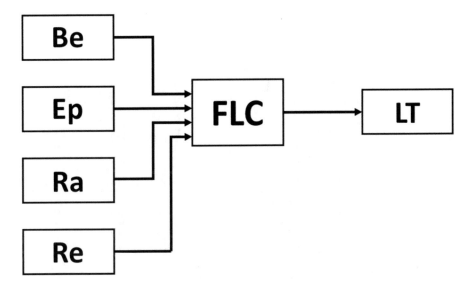

Fig. 3. Proposed system structure.

Table 1. Parameter and their term sets.

Parameters	Term set
Belief (Be)	Low (Lo), Medium (Me), High (Hi)
Experience (Ep)	Inexperience (In), Normal (N), Good (Gd)
Rationality (Ra)	Low (Lw), Medium (Md), High (Hg)
Reliability (Re)	Low (L), Medium (M), High (H)
Logical Trust (LT)	LT1, LT2, LT3, LT4, LT5, LT6, LT7, LT8, LT9

The membership functions are shown in Fig. 4 and Table 1 provides a listing of parameters and their corresponding term sets. While Table 2 shows the FRB, comprising of 81 rules. The syntax of a control rule follows the pattern: IF 'Condition' THEN 'Control Action'. For instance, Rule 81 can be interpreted as follows: IF Belief is High AND Experience is Good AND Rationality is High AND Reliability is High, THEN Logical Trust is Logical Trust level 9 (LT9).

Table 2. FRB.

Rules	Be	Ep	Ra	Re	LT	Rules	Be	Ep	Ra	Re	LT
1	Lo	In	Lw	L	LT1	41	Me	N	Md	M	LT5
2	Lo	In	Lw	M	LT1	42	Me	N	Md	H	LT6
3	Lo	In	Lw	H	LT2	43	Me	N	Hg	L	LT5
4	Lo	In	Md	L	LT1	44	Me	N	Hg	M	LT6
5	Lo	In	Md	M	LT2	45	Me	N	Hg	H	LT7
6	Lo	In	Md	H	LT3	46	Me	Gd	Lw	L	LT4
7	Lo	In	Hg	L	LT2	47	Me	Gd	Lw	M	LT5
8	Lo	In	Hg	M	LT3	48	Me	Gd	Lw	H	LT6
9	Lo	In	Hg	H	LT4	49	Me	Gd	Md	L	LT5
10	Lo	N	Lw	L	LT1	50	Me	Gd	Md	M	LT6
11	Lo	N	Lw	M	LT2	51	Me	Gd	Md	H	LT7
12	Lo	N	Lw	H	LT3	52	Me	Gd	Hg	L	LT6
13	Lo	N	Md	L	LT2	53	Me	Gd	Hg	M	LT7
14	Lo	N	Md	M	LT3	54	Me	Gd	Hg	H	LT8
15	Lo	N	Md	H	LT4	55	Hi	In	Md	L	LT4
16	Lo	N	Hg	L	LT3	56	Hi	In	Md	M	LT5
17	Lo	N	Hg	M	LT4	57	Hi	In	Md	H	LT6
18	Lo	N	Hg	H	LT5	58	Hi	In	Md	L	LT5
19	Lo	Gd	Lw	L	LT2	59	Hi	In	Md	M	LT6
20	Lo	Gd	Lw	M	LT3	60	Hi	In	Md	H	LT7
21	Lo	Gd	Lw	H	LT4	61	Hi	In	Hg	L	LT6
22	Lo	Gd	Md	L	LT3	62	Hi	In	Hg	M	LT7
23	Lo	Gd	Md	M	LT4	63	Hi	In	Hg	H	LT8
24	Lo	Gd	Md	H	LT5	64	Hi	N	Lw	L	LT5
25	Lo	Gd	Hg	L	LT4	65	Hi	N	Lw	M	LT6
26	Lo	Gd	Hg	M	LT5	66	Hi	N	Lw	H	LT7
27	Lo	Gd	Hg	H	LT6	67	Hi	N	Md	L	LT6
28	Me	In	Lw	L	LT2	68	Hi	N	Md	M	LT7
29	Me	In	Lw	M	LT3	69	Hi	N	Md	H	LT8
30	Me	In	Lw	H	LT4	70	Hi	N	Hg	L	LT7
31	Me	In	Md	L	LT3	71	Hi	N	Hg	M	LT8
32	Me	In	Md	M	LT4	72	Hi	N	Hg	H	LT9
33	Me	In	Md	H	LT5	73	Hi	Gd	Lw	L	LT6
34	Me	In	Hg	L	LT4	74	Hi	Gd	Lw	M	LT7
35	Me	In	Hg	M	LT5	75	Hi	Gd	Lw	H	LT8
36	Me	In	Hg	H	LT6	76	Hi	Gd	Md	L	LT7
37	Me	N	Lw	L	LT3	77	Hi	Gd	Md	M	LT8
38	Me	N	Lw	M	LT4	78	Hi	Gd	Md	H	LT9
39	Me	N	Lw	H	LT5	79	Hi	Gd	Hg	L	LT8
40	Me	N	Md	L	LT4	80	Hi	Gd	Hg	M	LT9
						81	Hi	Gd	Hg	H	LT9

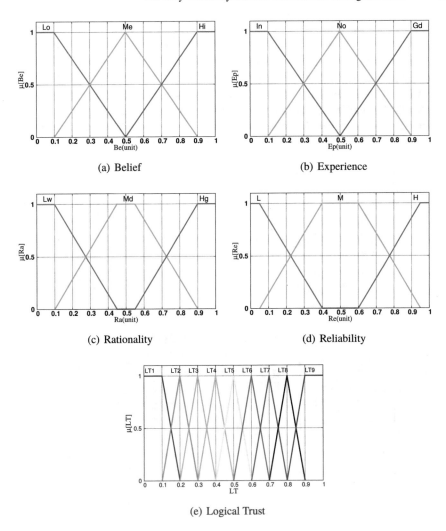

Fig. 4. Membership functions.

5   Simulation Results

In this section, we present the simulation results. We show the effect of four input parameters on LT. The results are presented in Fig. 5, Fig. 6, and Fig. 7. We show the relation of LT with Re by considering Be, Ep and Ra as constant parameters.

In Fig. 5, we keep Be 0.1 and increase Ra from 0.1 to 0.5 and then to 0.9. The LT is increased by 12% and 10%, when Ep is 0.1 and Re is 0.6. This shows that LT is increased with the increase of the Ra value. To investigate the influence of Ep on LT, we vary Ep value from 0.1 to 0.9 as shown in Fig. 5(a) and Fig. 5(b). When Ra is 0.5 and Re is 0.8, the LT value is increased 20%.

We can see the effect of Be on LT by comparing Fig. 5 and Fig. 6, where the Be value is increased from 0.1 to 0.5. When Ep is 0.1, Ra is 0.5, and Re is 0.3, the LT value is increased by 20%. This indicates that when Be increases, the LT also increases, which shows that the belief has a positive influence on LT.

In Fig. 7, we increase the Be value to 0.9. By comparing with Fig. 5 and Fig. 6, we see that the LT values are substantially increased. In case when Ep is 0.9, all LT values are higher than 0.5, indicating that people or devices are considered trustworthy.

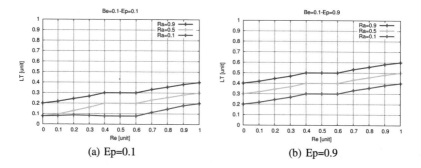

Fig. 5. Simulation results for Be=0.1.

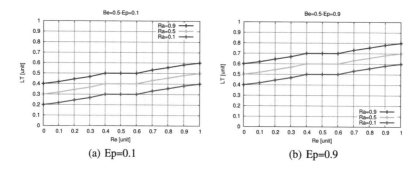

Fig. 6. Simulation results for Be=0.5.

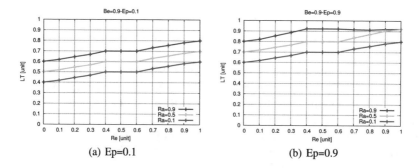

Fig. 7. Simulation results for Be=0.9.

6   Conclusions and Future Work

In this paper, firstly we provided a brief introduction to trust computing and FL. Then, we introduced our proposed FSALT system considering four input parameters. Based on the simulation results, we conclude as following.

- The LT values are increased when Re, Ep and Be values are increased, which demonstrates that they have a positive impact on LT.
- When Be and Ep are 0.9, for all Re and Ra values, the LT values are higher than 0.5, indicating that people or devices are considered trustworthy.

In the future work, we will consider new parameters and carry out extensive simulations to evaluate FSALT system.

References

1. Ting, H.L.J., Kang, X., Li, T., Wang, H., Chu, C.K.: On the trust and trust modeling for the future fully-connected digital world: a comprehensive study. IEEE Access **9**, 106–743 (2021). https://doi.org/10.1109/ACCESS.2021.3
2. Wang, D., Muller, T., Liu, Y., Zhang, J.: Towards robust and effective trust management for security: a survey. In: 2014 IEEE 13th International Conference on Trust, Security and Privacy in Computing and Communications, pp. 511–518 (2014)
3. Benzaïd, C., Taleb, T., Farooqi, M.Z.: Trust in 5G and beyond networks. IEEE Netw. **35**(3), 212–222 (2021)
4. Rahman, F.H., Au, T.-W., Newaz, S.S., Suhaili, W.S., Lee, G.M.: Find my trustworthy fogs: a fuzzy-based trust evaluation framework. Futur. Gener. Comput. Syst. **109**, 562–572 (2020)
5. Uslu, S., Kaur, D., Durresi, M., Durresi, A.: Trustability for resilient internet of things services on 5G multiple access edge cloud computing. Sensors **22**(24), 9905 (2022)
6. Cai, H., Li, Z., Tian, J.: A new trust evaluation model based on cloud theory in e-commerce environment. In: 2011 2nd International Symposium on Intelligence Information Processing and Trusted Computing, pp. 139–142 (2011)
7. Wang, Y., Vassileva, J.: Bayesian network-based trust model. In: Proceedings IEEE/WIC International Conference on Web Intelligence (WI 2003), pp. 372–378 (2003)
8. Zhou, P., Gu, X., Zhang, J., Fei, M.: A priori trust inference with context-aware stereotypical deep learning. Knowl.-Based Syst. **88**, 97–106 (2015). https://www.sciencedirect.com/science/article/pii/S095070511500307X
9. Zhang, D., Yu, F.R., Yang, R.: A machine learning approach for software-defined vehicular ad hoc networks with trust management. In: 2018 IEEE Global Communications Conference (GLOBECOM), pp. 1–6 (2018)
10. Jayasinghe, U., Lee, G.M., Um, T.-W., Shi, Q.: Machine learning based trust computational model for IoT services. IEEE Trans. Sustain. Comput. **4**(1), 39–52 (2019)
11. Hu, W.-L., Akash, K., Reid, T., Jain, N.: Computational modeling of the dynamics of human trust during human-machine interactions. IEEE Trans. Hum.-Mach. Syst. **49**(6), 485–497 (2019)
12. Zolfaghar, K., Aghaie, A.: Evolution of trust networks in social web applications using supervised learning. Procedia CS **3**, 833–839 (2011)
13. Kumar, S., Shah, N.: False information on web and social media: a survey (2018)

14. Braga, D.D.S., Niemann, M., Hellingrath, B., Neto, F.B.D.L.: Survey on computational trust and reputation models. ACM Comput. Surv. **51**(5), 1–40 (2018). https://doi.org/10.1145/3236008
15. Cho, J.-H., Chan, K., Adali, S.: A survey on trust modeling. ACM Comput. Surv. (CSUR) **48**(2), 1–40 (2015)
16. Jantzen, J.: Tutorial on fuzzy logic. Technical University of Denmark, Department of Automation, Technical Report (1998)
17. Zadeh, L.A.: Fuzzy logic. Computer **21**(4), 83–93 (1988)
18. Lee, C.-C.: Fuzzy logic in control systems: fuzzy logic controller. I. IEEE Trans. Syst. Man Cybern. **20**(2), 404–418 (1990)
19. Mendel, J.M.: Fuzzy logic systems for engineering: a tutorial. Proc. IEEE **83**(3), 345–377 (1995)

Source Code Vulnerability Detection Based on Graph Structure Representation and Attention Mechanisms

Yiran Sun[1], Ziqi Wang[1], Kewei Liu[2], and Baojiang Cui[1(✉)]

[1] Beijing University of Posts and Telecommunications, No.10 Xitucheng Road, Beijing, China
2021111153@bupt.cn, {wangziqi,cuibj}@bupt.edu.cn
[2] Tsinghua University, Tsinghua University, Haidian District, Beijing, China
kw-liu20@mails.tsinghua.edu.cn

Abstract. With the increasingly serious trend of information security, software vulnerability has become one of the main threats to computer security. How to accurately detect the vulnerabilities in programs is a key issue in the field of information security. However, the accuracy of existing static vulnerability detection methods decreases significantly when detecting vulnerabilities with inconspicuous characteristics. To solve this issue, a function-level vulnerability detection method based on code property graph and attention mechanism is proposed. The method firstly converts the program source code into a code property graph containing semantic feature information and slices it to eliminate redundant information not related to sensitive operations; secondly, the code property graph is encoded into numerical embedding vectors by using BERT; then, a large-scale feature dataset is used to train a neural network based on the attention mechanism; finally, the trained neural network is used to realize the vulnerability detection of the target program. The experimental results show that the accuracy of this method on SARD dataset and GitHub dataset reaches 95.3% and 94.2% respectively, which is a significant improvement compared with the baseline, and proves that this method can effectively improve the accuracy of detecting vulnerabilities with inconspicuous vulnerability features.

1 Introduction

With the development of information technology, computer software has become an indispensable part of life, and computer systems are widely used in all walks of life, including medical, educational, military, political and new retail fields. With the rapid development and popularization of computer systems, software vulnerabilities have become one of the major threats to the security of computer systems. Once a vulnerability is exploited by an attacker, it can lead to serious consequences.

However, the accuracy of existing static vulnerability detection methods generally decreases when detecting vulnerabilities with obscure vulnerability characteristics. Aiming at the above problem, this paper combines code property graph [10], attention mechanism and vulnerability detection, and proposes a function-level vulnerability detection method. The method firstly generates a code property graph and an expanded

L. Barolli (Ed.): EIDWT 2024, LNDECT 193, pp. 53–63, 2024.
https://doi.org/10.1007/978-3-031-53555-0_6

AST from the program source code to represent the semantic features of the code; secondly, the code graph is encoded into numerical embedding vectors by using a pre-trained BERT model [18]; lastly, a neural network model based on the attention mechanism is constructed, and is used to predict whether there is a vulnerability in the target program.

In order to verify the validity of this method, VULDEEPECKER [3], LIN et al. [1], μVULDEEPECKER [4], VUDDY [2], DEEPBUGS [6], DEVIGN [5], and FUNDED [13]are selected as the baseline of the method, and we conduct the experiments on the SARD dataset and the GitHub dataset. The experimental results show that the accuracy of the proposed method on the above two datasets reaches more than 92%, and all the metrics are higher than the baselines.

The main contributions of this paper are summarized as follows:

1. A method for modeling program source code based on code property graph is proposed, which first represents the program as a code property graph, and then encodes it into numerical embedding vectors by using a pre-trained BERT model, which can effectively avoid the loss of the semantic information of the program and provide support for the subsequent machine learning tasks.
2. A neural network model for source code vulnerability detection is proposed, which utilizes the attention mechanism to capture the key features related to the edge types while realizing the vulnerability detection.
3. By comparing the proposed method with SOTA on SARD dataset and GitHub dataset, the experimental results show that the proposed method has significant improvement over the comparison methods.

Section 2 introduces the background knowledge and related techniques. Section 3 describes the overall scheme and implementation details of the proposed vulnerability detection method. Sections 4 and 5 describes the experiments conducted in this paper. Section 6 summarizes the work of this paper and discusses the shortcomings and room for improvement of this method.

2 Related Work

This section introduces the background knowledge and related techniques involved in this paper.

2.1 Source Code Vulnerability Detection Techniques

Traditional vulnerability detection methods [7–9] mainly use predefined rules provided by human experts for code analysis, which is labor-intensive and inaccurate. Recently, researchers have proposed a number of deep learning (DL)-based vulnerability detection methods [1–5], which achieve state-of-the-art vulnerability detection performance by automatically learning the patterns of the vulnerable code without using human heuristics.

2.2 Application of Graph Structure Representation in Vulnerability Detection

Because the abstract graph structure of the program contains sufficient semantic information, the use of abstract graph structure for program analysis is currently a more popular trend. For example, TBCNN [11] models the program source code as an abstract syntax tree, and then performs tree-based convolution and pooling operations based on the structure of the abstract syntax tree and node types to classify the program. SecureSync [12] generates an abstract syntax tree and a program dependency graph from source code, and determines whether there are vulnerabilities in the source code through sub-graph isomorphism.

2.3 Attention Mechanism

Attention Mechanism (AM) [19] is a technique widely used in machine learning and natural language processing to model human attentional behavior. It allows a model to process sequential data by assigning different attentional weights to different parts of the input in order to pay more attention to the information relevant to the task at hand.

Self-Attention [20] is similar to Attention, they are both attention mechanisms. The difference is that Attention is source to target, the input source and output target are different. Self-Attention is source to source, is the Attention mechanism that occurs between the elements inside the source or between the elements inside the target.

3 Proposed Method

This section provides a detailed description of the vulnerability detection method based on source code graph structure representation and attention mechanism proposed in this paper. Firstly, the conversion method from source code to graph structure is introduced in Sect. 3.1, followed by the training and learning process of the model proposed in this paper in Sects. 3.2 and 3.3.

3.1 Conversion Method from Source Code to Graph Structure

3.1.1 Variable Name Replacement

To avoid interference with variable names during training, we use the compiler parser to rewrite variable names using a consistent naming scheme as a preprocessing step. This step ensures that trivial semantic differences in programs such as the choice of variable names do not affect the choice of token embeddings.

3.1.2 Generate Code Graph

After variable name substitution and code slicing of the source code, it is ready for code graph conversion. In this paper, two code graph representations are used: the extended AST proposed by Wang et al. [13] and the CPG generated by joern [21]. In this paper, we utilize the combination of the two graphs to comprehensively represent the semantic information of the code by storing the relationships of the extended AST and CPG in different relationship graphs - one relationship graph for each of the above nine relationships. We use the adjacency matrix to record the edge connections for each relation graph.

3.2 Graph Node Representations

In this paper, we use a pre-trained BERT model [18] to map the nodes (e.g., Stmt) and tokens of each program graph to numerical embedding vectors. The purpose of this is to construct a vector space such that words with similar context in the source code are close to each other in the vector space. To capture type information, we connect the embeddings of variables, constants, and (return) type of functions (e.g., int for integer variables) to the AST node name representations, and then go through a linear layer to obtain an initial representation of each node in the graph.

3.3 Model Training and Optimization Methods

The goal of a graph-supervised classification task is to learn a nonlinear mapping f from the input data to the output space,

$$f : X \mapsto Y \tag{1}$$

where X is the input data and Y is a set of graph labels.

Due to the use of non-injective aggregation functions, standard GNNs are unable to generalize to unknown graph structures, which limits the expressive power of such GNN architectures. We overcome this problem by using a graph isomorphism network (GIN) [14], which learns injective aggregations. This makes the model as powerful as the Weisfeiler-Lehman (WL) test [15] in graph classification tasks. As shown in Fig. 1, each layer of the GIN undergoes two main steps of neighborhood aggregation and non-linear transformation in passing and aggregating node feature information, which are described in detail next.

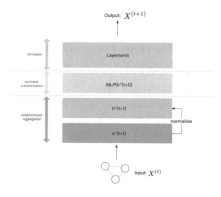

Fig. 1. Structure of A Single GIN Layer

3.3.1 Neighborhood Aggregation

The graph node representation obtained from BERT is aggregated with the features of the connected nodes through the aggregation function, but due to the different types of edges between the nodes, the weights for training should be different, so it is necessary to compute the weight matrix required for training in relation to the edge type, here the self-attention mechanism is used, the paradigm is as follows:

$$softmax\left(\frac{q*k^T}{\sqrt{d}}\right)*v \tag{2}$$

where both q and k are obtained from X by matrix operations:

$$q = X * W_q \tag{3}$$

$$k = X * W_k \tag{4}$$

where W_q and W_k are trainable parameters, the computed q and k are also n*768 dimensional matrices, whereas multiplying q and the transposed matrices of k gives the weight matrix of one type of edges (n*n), and then the weight matrices of the nine types of edges are summed:

$$E = \sum_{edge_{num}} q*k^T \tag{5}$$

When summing here, the adjacency matrix needs to be taken into account, adding 0 if the two points are not adjacent to each other, and adding the weight values if the two points are adjacent to each other. As the last to be processed by the softmax function, in order to ensure that the weight between the two points that are not adjacent be the weight of 0, we assigned the value of -1e9 to the element 0 of the matrix as the effect of $-\infty$.

The final weight matrix is calculated by substituting E into the paradigm:

$$E^{(t)} = softmax\left(\frac{E}{\sqrt{d}}\right) \tag{6}$$

This yields the n*n-dimensional weight matrix $E^{(t)}$. After that, the node embedding of the current (t+1)th level is computed using the aggregation and combination functions, and the weight matrix is used as an argument to the update function:

$$a^{(t+1)} = (1+\varepsilon)X^{(t)} + E^{(t)}X^{(t)} \tag{7}$$

where ε is a learnable parameter. Then the normalization operation needs to be performed on the computed results:

$$h^{(t+1)} = Layernorm(a^{(t+1)}) \tag{8}$$

Above is the operation of the neighborhood aggregation part of the model proposed in this paper.

3.3.2 Nonlinear Transformation

The vectors obtained from the neighborhood aggregation step are nonlinearly transformed by the MLP function:

$$MLP(h^{(t+1)}) = GELU(h^{(t+1)} * W + b) * W + b \tag{9}$$

where both W and b are trainable parameters. In order to avoid gradient explosion, the MLP function is processed using residual linking. Afterwards, it is normalized once again to be able to obtain the output of the (t+1)th layer vector:

$$X^{(t+1)} = Layernorm(h^{(t+1)} + MLP(h^{(t+1)})) \tag{10}$$

3.3.3 Graph Level Readout

In the graph-level readout, the node embeddings of each layer are summarized and then concatenated, and the final representation of the graph G combines the representations of the embeddings of all K layers:

$$h_G = CONCAT(\{SUM(X^{(k)})|k = 0, 1, 2, ..., K-1\}) \tag{11}$$

where K is the number of layers. A one-dimensional graph level vector representation is obtained using the global MLP (label prediction), which can be used directly for categorization tasks.

4 Experimental Setup

4.1 Evaluation Dataset

Table 1. Dataset For Evaluating Vulnerability Detection

Source	vul. types	samples	positive samples
SARD&NVD	30	90954	45477
GitHub	10	10400	5200

Table 1 gives the details of the FUNDED dataset selected for this paper, which contains a total of 101,354 function-level samples in the source language C. The data is not available in any other language. Half of this sample is positive (vulnerable) code samples. The dataset restricts the scope to the top 5 to the top 30 most dangerous software bugs (e.g., "buffer overflow", "out-of-bounds read/write", "NULL pointer dereference") as defined in CWE 2019. The publishers built this dataset from SARD [16], NVD [17], and open source projects hosted on GitHub.

4.2 Competitive Approaches

In terms of vulnerability detection, we compare the method proposed in this paper with seven related methods: VULDEEPECKER , μVULDEEPECKER , LIN et al. , VUDDY , DEEPBUGS, DEVIGN, and FUNDED; the first three methods are built on BiLSTM, VUDDY uses hash functions to discover vulnerable code clones, DEEP-BUGS utilizes feed-forward neural networks for vulnerability detection, DEVIGN uses standard GNN with untyped AST edges on the graph representation, and FUNDED uses GGNN with expanded AST to represent source code features.

4.3 Implementation

We implemented a GIN-based vulnerability detection model using PyTorch v.1.13, and the generation of graph structures using NetworkX. To construct the source code graph, we used joern to generate CPG14 and the FUNDED method proposed by Wang et al. to obtain the expanded AST. We trained and tested all the methods on a multi-core server equipped with 32-core 2.4 GHz Intel Xeon CPUs and Tesla V100 GPUs.

4.4 Evaluation Methodology

We use four higher-is-better metrics:

1. Accuracy: Ratio of correctly labeled cases to total number of cases tested.
2. Precision: The ratio of the number of samples correctly predicted to the total number of samples predicted to have a particular label. High precision indicates low false alarm rate.
3. Recall: The ratio of the number of correctly predicted samples to the total number of test samples belonging to a category. A high recall rate indicates a low false negative rate.
4. F1 score: The harmonic mean of precision and recall, calculated as $2 \times \frac{recall \times precision}{recall + precision}$. This metric is useful when the vulnerability types of the test data are unevenly distributed.

5 Experimental Results

5.1 Selection of Model Parameters and Graph Structure

We evaluate the performance of graph isomorphic networks on binary classification problems through several experimental setups. First, we consider the impact of different graph structures on model performance: extended AST alone, CPG alone, and a combination of both. In addition, we evaluate the impact of the number of model layers (K-hop neighborhood spans: 2, 3, 5, and 8) on model performance. Table 2, 3 presents the performance results expressed in terms of accuracy, precision, recall, and f1 score.

Table 2. Selection of Graph Structure's Impact on Model Performance

Graph	Accuracy	Precision	Recall	F1 Score
Extended AST	0.81	0.74	0.89	0.81
CPG	0.94	0.91	0.97	0.94
Combined	0.95	0.91	0.99	0.95

Table 3. Selection of Model Layer's Impact on Model Performance

Number of Layers	Accuracy	Precision	Recall	F1 Score
2	0.94	0.90	0.97	0.92
3	0.95	0.91	0.99	0.95
5	0.93	0.89	0.97	0.93
8	0.91	0.85	0.94	0.90

We chose the extended AST generated using the FUNDED tool [13] and the CPG generated using Joern [21] as the graph structure representations of the source code, and embedded them separately as vectors using the BERT model, comparing the impact on the performance of the binary classification task by using the different graph structures alone, and by splicing together the vector representations of the two graph structures. It can be found that the model trained using the extended AST alone has the worst classification performance, the model trained by CPG has improved performance, and the model trained by combining the two has the best results. This suggests that the extended AST proposed by Wang et al. does not provide a comprehensive abstraction of the source code semantics, but happens to be complementary to CPG to a certain extent, and thus we chose to splice the vector representations of the extended AST and CPG as inputs to the model in subsequent experiments.

We then compared the classification performance of 2, 3, 5 and 8 layers used in the GIN framework to find the optimal number of layers for graph-level classification. In all the same training setups, it is not the case that using a higher number of layers results in higher binary classification performance. This is because the more layers, the wider the range of neighborhoods that can be explored, but the binary classification task we chose is too simple for a complex model, which may lead to the appearance of gradient instability, and thus instead does not present as good results as a smaller number of layers. From the table, it can be seen that the GIN-3 model performs the best, so we choose the three-layer GIN model for the subsequent experiments.

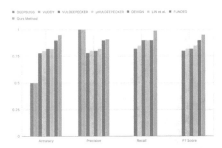

Fig. 2. Evaluation on standard vulnerability databases.

5.2   Evaluation on SARD and NVD

In the following experiment, we detect the CWE top-30 vulnerability types. We trained our detection model using training samples from the standard vulnerability database and samples collected by FUNDED's data collection tool.

As shown in the Fig. 2, the overall performance of vulnerability detection of the models proposed in this paper is the best. VUDDY and DEEPBUGS have lower detection accuracy due to the limitations of their detection models. VULDEEPECKER and μVULDEEPECKER using BiLSTM are effective for a few vulnerability types, but the detection accuracy for some types (e.g., CWE-469, CWE-676, and CWE-834) is less than 50% and may lead to a large number of false positives (i.e., low accuracy). By utilizing a rich manually labeled training dataset, FUNDED is the best performing competitive method. However, its pre-trained word2vec network may fail to detect certain vulnerable cases due to not capturing a certain linguistic keyword, e.g., sizeof. DEVIGN has the second best overall accuracy, but shows only a weak improvement of less than 2% compared to μVULDEEPECKER and is outperformed by FUNDED.

Overall, the detection model proposed in this paper is the only scheme with an average accuracy of more than 92% and has the best overall performance across all evaluation metrics.

5.3   Evaluation on GitHub

We extend our experiments to C functions collected from GitHub. In this experiment, we train on the SARD-NVD dataset and test on the GitHub dataset. Figure 3 reports the results for each metric. The accuracy on the GitHub code samples is degraded because the training data from the standard database cannot fully represent the vulnerability code samples in real programs. Overall, the model proposed in this paper performs best in terms of accuracy, recall, and F1 score. While VUDDY has the highest precision, the recall is much lower. This is because although VUDDY has the lowest false positives, it is too limited in detecting vulnerabilities - it misses more than 85% of the vulnerability test samples. In terms of accuracy, FUNDED is the best performing alternative model, but its F1 score is 20% lower than the model proposed in this paper, suggesting that the model proposed in this paper strikes a better balance between false positives and negative positives.

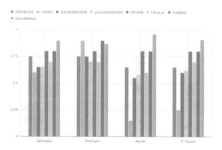

Fig. 3. Evaluation on GitHub samples.

6   Conclusions and Future Work

In this paper, to address the problem that the accuracy of existing static vulnerability detection methods generally decreases when detecting vulnerabilities with inconspicuous vulnerability features, a vulnerability detection method based on code property graph and attention mechanism is researched and implemented. The method represents the program source code as a code property graph, encodes the code property graph into embedding vectors using a pre-trained BERT model, trains a neural network model using the vectors, automatically learns the vulnerability patterns of the code, and predicts whether there are vulnerabilities in the target program using the trained model. In this paper, experiments are conducted on the standard vulnerability dataset composed of SARD and NVD, and the open source vulnerability dataset composed of GitHub, in which the F1 scores of the proposed method on the above two datasets reach 95.3% and 94.2%, which is a significant improvement compared with the baseline.

The vulnerability detection method proposed in this paper can effectively identify whether there is a vulnerability in a single function, which still has some room for improvement: first, this method detects vulnerabilities in a single function, without examining the context of the call, if the function to be detected does not have a vulnerability, but there is a vulnerability in the function it calls, the called function needs to be detected in order to find the vulnerability. Secondly, this method treats the vulnerability mining problem as a binary classification problem, so as to effectively predict the existence of vulnerabilities, but it can only predict whether there are vulnerabilities in the function, regardless of a single vulnerability or multiple vulnerabilities, and therefore the detection capability of the function when there are multiple vulnerabilities needs to be improved. Future research will be conducted to address these two issues.

References

1. Lin, G., et al.: Software vulnerability discovery via learning multi-domain knowledge bases. IEEE Trans. Dependable Secure Comput. **18**(5), 2469–2485 (2019)
2. Kim, S., et al.: VUDDY: a scalable approach for vulnerable code clone discovery. In: 2017 IEEE Symposium on Security and Privacy (SP). IEEE (2017)
3. Li, Z., et al.: Vuldeepecker: a deep learning-based system for vulnerability detection. arXiv preprint: arXiv:1801.01681 (2018)

4. Zou, D., et al.: μVulDeePecker: a deep learning-based system for multiclass vulnerability detection. IEEE Trans. Dependable Secure Comput. **18**(5), 2224–2236 (2019)

5. Zhou, Y., et al.: Devign: effective vulnerability identification by learning comprehensive program semantics via graph neural networks. In: Advances in Neural Information Processing Systems, vol. 32 (2019)

6. Pradel, M., Sen, K.: Deepbugs: a learning approach to name-based bug detection. Proc. ACM Program. Lang. **2**(OOPSLA), 1-25 (2018)

7. Cherem, S., Princehouse, L., Rugina, R.: Practical memory leak detection using guarded value-flow analysis. In: Proceedings of the 28th ACM SIGPLAN Conference on Programming Language Design and Implementation (2007)

8. Fan, G., et al.: SMOKE: scalable path-sensitive memory leak detection for millions of lines of code. In: 2019 IEEE/ACM 41st International Conference on Software Engineering (ICSE). IEEE (2019)

9. Heine, D.L., Lam, M.S.: Static detection of leaks in polymorphic containers. In: Proceedings of the 28th International Conference on Software Engineering (2006)

10. Yamaguchi, F., et al.: Modeling and discovering vulnerabilities with code property graphs. In: 2014 IEEE Symposium on Security and Privacy. IEEE (2014)

11. Mou, L., et al.: Convolutional neural networks over tree structures for programming language processing. In: Proceedings of the AAAI Conference on Artificial Intelligence, vol. 30, no. 1 (2016)

12. Pham, N.H., et al.: Detection of recurring software vulnerabilities. In: Proceedings of the 25th IEEE/ACM International Conference on Automated Software Engineering (2010)

13. Wang, H., et al.: Combining graph-based learning with automated data collection for code vulnerability detection. IEEE Trans. Inf. Forensics Secur. **16**, 1943–1958 (2020)

14. Xu, K., et al.: How powerful are graph neural networks?. arXiv preprint: arXiv:1810.00826 (2018)

15. Shervashidze, N., et al.: Weisfeiler-lehman graph kernels. J. Mach. Learn. Res. 12(9) (2011)

16. NIST. Software Assurance Reference Dataset Project. https://samate.nist.gov/SRD/. Accessed 1 June 2019

17. National Vulnerability Database (NVD). https://nvd.nist.gov. Accessed 1 May 2020

18. Devlin, J., et al.: BERT: pre-training of deep bidirectional transformers for language understanding. arXiv preprint: arXiv:1810.04805 (2018)

19. Vaswani, A., et al.: Attention is all you need. In: Advances in Neural Information Processing Systems, vol. 30 (2017)

20. Shaw, P., Uszkoreit, J., Vaswani, A.: Self-attention with relative position representations. arXiv preprint: arXiv:1803.02155 (2018)

21. Joern (2014). https://joern.readthedocs.io/en/latest/

22. Lipp, S., Banescu, S., Pretschner, A.: An empirical study on the effectiveness of static C code analyzers for vulnerability detection. In: Proceedings of the 31st ACM SIGSOFT International Symposium on Software Testing and Analysis (2022)

23. Lin, Y., et al.: Vulnerability dataset construction methods applied to vulnerability detection: a survey. In: 2022 52nd Annual IEEE/IFIP International Conference on Dependable Systems and Networks Workshops (DSN-W). IEEE (2022)

An Efficient Vulnerability Detection Method for 5G NAS Protocol Based on Combinatorial Testing

Siyuan Wang, Zhiwei Cui, Jie Xu[(✉)], and Baojiang Cui

School of Cyberspace Security, Beijing University of Posts and Telecommunications, Haidian District Xitucheng Road No. 10, Beijing, China
{wangsiyuan,zwcui,cheer1107,cuibj}@bupt.edu.cn

Abstract. With the rapid development and wide application of 5G networks, the security of 5G networks has become a widely concerned issue. Protocol security is the foundation of 5G network security. To protect 5G protocol security, we propose an efficient vulnerability detection method for 5G NAS protocol. In this work, we use a combinatorial testing algorithm to generate testing cases, which can detect the vulnerability caused by multi-parameter interaction. Furthermore, we define compliance constraints and semantic constraints to restrict the scale of input space and maximize the effectiveness of the testing cases. Finally, we implement a prototype system based on this method and then conduct practical vulnerability detection on 5G UE simulation environments UERANSIM and srsUE. Through experiments, we find five security vulnerabilities having both security and privacy implications and prove that our method has better performance in terms of protocol state coverage and the scale of testing cases.

1 Introduction

The fifth generation mobile network(5G) is the infrastructure for the internet of everything, which has been widely applied in various vertical industries. In order to satisfy the demands of application scenarios with large bandwidth, extremely low delay, and massive number of connections, 5G network has introduced many new technologies and services. As a result, 5G network has the characteristics of diverse types of terminals, virtualization of network functions, and openness of network. The traditional security boundaries are being eroded and the resource and environment of 5G face diverse security challenges. Signaling is the foundation for realizing the functionalities of 5G network. Therefore, the security of signaling is a prerequisite for ensuring the security of 5G network. The 3GPP standards serve as a guide for the implementation of 5G network. But due to the complexity and ambiguity of some parts of the standards, different device manufacturers may have variations in implementing the standards, during the production process, which may lead to security risks in 5G devices. Therefore, it is of great significance to conduct security vulnerability detection on signaling implementation to maintain 5G security.

5G network has strict requirements for the processing and verification of signaling. Therefore, applying traditional vulnerability detection methods for general internet protocols to 5G signaling may lead to low testing efficiency due to a large number of

L. Barolli (Ed.): EIDWT 2024, LNDECT 193, pp. 64–74, 2024.
https://doi.org/10.1007/978-3-031-53555-0_7

non-compliance signaling being discarded directly during testing. Additionally, using exhaustive methods to generate testing cases will cause an explosive growth in the scale of testing cases as the number of testing parameters grows.

Aiming at such problems, we propose an efficient vulnerability detection method for 5G NAS protocol based on combinatorial testing. Firstly, through analysis of 3GPP standards, we define signaling compliance constraints to reduce packet loss rate caused by decoding failures. Secondly, we define semantic constraints for field values to filter fields with high security risk and minimize the scale of testing cases. Finally, we use a combinatorial testing algorithm to address the problem of a significant increase in the scale of testing cases caused by using full permutation. This further improves the testing efficiency. We also conduct vulnerability detection on the 5G UE simulation platform and find multiple security vulnerabilities which can lead to leakage of UE privacy data, integrity protection bypass, and other implications.

In summary, the contributions of our work are as follows:

- We propose an efficient vulnerability detection method for 5G NAS protocol based on combinatorial testing, which can significantly improve testing efficiency and detect security vulnerabilities caused by the interaction of multiple parameters.
- We implement an efficient vulnerability detection prototype system for 5G NAS protocol and successfully discover five security vulnerabilities of 5G NAS protocol in srsUE and UERANSIM.
- The method we propose has better performance in terms of testing efficiency and coverage based on comparative experiments with traditional testing methods.

2   Related Work

Presently, there are two primary approaches for security analysis in the signaling of mobile networks.

One approach is to analyze whether the standard of the signaling has security vulnerabilities. This approach includes a Dolev-Yao-style adversary to enhance the normal protocol communication model. Subsequently, it uses symbolic model-checking tools to verify security-related properties and generate counterexample which could be a potential attack path. Representative research of this approach includes 4GInspector [3], 5GReasoner [4], and ProChecker [5], which have discovered potential attacks in RRC and NAS protocols such as authentication bypass, denial of service, and man-in-the-middle attacks. This approach overlooks the vulnerability detection of the signaling implementation. Furthermore, it solely concentrates on the vulnerabilities in handling signaling procedures and neglects the potential security vulnerabilities in more fine granularity, such as field processing.

The alternative approach uses fuzzing [6] to generate a substantial volume of abnormal structure messages for detecting potential memory corruption vulnerabilities in the implementation of signaling. The disadvantage of this method is that it does not fully consider the semantic characteristics of the signaling fields. As a result, it generates a significant scale of ineffective testing cases, therefore reducing the testing efficiency. Another way is to generate a large number of consistent and inconsistent testing cases to further discover potential security vulnerabilities due to violating the standards. Representative research of this approach includes LTEFuzz [7] and DoLTEst [8]. This approach does not consider carefully about constraints before generating testing cases, thus leading to the presence of redundant or invalid testing cases. Moreover, it overlooks the impacts caused by interactions among multiple parameters.

3   Background

3.1   Architecture of 5G System

As shown in Fig. 1, a standalone 5G network consists of three main components: user equipment (UE), radio access network (RAN), and the 5G core network (5GC) [1]. The UE usually comprises a mobile equipment(ME) and a universal subscriber identity module(USIM) card. The USIM card stores the critical parameters for authentication and security context generation. The ME processes received signaling from the air interface.

In the 5G standalone architecture, the next generation-radio access network(NG-RAN) is a network that provides communication services between UE and 5G base station(gNB), as well as between gNBs. And the NG-RAN establishes connectivity with the core network via the NG interface.

5G core network(5GC) implements a service-oriented architecture consisting of multiple network functions (NFs). The Access and Mobility Management Function(AMF) provides access service to the 5GC for NG-RAN and UE. AMF is capable of offering bidirectional authentication functions to ensure secure access and provides encryption and integrity protection for Non-Access Stratum (NAS) signaling. The Authentication Server Function (AUSF) and the Unified Data Management (UDM) generate and verify authentication and security protection vectors based on the 5G Authentication and Key Agreement (AKA) mechanism.

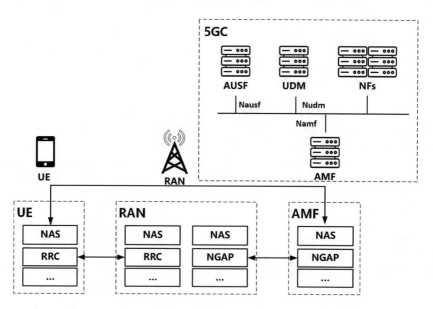

Fig. 1. Architecture of 5G system.

3.2 NAS Protocol

5G NAS protocol provides communication services between the UE and the AMF over the NG interface [2]. The 5G NAS protocol mainly consists of two functions, 5GMM (5G Mobility Management) and 5GSM (5G Session Management). The main registration procedure of the NAS protocol is depicted in Fig. 2. Initially, the UE sends a registration request message with the UE identifier. Subsequently, the core network triggers an authentication procedure to complete mutual authentication between the UE and the 5GC based on AKA protocol. Upon successful authentication, the core network commences a security mode control procedure to negotiate and establish a secure context between the UE and the 5GC. After this, the security protection of all subsequent signaling is activated. Finally, the UE responds with a registration complete message, indicating the successful completion of the registration process.

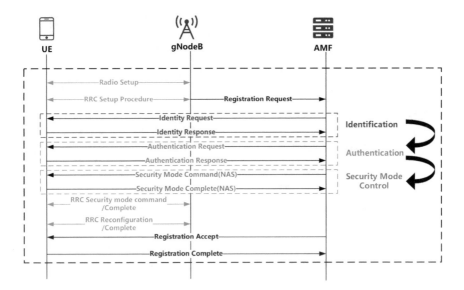

Fig. 2. UE registration procedure of NAS protocol.

3.3 Combinatorial Testing

Combinatorial testing (CT) is a testing case generation algorithm that generates testing cases for scenarios that involve multiple input parameters. Its purpose is to detect the impact of individual and multiple parameter interactions on the testing targets. Assuming the target being tested has k input parameters, denoted as $P = \{p_1, p_2, ..., p_k\}$, where parameter $p_i (i \leq k)$ can take discrete values from a finite set $V_i (i \leq k)$. The value range of the parameter set P is represented by $V = \{V_1, V_2, ..., V_k\}$. The parameter set P and value range V constitute the input space of the combinatorial testing (CT) model. Additionally, define the constraint set $CA = \{ca_1, ca_2, ..., ca_n\}$ to describe the dependencies among the parameters. The t-way combinatorial testing algorithm is an optimized method used for determining the necessary combinations for effective combinatorial testing. In this context, the value of $t(1 \leq t \leq k)$ represents the coverage strength. The t-way sequence refers to the t-dimensional combination of all values in the value range $V_{a1}, V_{a2}, ..., V_{at}$ for any t parameters $p_{a1}, p_{a2}, ...p_{at}$ in P. Every t-way sequence that adheres to the CA constraints is included at least once in the testing case set generated. The t-way combinatorial testing method strives to achieve a balance between the number of testing cases and the rate at which vulnerabilities are detected.

4   Proposed Approach

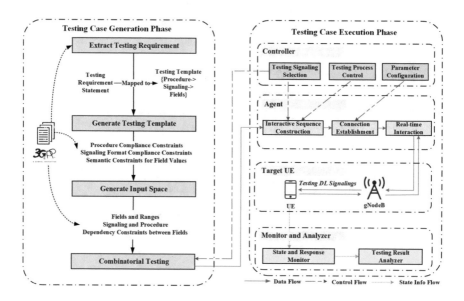

Fig. 3. Architecture of proposed method.

The framework of the efficient vulnerability detection method for 5G NAS protocol based on combinatorial testing is shown in Fig. 3. The method consists of two stages, testing case generation and testing case execution.

During the testing case generation stage, security-related testing requirements are derived through an analysis of the 3GPP standards. The testing requirement statements are then decomposed and mapped to specific NAS protocol process, signaling, and fields to construct the testing template. Subsequently, to improve the testing efficiency, we define constraints to restrict the input space of combinatorial testing algorithms. Finally, we use combinatorial testing algorithms to instantiate testing templates and generate testing cases that can detect the vulnerability caused by single-parameter effects and multi-parameter joint effects.

During the testing case execution phase, we implemented a vulnerability detection system for the NAS protocol. The system consists of three main modules. First and foremost, the control module configures the parameters necessary to establish a connection with the target UE and selects the signalings and procedures for testing. Based on the information provided by the control module, the agent module generates signaling interaction sequences to establish a connection with the target UE and engages in real-time interaction. The module for status monitoring and analysis monitors the status and response of the target UE by employing techniques like packet capture. Finally, the testing results are compared with the testing oracle, and obtain the testing conclusion, that is, whether there are potential security risks.

4.1 Testing Requirements Extraction and Testing Templates Generation

The 3GPP standard is described using natural language. The content and logic of the standard are highly complex. Therefore, we utilize the method of keyword divergence positioning to extract testing requirements. The 3GPP standards define the minimum requirement set for building 5G network, and according to ETSI regulations, requirement definitions in the standards should use modal verbs such as "shall", "shall not", "should", "should not", "may", "may not". Therefore, requirement statements can be further filtered through modal verbs. In this paper, we mainly extract the statements related to signaling and security function design in the standards. The security-related testing requirements are developed through the analysis of the contextual relationship among the requirement statements.

The extracted testing requirements from the standard only describe the signaling and security functions. It is necessary to form a testing template to guide the generation of executable testing cases. We analyze the characteristics of protocol and generate the testing template through field value selection and combination, signaling selection, and interaction procedure construction. Through the testing template, the testing requirements can be mapped to a specific execution stage in the actual communication process.

4.2 Input Space Minimization Based on Constraints

The constraints consist of the compliance constraints for signaling and the semantic constraints for field values.

4.2.1 Compliance Constraints

5G network implements stringent verification mechanisms to ensure reliable signaling reception and processing. Firstly, the signaling processing in 5G network is dependent on the interaction context and real-time status of two communicating devices. When the device is not in a state to receive the signaling, the testing message will be discarded directly. Furthermore, upon receiving the testing message, a failed decoding message will also be discarded promptly. To tackle these challenges, this paper establishes compliance constraints for signaling procedure and format to enhance the compliance of generated testing cases, allowing the testing cases to reach the target functional module and improve the testing efficiency.

Signaling Procedure Compliance Constraints. Primarily consist of contextual constraints within the signaling procedures, such as security context and session identifiers. This also involves constraints on the state of the device, which refer to the construction of the pre-signaling sequence that guides the testing target to reach the state of receiving testing cases.

Signaling Format Compliance Constraints. Classify signaling fields according to functions. Identification field: It is used to indicate the type of signaling, information element, and field. When generating testing cases, valid values are selected for identification fields based on the testing functions. Length field: During the process of decoding message, the message structure needs to be divided based on the length field. And then extract the field value correctly. Therefore, the length field needs to be modified again after adding or deleting bytes to the data packet. Check field: The NAS protocol has the functions of authentication and integrity protection of messages. In relevant tests that do not involve authentication and integrity protection, it is necessary to ensure the correctness of authentication vectors and MAC values to prevent interference with the testing results. Other types of data should be selected according to the corresponding value constraints based on the field data types.

4.2.2 Semantic Constraints for Field Values

The NAS protocol structure is relatively complex and contains many types of fields. If all possible values of the fields are tested, a large number of testing cases will be generated. Based on the semantic constraints for field values, we retain key and representative testing cases to maximize the effectiveness of testing cases and improve testing efficiency while ensuring testing coverage.

Through the analysis of the 3GPP standards, we locate fields and field values with higher risks. And use the relevance of fields to signaling security and the specificity of field values as risk assessment criteria. For example, the fields related to signaling security such as security header type, ngksi, etc. have higher risk. Additionally, the field values are divided into five categories: standard permitted values, standard prohibited values, reserved values, state mismatch values, special instruction values, and standard undefined values. Only one representative value is tested for each identical semantic based on the equivalent partition method. Through semantic constraints for field values, the scale of testing cases can be effectively reduced while ensuring the vulnerability detection rate.

4.3 Testing Cases Generation Based on Combinatorial Testing Algorithm

Security vulnerabilities are not necessarily triggered by a single abnormal parameter but may be triggered by a combination of multiple abnormal parameters. If the full permutation method is used to generate testing cases, for a large number of testing parameters with a wide range of parameter values, the number of testing cases will explode, which will consume a large amount of testing resources and increase the difficulty of testing. To solve this problem, we use t-way combinatorial testing algorithm to combine testing parameter values and generate testing cases with the chosen testing strength. IPOG algorithm [9] is an efficient t-way algorithm. The IPOG algorithm first generates combinations of all possible values for t parameters and forms an initial table of parameter combinations. Subsequently, starting from the t + 1 parameter, new parameter values are systematically employed to expand the table of parameter combinations both horizontally and vertically.

4.4 Testing Cases Execution

We implement a vulnerability detection system to execute the testing cases. The system mainly consists of three modules.

Controller. This module configures connection parameters to guide the agent module to establish a connection with UE. Select testing strength and testing template to guide the testing case generation. Additionally, it controls the execution of the testing system based on the information provided by monitor and analyzer.

Agent. This model establishes a connection with UE according to the configuration parameters. Before sending testing signaling, it constructs a pre-signaling sequence to enable the UE to reach the desired state of waiting for receiving the testing signaling, according to signaling compliance constraints. The module then interacts with the UE by sending the testing signaling sequence and receiving response data packets in real-time.

Monitor and Analyzer. This module is used to monitor the status of the testing UE and analyze its response to determine the presence of security vulnerabilities. If the corresponding process of the testing UE is abnormal, the status monitoring needs to pass the information to the controller. The controller will pause the testing process and wait for the UE to restart. Additionally, the module monitors the response of the UE to testing cases, compares the captured response with the testing oracle, and analyzes whether there are potential security vulnerabilities.

5 Experiment

5.1 Environment

The experimental environment of this paper is as follows.

(1) Hardware: X86 server (CPU: 11th Gen Intel(R) Core(TM) i7-11700. Memory: 32 GB, OS: Ubuntu18.04).
(2) Software: We take two 5G UE projects UERANSIM [10] and srsUE in srsRAN [11] as testing targets to verify the feasibility and effect of the proposed vulnerability detection method.

5.2 Vulnerabilities

This article selects five representative downlink NAS messages initiated by AMF for testing, namely identity request, authentication request, security mode command, registration accept, and configuration update command. Furthermore, we only focus on some security-related fields such as "security header type", "ngksi", "MAC" in all security protected messages and some other fields such as "identity type" in identity request, "ABBA" and "AUTN(SQN xor AK, AMF, MAC)" in authentication request, "selected integrity algorithm" in security mode command.

During the testing on srsUE and UERANSIM based on 5G NAS protocol, the vulnerability detection approach proposed in this paper has successfully discovered five

Table 1. Vulnerability

Vulnerability	Vulnerability Description	Notable Implications	srsUE	UERANSIM
User identity information leaked	When the security context establishment is not completed, UE responses the identity request whose identity type is other than SUCI, such as 5G-GUTI, IMEI, etc. (The standard states that only identity request with SUCI can be responded without security protection)	User identity information leaked	YES	YES
Multiplexing or error handling of ngKSI	When the value of ngksi in the authentication request message is the same as that in the registration request message, or the value of ngksi in the authentication request message is 7, which is a reserve value for UE, the UE continues the process without returning an error	The indication of security context conflicted	YES	NO
Authentication method recognition error	When the AMF identifier in AUTN is not 0x8000, the UE performs the 5G AKA authentication process normally	Implementing wrong authentication algorithm	NO	YES
Lack of verification of network identity	The UE does not verify the MAC value contained in the AUTN, but processes the SQN value in the AUTN. This means that the UE has not authenticated the 5GC	UE and the 5GC out of sync	NO	YES
Integrity protection bypass	When the selected integrity algorithm field takes a value of null algorithm or a reserved value, and the MAC takes a value of 0, the UE continues to the next procedure without integrity protection. (The standard states that for the UE, integrity protected signalling is mandatory for the 5GMM NAS messages.)	Integrity protection bypass	YES	NO

types of vulnerabilities in the NAS protocol implementation. As shown in Table 1. In summary, the cause of the vulnerabilities is that the implementation of NAS protocol is inconsistent with the definition of 3GPP standards.

5.3 Comparison Analysis

We compare the combinatorial testing algorithm based on constrained optimization proposed in this paper with the general combinatorial testing algorithm and the full permutation algorithm. Evaluate algorithm efficiency by comparing the number of testing cases needed to detect the same number of security vulnerabilities. The experimental results are shown in Fig. 4. We have found that the algorithm proposed in this paper can significantly reduce the number of testing cases and improve the testing efficiency compared with the other two algorithms.

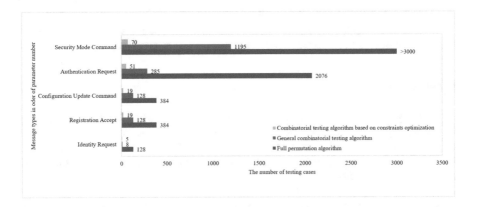

Fig. 4. Histogram of the number of testing cases generated.

Furthermore, we evaluate the impact of introducing signaling compliance constraints on the testing coverage of the protocol state. As shown in Fig. 5, using signaling compliance constraints can greatly increase the testing coverage of the protocol state.

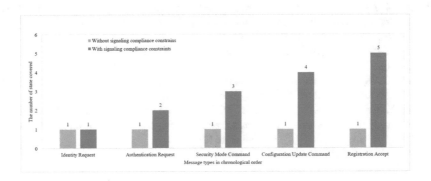

Fig. 5. Histogram of the number of states covered.

In summary, the vulnerability detection method proposed in this article has greatly improved the traditional security vulnerability detection method in terms of testing coverage and testing efficiency.

6   Conclusion

This article proposes an efficient vulnerability detection method for 5G NAS protocol based on combinatorial testing. By defining compliance constraints and semantic constraints for field values, this method improves testing coverage and minimizes the input space of the combinatorial testing algorithm, effectively reducing the scale of testing cases and improving the testing efficiency. In order to evaluate the effectiveness of the algorithm, we design and implement a prototype system, conduct actual tests on two open source UE projects, srsUE and UERANSIM, and find five security vulnerabilities. Furthermore, the experimental results prove that this method has better performance in testing efficiency and can cover as many protocol states as possible.

References

1. 3GPP. System architecture for the 5G System (5GS). TS 23.501 (2021). https://portal.3gpp. org/desktopmodules/Specifications/SpecificationDetails.aspx?specificatio-nId=3144
2. 3GPP. Non-Access-Stratum (NAS) protocol for 5G System (5GS); Stage3. TS 24.501 (2021). https://portal.3gpp.org/desktopmodules/Specifications/SpecificationDetails. aspx? specificationId=3370
3. Hussain, S., Chowdhury, O., Mehnaz, S., et al. LTEInspector: a systematic approach for adversarial testing of 4G LTE. Network and Distributed Systems Security (NDSS) Symposium (2018)
4. Hussain, S.R., Echeverria, M., Karim, I., et al.: 5G reasoner: a property-directed security and privacy analysis framework for 5G cellular network protocol. In: Proceedings of the 2019 ACM SIGSAC Conference on Computer and Communications Security, pp. 669-684 (2019)
5. Karim, I., Hussain, S.R., Bertino, E.: ProChecker: an automated security and privacy analysis framework for 4G LTE protocol implementations. In: 2021 IEEE 41st International Conference on Distributed Computing Systems (ICDCS), pp. 773-785. IEEE (2021)
6. Liu, X., Cui, B., Fu, J., et al.: HFuzz: towards automatic fuzzing testing of NB-IoT core network protocols implementations. Futur. Gener. Comput. Syst. **108**, 390–400 (2020)
7. Kim, H., Lee, J., Lee, E., et al.: Touching the untouchables: dynamic security analysis of the LTE control plane. 2019 IEEE Symposium on Security and Privacy (SP), pp. 1153–1168. IEEE (2019)
8. Park, C., Bae, S., Oh, B., et al.: DoLTEst: in-depth downlink negative testing framework for LTE devices. In: USENIX Security Symposium (2022)
9. Lei, Y., Kacker, R., Kuhn, D.R., et al.: POG: a general strategy for t-way software testing. In: Annual IEEE International Conference and Workshops on the Engineering of Computer-Based Systems (ECBS 2007), pp. 549–556. IEEE (2007)
10. srsUE. https://github.com/srsran/srsRAN
11. UERANSIM. https://github.com/aligungr/UERANSIM

Graph-Based Detection of Encrypted Malicious Traffic with Spatio-Temporal Features

Qing Guo, Wenchuan Yang$^{(\boxtimes)}$, and Baojiang Cui

Beijing University of Posts and Telecommunications, Beijing, China
{seven7,yangwenchuan,cuibj}@bupt.edu.cn

Abstract. With the continuous advancement of information technology, the concerns regarding privacy and network communication security are growing. Many applications have adopted encryption to ensure the confidentiality of network communication. However, the use of encryption has also provided opportunities for attackers. Attackers have begun to use encryption to conceal malicious activities, which poses a significant challenge for traffic detection. Traditional traffic detection methods primarily operate at the packet-level or session-level granularity and often neglect to consider the interrelationships between multiple sessions, thereby falling short of capturing comprehensive communication patterns exhibited by malware. In the paper, we propose a graph-based detection of encrypted malicious traffic, known as *HIG-RF*. It utilizes the GraphSAGE algorithm to generate embedding, comprehensively capturing the behavior patterns of hosts. And then we use the Random Forest model to comprehensively assess the probability of host infection. Our experiments show that HIG-RF achieves over 98% classification accuracy and over 98.5% recall, outperforming other advanced models.

1 Introduction

With the rapid development of the Internet, there is a growing demand among users for privacy protection and network communication security. The volume of encrypted traffic in global communication networks has shown an explosive growth trend. Going back nine years to December 2013, the Google Transparency Report [8] shows just 48% of worldwide web traffic was encrypted. Nowadays, the volume of encrypted web traffic tracked by Google is up to 95%.

While traffic encryption contributes to data privacy and security protection, it also serves as a sanctuary for malware. Attackers frequently exploit encrypted traffic to conceal their malicious activities. Malware has begun to utilize the TLS protocol to conceal their network activities. According to data from Zscaler Lab, by the end of 2022, the percentage of malware using the TLS protocol to hide their attack behaviors has reached a high of 85% [17]. Therefore, detecting encrypted malicious traffic has become a challenge that cannot be ignored.

Researchers have proposed various encrypted malicious traffic detection methods. However, since the payload of encrypted traffic is invisible, these research efforts mainly focus on two aspects: unencrypted TLS handshake message analysis and metadata analysis [13,15]. In these studies, traditional machine learning methods and emerg-

L. Barolli (Ed.): EIDWT 2024, LNDECT 193, pp. 75–86, 2024.
https://doi.org/10.1007/978-3-031-53555-0_8

ing deep learning models have been introduced into the detection of encrypted malicious traffic, and have achieved good results to some extent. However, these methods primarily rely on packet-level or session-level detection, resulting in incomplete and incomplete utilization of traffic information.

Due to the complexity of malware network activities, either packet-level detection or session-level detection is challenging and may not provide a comprehensive understanding of the encrypted communication activities of malware. Subsequent studies by Anderson et al. [2,3] have proposed using contextual background data, to augment available information. However, their primary aim is to complement missing features in individual encrypted sessions. While these efforts have led to improved detection performance, they have not thoroughly analyzed the spatio-temporal characteristics of sessions. Research indicates that malware network activities often follow specific access sequence rules [7]. Most malware tends to connect to different target servers in distinct stages, forming a connection sequence. For example, malware commonly conducts disguised innocuous network connectivity tests, which serve as preparatory actions before connecting to control servers, such as querying a host's public IP or downloading server certificates.

In response to the issues associated with the above detection methods, we propose a graph-based detection of encrypted malicious traffic with spatio-temporal features. First, we parse the traffic at the session granularity and extract statistical features from each session, including handshake features, certificate features, statistical features, and host behavior features. On this basis, we construct a heterogeneous graph called the Host Interactive Graph (HIG), which is used to model the complex network behavior of hosts. And then we employ GraphSAGE to relate sessions within the HIG graph, generating spatio-temporal features that comprehensively represent the behavior patterns of hosts. By utilizing GraphSAGE to correlate sessions in spatio-temporal context, we generate embedding representing the spatio-temporal characteristic of hosts, and then transform the problem of detecting encrypted malicious traffic into a host node classification task.

Our main contributions are summarised as follows:

1) We propose the Host Interactive Graph to capture the spatio-temporal characteristics of inner hosts. In HIG, nodes represent hosts or servers, and edges represent sessions. Every session between a host and a server forms an edge in the HIG.
2) We propose an information aggregation method based on GraphSAGE to generate embedding to characterize the host complex behaviour and identify potential infected hosts. And the problem of malicious encrypted traffic detection is transformed into host node classification problem, achieving efficient detection of malicious encrypted traffic.

2   Related Work

As the demand for security and privacy protection continues to grow, the volume of encrypted traffic has steadily increased. Identifying potential risks and threats within this vast encrypted traffic has become a central focus of current network security research. Related researches are mainly conducted from three perspectives: payload-based detection, feature engineering-based detection, and raw traffic-based detection.

2.1   Payload-Based Detection

The methods based on payload focus on identifying and matching the specific content of network traffic. These include deep packet inspection (DPI) and decryption methods through intermediate proxies. Traditional DPI [4, 16] are no longer effective because the encryption process transforms the original stream data into a pseudo-random and meaningless character sequence. To address the problem of DPI ineffectiveness after encryption, some industry has proposed using intermediate proxies to decrypt traffic for detection [11]. However, this approach severely compromises user privacy and impacts network performance.

2.2   Feature Engineering-Based Detection

The methods based on feature engineering focus on identifying encrypted malicious traffic without decryption. The key to this approach is to select highly discriminative features for training accurate detection models. In 2016, Cisco [1] found that flow metadata, packet lengths and timing sequences, byte distributions, and TLS packet header information are effective in identifying encrypted malicious traffic. Then they optimized detection results by utilizing related background traffic information, such as DNS request-response data. Futhermore, Anderson et al. [2] addressed the issues of noisy labels and non-stationarity in encrypted malicious traffic detection. They compared six machine learning models and found that the random forest is the most robust models for detecting encrypted malicious traffic.

2.3   Raw Traffic-Based Detection

The methods based on raw traffic treat binary network flows as a special type of text and transform traffic identification into a text classification task. They utilize CNN [15] or RNN [10] to learn semantic representations of traffic, eliminating the need for feature selection and extraction, achieving an end-to-end classification effect from source byte data to the final classification result. However, this method, relying on the detection of single flow features, lacks generalizability to unknown samples and may result in a significant number of unexplainable false positives [6].

Most previous methods detect traffic based solely on a snapshot of communication behavior at a specific moment, neglecting to consider the spatio-temporal relationships among traffic over a period of time. The issue of insufficient representative information in encrypted malware detection has not been effectively addressed. Therefore, our aim is to develop an algorithm that can holistically capture the spatio-temporal behavioral characteristics of hosts to identify the suspicious communication behaviour efficiently.

3   Methodology

3.1   Overview

As shown in Fig. 1, our detection system consists of five steps: traffic parsing, feature extraction, HIG construction, generate embedding through GraphSAGE, and detection. Below we elaborate on the functionality of each steps.

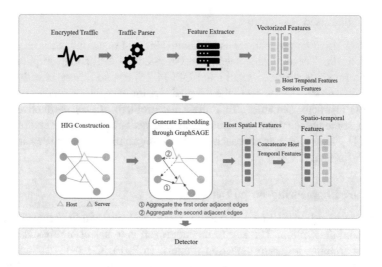

Fig. 1. Overview of the system

3.2   Traffic Parsing

We analyze encrypted traffic at the session-level granularity. In a session, packets share the same set of five-tuple values (source IP, source port, destination IP, destination port, and protocol), where source and destination IP addresses and ports can be interchangeable. By utilizing sessions, we can capture intricate interactions between clients and servers. Following this, we parse these sessions, extracting unencrypted protocol details, including DNS queries, TCP headers, TLS handshakes, and more. The parsed information will assist us in providing a more detailed characterization of these sessions.

3.3   Feature Extraction

Due to the invisibility of payload in encrypted traffic, the detection relies primarily on unencrypted handshake messages, certificate details, and side-channel statistical data of the traffic. In order to identify malicious traffic, we have meticulously designed a set of features to characterize various session attributes, as shown in Table 1. Furthermore, to gain a more comprehensive insight into host network activity, we have incorporated temporal features of hosts to capture their activities of hosts over a period, as detailed in Table 2.

Table 1. Features of session

Category	Feature
TLS handshake features	Client/Server TLS version
	Client Ciphersuite
	Number of Client Ciphersuites
	Server Chosen Ciphersuite
	Number of Client Extensions
	Number of Signatrue Algorithms
	Signature Algorithm
	Number of EC Point Formats
	Number of Elliptic Curve
	Present of SNI
Certificate features	Length of Certificate Chain
	Self-signed Certificate
	Certificate's Expiration
	Number of SAN
	Equivalence of SNI and CN
Statistical features	TLS Packet Length
	TLS Time Interval
	Number of Packets
	Max/Min/Average Packet Length

Table 2. Temporal features of host

Category	Feature
Duration	Max/Min/Average Duration
	Number of Short Session
Data	Mean/Median/Mode Length of Data Exchanged
	Mean/Median/Mode Length of Data Send/Received
	Mean/Median/mode Number of Packets Sent/Received
Port	Number of Non-standard DST Port
	Number of Non-standard DST Port in TLS Session
Connection	Number of Session
	Number of Failed or Rejected Attempts
	Number of Connections Per Second
	Number of Failed or Rejected Attempts Per Second

3.4 HIG Construction

In order to effectively conduct correlation analysis of encrypted traffic, we propose the Host Interaction Graph (HIG) to model the complex spatio-temporal relationships between encrypted sessions. The $HIG = (H, D, E, I, T, F)$ is a host-server bipartite graph designed to capture the interactions between internal hosts and external servers. We represent the internal hosts as a vertex set H, with each vertex uniquely identified by an IP address. Similarly, external servers are represented by a vertex set D, with each vertex identified by a server domain name or IP address. We use the edge set $E = \{e | e = (h_e, d_e), h_e \in H, d_e \in D\}$ to represent the sessions between hosts and servers. If a TLS session occurs between host h and server d, an edge $e = (h, d)$ connecting h and d is added to E. $I = \{i_e | e \in E\}$ represents the temporal feature of each edge, where i_e indicates the order of e associated with host h_e. $T = \{t_e | e \in E\}$ denotes the features extracted from the sessions as described in Table 1. $F = \{f_h | h \in H\}$ denotes the temporal features extracted from the behavior of internal hosts as described in Table 2.

3.5 Generate Embedding Through GraphSAGE

Node classification is one of the most widespread applications in the field of graph neural networks(GNN). Its purpose is to utilize labeled node information to represent new features for unlabeled nodes and classify them accordingly. The pioneering work in the realm of GNN, namely the Graph Convolutional Network (GCN) [9], excels in node classification tasks by capturing global graph information and training an independent embedding representation for each node. However, GCN employs a transductive learning approach, necessitating the involvement of all nodes in the training process to obtain node embeddings. As a result, adding a new node requires retraining all existing nodes, incurring substantial computational overhead.

Considering the real-time requirements of detection, we choose to use the Graph-SAGE with a neighbor sampling and aggregation mechanism as the node embedding algorithm. GraphSAGE is an inductive node learning model designed to learn an aggregation function that can aggregate vertex features by sampling local neighbors of vertices, rather than training separate embedding representations for each vertex. When new nodes are added to the graph, the GraphSAGE model can quickly obtain embeddings for these new nodes through the aggregation function.

In order to get the embedding of host nodes in HIG, we adopt the GraphSAGE algorithm to aggregate vertex features by locally sampling adjacent edges of host node n. The adjacent edge refers to the edge $N = \{e | e = (h_e, d_e), h_e = n\}$, which is connected to node n. The specific steps are described below.

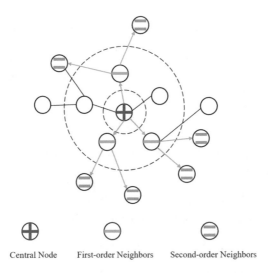

Fig. 2. Process of edge sampling and aggregation

Step 1:Sampling

If the model uses all adjacent edges to learn node embedding, the computational complexity becomes uncontrollable due to the uncertain size of the adjacent edge set. For each edge, GraphSAGE divides the entire graph structure data into k layers, starting from the innermost layer and moving outward. In this process, each edge at layer $l(0 \leq l \leq k)$ samples S_l adjacent edges. If the number of adjacent edges is less than S_l, all adjacent edges are used; otherwise, S_l edges are randomly sampled from the adjacent edges. By default, the value of k is set to 2, and it is necessary to ensure that $S_1 \cdot S_2 \leq 500$. Figure 2 illustrates the sampling process when $k = 2$, $S_1 = 3$, and $S_2 = 2$. In the Fig. 2, dashed lines are used for stratification. Nodes marked with a cross represent the central node, nodes with a single horizontal line represent the sampled first-order neighbor nodes, and nodes with two horizontal lines represent the sampled second-order neighbor nodes. Blank nodes represent those that were not sampled. The edges connecting the central node and first-order neighbor nodes, as well as the first-order neighbor nodes and the second-order neighbor nodes, are the objects of sampling, indicated by edges with arrows in the Fig. 2.

Step 2:Aggregation

The aggregation operation enables the integration of adjacent edge features, and it is essential for the aggregation process to be adaptive to the number of aggregated edges. Regardless of the number of neighbors a node has, the dimensions of the output after aggregation must remain consistent. During aggregation, the features of second-order edges are first aggregated into the corresponding first-order edges. Subsequently, the features of first-order edges are aggregated into the central node. Finally, classification is performed based on the aggregated central node features. The node aggregation process is depicted in Fig. 2, with arrows indicating the direction of aggregation along the edges between nodes.

Assuming that the edges are divided into k-th layers, the feature of edge e in the $(k-1)$-th layer is represented as t_e^{k-1}. The set of k-th layer adjacent edges is denoted as $N(e)$, the adjacent edge is denoted as e', and its feature as $t_{e'}^k$, where $\forall e' \in N(e)$. Let the aggregation function be denoted as Agg. The aggregation operation can be represented as the Eq. (1), $t_{N(e)}^{k-1}$ represents the aggregated feature obtained through adjacent edges of edge e:

$$t_{N(e)}^{k-1} = Agg[t_{e'}^k, \forall e' \in N(e)]$$

(1)

Then t_e^{k-1} and $t_{N(e')}^{k-1}$ are horizontally concatenated using the CONCAT function to form a new feature for edge e, which is used in the aggregation computation for the $(k-2)$-th layer. The process can be represented as the Eq. (2). $\bar{t}_{N(e)}^{k-1}$ represents the features learned through the model for edge e in the $(k-1)$-th layer, σ is the activation function, and W_k denotes the learnable parameters for the k-th layer.

$$\bar{t}_{N(e)}^{k-1} = \sigma[W^k CONCAT(t_e^{k-1}, t_{N(e)}^{k-1})]$$

(2)

After completing the aggregation for the 0-th layer, the host feature f_n and edge feature $\bar{t}_{N(e)}^0$ are horizontally concatenated to obtain the final host feature $\bar{f}_n$.

$$\bar{f}_n = CONCAT(f_n, \bar{t}_{N(e)}^0)]$$

(3)

Step3: Aggregation function design
In reality, each host may access numerous public services, generating a significant amount of traffic data. Aggregating the raw data without distinction may not provide insightful information about host behavior. Therefore, prior to information propagation, we calculate the importance of edges based on the frequency of servers being accessed and the order of session occurrences. We posit that if a server, denoted as m, is frequently accessed by many hosts, it is likely to be a target for host n. Furthermore, if host n accesses server m in the early stages of communication, relative to other servers, m may hold higher significance for n, as it could be the primary destination for n. Thus, we define the importance of the edge e, denoted as $\rho(e)$, with respect to its host, m, as follows:

$$\rho(e) = \frac{|\{e'|e' \in E \wedge d_{e'} = m\}|}{i_e \cdot |E|}$$

(4)

$|\{e'\}|$ represents the number of edges terminating at m, $|E|$ is the total number of edges, and i_e is the order in which edge e is connected to n.

As shown in Eq. (5), the aggregation function we designed aggregates based on the importance of adjacent edges, aiding in capturing complex relationships between sessions more accurately.

$$t_{N(e)}^{k-1} = \sum_{e'}^{N(e)} \rho(e')t_{e'}^k$$

(5)

3.6   Detection

Through node embedding, we compute spatio-temporal behavioral feature vectors for hosts and feed these vectors into a machine learning model to assess the probability of hosts being infected. Cisco [2] has pointed out that inaccurate foundational facts and highly unstable data distributions hinder the effectiveness of machine learning algorithms in malicious network traffic monitoring. In this context, the Random Forest (RF) excels, particularly in scenarios with noisy labels and unstable data distributions. Considering its ability to resist overfitting and its superior performance in comparative experiments, we ultimately chose the Random Forest regression algorithm as our detector. It is an ensemble learning method that makes predictions by aggregating the outputs of multiple decision trees. Based on the infection prediction values for each host, the detector provides a list of suspicious hosts along with their access information.

4 Experiment

4.1 Dataset

MTA [5]: This dataset originates from the Malware-traffic-analysis project, which is an open-source malware traffic initiative continuously updating various malware traffic samples since 2013. In this experiment, we collected traffic sample data publicly available from 2014 to 2022.

CTU-13 [14]: This dataset is obtained from CTU University, where specific malware were performed in 13 different scenarios to collect traffic samples. It is widely utilized as a benchmark dataset for encrypted malicious traffic detection tasks. The dataset includes both normal traffic samples from various scenarios and malware traffic samples.

In order to filter out the benign traffic among the malicious traffic samples, we used the Alexa Top 1 Million popular website list to filter out normal traffic from the captured malicious traffic files. After data preprocessing, we ultimately acquired 58,138 malicious encrypted sessions and 71,408 benign encrypted sessions. Among these, 70% are used for training malicious encrypted traffic detection models, 20% for testing, and 10% for validation.

4.2 Baselines

To validate the effectiveness of the proposed model, we compared it with various benchmark methods in the field of malicious traffic detection. Details of the baseline methods are as follows:

ETA [1]: It employs a random forest model to detect malware traffic. Specifically, ETA leverages flow features, including TLS handshake metadata, DNS context flows related to encrypted flows, and HTTP header information from HTTP context flows within a 5-minute window of the same source IP address.

1D-CNN [15]: It is a representation learning-based method for malicious traffic classification. It eliminates the need for manually extracting and selecting a feature set from network traffic, instead directly taking raw traffic as input data for a deep neural network. The entire process of traffic data representation learning is handled by the deep neural network.

HIG: It utilizes the GraphSAGE algorithm to compute the spatio-temporal features of hosts, which are then into the detector to acquire the probability of a host being infected. To comprehensively explore the best detector algorithms, we designed three models separately: **HIG-LR**, **HIG-SVM**, and **HIG-RF**, corresponding to the three typical machine learning algorithms of linear regression, support vector machine, and random forest, respectively.

4.3 Experimental Setup

Implementation: In the traffic analysis phase, we employed Zeek (version 5.0.10) [12] to aggregate and parse the traffic, generating diverse traffic logs. To implement the HIG graph representation module, we utilized NetworkX (version 2.5.1) for constructing a heterogeneous graph. Lastly, we implemented the detection algorithms using scikit-learn (version 0.23.2).

Parameter and Metrics: For the baseline models, we kept them at their default parameters. As for HIG, we set the node embedding dimension obtained through GraphSAGE to 256 and the learning rate to 0.001. For ease of experimental observation, we set the number of samples S_l equal for each layer and represented it as S. We also selected the best-performing hyperparameters on the validation set, where $S = 10$ and $K = 10$. Additionally, we used Accuracy, Precision, Recall, and macro F1 as evaluation metrics.

4.4 Evaluation Results

On the dataset, we conducted an extensive series of experiments to assess the performance of HIG and other baseline models. For each experiment, we repeated it ten times and took the averages of accuracy, recall, and micro-F1 as the final results. Table 3 demonstrates that the proposed HIG-RF model outperforms other baseline models across all evaluation metrics on the dataset. We made the following observations regarding different baseline methods:

1) In the case of HIG, through comparisons with various machine learning algorithms, we found that HIG-RF significantly outperforms other baselines across all evaluation metrics. This is primarily attributed to RF, which is a classic ensemble learning algorithm equipped with excellent resistance to overfitting and noise.
2) Regarding 1D-CNN, although it doesn't require manual feature engineering, its performance across all evaluation metrics falls short of ETA and HIG-RF. This is because it only captures text patterns in traffic and lacks a comprehensive multidimensional assessment, making it susceptible to overfitting when dealing with complex traffic, resulting in suboptimal model performance.
3) When compared to ETA, HIG-RF shows noticeable improvements in all evaluation metrics. This is mainly because ETA solely focuses on session-level features for detection and doesn't consider the spatio-temporal relationships between multiple sessions, thereby failing to comprehensively capture the behavioral characteristics of malware. HIG-RF, with the use of GraphSAGE, correlates sessions within a local network over a period of time, calculates the spatio-temporal features of hosts, providing a more comprehensive view of malware's behavioral patterns.

Table 3. Detection Performance of HIG-RF and Baselines

Model	Accuracy	Precision	Recall	F1-Score
ETA	0.9723	0.9713	0.9730	0.9721
1D-CNN	0.9658	0.9647	0.9664	0.9655
HIG-LR	0.9581	0.9591	0.9562	0.9575
HIG-SVM	0.9773	0.9766	0.9777	0.9771
HIG-RF	**0.9851**	**0.9844**	**0.9855**	**0.9849**

5   Conclusion

To comprehensively characterize the complex behavioral patterns of malicious software, we introduce a graph-based detection of encrypted malicious traffic with spatio-temporal features, known as HIG-RF. It models the complex communication behaviors of hosts as a heterogeneous graph and then transforms the problem of detecting encrypted malicious traffic into a node classification problem on this heterogeneous graph. Specifically, we employ the GraphSAGE algorithm to extract spatio-temporal characteristic of hosts from the graph and subsequently feed these features to a detector to calculate the probability of host nodes being infected. Experimental results demonstrate that HIG-RF outperforms various existing models on the dataset, achieving a classification accuracy of over 98% and a recall rate of over 98.5%.

References

1. Anderson, B., McGrew, D.: Identifying encrypted malware traffic with contextual flow data. In: Proceedings of the 2016 ACM Workshop on Artificial Intelligence and Security, pp. 35–46 (2016)
2. Anderson, B., McGrew, D.: Machine learning for encrypted malware traffic classification: accounting for noisy labels and non-stationarity. In: Proceedings of the 23rd ACM SIGKDD International Conference on knowledge discovery and data mining, pp. 1723–1732 (2017)
3. Anderson, B., Paul, S., McGrew, D.: Deciphering malware's use of TLS (without decryption). J. Comput. Virol. Hacking Tech. **14**, 195–211 (2018)
4. Creech, G., Hu, J.: A semantic approach to host-based intrusion detection systems using contiguousand discontiguous system call patterns. IEEE Trans. Comput. **63**(4), 807–819 (2013)
5. Duncan, B.: Malware traffic analysis (2023). https://malware-traffic-analysis.net/
6. Fu, C., Li, Q., Shen, M., Xu, K.: Frequency domain feature based robust malicious traffic detection. IEEE/ACM Trans. Networking **31**(1), 452–467 (2022)
7. Fu, Z., et al.: Encrypted malware traffic detection via graph-based network analysis. In: Proceedings of the 25th International Symposium on Research in Attacks, Intrusions and Defenses, pp. 495–509 (2022)
8. Google: google transparency report (2023). https://transparencyreport.google.com/
9. Kipf, T.N., Welling, M.: Semi-supervised classification with graph convolutional networks. arXiv preprint arXiv:1609.02907 (2016)
10. Marín, G., Caasas, P., Capdehourat, G.: DeepMAL - deep learning models for malware traffic detection and classification. In: Data Science – Analytics and Applications, pp. 105–112. Springer, Wiesbaden (2021). https://doi.org/10.1007/978-3-658-32182-6_16

11. Ponemon: hidden threats in encrypted traffic (2016). https://www.ponemon.org/local/upload/file/A10%20Report%20Final.pdf
12. Project, Z.: Zeek (2023). https://zeek.org/
13. Shekhawat, A.S., Di Troia, F., Stamp, M.: Feature analysis of encrypted malicious traffic. Expert Syst. Appl. **125**, 130–141 (2019)
14. University, C.T.: Ctu-13 (2023). https://www.stratosphereips.org/datasets-ctu13
15. Wang, W., Zhu, M., Wang, J., Zeng, X., Yang, Z.: End-to-end encrypted traffic classification with one-dimensional convolution neural networks. In: 2017 IEEE International Conference on Intelligence and Security Informatics (ISI), pp. 43–48. IEEE (2017)
16. Zhang, H., Papadopoulos, C., Massey, D.: Detecting encrypted botnet traffic. In: 2013 Proceedings IEEE INFOCOM, pp. 3453–1358. IEEE (2013)
17. Zscaler ThreatLabz: State of encrypted attacks 2022 report (2022). https://www.zscaler.com/blogs/security-research/2022-encrypted-attacks-report

Fuzzing IoT Devices via Android App Interfaces with Large Language Model

Wenxing Ma and Baojiang Cui[✉]

Beijing University of Posts and Telecommunications,
Haidian District Xitucheng Road No.10, Beijing, China
{mono,cuibj}@bupt.edu.cn

Abstract. Most of the current automated testing methods for IoT devices rely on firmware analysis and firmware emulation. However, due to the diverse architectures and structures of firmware across different vendors, these methods have limited applicability and cannot perform large-scale testing. To address this issue, we have designed a tool called FIAL, a novel IoT device fuzzing method based on accompanying app interfaces. FIAL allows for the discovery of firmware vulnerabilities without the need for firmware analysis. It leverages a large language model to analyze the accompanying app and extract the most effective function interfaces for fuzzing. We applied FIAL to analyze five popular devices and discovered a total of 14 bugs, including 5 new vulnerabilities. We conducted a comparison with two other network fuzzing tools, and the experiment showed that FIAL can uncover more exploitable vulnerabilities using fewer test cases.

1 Introduction

In recent years, the increasing number and expanding scale of IoT devices have provided attackers with numerous attack triggers and surfaces, resulting in a proliferation of attacks targeting IoT devices. To efficiently and effectively analyze the security of IoT devices, a significant amount of research on automated vulnerability detection for IoT devices has been conducted. However, the analysis of IoT devices often requires extracting the file system from the device firmware, and the firmware formats used by manufacturers vary greatly, making it difficult to extract firmware on a large scale through automated methods. Currently, there are also black box testing methods [5,20] specifically for IoT devices. However, these methods often have poor sample effectiveness due to a lack of knowledge about data formats and protocol formats [19]. Meanwhile, an increasing number of IoT manufacturers have developed companion apps for their devices. Compared to the devices themselves, these companion apps are easier to obtain online and subject to automated analysis. Therefore, there is an urgent need for an automated vulnerability detection method for IoT devices that relies solely on the accompanying applications instead of the firmware.

In this paper, we propose a method for black-box fuzzing of IoT devices based on fuzzing interfaces for accompanying applications. We leverage the powerful code comprehension capabilities of the large language model to identify the

L. Barolli (Ed.): EIDWT 2024, LNDECT 193, pp. 87–99, 2024.
https://doi.org/10.1007/978-3-031-53555-0_9

most suitable functions for dynamic hooking and mutation in the accompanying applications. Additionally, we develop a fuzzing framework to analyze and monitor two common vulnerability types in IoT devices: overflow vulnerabilities and injection vulnerabilities.

The main contributions of this paper are as follows:

1. We use the large language model combined with taint analysis to find the most suitable trigger functions for fuzzing in accompanying applications.
2. We construct a black-box fuzzing framework FIAL (Fuzzing IoT Devices via Android App Interfaces with Large Language Model) and verify its effectiveness.

The rest of this paper is organized as follows. Section 2 reviews the related work. Section 3 describes the principle and system design for FIAL. Section 4 represents the implementation and experimental result of FIAL. Section 5 concludes the paper and looks forward to the future development.

2   Related Work

2.1   Fuzzing for IoT Devices

The prevailing method for vulnerability discovery in IoT devices is currently fuzzing. Fuzzing techniques can be classified into white-box, grey-box, and black-box testing, based on the collection and analysis of feedback information obtained from program execution.

White-box testing often utilizes techniques such as symbolic execution or taint analysis to precisely gather program execution and state information, which guides the generation of test cases, as exemplified by QSYM [25] and FirmCorn [17]. Grey-box fuzzing employs lightweight monitoring techniques to efficiently discover vulnerabilities, such as the widely used AFL (American Fuzzy Lop) [26] and honggfuzz [15]. Due to the unavailability of source code from IoT device manufacturers, grey-box testing relies on device emulation for implementation. However, existing firmware emulation or simulation methods either rely on extensive manual labor [14] or have limited applicability.

Compared to the previous two methods, black-box testing does not require concerns regarding firmware acquisition and unpacking for IoT devices, making it more suitable for large-scale analysis. However, due to the lack of knowledge about data and protocol formats, the effectiveness of test samples in black-box testing is relatively poor [19]. IoTfuzzer [9] conducts device testing by utilizing accompanying application network send functions. However, it only defines a limited number of network interfaces manually, which may result in incomplete testing coverage. Diane [21] employs the method of constructing a real traffic environment to identify accompanying application function interfaces capable of producing traffic. Nevertheless, this approach is inefficient and prone to high false positive rates.

This paper utilizes LLM (Large Language Model) to assist in identifying the triggering functions and mutation parameters for fuzzing in accompanying applications. By leveraging the information from accompanying applications, it generates effective fuzzting samples, thereby enhancing the logical trigger depth of fuzzing.

2.2   Taint Analysis

Taint analysis technique can extract specific tainted source-to-sink propagation paths based on predefined taint propagation rules.

In the field of IoT device analysis, KARONTE [22] takes the paradigm of inter-process communication (IPC) between the web server and binary files of embedded devices as the starting point for taint analysis. However, the abundance of IPC interfaces leads to excessive analysis, resulting in a high false positive rate. SaTC [10], on the other hand, leverages the shared keywords between the frontend and backend as the starting point for taint analysis, taking into consideration the system design characteristics of IoT devices. This approach reduces the complexity of symbolic execution.

In the field of Android system analysis, TaintDroid [13] achieves system-level dynamic taint analysis by customizing and modifying the Android system. However, it is only compatible with older versions of Android, and the manual effort required to update it for newer versions is too high. FlowDroid [8], on the other hand, is a static taint analysis tool specifically designed for Android applications. It improves the efficiency and accuracy of taint analysis by building precise Android lifecycle models.

In this paper, we employ taint analysis technique to extract the function call chain from sensitive input sources to network sending functions in Android applications.

2.3   LLM-Assisted Vulnerability Detection

Large Language Models (LLMs) are extremely large deep learning models that are pretrained on massive amounts of data. In recent years, there have been several studies that have utilized LLMs in the field of vulnerability discovery.

For example, FuzzGPT [12] and TitanFuzz [11] utilize the code generation capability of LLMs for fuzzing DL libraries. KARTAL [23] leverages the multitasking and few-shot learning capabilities of LLMs to detect logical vulnerabilities in web applications. [24] applies LLMs for static analysis of code snippets in IoT software to identify vulnerability types and the lines of code that produce vulnerabilities. [18] utilizes the strong comprehension capability of LLMs to identify the sinks and sources of binary files, and performs data flow analysis to discover dangerous function call chains between sinks and sources, with the help of LLMs to identify potential vulnerabilities within the call chain.

However, while LLMs exhibit strong understanding of code functionality, their comprehension of logical security vulnerabilities is somewhat lacking. In contrast to the aforementioned studies, this paper employs LLMs to identify the most effective fuzz testing trigger functions in companion applications, rather than directly searching for vulnerabilities using LLMs.

3   Proposed Approach

FIAL is a framework for black-box fuzzing of IoT devices using accompanying Android applications. It identifies and utilizes the interaction interfaces between the app and the IoT devices to send test data to the target device. Overall, there are two main parts in FIAL: firstly, leveraging the power of large language models and conducting automated analysis of the accompanying app to find key functions that trigger the interaction interfaces; secondly, hooking these key functions and dynamically mutating the relevant parameters at runtime to monitor the IoT device's state and determine if the mutated data triggers any device vulnerabilities. Specifically, FIAL consists of four modules, as illustrated in Fig. 1:

1. First, FIAL extracts the Network Message Source. The APK file of the accompanying app is decompiled to obtain smali code, which is then converted into vector format using word embedding algorithms and stored in a vector database. Query statements are constructed and used as prompts along with the related code to be inputted into the large language model to output the Network Message Source.
2. Next, FIAL extracts the NetworkSender. Java-level network message sending functions are manually marked, and lightweight binary taint analysis techniques are used to analyze the relevant JNI function interfaces for native-level network message sending functions.
3. Subsequently, using the FlowDroid taint analysis tool, FIAL gets the taint propagation paths from the Network Message Source to the NetworkSender. The large language model is employed to analyze and identify more effective function call subchains between structure transformation related and authentication functions. The top-level functions in these subchains are identified as the hooked functions (*HF*s) during dynamic testing.
4. Finally, a real-world testing environment for IoT devices is set up, where the *HF*s extracted in the previous stage are dynamically hooked. Different field mutation and monitoring strategies are employed based on various vulnerability types, and the device's vulnerability triggering is determined based on the monitoring status.

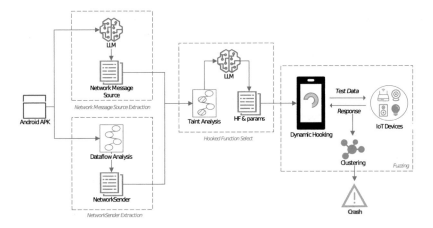

Fig. 1. Overview of FIAL

3.1   Network Message Source Extraction

The network messages exchanged between the app and the IoT device have a highly structured format. Mutating the parameters of the network send function directly leads to invalid packets that fail the early checks of the IoT device, thereby not triggering deeper logic. To address this, we employ a large language model to identify the various data elements constituting the structure of the network message and mutate them from a location close to their generation.

We define the network message source as the ultimate source of the data comprising the network messages sent by an application. This includes hard-coded data defined by the application, user input data, data obtained from other local or web services, and system-level data (e.g., MAC address information). In well-structured programs, these data are typically obtained using functions starting with words like "get," "load," or "request." Currently these metadata are often identified by manual tagging [16], which is time-consuming and labor-intensive. Moreover, many applications nowadays enable obfuscation protection such as proguard, so it is not possible to recognize the semantics of a function simply by its name. To overcome these challenges, we leverage the strong code understanding provided by a large language model (LLM) to identify functions related to fetching, loading, and requesting data, and designate them as network message sources.

Specifically, we first take an Android application as input and decompile the apk file into smali code using the decompiler tool apktool [1]. To address the contradiction between the large language model's restriction on token and the large amount of code in the project, we use the GPT Index tool [4] to convert the smali code file into an index file and use it for subsequent queries, i.e., we use the word embedding algorithm to convert the smali code into a vector format and save it in a vector database, and then convert the query statement into a vector format using the same algorithm and match it with the vector database during the subsequent queries. In the subsequent query, the query statement is converted into vector format using the same algorithm and matched with the vector database to retrieve the relevant code efficiently. Finally, we construct the prompt, which contains the query statement and the relevant code, so that the LLM can output the file and code location of those functions that fetch, load, and request data according to the function name and code semantic understanding in the code. In this way, we can get the network message source of the Android application.

3.2   NetworkSender Extraction

Next, we look for Java layer interfaces that provide data for the network message sending system calls, namely NetworkSender. This includes both the standard libraries, third-party libraries, and application-customized Java layer network message sending functions, as well as the JNI function interfaces that send the data to the native layer's custom network sending functions.

Java Layer Network Sending Functions. For the Java layer network sending functions, due to the small number, we manually collect network message sending functions from existing Java layer standard libraries and third-party libraries, and analyze whether the application has any custom Java layer network message sending functions through reverse analysis.

JNI Interfaces for Native Layer Network Sending Functions. We use lightweight binary taint analysis to get the required JNI function interfaces. In the native layer, all network messages are sent by the system library function send. We use send function as sink and all the JNI native methods as sources, and use lightweight taint analysis algorithm to find out all the sources that can reach the sink point as the extracted network message sending function. We take the intermediate representation in IDA Pro disassembler, Ctree, to assist in analyzing the program data flow and control flow, which can be applied to analyze library functions under different architectures. For each library file that calls the send function, perform Algorithm 1, first use Ctree to generate a control flow graph CFG, which is then used for taint propagation analysis. When a JNI local method parameter is found to have tainted the parameters of the send function, we consider the JNI interface corresponding to the Java layer function is counted as NetworkSender. By looking for these Java layer interfaces, we can then analyze the data flow in the Java layer and find the code path from the metadata to the network message sending function.

Algorithm 1: findSourcesReachSender

Input : JNISet: Set of JNI functions, Sender: send function, CFG: control flow generated by Ctree

Output: JNISenders: Set of JNI functions that can reach the Sender

1 **Function** findSourcesReachSender(*JNISet, Sender, CFG*):
2   **foreach** *sourceFunc* **in** *JNISet* **do**
3   *taintedSet* = {} ;   // Set to store tainted parameters
4   **foreach** *parameter* **in** *sourceFunc.parameters* **do**
5   *taintedSet*.add(*parameter*)
6   *worklist* = *CFG.getSuccessors*(*sourceFunc*)
7   **while** *worklist is not empty* **do**
8   *callSite* = *worklist*.pop() *callee* = *callSite.callee*
9   **if** *callee* == *Sender* **then**
10   *sourceFunc.isReachable* = *true* **break**
11   **foreach** *i* **in** *range(len(callSite.arguments))* **do**
12   *argument* = *callSite.arguments*[*i*] **if** *argument* **in** *taintedSet* **then**
13   *callee.parameters*[*i*].isTainted = **true**
          worklist.append(*CFG.getSuccessors*(*callee*))

14   *JNISenders* = {}
15   **foreach** *sourceFunc* **in** *JNISet* **do**
16   **if** *sourceFunc.isReachable* **then**
17   *JNISenders*.add(*sourceFunc*)
18   **return** *JNISenders*

3.3 NetworkSender Extraction

Subsequently, we are looking for the functions that hook during fuzzing and their parameters that need to be mutated. We use Network Message Source as source and NetworkSender as sink to search the taint propagation path using Flowdroid. Therefore, this taint propagation path is the path from the sensitive input source to the network message sender function.

In this function call chain, there may exist some data validation or structure transformation functions that will lead to the effectiveness of the fuzz test being reduced. Such functions include encryption and decryption functions, coding and decoding functions, and checksum verification functions. In order to solve this problem, we again utilize the comprehension power of the LLM to find the function call sub-chain that has encryption function and encoding function before the function call chain that has decryption function, decoding function, and validation function, and we use the function closest to the Network Message Source and the related parameter in this sub-chain as the fuzzing phase of the hooked function (*HF*).

For example, Listing 1 is part of a function chain to a obtained by taint analysis of the Orbi APP. The code marked in red is the taint propagation path for the password parameter. In the first function, after this app gets the password parameter from the UI, it first performs a validation and discards the data that does not meet the rule. After we detect the validation function using LLM, we remove it from the function chain and eventually choose the third function as the new chain starting point. The second function fails when decompiled using JADX [3], but the smali code is comprehensive, and we were able to identify its validation function based on the smali code using LLM. Eventually, the password parameter became part of the parameter event of the third-party network send function addEvent, which is highly structured and not suitable for fuzzing. Ultimately, our tool chooses sendUpdateAdminPassword as the *HF*.

Listing 1. Part of function chain to get the password from the UI and ultimately flow to a third party network sending function. The example is based on the Orbi APP.

```
protected ValidatorResult validateNewPassword() {
    String obj = this.etPassword.getText().toString();
    return this.validator.validateChangeAdminPassword(obj
        );
}

public static final ValidatorResult validateAdminPassword
    (Validator validator, String pwd, boolean flag,
    AndroidResourceProvider provider) {
    // unable to decompile
}

public void sendUpdateAdminPassword(String oldPassword,
    String newPassword, String enableRecovery, int
    securityQuestion1, String answer1, int
    securityQuestion2, String answer2) {
    ...
}

public void addEvent(final Event event) {
    ...
}
```

3.4   Fuzzing

In the fuzzing phase, we dynamically hook the extracted *HF*s and mutate the parameters, and then determine whether crashes occur in the target IOT devices based on the monitoring strategies. We detect the two most common types of vulnerabilities in IOT devices, i.e., overflow vulnerabilities and command injection vulnerabilities, and formulate different mutation and monitoring stratingegies for them.

Overflow Vulnerability. For overflow vulnerabilities, we choose to use different variation methods according to different field types:

1. For double, float, int and other numeric types, generate numeric boundary data and oversized data to test integer overflow and out-of-bound access vulnerabilities, such as 2^{31}, -1, 0 and other critical values.
2. For char, string and other character types, according to the continuous ASCII fields and binary fields to develop different strategies to test the overflow and out-of-bound access vulnerabilities. For text type data, arbitrary strings of different lengths are generated. The strings include English character case forms, Arabic numerals, and special symbols. For binary data, the data is inverted randomly and the length of the data is changed dynamically.
3. For object types, we follow the above rules to mutate their member variables in turn.

For monitoring strategies, when a vulnerability is triggered, the running state of the IoT device and its traffic behavior become abnormal. Abnormal traffic has characteristics that are different from normal traffic in terms of packet count, traffic volume, and other aspects. Therefore, we use the K-means algorithm to perform cluster analysis on the traffic to detect anomalous flow, thereby inferring the device's abnormal running state and recording the relevant samples that trigger overflow vulnerabilities.

Command Injection Vulnerability. For character-type data, we detect command injection vulnerabilities based on a library of vulnerability patterns. We use multiple command injection insertion forms of the reboot instruction, which is common to most architectures and systems, as the main variant samples (e.g., `;reboot;`, `&&{IFS}reboot;`), and also use other common commands as auxiliary judgments, such as SQL injection commands.

For monitoring strategy, we regularly send ping packets to the target IoT device to detect whether the device is running normally or not. If the device executes the reboot command, it will be unresponsive for a period of time, so that we can determine that the corresponding sample triggers the command injection vulnerability. Since the test period of the ping command is longer than the test packet sending period, when the device is unresponsive, we retest the test packets sent during this period and control the sending rate to determine which sample triggered the crash of the device.

4   Experiment and Result

In this section, we set up the environment and evaluate the results. In terms of system design, unlike traditional IoT device fuzzing frameworks, we leverage companion applications to analyze IoT devices. This allows us to conduct larger-scale and more comprehensive automated testing. Additionally, through automated analysis of companion applications, we are able to generate more effective fuzzing test cases.

4.1   Experiment

In terms of framework environment setup, in the network message source extraction phase, we utilize the LlamaIndex tool [4] to convert the decompiled code into vector format and process prompts with ChatGPT-4.0. In the NetworkSender extraction phase, we manually analyze a small amount of code to identify the custom network sending functions at the Java layer of the application. In the *HF* Extraction phase, we extract the function call chain from the Network Message Source to the NetworkSender based on Flowdroid [8] and again employ ChatGPT-4.0 to identify the most suitable hooked functions for the fuzzing phase. Lastly, in the fuzzing phase, we install the companion application on Google Pixel 2 and use the Frida tool [2] to hook all identified hooked functions. We perform sample generation, traffic monitoring and analysis on a single host (Ubuntu 20.04, 16GB RAM).

4.2   Result Evaluation

We conducted tests on five IoT devices from five different well-known vendors. The device model, firmware versions, companion applications, and the number of extracted HF are shown in Table 1. FIAL identified a total of 14 vulnerabilities, including 3 injection vulnerabilities and 11 overflow vulnerabilities, of which 5 are new vulnerabilities. These vulnerabilities have the potential to cause denial of service attacks or achieve privilege escalation and command execution effects. The vulnerabilities are listed in Table 2.

We compared FIAL with two popular open-source network protocol fuzzers, which are Sulley [7] and BED [6] as shown in Table 3. Under a 10-hour testing scenario, we recorded and compared the average number of test cases, crashes, and real vulnerabilities found for the 5 devices. It can be observed that FIAL was able to discover more real exploitable vulnerabilities with fewer test cases compared to the other fuzzers. Despite Sulley discovering more crashes, it was validated that the majority of them were found to be ineffective.

Table 1. Summary of Tested Devices

Vendor	Device Model	Firmware Version	Official Android Mobile APP	No. *HF*
Linksys	WRT54GL	4.30.18.006	com.cisco.connect.cloud	4
Netgear	RBR750	4.6.14.3	com.dragonflow.android.orbi	3
Tenda	AC9	15.03.06.42	com.tenda.router.app	3
Dlink	DIR-882	FW110B02	com.dlink.dlinkwifi	2
TPLink	IPC43AW	1.0.4	com.tplink.tether	2

[a] All the apps on the table can be downloaded from Google Play.

Table 2. Summary of Discovered Vulnerabilities

Device	Vulnerability Type	No. Issues	Zero Day
Linksys WRT54GL	Command Injection	2	
Netgear RBR750	Command Injection	1	
Netgear RBR750	Buffer Overflow	1	✓
Tenda AC9	Buffer Overflow	5	✓
Dlink DIR-882	Buffer Overflow	4	
TPLink IPC43AW	Buffer Overflow	1	✓

Table 3. Effectiveness comparison

	No. Test Cases	No. Crashes	No. Vulnerabilities
Sulley	146638	20	1
BED	15622	7	2
FIAL	5493	10	5

5   Conclusion and Future Work

In this paper, we present a porotype tool, FIAL, which can fuzz IoT devices via accompanying android app interfaces through power of large language model. This tool first uses LLM to extract the network message sources from the companion application and then utilizes a lightweight taint analysis algorithm to extract the NetworkSender. The extracted network message sources are treated as sources, and the NetworkSender functions are treated as sinks to extract the function call chains through taint analysis. LLM is then employed to analyze and identify the best interfaces for fuzzing. Finally, mutation strategies and monitoring strategies are devised based on the characteristics of different vulnerabilities. Through experimentation, we discovered 14 bugs, including 5 new vulnerabilities.

However, FIAL also has some limitations. For example, this tool cannot test interfaces of IoT devices other than those interacting with the companion application. Although the app covers most of the interfaces, there may be functionalities that do not require user control and are not present in the app. Additionally, this tool cannot detect other types of vulnerabilities such as encoding bypass and race conditions. Therefore, in the future, it is necessary to adapt more mutation and monitoring strategies to detect other types of logical vulnerabilities.

References

1. apktool. https://apktool.org/. Accessed 15 Nov 2023
2. frida. https://github.com/frida/frida/releases/. Accessed 15 Nov 2023
3. Jadx. https://github.com/skylot/jadx. Accessed 15 Nov 2023
4. Llamaindex. https://github.com/run-llama/llama_index. Accessed 15 Nov 2023
5. Peach. https://www.peachfuzzer.com/. Accessed 15 Nov 2023
6. Penetration testing tool: Bed. http://tools.kali.org/vulnerability-analysis/bed. Accessed 15 Nov 2023
7. Sulley. https://github.com/OpenRCE/sulley. Accessed 15 Nov 2023
8. Arzt, S., Rasthofer, S., Fritz, C., Bodden, E., Bartel, A., Klein, J., Le Traon, Y., Octeau, D., McDaniel, P.: Flowdroid: precise context, flow, field, object-sensitive and lifecycle-aware taint analysis for android apps. Acm Sigplan Notices **49**(6), 259–269 (2014)
9. Chen, J., et al.: Iotfuzzer: discovering memory corruptions in iot through app-based fuzzing. In: NDSS (2018)
10. Chen, L., et al.: Sharing more and checking less: Leveraging common input keywords to detect bugs in embedded systems. In: 30th USENIX Security Symposium (USENIX Security 21), pp. 303–319 (2021)
11. Deng, Y., Xia, C.S., Peng, H., Yang, C., Zhang, L.: Large language models are zero-shot fuzzers: Fuzzing deep-learning libraries via large language models. In: Proceedings of the 32nd ACM SIGSOFT International Symposium on Software Testing and Analysis, pp. 423–435 (2023)
12. Deng, Y., Xia, C.S., Yang, C., Zhang, S.D., Yang, S., Zhang, L.: Large language models are edge-case generators: Crafting unusual programs for fuzzing deep learning libraries. In: 2024 IEEE/ACM 46th International Conference on Software Engineering (ICSE), pp. 830–842. IEEE Computer Society (2023)
13. Enck, W., Gilbert, P., Han, S., Tendulkar, V., Chun, B.G., Cox, L.P., Jung, J., McDaniel, P., Sheth, A.N.: Taintdroid: an information-flow tracking system for realtime privacy monitoring on smartphones. ACM Trans. Comput. Syst. (TOCS) **32**(2), 1–29 (2014)
14. Feng, B., Mera, A., Lu, L.: {P2IM}: Scalable and hardware-independent firmware testing via automatic peripheral interface modeling. In: 29th USENIX Security Symposium (USENIX Security 20), pp. 1237–1254 (2020)
15. Google: honggfuzz. https://google.github.io/honggfuzz/. Accessed 15 Nov 2023
16. Gordon, M.I., Kim, D., Perkins, J.H., Gilham, L., Nguyen, N., Rinard, M.C.: Information flow analysis of android applications in droidsafe. In: NDSS, vol. 15, p. 110 (2015)
17. Gui, Z., Shu, H., Kang, F., Xiong, X.: Firmcorn: vulnerability-oriented fuzzing of iot firmware via optimized virtual execution. IEEE Access **8**, 29826–29841 (2020)
18. Liu, P., et al.: Harnessing the power of llm to support binary taint analysis. arXiv preprint arXiv:2310.08275 (2023)
19. Muench, M., Stijohann, J., Kargl, F., Francillon, A., Balzarotti, D.: What you corrupt is not what you crash: challenges in fuzzing embedded devices. In: NDSS (2018)

20. Pereyda, J.: boofuzz documentation. THIS REFERENCE STILL NEEDS TO BE FIXED (2019)
21. Redini, N., et al.: Diane: identifying fuzzing triggers in apps to generate under-constrained inputs for iot devices. In: 2021 IEEE Symposium on Security and Privacy (SP), pp. 484–500. IEEE (2021)
22. Redini, N., et al.: Karonte: detecting insecure multi-binary interactions in embedded firmware. In: 2020 IEEE Symposium on Security and Privacy (SP), pp. 1544–1561. IEEE (2020)
23. Sakaoglu, S., et al.: Kartal: web application vulnerability hunting using large language models (2023)
24. Yang, Y.: Iot software vulnerability detection techniques through large language model. In: International Conference on Formal Engineering Methods, pp. 285–290. Springer (2023)
25. Yun, I., Lee, S., Xu, M., Jang, Y., Kim, T.: QSYM : A practical concolic execution engine tailored for hybrid fuzzing. In: 27th USENIX Security Symposium (USENIX Security 18), pp. 745–761. USENIX Association, Baltimore, MD, August 2018. https://www.usenix.org/conference/usenixsecurity18/presentation/yun
26. Zalewski, M.: American fuzzy lop. 2014 (2014)

Towards the Automated Population of Thesauri Using BERT: A Use Case on the Cybersecurity Domain

Elena Cardillo[1]($\boxtimes$), Alessio Portaro[1], Maria Taverniti[1], Claudia Lanza[2], and Raffaele Guarasci[3]($\boxtimes$)

[1] Institute of Informatics and Telematics, National Research Council, Rende, Italy
`elena.cardillo@iit.cnr.it` , `maria.taverniti@cnr.it`
[2] University of Calabria (UNICAL), Rende, Italy
`claudia.lanza@unical.it`
[3] Institute for High Performance Computing and Networking, National Research
Council of Italy (CNR), Rende, Italy
`raffaele.guarasci@icar.cnr.it`

Abstract. The present work delves into innovative methodologies leveraging the widely used BERT model to enhance the population and enrichment of domain-oriented controlled vocabularies as Thesauri. Starting from BERT's embeddings, we extracted information from a sample corpus of Cybersecurity related documents and presented a novel Natural Language Processing-inspired pipeline that combines Neural language models, knowledge graph extraction, and natural language inference for identifying implicit relations (adaptable to thesaural relationships) and domain concepts to populate a domain thesaurus. Preliminary results are promising, showing the effectiveness of using the proposed methodology, and thus the applicability of LLMs, BERT in particular, to enrich specialized controlled vocabularies with new knowledge.

Keywords: Thesauri · Domain-specific language modeling · Semantic analysis · Knowledge Extraction · LLMs

1 Introduction

The field of Natural Language Processing (NLP) has experienced remarkable growth, largely driven by the emergence of Large Language Models (LLMs). The advent of Transformer-based models, starting with the release of models like BERT [5], has marked a new era in NLP approaches, consistently elevating

Although the authors have cooperated in the research work and in writing the paper, Elena Cardillo contributed to the Conceptualization of the paper, and an overall revision of the paper; Alessio Portaro formalized the experimental pipeline and carried out the experiments described in Sect. 4; Maria Taverniti contributed to conceptualization and revision; Claudia Lanza specifically dealt with Sect. 3.1.; Raffaele Guarasci contributed to the Conceptualization of the paper, and the Methodology.

L. Barolli (Ed.): EIDWT 2024, LNDECT 193, pp. 100–109, 2024.
https://doi.org/10.1007/978-3-031-53555-0_10

the standards for achievable results across various tasks such as text classification [1,8,31], sentiment analysis [17], and anaphora and coreference resolution [7,10,22,29]. The profound potential of LLMs to encode linguistic knowledge has become a focal point of scholarly interest [14]. However, within this landscape, a growing discrepancy between high-resource and low-resource languages has become evident. To address this imbalance, the exploration of transferring linguistic knowledge from one language to another has become a prominent topic in the NLP field [11,12].

This work starts from the need to propose innovative methods that have been shown to improve performance in various areas of the NLP field, to support the creation and integration of controlled vocabularies such as Thesauri, by identifying specific (non-explicit) relations and domain concepts relevant for indexing. In particular, we propose a NLP-inspired pipeline based on the wide-popular LLM BERT to extract the knowledge graph from a specialized domain thesaurus and then apply a natural language inference (NLI) method to extract potential relationships. The resource chosen for this purpose is the cybersecurity thesaurus published on the Italian Cybersecurity Observatory (OCS) web platform[1], containing more than 200 Italian/English indexing terms.

The paper is organized as follows: Sect. 2 offers an overview of recent related works. Following that, Sect. 3 describes in details the materials and the research methodology, encompassing information on the structural probe, the chosen LLM, and the dataset employed. Section 4 outlines the experimental evaluation and subsequently presents and discusses some preliminary results. Lastly, Sect. 5 provides some conclusions and alludes to potential future developments.

2   Related Works

Earning knowledge graphs has garnered significant attention recently, with advancements driven by LLMs. Early approaches to knowledge graph extraction primarily relied on rule-based systems and information retrieval techniques [21]. These methods often faced challenges in handling unstructured or ambiguous textual data and required extensive manual revision by experts. The release first of word embeddings [20] and then of large neural models of the language [5] shifted the paradigm by allowing unsupervised NLP techniques to be used in KG extraction as well. In particular, several studies have explored BERT's effectiveness in entity recognition and relationship extraction. [34,36] show how fine-tuning BERT models on specific entity and relation extraction tasks significantly improve extraction accuracy. Moreover, BERT's ability to understand context and relationships is leveraged in classification tasks. Techniques such as attention mechanisms and multi-instance learning [35] showcase improvements in identifying and classifying complex relationships within textual data.

Concurrently, NLI has emerged as a complementary approach for knowledge graph construction. By leveraging BERT's contextual embeddings, NLI models aid in reasoning about entailment and contradiction, facilitating the extraction of

---

[1] https://www.cybersecurityosservatorio.it/it/Services/thesaurus.jsp.

implicit relationships within the text [3]. In recent years, studies have explored joint learning approaches that simultaneously address entity recognition and relation extraction tasks. As suggested by [18], BERT-based models exhibit the capacity to enhance both aspects concurrently, leading to more coherent and accurate knowledge graph construction.

3   Materials and Methods

This section describes in detail the resources and the neural model of the language (NLM) used in the experiment. In particular, the resources used are the following: *i)* the bilingual (Italian-English) cybersecurity thesaurus published on the mentioned Italian OCS website, namely the OCS Thesaurus, which is a controlled vocabulary offering a structured representation of the domain knowledge through semantic relationships between different indexed terms belonging to the field of study; *ii)* a corpus of cybersecurity regulations extracted from the OCS website using automated tools. Concerning the NLM method, we take into account the wide-popular BERT [5]. The choice was driven by the enormous popularity and versatility of the model, applied in recent years to a wide variety of tasks.

3.1   The OCS Thesaurus

Considering the specialized domain taken into account, that is Cybersecurity, characterized by a highly technical terminological distribution in the documentation [4], a preliminary task of this research activity was concentrated on the consultation of already existing resources aimed at formalizing this knowledge-domain. A first choice fell on the mentioned OCS bilingual Thesaurus realized as part of this project and built by taking into account an heterogeneous set of source texts from which a list of technical terms arranged through a semantic relationship structure were extracted (see [23,26]). According to the ISO 25964:2013 standard (2013, p. 12) [25] a thesaurus is a "Controlled and structured vocabulary in which concepts are represented by terms organized so that relationships between concepts are made explicit and preferred terms are accompanied by lead-in entries for synonyms or quasi-synonyms". The main purpose assigned to such a semantic resource is that of organizing the terminology characterizing a given sector-oriented domain of study in a way that supports the indexing operations and targeted knowledge discovery [2], as well as a terminological control over the information retrieval tasks (see [6]). Indeed, as Lykke (2001, 778) [19] argued, "the thesaurus is a tool that helps individual users to get an understanding of the collective knowledge domain". The terms reflecting the knowledge domain are semantically connected with each other following the rules provided by the ISO standards 25964-1:2011 [24] and 25964-2:2013 [25] according to which there are three main kinds of semantic relationships covered within thesauri:

1. Hierarchical relationship, marked with the tags Broader Term (BT) and Narrower Term (NT), points to the specificity connection between terms. This includes relationships of types "generic" (i.e., class-member relationship "IS-A"), "instance" and "partitive";

2. Equivalence relationship, marked with the tags Used (USE) and Used For (UF), manages the synonymy link between terms representing the same concept;
3. Associative relationship, marked with the tag Related Term (RT), denotes the coordination of terms belonging to the same category or to others.

These relationships are aimed at managing the conceptual framework of a specialized domain of study in order to guarantee a semantic tool able to normalize the information and avoid ambiguity in treating sector-oriented information [2]. Given these premises, constructing a semantic tool representing specialized fields of knowledge under the lens of a terminological network requires to rely on as much technical documentation as possible in order to detect the right amount of sector-oriented terms. The specialized corpus employed to populate the OCS Thesaurus and, above all, to shape its structure consists of 57 documents coming from authoritative legislative (e.g., national and international regulations, best practices) and popular sources (scientific journals on the subject), as well as glossaries on the domain (e.g., the NIST *Glossary of key information terms* [28] and ISO 27000:2016 Information technology - Security techniques - Information security management systems - Overview and vocabulary) [13,15]. The number of terms in the OCS Thesaurus, extracted and selected alongside the supervision of domain experts and through the Term Frequency/Inverse Document Frequency (TF/IDF) statistical measure [33] is 238. To test the methodology proposed in this work we selected a sample of 5 documents in English, already part of the corpus used for the construction of the OCS Thesaurus, namely the legislative framework documentation on an European level and a specific set of relationships characterizing one Top Term (TT) of the resource, i.e., *Cybersecurity* consisting of 153 connecting terms. For instance, an example of a set of semantic relations taken into account is the following:

- **Hierarchy**: Cybersecurity *NT* Cyber risk management;
- **Equivalence**: Cybersecurity *UF* Information Security;
- **Association**: Cybersecurity *RT* Privacy.

3.2   Neural Language Model

Presently, among Transformer-based Neural Language Models (NLMs) used in NLP, BERT [5] holds a prominent position, with its efficiency and high performance well-established in the literature. The *BERT-base* version of this deep neural network architecture comprises 12 layers of decoder-only Transformers, each with 768-hidden dimensional states and 12 attention heads, totaling 110 million parameters. Conceptually rooted in the Transformer encoder architecture [32], BERT employs a multi-layer bidirectional design, pre-trained on extensive unlabeled text through two primary objectives: masked language modeling and next sentence prediction.

The versatility of BERT lies in its ability to provide robust context-dependent sentence representations, subsequently adaptable to diverse NLP tasks through a fine-tuning process tailored to specific requirements. This fine-tuning necessitates the adjustment of various hyperparameters, directly impacting the achievable outcomes. BERT's pre-training methodology centers on masked language modeling, where a portion of words in the training corpus is randomly masked. This allows the model to learn from both sentence directions while predicting the masked words. The choice between *cased* and *uncased* input vocabularies leads to two distinct pre-trained models. BERT's bidirectional analysis maintains significant generative capacity in deep constituent network layers, with outer layers adapted for task-specific fine-tuning. This dual-layered approach has established BERT as a benchmark model in recent literature. In BERT, each input word sequence begins with a special *[CLS]* token, generating a vector of size H (hidden layer size) as the output, representing the entire input sequence. Additionally, a unique *[SEP]* token must conclude each sentence within the input sequence.

Given an input sequence of words $t = (t_1, t_2, \ldots, t_m)$, BERT's output is denoted as $h = (h_0, h_1, h_2, \ldots, h_m)$, where $h_0 \in \mathbb{R}^H$ serves as the final hidden state of the *[CLS]* token, offering a pooled representation for the entire input sequence.

Subsequently, $h_1, h_2, \ldots, h_m$ represent the final hidden states of the remaining input tokens. In the context of fine-tuning BERT for classifying input sequences into K distinct text categories, the utilization of the final hidden state h_0 facilitates feeding a classification layer, followed by a softmax operation to convert category scores into likelihoods, as expounded by Sun et al. [30]:

$$P = softmax(CW^T) \tag{1}$$

where $W \in R^{K x H}$ is the parameter matrix of the classification layer.

4   Experimental Assessment

The proposed methodology starts from the work proposed in [3]. It is divided in two macro-steps as shown in Fig. 1:

1. **Concepts Identification and Extraction**: The first layer of the application aims to extract concepts and their relationships in the form of a Knowledge Graph (KG). As mentioned in Sect. 3, the corpus of documents used to run the experiments comprises legislative texts containing cybersecurity regulations. In order to have more semantically relevant knowledge graphs, a smaller set of documents was later selected by legal experts of the team from the initial texts' corpus.

   For the extraction of the KG we used part of the software developed by [27] within the scope of the Interlex project [16] . This tool first extracts concepts from the text in PDF files, leveraging the capabilities of the **SpaCy**[2] library which can infer PoS-tagged dependency tree (DPT). Using the DPT, the Interlex module extracts relevant concepts navigating nodes tagged as noun

---

[2] https://spacy.io.

phrases. Once the concepts' set is created, it can search the tree for those tokens connecting concepts. The latter is used to describe a relation between the two concepts. This relation is simply in the same form as the sentence from which it was extracted, with the only difference of having two blanks as placeholders for the subject and the object of the sentence. This method for relations extraction has the apparent advantage of preserving the natural language structure, that is very useful when the resulting data is fed into an LLM. The Interlex module was preferred among other alternatives for KG extraction since triples composed of the two concepts (subject and object) and the relation connecting the two (predicate) come together to form a KG.

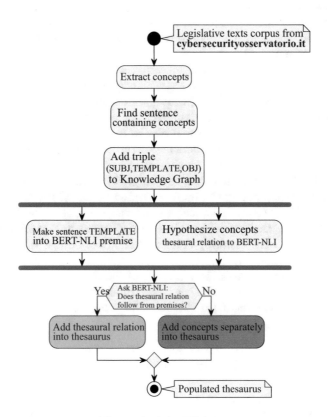

Fig. 1. Activity Diagram

2. **Natural Language Inference using KG**: In the second layer, the extracted KG populates a Thesaurus, in this case the OCS Thesaurus mentioned in Sect. 3.1 using LLMs. The KG model is known to have much redundant semantic information because many instances of the same relation can appear in the same graph. Furthermore, the relations in a KG are typically subject-object relations, whereas the relations in a thesaurus are more linguistic correlations between concepts (as mentioned above, can be hierarchical, equiva-

lence and associative relationships). To resolve this redundancy issue, we need to identify said linguistic relations. An expert would identify and classify the thesaural relationships using a classic approach to build a thesaurus. However, our scope was to automate this process, leveraging the enormous linguistic capabilities of modern LLMs like BERT. More specifically, we used a fine-tuned version of the BERT model further trained for NLI tasks. In particular, the model used is **HuggingFace**'s mDeBERTa-v3-base-xnli-multilingual-nli-2mil7 [16]. The classic input for an NLI model is composed of a premise and a hypothesis, which the model can classify in three ways:

- **Entailment**, meaning the hypothesis that can be inferred form the premises;
- **Contradiction**, meaning the hypothesis that contradicts the premises;
- **Neutral**, meaning the hypothesis that neither descends or contradicts the premises;

Since the hypothesis we wanted to verify is a series of questions about the relations between each pair of concepts, the premise consisted in some context about the two concepts. For this task, we decided to include every relation in the KG connected to each pair concept as context. The relations were extracted as a template with some blanks for the subject and object of the sentence, so we replaced the blanks with the concepts that the said relation connects in the KG. Thanks to the Interlex library, once this operation is completed, the resulting sentence can be perfectly understandable in natural language, perfectly fitting the input expected by an NLP transformer-based model like BERT. Given a context and a hypothesis, the NLI model predicts a class. So in our task, when the class corresponded to **Entailment**, we created a relation in the resulting domain thesaurus between the two concepts. The type of relation matched the one we hypothesized as input for the NLI model. To give an example, from a triple of the KG extracted having as subject "Security" and object "Privacy", we could identify the type of thesaural relationships among them (in this case RT) and verify its presence in the Thesaurus.

5   Conclusion and Future Works

This paper showed the preliminary steps of an approach that exploits modern neural models of language to extract/add information to a controlled vocabulary. Specifically, BERT was used to validate a pipeline that extracts concepts from a thesaurus on cybersecurity to conduct an NLI task. Exploiting BERT embeddings, the approach is able to extract information from a corpus of Cybersecurity-related documents. The proposed NLP-inspired pipeline seamlessly integrates NLMs, knowledge graph extraction, and NLI to discern implicit relations and domain concepts, enriching a domain thesaurus.

Preliminary findings demonstrated the robustness of the proposed methodology, highlighting the applicability of state-of-the-art Large Language Models in augmenting specialized controlled vocabularies with new knowledge. The outcomes underscore the potential for integrating BERT-based techniques to enhance the semantic richness and utility of domain-oriented thesauri, without relying on outdated lexicons. This research contributes to the advancement of methodologies for constructing and enriching, thus updating controlled vocabularies, providing a contemporary framework for knowledge extraction and relationship identification in domain-specific contexts. Even in a preliminary stage, the promising results obtained open up for further exploration and application in the rapid growing landscape of domain-specific knowledge management. For future work, given the great versatility of BERT and its availability in multilingual versions, the approach here proposed will be tested taking into account Italian terms contained in the thesaurus. It could open up numerous research possibilities regarding parallel analyses on the enrichment of the translations provided in the resource.

Furthermore, new approaches will be tested that promise to exceed the current performance achievable with NLMs, in order to be able to identity and extract, for example, precise types of hierarchical relationships (i.e., partitive, instance) to correctly connect the concepts extracted and properly shape/update the thesaurus structure. Moreover, novel perspectives on analysis have opened up with the emergence of Quantum NLP, or that sub-branch of NLP that uses methods derived from quantum theory to increase performance [9]. This could be investigated to improve the results in terms of accuracy and quantity of new knowledge that can be identified to populate the thesaurus.

Acknowledgements. This work was partially supported by project SERICS (PE00000014) under the MUR National Recovery and Resilience Plan funded by the European Union - NextGenerationEU.

References

1. Bonetti, F., Leonardelli, E., Trotta, D., Guarasci, R., Tonelli, S.: Work hard, play hard: Collecting acceptability annotations through a 3d game, pp. 1740–1750 (2022)
2. Broughton, V.: Essential Thesaurus Construction. Facet (2006). https://doi.org/10.29085/9781856049849
3. Chen, W., Ji, H.: Infer: Capturing implicit entity relations for knowledge graph completion using contextualized language models. arXiv preprint arXiv:2006.05295 (2020)
4. Claudia, L., Elena, C., Maria, T., Roberto, G.: Terminology management in cybersecurity thought knowledge organization systems: an Italian use case. Int. J. Adv. Secur. **1–2**, 17–27 (2020)
5. Devlin, J., Chang, M.W., Lee, K., Toutanova, K.: BERT: pre-training of deep bidirectional transformers for language understanding. In: Proceedings of the 2019 Conference of the North American Chapter of the Association for Computational Linguistics: Human Language Technologies, Volume 1 (Long and Short Papers), pp. 4171–4186. ACL, Minneapolis, Minnesota (2019). https://doi.org/10.18653/v1/N19-1423

6. Gabler, S.: Thesauri - a toolbox for information retrieval. Bibliothek Forschung und Praxis **47**(2), 189–199 (2023). https://doi.org/10.1515/bfp-2023-0003

7. Gargiulo, F., et al.: An electra-based model for neural coreference resolution. IEEE Access **10**, 75144–75157 (2022). https://doi.org/10.1109/ACCESS.2022.3189956

8. Guarasci, R., Damiano, E., Minutolo, A., Esposito, M., De Pietro, G.: Lexicon-grammar based open information extraction from natural language sentences in italian. Expert Syst. Appl. **143**, 112,954 (2020). https://doi.org/10.1016/j.eswa.2019.112954

9. Guarasci, R., De Pietro, G., Esposito, M.: Quantum natural language processing: Challenges and opportunities. Appl. Sci. (Switzerland) **12**(11) (2022). https://doi.org/10.3390/app12115651

10. Guarasci, R., Minutolo, A., Damiano, E., De Pietro, G., Fujita, H., Esposito, M.: ELECTRA for neural coreference resolution in italian. IEEE Access **9**, 115,643–115,654 (2021). https://doi.org/10.1109/ACCESS.2021.3105278

11. Guarasci, R., Silvestri, S., De Pietro, G., Fujita, H., Esposito, M.: Bert syntactic transfer: a computational experiment on Italian, French and English languages. Comput. Speech Lang. **71**, 101,261 (2022)

12. Guarasci, R., Silvestri, S., De Pietro, G., Fujita, H., Esposito, M.: Assessing bert's ability to learn Italian syntax: a study on null-subject and agreement phenomena. J. Ambient. Intell. Humaniz. Comput. **14**(1), 289–303 (2023)

13. Hazem, A., Daille, B., Claudia, L.: Towards automatic thesaurus construction and enrichment. In: B. Daille, K. Kageura, A.R. Terryn (eds.) Proceedings of the 6th International Workshop on Computational Terminology, pp. 62–71. European Language Resources Association, Marseille, France (2020)

14. Jawahar, G., Sagot, B., Seddah, D.: What does BERT learn about the structure of language? In: Proceedings of the 57th Annual Meeting of the Association for Computational Linguistics, pp. 3651–3657. ACL, Florence, Italy (2019). https://doi.org/10.18653/v1/P19-1356

15. Lanza, C.: Semantic control for the cybersecurity domain: investigation on the representativeness of a domain-specific terminology referring to lexical variation. CRC Press (2022). https://doi.org/10.1201/9781003281450citation-key

16. Laurer, M., Atteveldt, W.v., Casas, A.S., Welbers, K.: Less Annotating, More Classifying - Addressing the Data Scarcity Issue of Supervised Machine Learning with Deep Transfer Learning and BERT - NLI. Preprint (2022). Publisher: Open Science Framework

17. Li, W., Zhu, L., Shi, Y., Guo, K., Cambria, E.: User reviews: sentiment analysis using lexicon integrated two-channel cnn-lstm family models. Appl. Soft Comput. **94**, 106,435 (2020). https://doi.org/10.1016/j.asoc.2020.106435

18. Liu, Y., Ott, M., Goyal, N., Du, J., Joshi, M., Chen, D.: Joint entity recognition and relation extraction as a multi-head selection problem. arXiv preprint arXiv:2201.10208 (2022)

19. Lykke, M.: A framework for work task based thesaurus design. J. Documentation **57**, 774–797 (2001). https://doi.org/10.1108/EUM0000000007100

20. Mikolov, T., Chen, K., Corrado, G., Dean, J.: Efficient estimation of word representations in vector space. arXiv preprint arXiv:1301.3781 (2013)

21. Miller, G.A.: Introduction to wordnet: an on-line lexical database. Int. J. Lexicogr. **3**(4), 235–244 (1990)

22. Minutolo, A., Guarasci, R., Damiano, E., De Pietro, G., Fujita, H., Esposito, M.: A multi-level methodology for the automated translation of a coreference resolution dataset: an application to the italian language. Neural Comput. Appl. **34**(24), 22,493 - 22,518 (2022). https://doi.org/10.1007/s00521-022-07641-3

23. Nielsen, M.L.: Thesaurus construction: key issues and selected readings. Cataloging Classification Quarterly **37**(3–4), 57–74 (2004). https://doi.org/10.1300/J104v37n03_05

24. Organization, I.S.: ISO 25964-1:2011 Information and documentation - Thesauri and interoperability with other vocabularies - Part 1: Thesauri for information retrieval (2011)

25. Organization, I.S.: ISO 25964-2:2013 Information and documentation - Thesauri and interoperability with other vocabularies - Part 2: Interoperability with other vocabularies (2013)

26. Zadeh, B.Q., Handschuh, S.: The ACL RD-TEC: a dataset for benchmarking terminology extraction and classification in computational linguistics. In: Proceedings of the 4th International Workshop on Computational Terminology (Computerm), pp. 52–63. Association for Computational Linguistics and Dublin City University, Dublin, Ireland (2014). https://doi.org/10.3115/v1/W14-4807

27. Sovrano, F., Palmirani, M., Vitali, F.: Legal knowledge extraction for knowledge graph based question-answering. In: Legal Knowledge and Information Systems, pp. 143–153. IOS Press (2020)

28. National Institute of Standards: Glossary of key information security terms. Tech. rep., NIST Interagency or Internal Report (NISTIR) 7298 Rev. 2, May 2013

29. Sukthanker, R., Poria, S., Cambria, E., Thirunavukarasu, R.: Anaphora and coreference resolution: a review. Inf. Fusion **59**, 139–162 (2020). https://doi.org/10.1016/j.inffus.2020.01.010

30. Sun, C., Qiu, X., X.Y., X., H.: How to fine-tune bert for text classification? In: China National Conference on Chinese Computational Linguistics, pp. 194–206. Springer (2019)

31. Trotta, D., Guarasci, R., Leonardelli, E., Tonelli, S.: Monolingual and cross-lingual acceptability judgments with the Italian CoLA corpus. In: M.F. Moens, X. Huang, L. Specia, S.W.t. Yih (eds.) Findings of the Association for Computational Linguistics: EMNLP 2021, pp. 2929–2940. Association for Computational Linguistics, Punta Cana, Dominican Republic (2021). https://doi.org/10.18653/v1/2021.findings-emnlp.250

32. Vaswani, A., Shazeer, N., Parmar, N., Uszkoreit, J., Jones, L., Gomez, A.N., Kaiser, L., Polosukhin, I.: Attention is all you need. In: Advances in Neural Information Processing Systems 30: Annual Conference on Neural Information Processing Systems 2017, pp. 5998–6008. Long Beach, CA, USA (2017)

33. Wu, H.C., Luk, R.W.P., Wong, K.F., Kwok, K.L.: Interpreting tf-idf term weights as making relevance decisions. ACM Trans. Inf. Syst. **26**, 13:1–13:37 (2008)

34. Wu, S., Dredze, M.: Beto, Bentz, Becas: The surprising cross-lingual effectiveness of BERT. In: Proceedings of the 2019 Conference on Empirical Methods in Natural Language Processing and the 9th International Joint Conference on Natural Language Processing (EMNLP-IJCNLP), pp. 833–844. Association for Computational Linguistics, Hong Kong, China (2019). https://doi.org/10.18653/v1/D19-1077

35. Zhang, S., Wang, Z., Tang, J.: Bert for joint entity and relation extraction via context-aware coreference resolution. Inf. Process. Manage. **58**(5), 102,356 (2021)

36. Zhang, Y., Zhang, Y., Ji, D.: Kg-bert: Bert for knowledge graph completion. arXiv preprint arXiv:2002.00388 (2020)

Network Scanning Detection Based on Spatiotemporal Behavior

Pengyuan Zhang and Baojiang Cui[✉]

School of Cyberspace Security, Beijing University of Posts and Telecommunications,
Haidian District, Xitucheng Road No.10, Beijing, China
{zhangpengyuan,cuibj}@bupt.edu.cn

Abstract. Network scanning stands as a crucial tactic for attackers to identify and access network assets. Timely recognition of network scanning behavior facilitates early detection of attackers' intentions, enabling proactive defensive measures. However, network scans often involve attempts to access specific pathways, and the traffic itself doesn't exhibit overt malicious traits, making it challenging to discern whether a single traffic instance represents scanning behavior. This paper devises a spatiotemporal behavior detection method that integrates temporal features of traffic content and network node access relationship features. Leveraging LSTM, the paper amalgamates network traffic data over a period, resulting in a graph data structure encompassing node features, edge features, and node access relationships. The EdgeGAT network is employed for classifying and identifying graph data. Experimental findings indicate the positive impact of node access relationships within traffic data on identifying asset scanning behavior, thereby enhancing the effectiveness of traditional traffic identification methods.

1 Introduction

Scanning the assets of a target network to gather precise and detailed information is a crucial preliminary step for attackers before initiating a network assault. To obtain critical data such as network device versions, website backend information, or backend framework details, attackers conduct asset reconnaissance scans on the target network. Subsequently, they assess the vulnerability of target devices based on the device manufacturer and version number, employing corresponding vulnerabilities to launch attacks. Throughout this process, attackers typically extract software models, firmware, and other information by scrutinizing keywords within device interaction protocol banners, checking for specific device configuration pages, crafting error messages from specific devices, or probing management backend login pages.

However, these reconnaissance methods employed in network traffic data do not inherently contain explicit malicious attack instructions. Each individual instance of scanning traffic, when observed in isolation, may not manifest overtly malicious traits. Yet, collectively, this scanning behavior constitutes malicious intent.

Addressing the need to analyze aggregated traffic data over a period, this paper proposes a spatiotemporal correlation analysis-based scanning behavior detection technique. This approach integrates the functionalities of Graph Neural Networks (Edge-GAT) and Long Short-Term Memory Networks (LSTM), enabling a comprehensive

L. Barolli (Ed.): EIDWT 2024, LNDECT 193, pp. 110–117, 2024.
https://doi.org/10.1007/978-3-031-53555-0_11

analysis of network traffic data across spatial structures and temporal dynamics. By constructing a network traffic graph, the EdgeGAT network extracts interaction patterns among nodes and potential anomalous relationships. Simultaneously, the LSTM network interprets time series data, capturing long-term behavioral dependencies and periodic anomalies. The fusion of temporal and spatial relationships provides multidimensional insights, enabling the identification of node access behaviors that exhibit anomalies across both time and space dimensions. This enhances the model's precision in scanning pattern detection, facilitating more effective formulation and adjustment of defensive strategies.

This paper aims to explore the detection of network scans by leveraging spatiotemporal behaviors within network traffic. To comprehensively delve into this subject, the paper is divided into the following sections: the initial section will review relevant literature, discussing the latest research findings and viewpoints in the field regarding the utilization of spatiotemporal correlation analysis for advancing malicious traffic detection. The second section will elaborate on our proposed model methodology based on the EdgeGAT + LSTM network. We will elucidate our approach and its application within the research context. The third section will present our research findings and their analysis. We will showcase and interpret the obtained data, experimental results, or investigative discoveries. Finally, in the conclusion, we will summarize the main discoveries of this paper, discussing implications for future research and practice. We will offer suggestions, outline potential research directions, and emphasize the contributions and significance of this paper.

2 Related Work

Experts and scholars in the field of cybersecurity have consistently proposed methods utilizing spatial and temporal relationships to defend against and detect network attacks. These methods employ diverse detection perspectives, analyzing and scrutinizing entities or behaviors within systems.

The application of Graph Neural Networks (GNNs) has garnered considerable attention within this domain. These networks effectively model network structures, enabling a better understanding and analysis of intricate network data. Specifically, research in dynamic and static graphs constitutes two significant directions within the realm of graph neural networks. For static graphs, various models have been proposed to handle network data. Graph Convolutional Networks (GCNs) [1] and Graph Attention Networks (GAT) [2] are widely used models for tasks such as node classification, link prediction, and graph representation learning. GCN performs admirably in handling static graphs by aggregating information from nodes and their neighbors to learn node representations. On the other hand, GAT introduces an attention mechanism, dynamically assigning different weights to neighboring nodes, allowing a more flexible capture of relationships among nodes. Research in both static and dynamic graphs continuously explores leveraging graph neural network models to better understand and tackle challenges within the cybersecurity domain. While models like GCN for static graphs have excelled in tasks like node classification and graph representation learning, studies on GAT and similar networks for dynamic graphs aim to capture temporal features during network evolution, enhancing the identification and prediction of network attacks and anomalous behavior.

Previous studies [3–5] have utilized graph-based approaches to analyze connection behaviors between hosts and DNS servers, extracting domain features through rules to detect the establishment of Command and Control (C&C) channels in early-stage intrusions. Other studies [6–8] have analyzed entity behaviors, constructed association graphs among lower-level entities, and employed rule matching, feature analysis, and attack scenario reconstruction to identify malicious entities for attack detection. Furthermore, some research [9, 10] trained multiple identity authentication features and monitored alerts' associations using machine learning to detect compromised hosts and attack chains. Additional studies [11, 12] combined graph methods and neural networks to model homogenous or heterogeneous entities, employing meta path analysis or graph-matching neural networks to identify abnormal behavior between entities. One study [13] constructed a user behavior heterogeneous graph based on ten rules and used clustering methods to detect abnormal behavior, thereby revealing differences in invaded users' daily behavior compared to their typical conduct. Another proposal, the MalRank algorithm [14], relied on graph-based inference to infer malicious scores of nodes based on their associations with other entities in a knowledge graph, issuing alerts. However, this method heavily relied on threat intelligence, and the computation of malicious scores exhibited subjectivity. Research [11] modeled identity authentication behavior within enterprise-level networks and introduced unsupervised graph learning techniques to identify lateral movement events. Additionally, a study [15] analyzed extensive network traffic, evaluating the temporal changes of internal hosts in a multidimensional feature space to analyze the suspicion of Advanced Persistent Threat (APT) activities, ranking suspicious internal hosts for expert analysis.

These research methodologies primarily focus on analyzing entity behaviors and utilize graph methods or deep learning to predict normal and abnormal behaviors. However, they commonly overlook the consideration of network node spatiality. During attacks or scanning processes, there exist connections between the attacker and the victim hosts, connections that attackers manifest and find challenging to completely conceal. Approaches lacking spatial considerations miss out on many edge details that complement each other among host characteristics. Therefore, this paper proposes a novel detection method that concurrently considers the spatial characteristics of network nodes and the temporal features of traffic, aiming to address this limitation.

3 Model Methodology

To address the spatiotemporal characteristics of scanning traffic, this paper introduces a detection model based on the spatiotemporal behavior of scanning traffic. Initially, the model transforms raw logs into graph-structured data, preserving multiple records between nodes over a period. This approach enables a more comprehensive capture of dynamic changes in network activities. Through encoding, the model effectively extracts subnet information and access data between nodes, transforming them into node and edge features, subsequently structured into a dataset suitable for graph networks.

The model comprises two key components: an LSTM layer for feature extraction and a GAT layer for graph classification. Initially, the LSTM layer processes the original graph data, extracting a unified dimension of edge feature data from the HTTP dataset.

The model updates these new edge feature data onto the graph and employs a GAT network to classify the entire connected graph. This methodology maximizes the utilization of graph connectivity, enabling the model to comprehensively comprehend and classify the characteristics of the entire network graph, and facilitating efficient identification and classification of network activities. The overall logic is depicted in Fig. 1, while detailed logic and processes will be expounded in subsequent sections.

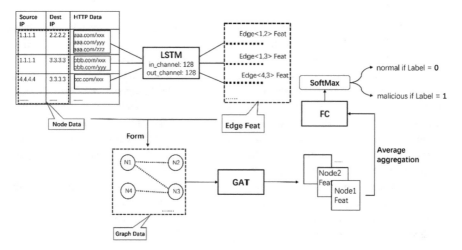

Fig. 1. Method Framework

3.1 Data Preprocessing

Before training the model, it's necessary to convert regular traffic parsing logs into the format of a graph dataset, consisting of three key components: node features, node interaction relations, and edge features. Following the earlier description, the initial step involves selecting a time window and gathering all traffic data within that time frame. Each IP address is regarded as a node within the graph, while the traffic data records between two IP addresses constitute the node interaction relations. The initial feature set for each node is defined as the node's in-degree and out-degree, denoted as (in-degree, out-degree).

The edge feature between any two nodes constitutes an array encompassing all encoded traffic data content between those nodes. Primarily, we utilized a character-level tokenizer to tokenize the traffic log text. Limits were set for a maximum of 20,000 characters and a maximum sequence length of 128. The Tokenizer was used to convert the original text into integer sequences, encoding each character based on its frequency in the dataset. To ensure consistent input lengths for the neural network, the pad_sequences method from the Keras library was applied to either pad or truncate integer sequences to a uniform length of 128.

These preprocessing steps provided a data structure suitable for input into the LSTM network, enabling the model to better comprehend and learn patterns and features within the traffic log. The data was handled in a way that retained its original features without substantial modification or filtering, ensuring data integrity and reliability.

Finally, the graph was labeled with its classification attribute, either as scan type or normal type. The designed model focuses on the overall graph classification and prediction rather than specific edges and nodes within the graph.

3.2 Model Design

The model devised in this approach comprises two main components: an LSTM network and an EdgeGAT network. Following the description in the data preprocessing section, the feature content between two nodes constitutes an array of traffic data features. The LSTM network processes this feature array to extract a standardized sequence feature representing the relationship between the two nodes. Utilizing the standardized edge features, a graph neural network performs global pooling operations to extract features from the entire graph, facilitating the classification of the entire graph data. A series of graph neural network models, exemplified by GCNs, have demonstrated good performance in tasks such as node classification, graph classification, and link prediction within the graph domain. Most of these models are constructed based on the message-passing approach, which involves aggregating neighbor information and updating node self-states. Assuming features $x_v \in \mathbb{R}^{d_1}$ are present on node v and features $w_e \in \mathbb{R}^{d_2}$ are present on edge (u, v), the message-passing paradigm defines computations on a per-node and per-edge basis as follows:

$$m_e^{(t+1)} = \phi\left(x_v^{(t)}, x_u^{(t)}, w_e^{(t)}\right), (u, v, e) \in E \tag{1}$$

$$x_v^{(t+1)} = \psi\left(x_v^{(t)}, \rho\left(\left\{m_e^{(t+1)} : (u, v, e) \in \mathcal{E}\right\}\right)\right) \tag{2}$$

In the equation above, ϕ is the message function defined on each edge, generating messages by combining edge features with the features of the two nodes it connects. The aggregation function ρ consolidates the messages received by nodes. The update function ψ combines the aggregated messages with the node's features to update the node's features.

In this paper's graph data, there is rich information on the edges. Referencing citation [16], the EdgeGAT layer with edge features is utilized in the graph neural network section. LSTM and EdgeGAT are trained together. Initially, each edge's feature undergoes LSTM processing, where the input and hidden state dimensions for each edge feature are set at 128. After processing through LSTM units and fully connected layers, updated edge features are obtained. The EdgeGATConv layer is employed to extract node and edge feature information from the graph data. Its output includes updated node and edge features. All edge features are normalized for consistency and, along with node features, fed into the EdgeGATConv layer to derive new node features. Post pooling operation on these new node features, a final graph-level representation is obtained through another fully connected layer, producing an output dimension of 2, used for graph-level prediction tasks. The model updates parameters via a cross-entropy loss function and gradient descent and employs softmax to derive the ultimate classification results.

4 Experiment Design and Results

4.1 Data Collection and Processing

The experimental dataset in this paper is based on traffic data from the Beijing University of Posts and Telecommunications campus network and traffic data from an actual attack-defense competition conducted in the campus network environment. The attack-defense competition was the Huawei Cup Graduate Network Security Innovation Competition hosted by the Beijing University of Posts and Telecommunications in November 2022. Twenty-four teams conducted penetration attacks on the IT facilities within the campus network, and all traffic data was entirely preserved through traffic mirroring. The competition spanned two days, yielding a total of 800,000 traffic log entries, encompassing various scanner-generated traffic logs used by the attacking teams.

From the identified scanning traffic logs, the data was grouped using fixed time windows (e.g., 30s, 60s, etc.), constructing each time window's data into independent graph-type data. Eventually, a total of 1,163 graph data samples were obtained, representing scanning-type traffic samples. Similarly, for normal campus network traffic data within the same time periods, normal traffic samples were constructed using the same methodology, resulting in a total of 1,422 normal traffic sample graphs. All types of graph data contain complete access relationships between all nodes within the graph and full HTTP protocol data exchanged between nodes during access.

4.2 Experimental Results

The experiment in this paper was conducted in the PyTorch 2.0 and Python 3.8 environment, leveraging the DGL library version 1.1.2 to assist in constructing the graph neural network. The experiment comprised stages such as dataset preprocessing, construction of a DGL graph-type dataset, neural network architecture design, and training. Model evaluation was based on three metrics: accuracy, precision, and recall. Accuracy represents the percentage of correctly predicted graph data among all graph data, precision signifies the ratio of correctly predicted scanning-type graphs to the total predicted scanning-type graphs, while recall denotes the ratio of correctly predicted scanning-type graphs to all actual scanning-type graphs.

The first experiment aimed to verify whether incorporating network node access relationships improves the ability to classify malicious sequence collections. For this experiment, a time window size of 30 s was selected. Initially, the neural network's recognition capability was tested without utilizing network node relationships, solely based on traffic log content. In this scenario, the model employed only the LSTM module within our network, taking multiple traffic data contents within the time window as inputs, and training the LSTM network to identify whether this set of traffic data is scanning-type traffic. Subsequently, the experiment involved employing our designed model containing both the EdgeGAT and LSTM modules for scan identification. The complete node access relationships and traffic data content were used as inputs to test the model's recognition capability. The experimental results based on the evaluation metrics of accuracy, precision, and recall are presented in Table 1.

Table 1. Results of Comparative Experiments on Model Structure

Model	Accuracy	Precision	Recall
LSTM	89.2%	91.8%	85.2%
LSTM + EdgeGAT	94.7%	95.1%	90.3%

From the experimental results, it is evident that the EdgeGAT + LSTM network exhibits varying degrees of improvement in accuracy, precision, and recall compared to using the LSTM network alone. This indicates that leveraging the network node access relationships enhances the ability to discern network scanning traffic. Simultaneously, it's observed that using the LSTM network alone also demonstrates a high recognition capability for scanning traffic, suggesting substantial differences between scanning traffic collections and normal traffic content itself.

The second experiment aimed to test the impact of the selected time window size on network performance. Time window sizes of 30 s, 60 s, and 120 s were chosen, and the model's various metrics were evaluated. The experimental results based on the evaluation metrics of accuracy, precision, and recall are presented in Table 2.

Table 2. Experimental Results of Time Window

Time Window Size	Accuracy	Precision	Recall
30 s	94.7%	95.1%	90.3%
60 s	96.2%	96.8%	94.5%
120 s	97.1%	98.4%	96.3%

From the experimental results, it's evident that as the time window expands, the ability to identify scanning traffic shows an increasing trend. As expected, larger time windows can acquire richer network node access information and traffic data, while potentially reducing the problem of missing access relationships in graphs due to overly short time spans. However, in actual application scenarios, it's crucial to strike a balance between detection performance and capability, avoiding blindly expanding the time window.

5 Conclusion

To address the challenge of inconspicuous malicious features and delayed identification in the network scanning phase, this paper proposes a combined approach utilizing sequence feature extraction and graph neural network classification to identify network scanning behavior. Constructing a graph classification neural network incorporating EdgeGAT and LSTM, this study conducted tests on the Beijing University of Posts and Telecommunications campus network data containing scanning traffic. The results demonstrate the superiority of this network design approach over traditional sequential classification methods, while also validating the role of network node spatial relationships in detecting scanning behavior. Associative analysis based on temporal and spatial relationships holds significant importance in the field of malicious traffic detection, warranting further exploration and experimentation.

References

1. Velikovi, P., et al.: Graph attention networks (2017)
2. Kipf, T.N., Welling, M.: Semi-supervised classification with graph convolutional networks (2016)
3. Rahbarinia, B., Perdisci, R., Antonakakis, M.: Efficient and accurate behavior-based tracking of malware-control domains in large ISP networks. ACM Trans. Priv. Secur. (TOPS) **19**(2), 1–31 (2016)
4. Khalil, I., Yu, T., Guan, B.: Discovering malicious domains through passive DNS data graph analysis. In: Computer and Communications Security ACM (2016)
5. Shi, Y., Chen, G., Li, J.: Malicious domain name detection based on extreme machine learning. Neural Process. Lett. **48**, 1347–1357 (2018)
6. Milajerdi, S.M., et al.: Holmes: real-time apt detection through correlation of suspicious information flows. In: 2019 IEEE Symposium on Security and Privacy (SP). IEEE (2019)
7. Qingqiang, W.U.: Overview of network user behavior analysis and modelling. Digit. Libr. Forum (2015)
8. Pei, K., et al.: HERCULE: attack story reconstruction via community discovery on a correlated log graph. In: Proceedings of the 32Nd Annual Conference on Computer Security Applications (2016)
9. Bian, H.: Detecting network intrusions from authentication logs. MS thesis. University of Waterloo (2019)
10. Ghafir, I., et al.: Detection of advanced persistent threat using machine-learning correlation analysis. Fut. Gener. Comput. Syst. **89**, 349–359 (2018)
11. Bowman, B., et al.: Detecting lateral movement in enterprise computer networks with unsupervised graph {AI}. In: 23rd International Symposium on Research in Attacks, Intrusions and Defenses (RAID 2020) (2020)
12. Wang, S., et al.: Heterogeneous graph matching networks. arXiv preprint: arXiv:1910.08074 (2019)
13. Liu, F., et al.: Log2vec: a heterogeneous graph embedding based approach for detecting cyber threats within enterprise. In: Proceedings of the 2019 ACM SIGSAC Conference on Computer and Communications Security (2019)
14. Najafi, P., et al.: MalRank: a measure of maliciousness in SIEM-based knowledge graphs. In: Proceedings of the 35th Annual Computer Security Applications Conference (2019)
15. Marchetti, M., et al.: Analysis of high volumes of network traffic for advanced persistent threat detection. Comput. Netw. **109**, 127–141 (2016)
16. Monninger, T., et al.: Scene: reasoning about traffic scenes using heterogeneous graph neural networks. IEEE Robot. Autom. Lett. **8**(3), 1531–1538 (2023)

IoTaint: An Optimized Static Taint Analysis Method in Embedded Firmware

Huan Yu and Baojiang Cui[✉]

School of Cyberspace Security, Beijing University of Posts and Telecommunications,
Beijing, China
{littlewhitehat,cuibj}@bupt.edu.cn

Abstract. While IoT devices have created immense value for human life, they have also introduced unavoidable security risks. In recent years, attacks targeting IoT devices have become increasingly common, making the use of efficient and automated methods for discovering vulnerabilities in IoT devices a popular research direction. However, current vulnerability detection techniques face issues such as high false positive rates and huge time costs. Therefore, this paper introduces a prototype system for IoT device vulnerability detection, IoTaint, which is based on an optimized taint analysis method. IoTaint identifies tainted data sources by analyzing shared keywords between front-end and back-end files, tracks taint analysis across border binary and inter-files, and checks sink points of dangerous data. With low latency and low false positive rates, it is achieved by optimization strategies to efficiently identify vulnerabilities in IoT device firmware. Not only does IoTaint perform well in detecting Nday and 1day vulnerabilities, but it is also capable of discovering 0day vulnerabilities, for which are confirmed by CVE/CNNVD.

1 Introduction

The development of IoT (Internet of Things) device technology has brought unprecedented convenience to people's lives, as well as risks and challenges. Research indicates that by 2020, there were about 5.8 billion IoT endpoints in use [12]. However, the vast number of IoT device programs are troubled by medium and high-risk vulnerabilities, making them remotely controlled by malicious software. Report shows [13] that 57% of IoT devices worldwide are subject to moderate or severe attacks, with devices like edge gateways and smart cameras becoming potential security threats to homes and companies.

Among all IoT devices, those based on Linux embedded systems are most vulnerable to attack [14]. This is because the ELF (Executable and Linkable Format) binary programs in the backend, when processing parameters sent from the frontend, are prone to issues like overflow and injection. Unfortunately, existing methods still fall short of accurately and efficiently detecting and uncovering vulnerabilities in IoT devices. Current technologies primarily rely on dynamic analysis techniques like fuzzing [1,2] and emulation [3,4] and static analysis techniques like taint analysis. Dynamic analysis faces significant challenges in adapting to different environments, requiring extensive preparation and facing low success rates in emulation. Additionally, fuzz testing,

L. Barolli (Ed.): EIDWT 2024, LNDECT 193, pp. 118–129, 2024.
https://doi.org/10.1007/978-3-031-53555-0_12

a black-box approach often used on firmware programs, requires consideration of code coverage. However, much of the code is internal and does not interact with external inputs, leading to limited coverage of interactive code parts and difficulty in obtaining effective results within constrained conditional branches. On the other hand, static taint analysis techniques have high false negative and false positive rates. The high rate of false negatives is due to the neglect of indirect calls in static taint analysis, while the high rate of false positives arises from the redundant business code which is reused across different firmware models in IoT device programs, making it challenging to identify the sources of taint.

Therefore, to uncover IoT device vulnerabilities more accurately and efficiently, we propose an optimized taint analysis technique. Firstly, we support taint tracking for parameters that propagate across files, not limited to the binary itself. Specifically, we focus on script files within the filesystem of Linux-based embedded firmware, searching taint paths where front-end parameters are passed to script files via back-end programs. Secondly, we add a heuristic pruning operation before the existing taint-style vulnerability discovery step. This allows the taint engine to effectively reduce unnecessary static analysis during runtime, significantly cutting down redundant paths in the path taint tracking process and focusing on more problematic program or script call paths.

In summary, our major contributions are as follows:

1. We utilize a shared keyword approach to swiftly identify border binary files, namely the back-end interface programs interacting with front-end web programs, and apply the same method to determine the caller programs for RPC calls.
2. We employ a cross-file parameter transmission filtering algorithm to categorize and filter the script files involved in back-end RPC calls. This process helps us detect bugs in potential script call chains and assess the security of the scripts themselves, including languages like shell, Perl, and Python.
3. We enhance the traditional static symbolic execution taint analysis engine by adding an expanded feature for script file parameter taint analysis. We also delve deeper into heuristic pruning method and symbolic execution optimization strategies applied prior to the binary symbolic execution process. This approach opens up new avenues in vulnerability detection, thereby improving overall efficiency and accuracy.

The rest of the paper is organized as follows. We first summarize related work in Sect. 2 and present the overall design of IoTaint in Sect. 3. Then we demonstrate the accuracy and efficiency of IoTaint through experiments and present vulnerability detection result in Sect. 4. At last Sect. 5 concludes this paper.

2   Related Work

Our research focuses on IoT device firmware. Traditional analysis for IoT firmware is categorized into dynamic and static methods. Each of them has its own advantages and limitations.

Dynamic execution analysis techniques have always been a hot research direction in IoT device vulnerability detection, but they have struggled to resolve issues effectively. In terms of emulation, Firmadyne [3] managed to emulate over 1,900 images, but it only emulated file system images and repaired the NVRAM hardware environment variables issues. SRFuzzer [2], an automatic fuzz testing framework for SOHO routers, emulates data processing and communication methods of back-end binary programs, continuously generating constrained test cases. FirmFuzz [7] emulates firmware using QEMU [10] system mode, relying on a proxy server to obtain traffic and data in the network environment, thereby capturing packets directly for fuzz testing input. FIRM-AFL [4] optimizes the huge overhead brought by full-system emulation through enhanced process emulation technology and uses the grey-box fuzz tool AFL(American Fuzzy Lop) [15], guided by path coverage, for efficient and high-speed detection of Nday, 1day vulnerabilities, and 0day vulnerability discovery in IoT programs. However, all these methods are either trapped in dependency repair work for service program startups, hindered by the huge overhead of full-system emulation, or limited to obtaining fuzz inputs only from front-end parameter. This inability to discover hidden interfaces and parameters ultimately leads to minimal gains from long-running and high-memory consumption processes.

Therefore, static analysis techniques have shown better results in the work of detecting vulnerabilities in Linux-based IoT device firmware recently. In determining the target binary program, Karonte [6] uses a method of finding inter-process communication functions to map the interaction relationships between binaries with a graph network structure. SATC [8] analyzes front-end files to obtain web parameter key-value pairs, using the key names to locate the corresponding back-end binary program functions, thereby determining the starting address for back-end data processing. Emtaint [9] focuses on analyzing the relationship between assembly instructions in binary programs and indirect calls, ensuring low false negative rates in uncovering taint-style vulnerabilities. However, the false positive rates of these prototype systems are very high. This could be due to, on one hand, the path exploration in symbolic execution being lengthy and complicated, with extremely long constraint expressions and tens of thousands of branch states, leading to memory explosions and analysis failures. On the other hand, there's a lack of pruning in the paths leading to sink functions, still requiring extensive manual filtering.

In summary, current dynamic analysis techniques face issues with high costs of environmental configurations and limitations in fuzz testing inputs, while traditional static analysis techniques are constantly troubled by path explosions in symbolic execution leading to analysis failures.

3   Methodology

Based on the issues above, we propose IoTaint, an optimized static taint analysis prototype system tailored for IoT devices, specifically for firmware based on the Linux filesystem.

3.1   Overview

The prototype system starts by receiving a firmware binary file as input, which it unpacks and divides into three categories.

1. Static Files. These include front-end web program files like HTML, JS, XML, etc. They are not included in the vulnerability discovery process but are essential for understanding the front-end interface.
2. Binary Files. These are binary programs within the filesystem, which are the key targets for vulnerability detection. They include the back-end executable files that the system runs.
3. Backend Script Files. This category comprises scripts like shell, Perl, Python, etc. Notably, this does not include JS scripts. As many backend systems' initialization work or inter-process communication require scripts like shell scripts. These script files will be used to explore RPC calls and script call chains in the backend.

The front-end files and the binaries undergo a matching process under a shared keyword filter to identify border binaries, which are back-end interface programs directly receiving data from the frontend. These border binary files might contain vulnerabilities triggered by malicious user inputs. What's more, they themselves might be RPC callers triggering injection vulnerabilities in lower-level RPC-registered scripts. Simultaneously, the back-end script files are selected out by a cross-file parameter transmission filtering algorithm for RPC Scripts and Inter Scripts, which are then used for taint tracking by the taint engine.

Finally, the border binaries are processed in the taint engine, tracking from taint source functions to artificially set taint sinks. Through forward and backward analysis, multiple viable taint paths are identified. Utilizing heuristic pruning method and optimized static symbolic execution strategies, the prototype system aims to complete the vulnerability detection work more accurately, with lower false positives and higher efficiency. The overall architecture is illustrated in Fig. 1.

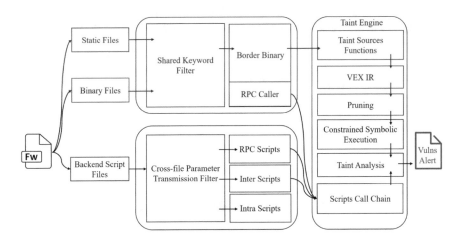

Fig. 1. Structure of IoTaint

3.2   Firmware Extracting

IoTaint takes a Linux-Based firmware file as input. We obtain the latest firmware by downloading it from the device's official website or extracting the complete image from an actual device using a heat gun for desoldering and a programmer for reading the compressed package. As the first step, it uses binwalk to decompress the firmware and access the root filesystem.

3.3   Target File Filtering

The target files for vulnerability detecting are the border binaries and RPC scripts. We obtain these target files through shared keyword filtering method and cross-file parameter transmission filtering method.

Files in the filesystem are categorized into static files, binary files, and back-end script files. After parsing static files and filtering RPC-related scripts from the back-end script files, we perform an intersection on the binary file set, filtering it based on the shared keyword method. This implies that an attacker's dangerous input can be processed by the binary program and might even flow into back-end script files involving remote calls.

We adopt the same method as SATC to obtain key information from the front-end static files, relying on shared keywords. However, the difference is that our focus is on whether there are indications of RPC calls in web interface applications and border binary programs. Specifically, we use a JS parser to capture some actions in static files that trigger HTTP requests. These requests contain necessary interface access information, such as parameter information. For instance, Fig. 2 shows a request to the interface *setMacFilterCfg*, with the request body *macFilterType = black&deviceList =*. Thus, we can use these three keywords, *setMacFilterCfg*, *macFilterType*, *deviceList* to identify which binary contain the same string, thereby confirming the correct border binary.

```
POST /goform/setMacFilterCfg HTTP/1.1
Host: 192.168.1.1
Accept: */*
Content-Type: application/x-www-form-urlencoded; charset=UTF-8
Accept-Encoding: gzip, deflate
Cookie: password=25d55ad283aa400af464c76d713c07aduyh1qw

macFilterType=black&
deviceList=
```

```
void __fastcall formSetMacFilterCfg(int a1)
{
    void *Var; // [sp+24h] [+24h]
    void *v5; // [sp+28h] [+28h]
    // get one param
    Var = websGetVar(a1, "macFilterType", &byte_4EC06C);
    // get another
    v5 = websGetVar(a1, "deviceList", &byte_4EC06C);
    v3 = sub_47BB34(Var, v5);
    ...
}
```

Fig. 2. Shared Keyword Mapping Process

On the other hand, we also consider back-end script files related to RPC calls, such as shell, Perl, and Python scripts. Because some devices' business process are handled with help of RPC (Remote Procedure Call), like jsonrpc or RESTful. Jsonrpc packages the method name and parameters to be called in a json format and transmits the data to another process via an HTTP request. As shown in Fig. 3, we can find scripts and border binary files using the same keywords, as well as RPC Callers and corresponding RPC scripts. For example, the request packet content of an RPC Caller is the result processed by a border binary, and its corresponding RPC script is *system-reboot*. This allows the backend to get front-end data naturally and clearly by using jsonrpc methods. Through this shared keyword method, we can directly locate the RPC call part in the border binary and corresponding RPC script which is helpful to obtain the RPC script call chain.

```
POST /jsonrpc HTTP/1.1
Host: 192.168.1.1
Accept: application/json, text/plain, */*
Cookie:sessionid= xxx

{
  "jsonrpc": "2.0",
  "method": "action",
  "params": {
    "rpc":"system-reboot"
}
```
RPC Caller

```
/* system-reboot*/
...
reboot to image() {
    img_selected="$1"
    active_firmware=`uci get firmware.firminfo.version`
    inactive_firmware=`uci get firmware.firminfo.inactive_version`
    logger -t system -p local0.alert "Rebooting the system .."
    ...
}
```
RPC Script

Fig. 3. RPC Caller and RPC Scripts

The shared keyword filtering method can only capture the RPC caller and their RPC scripts that are called in regular operations, but these are not all the RPCs in the firmware environment. To track the entire file system's RPC call chain, we designed an algorithm called Cross-File Parameter Transmission Filtering Method to filter common back-end scripts, i.e., RPC scripts and Inter-scripts, separating useless Intra-scripts, and generating a script call chain. We define such a script relationship where RPC scripts serve as the source of taint. The Inter-scripts they call can form branches of a call tree, each branch recording a call chain from an RPC script as the root node.

When it comes to the algorithm of Cross-File Parameter Transmission Filtering, after we collect the script paths in the file system, we start by obtaining the specific script path of the RPC script from the configuration file. We create a call tree with the collected RPC scripts and other Inter-scripts, which are the callee of RPC scripts, through file's content reading and depth-first search. The obtained script call tree will be used in the taint analysis engine for script security auditing and parameter taint tracking, checking for system-level command injection vulnerabilities. The Algorithm 1 is shown below.

Algorithm 1: RPC Scripts Filter

Function *RPCScriptsFilter(dirPath)*:
    configFile ← GetConfigFile(*dirPath*);
    RPCScripts ← GetRPCPaths(*configFile*);
    scriptList ← GetAllScriptsPath(*dirPath*);
    callTrees ← [];
    foreach *path* ∈ *RPCScripts* **do**
        treeNode ← ConstructCallTree(*path*);
        callTrees.add(*treeNode*);
    end
    return *callTrees*
end
Functon *ConstructCallTree(path)*:
    root ← CreateNode(*path*);
    stack ← [*root*];
    visited ← set();
    while *stack* ≠ ∅ **do**
        curr ← *stack*.pop();
        foreach *called* ∈ *GetScriptEachCall(curr.name)* **do**
            if *called* ∉ *visited* **then**
                child ←CreateNode(*called*);
                curr.AddChild(*child*);
                stack.push(*child*);
                visited.add(*called*);
            else
        end
    end
    return *root*
end

3.4 Optimized Taint Analysis

The taint engine considers two parts, one is taint analysis of script call chains, and the other is static taint analysis of border binary programs.

We assign the task of detecting vulnerabilities in the script call chain to the taint analysis engine. First, each script is statically analyzed, and taint source statements, i.e., statements that read external inputs, are marked. In the script call chain, starting from the taint source RPC script, we traverse in depth-first search order. When a Inter-script is encountered, we first check if it is called by a taint source statement. If so, the tainted parameters of the taint source are recorded at the script's entry. Then, data flow analysis is conducted from the Inter-script's entry, deducing how tainted parameters propagate in the script and recording cases where parameters are passed to other inner-functions. Our strategy targets various script languages to detect whether there are command execution functions and behaviors of obtaining external argv parameters. For example, we need to check functions like *system*, *eval*, etc., in the scripts, which can start a new process or thread through a system call and execute commands via shell, leading to injection. Additionally, regarding external parameters, Perl scripts often obtain command-line arguments through the @*ARGV* array, and there might also be socket inputs, file inputs from *open*, etc. We have collected script languages' external parameter inputs and command execution methods, written corresponding regular expression patterns for regex matching.

We start tracking from special taint source functions in IoT device border binary, such as *WebsGetVar*, *getenv*, which are faster and more accurate than starting from functions like *recv*, *fgets*, etc. Additionally, we have set up a heuristic pruning strategy specifically for IoT programs to exclude unnecessary paths as much as possible, aiming for quicker and more accurate discovery of overflow and injection vulnerabilities.

In order to perform cross-architecture binary program analysis, we use an IR after lifting binary code. Common forms of intermediate languages include LLVM, BAP, REIL, VEX, etc. Currently, VEX IR is the best choice, as it supports the original project Valgrind and is specialized for program analysis. VEX IR represents our target program's assembly code in IR and substitutes control flow graph basic blocks with unique IRSBs (IR Super-Blocks), creating the program's CFG and DFG. We implement IoTaint using Angr [11]. Angr's inherited program analysis interface functions are used to convert border binary into VEX IR.

Taint Sources. Taint sources are the starting points where tainted data is introduced. For IoT border binary programs, besides some proprietary custom functions that can be directly considered as returning pointers to tainted data structures, there are also regular C library function taint sources like the *recv* in socket, and *read*, *fgets*, etc. The pointer variables returned by these functions point to tainted data, and after data reception, the buffer becomes tainted.

Taint Propagation. General taint propagation analysis uses forward analysis, which involves enumerating all paths from the taint source functions to sink functions, but this is generally infeasible [5]. Therefore, to alleviate the difficulty of finding multiple sources and sinks in path searching, we use a static taint engine that combines forward analysis with backward analysis. For example, we guide forward taint analysis from each program point that contains a source (e.g., *recv*) and backward taint analysis from each program point that receives (e.g., *strcpy*, *sprintf*). Table 1 shows our data propagation rules for forward and backward analysis supported by VEX IR.

Table 1. IR Taint Propagation Rules

IR statement	Forward taint rules	Backward taint rules	Description
$t_i = \text{get}(r_j)$	$r_j \to t_i$	$t_i \mapsto r_j$	assign to taint var
$\text{put}(r_j) = t_i$	$t_i \to r_j$	$r_j \mapsto t_i$	assign to register
$t_i = t_j$	$t_j \to t_i$	$t_i \mapsto t_j$	assign t_j to t_i
$t_i = \text{OP}(a,b)$	$a \to t_i, b \to t_i$	$t_i \mapsto a, t_i \mapsto b$	OP ::= $+,-,*,/,<<,>>,<,>,<=,>=,!=,==$
$t_i = \text{ITE}(t_j,a,b)$	$a \to t_i$ or $b \to t_i$	$t_i \mapsto a$ or $t_i \mapsto b$	if t_j is True,then $t_i = a$ else $t_i = b$
$t_i = \text{Load}(t_j)$	$t_j \to t_i, *t_j \to t_i$	$t_i \mapsto t_j, t_i \mapsto *t_j$	assign value loaded from address t_j to t_i
$\text{Store}(t_j) = t_i$	$t_i \to *t_j$	$*t_j \mapsto t_i$	Store t_i to the memory at address t_j

$t_i \to t_j$ denotes that the taint propagate from t_i to t_j.

$t_i \mapsto t_j$ denotes that if t_i is tainted, check t_j.

Taint Sink. The taint sink is a point where tainted data converges and researchers use it to check for memory corruption or command injection vulnerabilities. Common taint sinks in IoT device programs in libc are like *strcpy*, *strcat*, *fgets*, *scanf*, *sscanf*, *sprintf*, *memcpy*, *strncpy*, *snprintf*, *system*, *popen* and so on.

What's more, We assign a heuristic pruning method to optimize target paths in IoT devices that can reach sinks. If there are operations that filter or truncate taint data along a path, we abandon that path and release all resources of the symbolic execution process. We establish a feature function library by summarizing and cataloging C library functions and length judgment functions of different brand manufacturers, collecting function feature assembly code snippets as feature codes. In taint analysis, if a path's function has the same name or contains the same feature code, we terminate the taint analysis flow. For example, in a path from *WebsGetVar* to *sprintf*, if there are functions like *strlen* or custom filter characters, we clear the taint if the tainted data is passed as their arguments. This approach avoids constraint expressions resolving. Furthermore, symbolic execution often encounters path explosion and crashes due to insufficient memory, with loops being a primary cause. However, there are almost no loops with a large number of iterations for processing flows in the border binary. Therefore, we can directly set the iteration number to a default value of 2 to avoid the pitfalls of loops.

4   Evaluation

We have developed a prototype system implemented in Python, specifically designed for vulnerability detection in IoT device firmware. This system is capable of detecting cross-file parameter transmission injection vulnerabilities arising from RPC, as well as buffer overflow and command injection vulnerabilities in border binaries. We utilized shared keyword filtering and cross-file parameter transmission filtering method to select target files. The Angr static analysis framework was employed to transform assembly code into VEX IR, generating CFGs and DFGs for binary programs. Combined with taint analysis and data flow analysis, our system infers vulnerabilities with the help of heuristic pruning and strategies optimization.

Our experimental subjects included popular SOHO routers, enterprise-level routers, and smart cameras available in the market. The experiments were conducted on an Ubuntu 18.04.4 LTS operating system, equipped with a 64-bit 8-core RYZEN 5000 CPU and 16 GB of RAM. Compared to SATC, our system employs a similar method in acquiring front-end interfaces and border binary. However, significant differences lie in the backend program taint path analysis and injection detection in file system scripts.

As shown in Table 2, our system was compared with SATC in detecting vulnerabilities in the same border binary across various firmware. *OFPaths* represents the number of detected buffer overflow paths from taint sources to sinks in the program. While *CIPaths* denotes the number of detected command injection paths. Our data indicates that the number of buffer overflow paths detected by our system is generally less than half of those detected by SATC's taint paths, a result attributed to our heuristic pruning method, representing a lower false positive rate. Moreover, IoTaint detects slightly more paths in command injection vulnerability, due to injection vulnerabilities in system scripts or cross-file parameter transmission, all accounted for in *CIPaths*.

Table 2. Experimental Data

ID	Vendor	Product	Type	Arch	Binary	IoTaint			SaTc		
						OFPath	CIPath	Time(s)	OFPath	CIPath	Time(s)
1	Tenda	AC9	SOHO	MIPS	httpd	86	20	2162	180	24	2891
2	Tenda	AC10	SOHO	MIPS	httpd	45	0	958	142	0	1464
3	Tenda	G1/G3	VPN	ARM	httpd	178	12	3295	329	10	4167
4	Cisco	RV345	VPN	ARM	jsoncmd	2	4	225	5	0	127
5	Dlink	878	SOHO	MIPS	prog.cgi	37	0	762	50	0	1136
6	Dlink	882	SOHO	MIPS	prog.cgi	41	1	879	6	0	1174
7	ToTolink	AC2600	SOHO	MIPS	cste.cgi	45	3	260	60	0	378
8	TPlink	IPC43AW	Camera	ARM	dsd	0	0	15	0	0	27

Another reason for SATC's lengthy runtime is lack of enough pruning. This can lead to interruption or timeout due to path explosion and memory crash, making it suitable only for short call chains. In contrast, our optimized method employs optimized symbolic execution strategy, conducting heuristic pruning before symbolic constraint resolving. This approach reduces the likelihood of null constraint formulas in tainted variables, thus enabling more precise path verification. Additionally, our method has proven to be time-efficient. Notably, the command injection path detection for RV345 took longer due to RPC scripts and Inter-scripts in the device. The data reveals that SATC's taint paths are numerous and prone to high false-positive rates, whereas our pruning optimization strategy significantly reduces unnecessary symbolic execution processes, saving considerable analysis time.

In summary, our prototype system, IoTaint, achieved efficient and automated vulnerability mining, aided by manual verification to identify several CNNVD and CVE 0day vulnerabilities. The system also facilitated efficient replication of complex 1day vulnerabilities, quickly pinpointing vulnerability sink points and obtaining comprehensive taint propagation paths. We conducted specific research on the Cisco RV340/345 enterprise routers and were able to identify and replicate three high-risk vulnerabilities caused by script issues, as shown in Table 3.

Table 3. BugID Result

ID	Vendor	Product	Vuln Type	Bug ID	0day/1day
1	Tenda	AC9	Command Injection	CNNVD-202109-1174	0day
2	Tplink	IPC43AW	Buffer Overflow	CNNVD-202109-1175	0day
3	Blink	X12	Command Injection	CNNVD-202011-1320	0day
4	Ruijie	RG-EW1200	Command Injection	CNNVD-202112-2841	0day
5	Tenda	AC10	Buffer Overflow	CVE-2023-45482	0day
6	Tenda	AC10	Buffer Overflow	CVE-2023-45483	0day
7	Tenda	AC10	Buffer Overflow	CVE-2023-45484	0day
8	Cisco	RV340/345	Command Injection	CVE-2022-20706	1day
9	Cisco	RV340/345	Command Injection	CVE-2022-20707	1day
10	Cisco	RV340/345	Command Injection	CVE-2022-20708	1day

5   Conclusion

In this paper, we proposed an IoT firmware vulnerability detecting prototype system, IoTaint, which utilizes static taint analysis for both binary program and cross-file parameter transmission vulnerabilities. Considering that the latter often involves binary and script files in the transmission chain, we conducted taint analysis on the RPC script call chain. The overall scheme employed shared keyword filtering and cross-file parameter transmission filtering method to obtain border binary and script call chains for taint engine tracking. Simultaneously, static analysis of binary was conducted using Angr and VEX IR in taint engine, complemented by heuristic pruning and optimization strategies to reduce the cost of symbolic execution for unnecessary paths. Compared horizontally with SATC, IoTaint was able to detect buffer overflow and command injection vulnerabilities more quickly and accurately, proving more efficient and with a lower false positive rate. IoTaint helped us discover several 0day and complex 1day vulnerabilities.

References

1. Chen, J, et al.: IOTFUZZER: discovering memory corruptions in iot through app-based fuzzing. In: Network and Distributed Systems Security (NDSS) Symposium 2018. Network and Distributed Systems Security (NDSS) Symposium 2018, San Diego, California, United States, 18 February 2018
2. Zhang, Y., et al.: SRFuzzer: an automatic fuzzing framework for physical SOHO router devices to discover multi-type vulnerabilities. In: Proceedings of the 35th Annual Computer Security Applications Conference, pp. 544–556, December 2019
3. Chen, D.D., Woo, M., Brumley, D., Egele, M.: Towards automated dynamic analysis for Linux-based embedded firmware. In: NDSS, vol. 1, p. 1, February 2016
4. Zheng, Y., Davanian, A., Yin, H., Song, C., Zhu, H., Sun, L.: FIRM-AFL: high-throughput greybox fuzzing of IoT firmware via augmented process emulation. In: 28th USENIX Security Symposium (USENIX Security 2019), pp. 1099–1114 (2019)
5. Cadar, C., Sen, K.: Symbolic execution for software testing: three decades later. Commun. ACM **56**(2) (2013)
6. Redini, N., et al.: KARONTE: detecting insecure multi-binary interactions in embedded firmware. In: 2020 IEEE Symposium on Security and Privacy (SP), pp. 1544–1561. IEEE, May 2020
7. Srivastava, P., Peng, H., Li, J., Okhravi, H., Shrobe, H., Payer, M.: FirmFuzz: automated IoT firmware introspection and analysis. In: Proceedings of the 2nd International ACM Workshop on Security and Privacy for the Internet-of-Things, pp. 15–21, November 2019
8. Chen, L., et al.: Sharing more and checking less: leveraging common input keywords to detect bugs in embedded systems. In: 30th USENIX Security Symposium (USENIX Security 2021), pp. 303–319 (2021)
9. Cheng, K., et al.: Finding taint-style vulnerabilities in Linux-based embedded firmware with SSE-based alias analysis (2021). arXiv preprint arXiv:2109.12209
10. Bellard, F.: QEMU, a fast and portable dynamic translator. In: USENIX Annual Technical Conference, FREENIX Track, vol. 41 (2005)
11. Wang, F., Shoshitaishvili, Y.: Angr-the next generation of binary analysis. In: 2017 IEEE Cybersecurity Development (SecDev). IEEE (2017)

12. Gartner Says 5.8 Billion Enterprise and Automotive IoT Endpoints Will Be in Use (2020). https://www.gartner.com/en/newsroom/press-releases/2019-08-29-gartner-says-5-8-billion-enterprise-and-automotive-io
13. Palo Alto Networks: 2020 Unit 42 IoT Threat Report (2020). https://iotbusinessnews.com/download/white-papers/UNIT42-IoT-Threat-Report.pdf
14. Khandelwal, S.: Thousands of MikroTik Routers Hacked to Eavesdrop on Network Traffic (2018). https://thehackernews.com/2018/09/mikrotik-router-hacking.html
15. Michał, Z.: American Fuzzy Lop. Retrieved 1 September 2022. https://lcamtuf.coredump.cx/afl/

An Innovative Approach to Real-Time Concept Drift Detection in Network Security

Federica Uccello[1]([⊠])(iD), Marek Pawlicki[2], Salvatore D'Antonio[1](iD), Rafał Kozik[2], and Michał Choraś[2]

[1] University of Naples 'Parthenope' Centro Direzionale, Isola C4, 80133 Naples, Italy
federica.uccello@assegnista.uniparthenope.it,
salvatore.dantonio@uniparthenope.it
[2] Bydgoszcz University of Science and Technology, PBS, Bydgoszcz, Poland
chorasm@pbs.edu.pl

Abstract. In the realm of cybersecurity, the detection of Concept Drift holds the potential to improve the adaptability and effectiveness of security systems. In particular, Security Information and Event Management (SIEM) frameworks can benefit from real-time Drift Detection, enabling prompt detection of changing attack patterns, and consequent update of the detection criteria. To explore such an opportunity, the proposed approach extends a previously introduced SIEM solution with Concept Drift Detectors. An experimental evaluation is presented using two well-known unsupervised detectors on a merged dataset featuring Concept Drift, taking into consideration metrics such as Error Rate, Precision, Recall, and Window Average Error Rate. The results demonstrate that the integrated mechanism successfully identifies Concept Drift, triggering SIEM alerts and prompting timely updates to correlation rules. The experiment's implications, limitations, and future directions are discussed, emphasizing the importance of continuous improvement in cybersecurity measures.

Keywords: Concept Drift · SIEM · NIDS · Artificial Intelligence

1 Introduction

Within the domain of data analysis applications, the challenge of coping with changing data, known as Concept Drift, is pervasive. The notion of Concept Drift entails alterations in the characteristics of a target variable, leading to performance decay of Machine Learning (ML) models [1]. In the realm of network security, the timely detection and adaptation to these shifts are crucial for preserving the efficacy of security systems [2]. Among various applications, Network Intrusion Detection Systems (NIDS) and Security Information and Event Management (SIEM) solutions stand to gain significantly from the prompt detection of Concept Drift [3]. In fact, the increasing sophistication of cyberthreats

L. Barolli (Ed.): EIDWT 2024, LNDECT 193, pp. 130–139, 2024.
https://doi.org/10.1007/978-3-031-53555-0_13

necessitates security measures that can dynamically respond to emerging attack patterns. Concept Drift Detection plays a crucial role in alerting users promptly when the underlying data distribution, indicative of potential threats, undergoes substantial changes [4]. This paper underscores the significance of real-time Concept Drift Detection in the context of cybersecurity, presenting a solution that integrates this capability into a rule-based SIEM framework. The proposed approach extends a rule-based SIEM solution introduced in our previous research [5]. The framework is enhanced by integrating real-time Concept Drift Detection mechanisms, allowing the SIEM to adapt in real-time to dynamic cyberthreats. The study adopts two distinct approaches: a static analysis to assess experiment viability and a real-time simulation designed to replicate operational scenarios more accurately.

The structure of this paper is organized as follows. Section 2 overviews related research. Section 3 provides a comprehensive overview of the proposed methodology, explaining in detail the experimental setup and the integration of Concept Drift Detection with the SIEM solution. Section 4 delves into the novel methodology for Concept Drift Detection, delineating the datasets used, the performance metrics employed, and the specific detectors implemented. Section 5 presents the experimental results, providing an analysis of drift detection methods in both static and real-time approaches.

2   Related Work

In the cybersecurity domain, Concept Drift and Concept Drift Detection have been the subject of attention and discussion in the literature, in different application domains. Among others, the Internet of Things (IoT) [6,7] and Industrial Internet of Things (IIoT) [8] were taken into consideration.

Qiao et al. [6] focuses on cyberattack detection in IoT scenarios, emphasizing the impact of Concept Drift. The authors propose a dynamic sliding window method based on residual subspace projection, offering a data-based Concept Drift Detection strategy. Notably, the approach avoids reliance on labeled information or statistics, presenting a cost-effective and efficient solution for detecting cyberattacks in IoT environments.

Similarly, Yang et al. [7] addressed Concept Drift challenges in IoT data analytics. The authors propose an adaptive framework utilizing optimized Light-GBM and a drift adaptation method, Optimized Adaptive and Sliding Windowing (OASW). The approach demonstrates high accuracy and efficiency in continuous learning and drift adaptation on IoT data streams without human intervention.

Focused on Industrial IIoT, [8] uses statistical drift detection to counter cybersecurity concerns in interconnected industrial systems. The authors propose a method to detect changes in data patterns caused by malware intrusions and train ML classifiers accordingly. Results indicate high accuracy, emphasizing the approach's success and ease of adoption.

Addressing the vulnerabilities of traditional state estimators in smart grids, the work presented in [9] analyzes the differences between physical grid changes and data manipulation changes. The proposed approach employs statistical hypothesis testing and dimensionality reduction to detect false data injection attacks under Concept Drift.

The potential of Concept Drift Detection in Intrusion Detection has also been explored in past research. The approach shown in [10] employs iP2V, an incremental feature extraction method, and a stable sub-classifier generation module to achieve high efficiency and adaptability to various real-world drifts, with promising experimental results.

Rajeswari et al. [11] propose an adaptive window with Support Vector Machine (ADWIN-SVM) for large-scale intrusion detection with multiple attributes. The approach employs data cleaning and normalization, with ADWIN algorithm detecting drift and SVM classifying data as normal or attack. Comparative analysis with various techniques is also provided.

Focusing on Denial of Service (DoS) attack detection, [12] proposes a methodology for monitoring changes in network traffic from individual source nodes based on concept drift. The framework applies ML techniques to dynamically define lower and upper bounds for each node, effectively discriminating abrupt changes from gradual changes in traffic magnitude. Real-world data testing validates the methodology's effectiveness. In comparison with the existing literature, this work presents a novel methodology for real-time Concept Drift Detection within the context of SIEM solutions. While existing literature primarily focuses on detection strategies for cyber-attacks, the present contribution integrates real-time Concept Drift Detection into a rule-based SIEM framework. This approach allows for the dynamic adaptation of cybersecurity measures in response to evolving attack patterns.

3   Overview of the Proposed Approach

This section delineates the proposed methodology for real-time Concept Drift Detection. Two distinct approaches are employed and tested in the conducted experiments: a static approach, utilized to assess the viability of the experiment, and a real-time approach, designed and implemented to simulate a realistic scenario.

Both approaches leverage the SIEM solution proposed by the authors in previous research [5], and three different Concept Drift Detectors. The dataset used is a combination of two versions of the same dataset ([13]), CIC-IDS, featuring identical features and labels, but collected in different years. The primary objective of this research is to enhance the proposed rule-based SIEM solution, making it more resilient to changes in attack patterns. Concept Drift Detection is employed to raise a SIEM alert, prompting the user to update the correlation rule once a change in the attack pattern is detected. For this purpose, the two datasets have been merged as explained in Sect. 4, creating a new dataset D_{12}, which is used in both approaches as described below. The proposed approaches use unsupervised Concept Drift Detector since in a realistic scenario there would not be access to the ground truth labels, but exclusively to network traffic. A general pipeline for the proposed methodology is depicted in Fig. 1.

3.1 Static Concept Drift Detection

In the static approach, dataset D_{12} has its original labels removed and it is used to feed the Concept Drift Detectors. Then, The Detectors analyze batches of D_{12} to determine the presence or absence of Concept Drift throughout it.

3.2 Real-Time Concept Drift Detection

Dataset D_{12} has its original labels removed, and it is used to generate simulated network traffic. The traffic is processed by the SIEM, which labels the samples as Benign or malicious based on its correlation rule. At the same time, Concept Drift Detectors analyse the traffic to determine the presence or absence of Concept Drift in the dataset. Upon detection, an alert is raised, detailing the index of the triggering sample and the involved detector. The alert is then dispatched to the SIEM message broker and visualized on the dedicated Dashboard, informing the security personnel that the SIEM rule is now outdated and an update is required. Further details are provided in the following Section.

4 Proposed Novel Methodology for Concept Drift Detection

In this section, the proposed approach for Concept Drift Detection is detailed, providing insights on the datasets employed, the performance metrics extracted, and the detectors implemented.

4.1 Dataset Description

The experimental setup incorporates two datasets, namely CIC-IDS2017 and CSE-CIC-IDS2018 [13]. Both datasets adhere to the same features and labels, facilitating a comprehensive analysis of the SIEM rule's adaptability over time. To ensure the presence of concept drift in the final dataset D_{12} used for experiments, the SIEM rule was tested on both datasets. A notable observation of significant performance degradation is presented in Table 1, indicating a discernible change in attack patterns and the occurrence of concept drift.

Table 1. Performance metrics for the SIEM predictions with respect to D_1 and D_2, using the same correlation criteria. Significant performance loss has been detected, denoting attack pattern change and Concept Drift.

Dataset	Precision		Recall		Accuracy
	Benign	Malicious	Benign	Malicious	
D_1	0.88	1	1	0.87	0.93
D_2	0.01	0.20	0.01	0.24	0.12

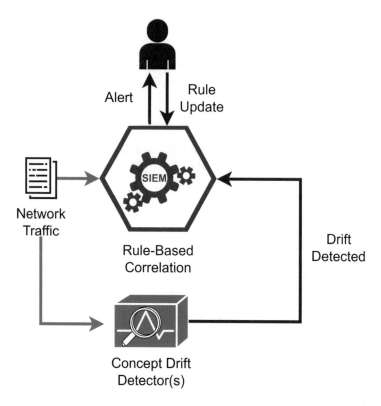

Fig. 1. High-level schema of the proposed approach for Drift Detection. Based on the defined Correlation Rule, the SIEM labels traffic as Benign or malicious. The traffic is also fed to the Concept Drift Detector. In case Concept Drift is Detected, the SIEM is notified, and an Alert is raised to the User, indicating that the Correlation Rule is obsolete and an update is required to face the change in the attack pattern.

The datasets were merged systematically by injecting chunks of samples from D_2 after the first n samples of D_1. The transition to using D_2 exclusively marked a deliberate injection of concept drift into the dataset, creating the merged dataset D_12, used for the actual experiments.

4.2   Performance Metrics

The performance metrics utilized for drift detection include the error rate (and its complement, accuracy), precision, and recall. These metrics are calculated as shown in the following Equations, and fed to the detectors as described previously. An additional parameter is also calculated and feed to the detectors: the Window Average Error Rate (WAER) parameter, which represents the average error rate calculated over a specific window size. It provides a smoothed measure of the error rate, helping to mitigate the impact of short-term fluctuations in performance metrics. This parameter is particularly useful in real-time concept drift detection, where it allows for a more stable and reliable indication of changes in the data distribution. The WAER serves as an additional metric for assessing the overall performance and stability of the drift detection system.

$$\text{Precision} = \frac{\text{True Positive}}{\text{True Positve} + \text{False Positive}} \tag{1}$$

$$\text{Recall} = \frac{\text{True Positive}}{\text{True Positive} + \text{False Negative}} \tag{2}$$

$$\text{Accuracy} = \frac{\text{Number of Correct Predictions}}{\text{Total Number of Predictions}} \tag{3}$$

$$\text{Error Rate} = 1 - \text{Accuracy} \tag{4}$$

4.3   Drift Detection Techniques

The present research employed three different unsupervised Concept Drift Detectors among the most well-known ones proposed to monitor and adapt to changes in data distribution: Page-Hinkley Test (PHT) [14] and Adaptive Windowing (ADWIN) [15]. PHT is a sequential analysis method that serves as a concept drift detector: it computes observed values, such as the actual accuracy of the classifier, and their mean up to the current moment. In the event of a concept drift, the base learner may fail to correctly classify incoming instances, leading to a decrease in actual accuracy. PHT detects this change by monitoring the cumulative difference between observed values and their mean. The comparison of this difference with a minimum threshold determines if a significant distribution change has occurred. The flexibility of adjusting the threshold allows for balancing between false alarms and the potential to detect actual changes. ADWIN employs sliding windows of variable size that are recalculated online based on observed changes in the data. ADWIN dynamically adjusts the window size, enlarging it when no apparent change is detected and shrinking it when a change is identified. The algorithm aims to identify two sub-windows within the larger window that exhibit distinct averages. A significant difference in averages suggests a change in the underlying data distribution, leading to the removal of the older portion of the window. ADWIN provides rigorous performance guarantees, including limits on false positives and false negatives, ensuring reliable and efficient drift detection. The maximum length of the window is determined by statistical consistency, offering a principled approach to handling changes in data patterns.

5   Experimental Results

In this section, a comprehensive analysis of the experimental results is presented, examining the performance of different drift detection methods in both static and real-time approaches. The experiments were conducted using two distinct strategies: one involved batch processing of the dataset, with fixed batch size, and the other one involved stream processing for real-time detection. Both cases had identical set up in terms of detector sensitivity. The results, as shown in Table 2 and 3 for the static and real-time approaches respectively, indicate a consistent overlap in the performance of drift detectors across both approaches. Notably, the static detector demonstrated slightly early detection of drift. Both ADWIN and PHT exhibited faster response times in the static approach. However, in the real-time scenario, the detectors exhibited a marginally slower sensitivity to the gradient drift, detecting it at a later stage.

Table 2. Drift Detection summary for the static approach. The metrics include precision, recall, error rate, window average error rate, and the index of the first sample in the chunk where drift was detected.

Detector	Precision	Recall	Error Rate	WAER	Sample Index
PHT	0.24	0.25	0.86	0.06	155500
ADWIN	0.22	0.39	0.83	0.064	156400

Table 3. Drift Detection summary for the real-time approach. The metrics include precision, recall, error rate, window average error rate, and the index of the first sample in the chunk where drift was detected.

Detector	Precision	Recall	Error Rate	WAER	Sample Index
PHT	0.18	0.21	0.889	0.28	219000
ADWIN	0.16	0.2	0.898	0.29	221000

Listing 1 illustrates a SIEM alert generated upon detecting concept drift in the real-time scenario. The alert includes essential information such as the metrics for the latest chunk of network traffic, the involved detector (in this case, PageHinkley), and the index of the first sample in the chunk where drift was detected. This structured alert format facilitates prompt user awareness and intervention.

```
1   {
2       "concept_drift_detected": true,
3       "error_rate": 0.889,
4       "window_average_error_rate": 0.28306363636363635,
5       "precision": 0.18456375838926176,
6       "recall": 0.21442495126705652,
7       "detector": [
8           "PageHinkley"
9       ],
10      "sample_index": 219000,
11      "Alert": "SIEM rule outdated: please update"
12  }
```

Listing 1: Example of alert format for Concept Drift. The alert provides a summary of the metrics and displays relevant information.

6   Discussion

The experimental results indicate that the proposed real-time Concept Drift Detection mechanism effectively integrates into the SIEM solution, demonstrating its capability to adapt to changes in attack patterns. The static approach, which serves as a benchmark for the experiment, shows that the drift detectors (PHT and ADWIN) respond consistently to the induced concept drift in the dataset. The detectors, while exhibiting slightly delayed detection times in the real-time scenario, successfully identify the drift, triggering SIEM alerts and informing security personnel of the need to update the correlation rule promptly. The presented experiment holds potential for enhancing traditional rule-based SIEM: by promptly detecting shifts in attack patterns, security professionals can update correlation rules in a timely manner, ensuring that the SIEM solution remains resilient to evolving threats. The structured format of SIEM alerts facilitates clear communication and aids security personnel in understanding the nature of the drift, enabling informed decision-making. While the experimental results are promising, it is essential to acknowledge certain limitations inherent in the conducted study. First and foremost, the experiment employs Concept Drift introduced by merging two different versions of the same dataset. While this provides a controlled environment for testing, real-world concept drift might manifest differently and could be influenced by a broader range of factors. Therefore, further validation with real-world datasets and diverse network conditions is necessary to assess the robustness of the proposed approach. Additionally, the evaluation metrics used in the experiment, including precision, recall, error rate, and window average error rate, provide a quantitative assessment of the detectors' performance. However, the qualitative aspects of the SIEM alerts, such as the interpretability of the provided information and the user's ability to respond effectively, need further exploration. User studies and feedback from security professionals can provide valuable insights into the practical utility and user-friendliness of the proposed real-time Concept Drift Detection mechanism.

7   Conclusion

The present research introduces a pioneering approach to real-time Concept Drift Detection within the context of a rule-based SIEM framework. The experiment showcases the effectiveness of the proposed mechanism in detecting shifts in attack patterns, enabling the SIEM solution to adapt dynamically to emerging threats. While the results are promising, acknowledging the experiment's limitations is crucial. Future work involves further validation with real-world datasets and user studies to assess practical applicability and user-friendliness. By providing a proactive defense mechanism against evolving cyber threats, the proposed approach contributes to the continuous improvement of cybersecurity posture. As the threat landscape evolves, the adaptability introduced through real-time Concept Drift Detection becomes increasingly crucial for maintaining the resilience and effectiveness of security systems.

Acknowledgment. This research has received funding from the European Union's Horizon 2020 Research and Innovation Programme under Grant Agreement No 101020560 CyberSEAS, and from the Industrial Cyber Shield (ICS) No C83C22001460001 project funded by INAIL. The content of this publication reflects the opinion of its authors and does not, in any way, represent opinions of the funders. The European Commission and INAIL are not responsible for any use that may be made of the information that this publication contains.

References

1. Žliobaitė, I., Pechenizkiy, M., Gama, J.: An overview of concept drift applications. In: Japkowicz, N., Stefanowski, J. (eds.) Big Data Analysis: New Algorithms for a New Society. SBD, vol. 16, pp. 91–114. Springer, Cham (2016). https://doi.org/10.1007/978-3-319-26989-4_4

2. Kuppa, A., Le-Khac, N.-A.: Learn to adapt: robust drift detection in security domain. Comput. Electr. Eng. **102**, 108239 (2022)

3. Mendes, C., Rios, T.N.: Explainable artificial intelligence and cybersecurity: a systematic literature review. arXiv:2303.01259 (2023)

4. Salman, N.S., Abdulrahman, A.A.: Survey on intrusion detection system based on analysis concept drift: status and future directions. Int. J. Nonlinear Anal. Appl. **14**(1), 299–307 (2023)

5. Coppolino, L., et al.: Detection of radio frequency interference in satellite ground segments. In: 2023 IEEE International Conference on Cyber Security and Resilience (CSR), pp. 648–653. IEEE (2023)

6. Qiao, H., Novikov, B., Blech, J.O.: Concept drift analysis by dynamic residual projection for effectively detecting botnet cyber-attacks in IoT scenarios. IEEE Trans. Ind. Inf. **18**(6), 3692–3701 (2022)

7. Yang, L., Shami, A.: A lightweight concept drift detection and adaptation framework for IoT data streams. IEEE Internet Things Mag. **4**(2), 96–101 (2021)

8. Amin, M., Al-Obeidat, F., Tubaishat, A., Shah, B., Anwar, S., Tanveer, T.A.: Cyber security and beyond: detecting malware and concept drift in AI-based sensor data streams using statistical techniques. Comput. Electr. Eng. **108**, 108702 (2023)

9. Mohammadpourfard, M., Weng, Y., Pechenizkiy, M., Tajdinian, M., Mohammadi-Ivatloo, B.: Ensuring cybersecurity of smart grid against data integrity attacks under concept drift. Int. J. Elect. Power Energy Syst. **119**, 105947 (2020)

10. Wang, X.: ENIDrift: a fast and adaptive ensemble system for network intrusion detection under real-world drift. In: Proceedings of the 38th Annual Computer Security Applications Conference, pp. 785–798 (2022)

11. Rajeswari, P.V.N., Shashi, M., Rao, T.K., Rajya Lakshmi, M., Kiran, L.V.: Effective intrusion detection system using concept drifting data stream and support vector machine. Concurrency Comput. Pract. Experience. **34**(21), e7118 (2022)

12. Rajeswari, P., Shashi, M.: Concept-drift based identification of suspicious activity at specific IP addresses using machine learning. Int. J. Recent Technol. Eng. **8**(3), 6651–6655 (2019)

13. Sharafaldin, I., Lashkari, A.H., Ghorbani, A.A.: Toward generating a new intrusion detection dataset and intrusion traffic characterization. ICISSp **1**, 108–116 (2018)

14. Page, E.S.: Continuous inspection schemes. Biometrika **41**(1/2), 100–115 (1954)

15. Bifet, A., Gavalda, R.: Learning from time-changing data with adaptive windowing. In: Proceedings of the 2007 SIAM International Conference on Data Mining, pp. 443–448. SIAM (2007)

Formal Analysis of 5G EAP-TLS 1.3

Naiguang Zhu, Jie Xu[(✉)], and Baojiang Cui

School of Cyberspace Security, Beijing University of Posts and Telecommunications,
Haidian District Xitucheng Road No. 10, Beijing, China
{naiguangzhu,cheer1107,cuibj}@bupt.edu.cn

Abstract. As the fifth-generation mobile network communication technology, 5G supports a wider range of application scenarios with faster speed and lower latency, so the importance of its security cannot be underestimated. Authentication protocols play a crucial role in security as the first safeguard of 5G networks. We provide a formal model of the 5G EAP-TLS 1.3 authentication protocol and perform an automated security analysis using ProVerif, a symbol-based formal verification tool, and our results show that there are some potential flaws in the current protocol design that will cause the protocol to violate the security goals that need to be guaranteed in the standard. To the best of our knowledge, this is the first formal verification study on the 5G EAP-TLS 1.3 protocol.

1 Introduction

With the development of communication technology, mobile networks have gone through multiple generations in the past decades, and 5G brings faster data transmission speeds, lower communication latency, higher connection density. The benefits brought by 5G attract many people to become its users, which puts forward higher requirements for the security of 5G networks. Among all the security mechanisms, the authentication protocol is a noteworthy aspect, which provides an important foundation for protecting user privacy and secure communication, and the rationality of the design of the authentication protocol directly affects the security capability of the protocol system.

There are mainly four authentication protocols in 5G networks, distributed in different standard protocols, 5G AKA and EAP-AKA protocols are described in the standard [1], 5G EAP-TLS is introduced in the standard [2], and the standard [3] provides a detailed description of 5G EAP-TLS 1.3. The first two authentication protocols are based on shared keys to achieve authentication and key negotiation, and the latter two authentication protocols are based on public key cryptographic algorithms. In the standard, the document uses English to describe the protocol flow and security goals, which lacks strong security proofs, and formal analysis effectively compensates for this shortcoming. Formal analysis transforms the protocol into a mathematical model, attributes the security goals, and uses logical inference rules and proof techniques to verify that the protocol system satisfies the required security properties, and if there exists a state that violates the security properties, the corresponding counterexample is given to help locate the problem.

In this work, we read the contents of several standard documents, portray the normal flow and session resumption flow of the 5G EAP-TLS 1.3 protocol, extract the security goals of the protocol, model and analyze the protocol using ProVerif [4], a formal analysis tool based on applied pi calculus, and find that there are potential flaws in the protocol that make it vulnerable to privacy attacks. To the best of our knowledge, this is the first formal analysis of the 5G EAP-TLS 1.3 authentication protocol, and we hope that our work can provide some help for related research.

The rest of this paper is organized as fellow. Related research work is presented in Sect. 2. The architecture of 5G and the 5G EAP-TLS 1.3 protocol flow will be illustrated in Sect. 3. We present the protocol security goals in Sect. 6. Our formal model of the protocol is shown in Sect. 5, and In Sect. 6, we illustrate the validation results and analyze them. Finally, it is summarized in Sect. 7.

2 Related Work

David et al. [5] provided the first faithful 5G AKA protocol model using the formal analysis tool Tamarin [6] and determined the minimum requirements for each security goals to be guaranteed, in addition, they identified a vulnerability of the 5G AKA protocol regarding privacy protection, they made recommendations for the improvement of the protocol, and they proposed a provable security fix for the identified flaws.

C. Cremers et al. [7] proposed a fine-grained formal model of the 5G AKA protocol containing user equipment, serving network, key network element entities in the home network. They found that the security of 5G AKA critically relies on unspecified assumptions about the underlying communication channel. In particular, they found an attack that exploits potential race conditions, and they also proposed a modification that proved to prevent the attack and reported their findings to the 3GPP.

E.K.K. Edris et al. [8] modeled the EAP-AKA' protocol and performed a security assessment according to the 5G specifications to determine whether the protocol meets the security requirements. In addition, the authors evaluated and simulated the performance of the EAP-AKA' and 5G AKA protocols with the help of the mathematical modeling method of Markov chains and the NS-3 network simulator to compare the communication costs of the two protocols, including metrics such as elapsed time, throughput.

Jingjing et al. [9] presented a formal model of the 5G EAP-TLS 1.2 protocol and conducted a comprehensive analysis based on the ProVerif formalization tool. The authors identified several potential flaws in the protocol design that may have an impact on the security goals and lead to security vulnerabilities during the implementation of the actual protocol system. This is the first relevant study on formal verification of the 5G EAP-TLS 1.2 protocol.

3 5G Authentication Protocol

This section describes the main flow of the 5G EAP-TLS 1.3 authentication protocol with reference to the 3GPP TS 33.501 [1], RFC 8446 [3], and RFC 9190 [10] standards. The flow is described in a simplified manner for readability, and interested readers can refer to the relevant standards directly. We first describe the 5G architecture as a whole, and then introduce the EAP-TLS 1.3 authentication protocol.

3.1 Architecture

The 5G network architecture shown in Fig. 1 consists of three main components: User Equipment (UE), Serving Network (SN), and Home Network (HN). User equipment usually represents a smartphone or IoT device that contains a Universal Subscribe Identity Module (USIM) with a unique Subscription Permanent Identifier (SUPI), the public key of the home network element and other information involved in the authentication process. The home network consists of various network elements and is where the main functions are located, providing identity management, session management and other functions for UE. The 5G EAP-TLS 1.3 authentication protocol process mainly involves the AUSF and the UDM network elements, which are responsible for the management and maintenance of the user identity data as well as the execution of the authentication and authorization process of the user. The service network, usually referred to as the base station, plays a key role in connecting the UE to other network elements in the core network. The service network and user devices communicate with each other through the air, and messages are transmitted between the service network and home network elements through authentication channels.

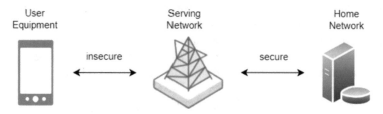

Fig. 1. 5G network architecture

3.2 Authentication Protocol

3.2.1 Initialization Protocol

Once the service network triggers the authentication of the UE, the UE converts SUPI into Subscription Concealed Identifier (SUCI) after randomization and encryption, $aenc(*, PKudm)$ denotes the asymmetric encryption of the message using the public key of the UDM network element, and R is a random number generated by the UE. The home network receives the SUCI, decrypts it to get the user permanent identity SUPI, and selects an authentication method for authentication. Figure 2 describes the EAP-TLS 1.3 authentication protocol process.

$$SUCI = aenc((SUPI, R), PKudm) \tag{1}$$

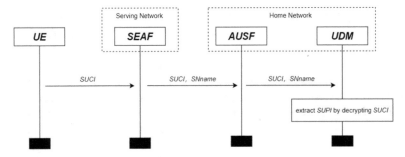

Fig. 2. Initiation of Authentication

3.2.2   5G EAP-TLS 1.3 Protocol

Normal

The 5G EAP-TLS 1.3 normal process is shown in Fig. 3 and key derivation in Fig. 4. The process covers key exchange, parameter generation, and mutual authentication phases. The standard is so extensive that we will describe the processes related to security. Interested readers can refer to the specific standards.

When the UE is initially accessed, a ClientHello message is sent containing a random number, legacy session id, cipher suite, and extensions. The ClientHello message participates in the generation of the handshake key, and after the generation of the handshake key, the handshake message will be encrypted and protected using the handshake key. The versions prior to EAP-TLS 1.3 support the function of session resumption, and this function is merged with pre-shared key in EAP-TLS 1.3. UE that has a cached session id set by a version of TLS prior to TLS 1.3 should set the legacy session id to this value. The key suite describes a set of cryptographic algorithms supported by UE, including encryption, hashing and other algorithms for the server to negotiate and select. The client requests the extension function from the server through the extension field, which contains the scope group, the value g^x that UE selected secret x acts on the scope group, and a field called pskExtention. Since it is a normal process, the pskExtention is set to 0, and the value of this field will be replaced with a series of tickets in the resumption process.

When the home network receives a ClientHello message, it sends ServerHello as a response, containing a random number, legacy session id echo, cipher suite, and extension. The legacy session id echo field is populated with the legacy session id value from the ClientHello; the Cipher Suite displays a single cipher suite selected from the ClientHello cipher suite list. With the ClientHello and ServerHello messages, the handshake key server handshake traffic secret can be computationally derived, and thereafter handshake messages are cryptographically protected, Fig. 4 shows the derivation of keys involved in the EAP-TLS 1.3 process. After ServerHello, the cryptographically protected CertificateRequest, Certificate, CertificateVerify, and Finished messages follow. The home network sends a CertificateRequest requesting the UE to send its certificate for subsequent authentication, sends its own encrypted certificate information, Certificate, to facilitate subsequent authentication of the UE to the home network, and sends a CertificateVerify message to prove that the home network possesses the private key corresponding to the certificate and to ensure the integrity of the handshake message up to this point. $Senc(*, Kshsts)$ denotes symmetric encryption using the handshake key computed by the home network, $Sign(*, SSKausf)$ denotes signing using the private key of the AUSF network element, and $Hash(*)$ denotes the computation of a hash value for the content. In the session resumption process, the pre-shared key will be used for mutual authentication, at that time there is no need to send CertificateRequest, Certificate, CertificateVerify messages. The Finish message is the last message in the authentication block, which is crucial for the verification of handshake messages and keys. The key Kfinish is derived from the handshake key Kshsts and a constant string.

$$hash1 = Hash(ClientHello, \ldots, Certificate) \tag{2}$$

$$CertificateVerify = Senc(Sign(hash1, SSKausf), Kshsts) \tag{2}$$

$$hash2 = Hash(ClientHello, \ldots, CertificateVerify) \tag{3}$$

$$Finished = Senc(HMAC(hash2, Kfinish), Kshsts) \tag{4}$$

When the UE receives the above message from the home network, it verifies the Certificate information and checks the CertificateVerify and Finished information; if the verification passes, the UE authenticates the home network successfully and sends its own encrypted Certificate, CertificateVerify and Finished messages so that the home network can authenticate the UE to realize mutual authentication. For the UE, the content of the hash contains the authentication information sent by the home network.

When the home network receives the authentication message sent by the UE and the verification passes, then the home network successfully authenticates the UE. The home network sends a NewSessionTicket message, which is encrypted using the computed server application traffic secret Ksats, and the message contains a identifier ticket id and a random ticket nonce. The message creates a unique association between a ticket and a pre-shared key derived from the resumption master secret. UE can use pre-shared keys for authentication for future resumption processes. The home network can send multiple tickets in a single connection. Multiple tickets help the user device to open multiple parallel HTTP connections.

$$PSK = HKDF(resumption\_master\_secret, "resumption", ticket\_nonce) \tag{5}$$

Resumption

The 5G EAP-TLS 1.3 resumption process is shown in Fig. 5. The session resumption process, in comparison to the above process, revolves around the pre-shared key PSK for mutual authentication. While the PSK can be established out-of-band, PSKs can also be generated in previous process. Each ticket id maps a PSK, which is used to enable mutual authentication for the session resumption process. The UE populates the ClientHello pskExtension field with a series of ticket ids, and the home network selects one of the ticket ids to negotiate the selection of the corresponding PSK for mutual

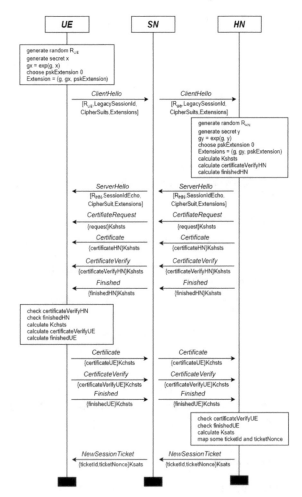

Fig. 3. 5G EAP-TLS 1.3 normal process.

authentication. The use of a pre-shared key eliminates the need for certificates and other messages involved, reducing the number of messages transmitted between networks. The PSK derives the handshake key server handshake traffic secret for the new connection, and subsequent handshake messages are encrypted and protected by this key.

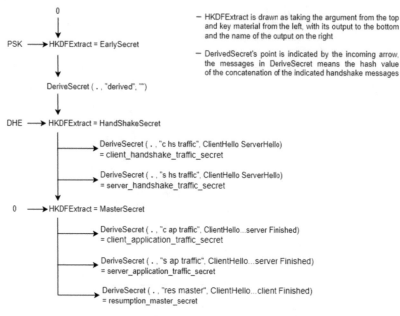

Fig. 4. Key derivation.

4 Security Goals

In this section, we extract protocol security goals from the 3GPP standard documents, and our formal definition are base on the textual descriptions of these security goals.

4.1 Authentication Properties

For the authentication property, Lowe [11] categorizes it into four classes: aliveness, weak agreement, non-injective agreement, and injective agreement, which require progressively higher degrees of authentication. During our analysis, we consider the injective agreement, determining that there are uniquely correctly paired UE and home networks communicating with each other during protocol operation and preventing replay attacks.

> [TS 33.501, Sec 5.1.2]
>
> **UE authorization**: The serving network shall authorize the UE through the subscription profile obtained from the home network. UE authorization is based on the authenticated SUPI.
>
> **Serving network authorization by the home network:** Assurance shall be provided to the UE that it is connected to a serving network that is authorized by the home network to provide services to the UE.
>
> **Access network authorization**: Assurance shall be provided to the UE that it is connected to an access network that is authorized by the serving network to provide services to the UE.

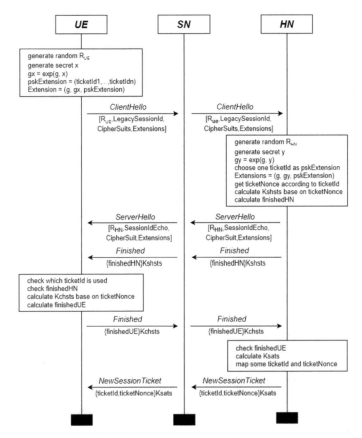

Fig. 5. 5G EAP-TLS 1.3 resumption process.

4.2 Confidentiality Properties

Confidentiality refers to ensuring that information is only accessed or read by authorized entities and is invisible or incomprehensible to unauthorized entities. Confidentiality is crucial in terms of protecting personal privacy, and it is only by ensuring that information remains confidential that sensitive information can be prevented from being leaked or maliciously abusing by illegals. The SUPI belongs to the private information and its confidentiality needs to be safeguarded. After the handshake key is negotiated between the UE and the home network, all subsequent messages need to be encrypted and protected by this key.

> [TS 33.102, Sec 5.1.1]
> **user identity confidentiality:** the property that the permanent user identity of a user to whom a services is delivered cannot be eavesdropped on the radio access link

4.3 Privacy Properties

With respect to privacy, we focus on the unlinkability of the UE. Unlinkability requires the protocol to ensure that the user's identity cannot be associated with a specific identifier, context, etc., and that an attacker cannot determine whether it is the same user from any two services. The protocol takes many measures to ensure the unlinkability of the UE, such as using SUCI in the transport instead of SUPI, certificate information associated with the user's identity is also encrypted with a handshake key, and the UE should not reuse a ticket across multiple connections. Despite all these measures to ensure the unlinkability, during our formal analysis and validation, we found some vulnerabilities that prevented the protocol from achieving this security goal.

> **[TS 33.102 5.1.1]**
> **user untraceability:** the property that an intruder cannot deduce whether different services are delivered to the same user by eavesdropping on the radio access link.

5   Formal Models

In this section, we describe how to model security properties using ProVerif. We modeled and analyzed the normal process and the resumption process in the same system, and the complete code is available online [12].

Proving reachability is the most basic function of ProVerif, and verifying the confidentiality of a term M essentially solves a reachability problem, i.e., whether an attacker can reach a state where a term M can be available. The statement to query the confidentiality of the term M is as follows:

```
query attacker(M)
```

Authentication is verified in ProVerif using correspondence assertions in conjunction with events. An event is expressed as event e(M1,…,Mn), M is a term constructed from the variables and constructor. We verify a more stringent injective agreement of the following form:

```
query x1:t1,…,xn:tn;
inj-event(e(M1,…,Mi)) => inj-event(e'(N1,…,Nj))
```

Injective agreement requires that the two events before and after correspond uniquely, i.e., for every event e(M1,…Mi) that occurs there is a unique corresponding event e'(N1,…,Nj) that occurs before it, effectively checking for the impact caused by replay. In the protocol model, we examine whether the occurrence of an event in which one party succeeds in examining an authentication message is preceded by the occurrence of a unique event in which the other party sends an authentication message.

Table is provided in ProVerif to support the storage of data, and in conjunction with table, we propose a new modeling approach for unlinkability verification in ProVerif. We create the linkability process as a complement to the attacker, which stores key information that may be associated with the user's identity during the transmission of messages. In the subsequent message interaction between the UE and the home network, the process determines whether there is any information in the interactive messages that is correlated with the stored information. Based on the above, we identified two vulnerabilities where the protocol fails to guarantee unlinkability.

6   Security Analysis

During the handshake between the UE and the home network, the ClientHello message is the first message sent by the UE. In protocol versions prior to EAP-TLS 1.3, the protocol supported the use of session id for session resumption, and switched to a pre-shared key mechanism in version 1.3. To guarantee compatibility, 5G EAP-TLS 1.3 sets a legacy session id field in the ClientHello message, which should be set to the value if the client has a cached session id set by a server prior to 1.3. Therefore, if the 5G EAP-TLS 1.2 protocol was originally used for mutual authentication, the UE retains the session id of the session, and then due to the protocol version upgrade, the UE carries the legacy session id in the message. An attacker can listen to the ClientHello message in the air, and if it finds any two ClientHello messages have the same value in the legacy session id field, it means that these messages come from the same UE, which obviously violates the security requirement of unlinkability as shown in Fig. 6.

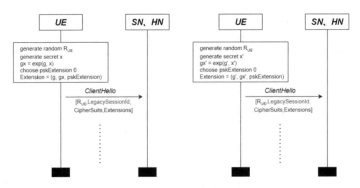

Fig. 6. Vulnerability I

The session resumption process of EAP-TLS 1.3 protocol requires a pre-shared key mechanism to realize mutual authentication. The protocol introduces the NewSessionTicket message, which is sent from the home network to the UE after the execution of the normal authentication process, and the home network can send multiple NewSessionTicket messages in one connection. Each NewSessionTicket message contains a unique ticket nonce and ticket id, which are mapped to each other. The ticket nonce is used for the generation of the pre-shared key between the two parties. The ClientHello message contains a structure called Extention, which has the field pskExtention. The pskExtention is populated by a series of ticket ids for the home network to select one to negotiate the pre-shared key generation. As a result, some of the ticket ids in the ClientHello messages of multiple session resumption processes for the same UE will be the same, and an attacker can determine whether multiple session resumption processes are initiated by the same UE as shown in Fig. 7.

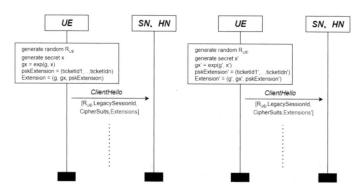

Fig. 7. Vulnerability II.

7   Conclusions

In this work, we analyze the EAP-TLS 1.3 protocol in 5G networks. We referenced several protocol standards and modeled the flow and security goals of the protocol with the help of a formal analysis tool ProVerif, and we found that the unlinkability of the protocol is not guaranteed even in the least-threat-capable attack scenarios, which are completely realisitc.

In the future, we would like to refine the modeling analysis so that it is fully automated and scripted, the current assumptions for each attack scenario require manual modification of the modeling code. In addition, the formal analysis considers whether the protocol design is safe and reliable, the protocol design is reliable does not mean that the security is guaranteed, the implementation of the protocol may also have unexpected potential vulnerabilities, we would like to further use fuzz and other techniques to study the correctness of the protocol implementation.

References

1. 3GPP. Security architecture and procedures for 5G system. TS 33.501, V17.7.0 (2022)
2. Simon, D., Aboba, B., Hurst, R.: RFC 5216: The EAP-TLS Authentication Protocol (2008)
3. Rescorla, E.: RFC 8446: The transport layer security (TLS) protocol version 1.3. (2018)
4. Blanchet, B.: Modeling and verifying security protocols with the applied pi calculus and ProVerif. Found. Trends® Priv. Secur. **1**(1–2), 1–135 (2016)

5. Basin, D., et al.: A formal analysis of 5G authentication. In: Proceedings of the 2018 ACM SIGSAC Conference on Computer and Communications Security (2018)
6. Meier, S., Schmidt, B., Cremers, C., Basin, D.: The TAMARIN prover for the symbolic analysis of security protocols. In: Sharygina, N., Veith, H. (eds.) Computer Aided Verification. CAV 2013. LNCS, vol. 8044, pp 696--701. Springer, Berlin (2013). https://doi.org/10.1007/978-3-642-39799-8_48
7. Cremers, C., Dehnel-Wild, M.: Component-based formal analysis of 5G-AKA: Channel assumptions and session confusion. In: Network and Distributed System Security Symposium (NDSS), Internet Society (2019)
8. Edris, E.K.K., Aiash, M., Loo, J.: Formalization and evaluation of EAP-AKA'protocol for 5G network access security. In: Array 16, p. 100254 (2022)
9. Zhang, J., et al.: Formal analysis of 5G EAP-TLS authentication protocol using proverif. IEEE Access **8**, 23674–23688 (2020)
10. Preuß Mattsson, J., Sethi, M.: RFC 9190: EAP-TLS 1.3: using the extensible Authentication Protocol with TLS 1.3 (2022)
11. Lowe, G.: A hierarchy of authentication specifications. In: Proceedings 10th Computer Security Foundations Workshop, IEEE (1997)
12. Naigaung Z.: Proverif model, 5G-EAP-TLS-1.3 (2023). https://github.com/Memory1111/5G-EAP-TLS-1.3-ProVerif

Perceiving Human Psychological Consistency: Attack Detection Against Advanced Persistent Social Engineering

Kota Numada, Shinnosuke Nozaki, Takumi Takaiwa, Tetsushi Ohki, and Masakatsu Nishigaki[✉]

Shizuoka University, 3-5-1 Johoku, Naka, Hamamatsu, Shizuoka 432-8011, Japan
nisigaki@inf.shizuoka.ac.jp

Abstract. Social engineering involves manipulating a target to make it act to achieve the attacker's objectives. Currently, direct and short-term attacks are common in the mainstream. However, with the development of science and technology, there has been a shift toward extended and long-term attacks using VR and other technologies. We define advanced persistent social engineering (APSE) as social engineering in which a sophisticated intrusion is continuously made in the psychological aspect of the target so that the target voluntarily behaves as intended by the attacker. Countermeasures against APSE attacks have not been studied yet. We propose a method to detect APSE attacks by checking for inconsistencies in personality because the principles of human behavior (personality) do not change easily but change when others intrude.

1 Introduction

A social engineering attack is the act of manipulating a target to make them take action (e.g., divulging confidential information or transferring money) to achieve the attacker's objectives. The amount of damage caused by social engineering attacks is increasing every year and is one of the most serious types of attacks [1]. In a social engineering attack, an attacker uses sophisticated psychological manipulation techniques to influence the target. Consequently, the target behaves as intended by the attacker without their knowledge.

Current typical social engineering attacks include phishing scam advertisements and communication frauds ("It's me" fraud). These methods involve the attacker cleverly providing "direct instructions" to achieve their goal by gaining control over the target in a "short period of time." Technical countermeasures such as denylisting [2] and anomaly detection [3] have been used. Additionally, there have been numerous reports on phishing, scam advertisements, and communication fraud, leading to increased social awareness of these attacks and their damages. Therefore, it is possible to develop educational countermeasures based on past cases and implement effective strategies against such "direct and short-term social engineering" by combining them with technical countermeasures.

Conversely, social engineering considers not only short-term attacks but also those that gradually control the target through "long-term intrusion." Additionally, with the development of science and technology, methods of intrusion by attackers are likely to "extend" [4]. Specifically, it is not difficult to imagine that sophisticated intrusion using XR (AR/VR/MR) devices will become feasible, and intrusion through brain-machine interface (BMI) devices will become possible in the near future. Since XR and BMI devices are head-mounted user interfaces, it is challenging for others to observe the information being sent and received. Therefore, an attacker can target a victim for a prolonged period without being noticed by others. Consequently, an attacker can conduct "extended and long-term social engineering attacks" on the target.

It is possible to intentionally limit the information shown to the target using VR devices or display information with enhanced dramatic effects using MR devices. BMI devices can directly interfere with the target's brain and rewrite information. The target may become intoxicated by the worldview presented by the attacker through effective presentation and may shift into a psychological state in which they behave as the attacker intends. The intoxicated target may provide information or money without limits, exacerbating the damage. Because the intoxicated target is not aware of the damage, and the attack does not become apparent, knowledge about the attack cannot be gathered, and countermeasures cannot be implemented by using a denylist. These are the problems that have arisen from the extension of social engineering attacks.

When attacks are conducted over a long period, attackers can gradually intrude to avoid detection as a major anomaly, making existing anomaly detection more challenging. Additionally, long-term intrusion can lead to a situation in which the target become enamored with an attacker, causing the target to relax their vigilance. The above are the problems that have arisen from the prolongation of social engineering attacks.

Such advanced social engineering attacks can benefit attackers seeking information and money. As countermeasures against direct and short-term social engineering have been implemented, it is possible that the trend of attacks will shift to "extended and long-term social engineering" in the future. To the best of our knowledge, countermeasures against extended and long-term social engineering attacks have not yet been studied. In this study, we introduce the term "advanced persistent social engineering (APSE)" (Fig. 1), which is defined as social engineering in which a target is subjected to continuous and sophisticated psychological intrusions so that they voluntarily behave as the attacker intended. Section 2 introduces the prerequisite knowledge, including related studies. Section 3 discusses APSE attacks, which are expected to become a threat in the future, and examines countermeasures against APSE attacks. Section 4 explains the concept of the proposed countermeasure, and Sect. 5 summarizes the study.

2 Prerequisite Knowledge

2.1 Current Mainstream of Social Engineering

Social engineering aims to gain profits or harm a target and its surroundings. The attacker cleverly uses psychological manipulation techniques to intrude unnoticed by the target and induces the target's behavior toward the attacker's intended actions. People exhibit psychological characteristics such as obedience to authority and impatience when given a

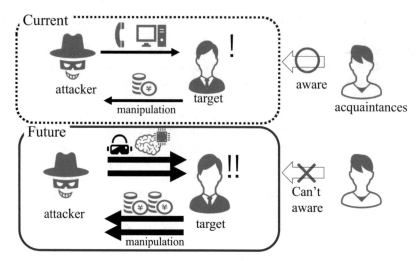

Fig. 1. Social-engineering attack

time limit [5]. The attacker exploits these characteristics and attempts to intrude through psychological manipulation techniques that skillfully use verbal or visual information.

2.2 Generative AI and XR Technologies

The use of generative AI enables the rapid and infinite generation of misinformation. Generative AI can also conduct automated attacks. Extended reality (XR) is a technology that merges real physical and virtual cyber spaces and can modify, complement, or extend "real" using virtual information. It can not only rewrite sensory information in a real physical space [6] but also replace the senses of others with their own [7]. The exploitation of these technologies has provided attackers with more psychological manipulation techniques than ever. An attacker can intentionally limit the information shown to the target using VR devices or display it with enhanced effects using MR devices. VR devices can also allow the target to share or experience the attacker's sensations. With these capabilities, the attacker may induce the target to perform the actions intended by the attacker.

2.3 BMI Technology

BMI has high expectations as a communication device for physically disabled people, such as the elderly and the handicapped, and technological development for social implementation is underway [8]. The misuse of BMI technology provides an attacker with the means to directly manipulate the brain and the five senses. If a subject communicates with an attacker's computer through a BMI device, there is a risk of malicious data being directly implanted into the brain. It is challenging for a third party to detect intrusion from an attacker or a behavioral change in the target. Therefore, mutual inspection among acquaintances (i.e., when acquaintances become aware that an attacker is intervening or manipulating friends or family members, they warn each other) cannot be expected.

2.4 Lifestyle Authentication

Lifestyle authentication is a user authentication technology based on user behavior logs [9]. It achieves user authentication by applying template matching (confirming a person's identity by matching their registered patterns) or anomaly detection (detecting deviations from normal patterns as "abnormalities") to human lifestyles. Continuous authentication can be performed using all types of lifelog data such as user location and Wi-Fi connection information. Humans differ in their daily activities. However, these activities can vary depending on factors like whether it is a weekday or holiday, whether one is at work or on vacation, whether one is alone or with friends, and even preferences like what one wants to eat that day. Such fluctuations in daily life can significantly affect authentication accuracy. Additionally, monitoring human behavior raises privacy concerns.

2.5 Big Five

The Big Five is a personality assessment scale known as the "Major Five-Factor Model," proposed by Lewis R. Goldberg [10]. It represents a person's personality based on "emotional instability," "diplomacy," "openness," "congeniality," and "integrity." This model offers a comprehensive and clear framework for understanding personality and finds application in various areas, including medical care, consumption preference surveys, and business performance analysis.

2.6 Personality Insights

Personality Insights [11] is an AI system that generates a user's Big Five personality profile based on their texts, such as tweets and blog posts. Personality Insights is implemented using machine learning to determine the relationship between the results of personality (Big Five) tests conducted on thousands of people and their texts. It expresses personality through 30 personality factors, each of which is further divided into six elements, based on the Big Five personality factors.

3 APSE Attacks and Countermeasures

3.1 Advanced Persistent Social Engineering

Current social engineering attacks primarily involve "direct and short-term social engineering," but it is conceivable that social engineering will transition to "extended or long-term social engineering" with the development of science and technology in the future. It is not difficult to imagine that intrusion using XR (AR/VR/MR) devices and intrusion through BMI devices will become possible in the near future. XR and BMI devices will not only provide an augmented means of intrusion to the attacker but will also be head-mounted user interfaces, making it difficult for others to see the information being sent and received. By effectively using augmentative effects to captivate the target, the attacker can compel the target to give away money or information without limit. The ability to intrude over a long period of time not only provides a method of gradual intrusion to avoid detection as a major anomaly but also relaxes the target's vigilance, helping the target become enamored with the attacker.

As a concrete example, consider the case of communication fraud ("It's me" fraud) or a pyramid scam, both of which can manifest in the metaverse space. In communication fraud, the attacker deceives the target by pretending to be the target's relative. Additionally, fraudulent groups play various roles, such as relatives, police, lawyers, etc., to commit theatrical crimes. However, in metaverse fraud, an avatar that mimics the target's relative can be used [12]. Moreover, the metaverse can create a realistic representation of the theater space. In a pyramid scheme fraud, attackers fabricate a fictitious company that garners attention from various fictitious investors, offering an attractive investment opportunity to the target. In metaverse fraud, it becomes effortless to make such schemes seem real. The attacker can strongly encourage the target's gambling spirit using various communication methods. Wearing a head-mounted display, the target becomes immersed in the metaverse space, allowing the attacker to intrude repeatedly without detection by others. As the intrusion progresses gradually, the target develops a desire to provide money to the attacker, eventually engaging in such actions without discomfort.

In this study, we define APSE as social engineering in which the attacker consistently and persistently intrudes into the target's mind, leading the target to voluntarily implement the attacker's intended actions. An APSE attack is believed to involve the following steps:

(1) The attacker continuously provides the target with "sensations" or "experiences" that lead to the attacker's intended behavior (intrusion).
(2) Through these stimuli, a specific memory is implanted in the target and the behavioral principles formed by this memory undergo a gradual transformation (convincement).
(3) With a change in behavioral principles, the target's actions eventually align with those intended by the attacker (manipulation).

3.2 Countermeasures

As APSE attacks unfold in the order of "intrusion," "convincement," and "manipulation," countermeasures against such attacks can be classified into three categories: intrusion, convincement, and manipulation.

1. Detection of intrusion: Typically, an attacker employs some form of psychological manipulation technique (e.g., Cialdini's law [5]) when attempting to intrude on a target. However, as humans are "social animals," seeking attention from others is a natural behavior (e.g., Chardeni phrases abound in everyday sales discussions). Therefore, even if psychologically manipulative phrases in conversations could be identified using natural language processing technology, distinguishing between an ordinary conversation (sales talk) and an APSE attack would pose a significant challenge.
2. Detection of convincement: There exists a relationship between behavior and personality [13], and alterations in personality often manifest as behavioral changes. While personality evolves over time, such changes are gradual and consistent [14]. It is acknowledged that personality undergoes significant shifts when influenced by external factors, such as mind control [15]. Therefore, it is potentially expected that the presence or absence of "mind control" can be tested by confirming the consistency of the target's behavioral principles (personality).

3. Detection of manipulation: A monitoring system can be implemented to detect a target's duped behavior and prohibit it using allow or deny lists. However, human behavior is influenced by various factors (e.g., weekdays versus holidays, being at work or on a break, being alone or with friends, and dietary preferences). Consequently, setting appropriate categories for permission or prohibition becomes challenging. Similarly, anomalies in a target's behavior can be detected through lifestyle authentication [6]. However, as previously mentioned, human behavior is subject to variation depending on various factors. Therefore, even if changes in a target's behavior can be detected using behavior monitoring technology, discerning whether these changes result from fluctuations in daily life or manipulation due to an APSE attack poses a challenge.

From the above discussion, detecting "convincement" emerges as a realistic countermeasure. Therefore, this study proposes an APSE attack detection system that focuses on the consistency of a target's behavioral principles (personality).

3.3   APSE Attack Detection Based on Personality Consistency Checks

In social engineering, the goal of an attacker is to change the "behavior" of the target for the attacker's benefit. To rephrase, it is the "behavior" of the target that undergoes change through social engineering. However, people's behaviors are diverse and can change voluntarily. For example, people may not consistently visit the same websites every day. While it is possible that an APSE attacker may have manipulated someone into visiting different websites than usual, it is also possible that the individual changed their websites preferences depending on their daily mood. Consequently, determining whether a person is being "manipulation" by an attacker just because their behavior is inconsistent is challenging, given the numerous possible factors contributing to such variability.

Therefore, this study focuses on the "behavioral principle" that is the underlying source of human behavior. Humans perform a wide variety of daily actions, but regardless of the kind of action, they follow their own behavioral principles. Behavioral principles are internal information, making them challenging to observe concretely. However, personality, one of the components of behavioral principles, has been predominantly studied in the field of psychological research, where methods of personality testing and the results of psychological or behavioral analyses based on personality have accumulated. It is known that personality is consistent [14], but it changes when subjected to intrusions from others [15]. Therefore, it is possible to test for the presence or absence of "convincement" by examining the consistency of a target's behavioral principles (personality).

The procedure for detecting personality consistency is illustrated as follows (Fig. 2):
[Preliminary Preparation Phase]

1. Recruit numerous collaborators and have them take the personality (Big5) test.
2. Request them to provide daily behavior logs, along with their personality.
3. Conduct statistical analysis of the relationship between behavior logs and personality
4. Implement a system that can diagnose personality using behavior logs.

[Enrollment Phase]

1. Monitor user behavior and collect behavior logs during the enrollment phase.
2. Diagnose personality from behavior logs. (Alternatively, users can take personality tests directly instead of providing behavior logs.)
3. Save the personality in the non-intrusive state as a "template."

   [Authentication Phase]

1. Constantly monitor user behavior and collect behavior logs.
2. Diagnose personality with each accumulation of a certain number of logs collected.
3. Compare it with the template and check whether the personality has changed.
4. Alert the user to a possible APSE attack when there is a change in personality.

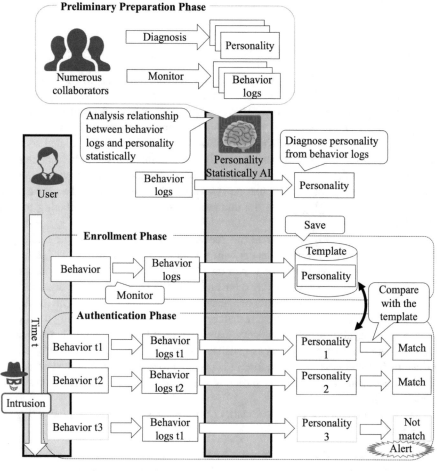

Fig. 2. Procedure for detecting personality consistency

3.4 Related Research

Research on personality traits and security can be divided into three main categories: (1) Investigation of the influence of personality traits on security awareness (e.g., [16]), (2) Promotion of security awareness considering personality traits (e.g., [17]), and (3) Enhancement of security measures considering personality traits (e.g., [18]).

The literature [16] applies personality psychology to analyze the impact of social engineering. It investigates the relationship between user personalities and their vulnerability to social engineering attacks.

The literature [17] applies personality psychology to social engineering awareness. The types of messages that increase sensitivity (appealing power) differ depending on the user's personality. Based on the user's personality, they generate "appealing security messages" for each user and present such messages to increase the user's security awareness.

The literature [18] applies personality psychology to countermeasures against social engineering. The effect of Cialdini's six laws (the psychological manipulation effect) differs depending on the user's personality. Based on the user's personality, the authors of the literature [18] identify "Cialdini phrases with a high psychological manipulation effect" for each user and protect users by alerting them when they receive emails containing such phrases.

To the best of our knowledge, there is no method that uses personality consistency for security measures, as presented in this study.

4 PoC System Based on Personality Insights

4.1 Implementation of Personality Consistency Testing

As a proof of concept (PoC) for the system proposed in this study, we implemented an APSE attack detection system, assuming there is an AI model that diagnoses personalities based on users' blog posts. The AI model responsible for personality diagnosis is Personality Insights [10], as described in Sect. 2.6. We assume that users regularly post blog articles daily. The proposed system consistently performs personality assessments based on daily blog posts. The user's usual personality (in the case of Personality Insights, the values of 30 personality factors based on the Big Five) is stored as a template, and when a change in personality is detected, the system notifies the user of the potential occurrence of an APSE attack.

4.2 Discussions

APSE attacks involve close communication between the attacker and target. As the target may become enamored, no reports will be filed, making it challenging to gather information about the attacks. In other words, countermeasures cannot keep up with attackers as long as we focus on the approaches from the attacker (detection of "intrusion"). Additionally, by the time the attacks are discovered, the attackers will have changed their attack patterns. In other words, countermeasures cannot keep up with attackers as long as we focus on the behavior of the target (detection of "manipulation"), either. In contrast, countermeasures that focus on the target's personality (detection of "convincement") have the potential to detect APSE attacks even if knowledge of the attack is not gathered. Therefore, we believe that the proposed system is a solution that can outsmart the cat-and-mouse games with the attacker and provide a defensive advantage.

Although we used Personality Insights to diagnose personality based on linguistic information, numerous studies have linked behavioral characteristics to personality [19]. We believe that our proposed method can be executed with other approaches. Additionally, the types of metabolites secreted by somatic cells, depending on the person's physical condition and mood [21], are being elucidated. In the future, it is expected that the metabolites of somatic cells dissolved in the blood, etc., will discover not only the physical condition of a person but also their mood. Therefore, it may become possible to examine the consistency of a user's personality not only from behavior logs but also from information such as blood metabolites.

User behavior logs contain highly private information, and many people may be reluctant to use them. However, personality diagnoses can be performed using non-sensitive information (e.g., blog posts), as in the case of Personality Insights. Additionally, while information about "personality" also has an identity, it is not considered to have linkability that uniquely determines the user. Therefore, the proposed method that uses personality traits mitigates privacy concerns.

4.3 Limitations

People may adopt different personalities depending on their opponents and the situation. However, it is thought that their responses to the same opponent in the same situation remain consistent [20]. Measures, such as testing the consistency of personality for each opponent or situation, are necessary.

In addition, the relationship between user behavior and personality is not one-to-one. An attacker can collect various behaviors diagnosed as having the same personality. If the attacker can change the user's behavior in a way that does not change their personality, the consistency of the personality will not be broken, and the APSE attack will not be detected.

In the future, with the development of large language models (LLM) such as ChatGPT, people are likely to write blog posts or choose their own actions with the support of AI. However, instead of considering the consistency of the behavior of the user alone or the AI alone, it would be possible to consider the user and the AI as one "individual" and extend our idea to check the consistency of behavior in the unit of "person and AI (user supported by AI)."

In Sects. 3.2 and 3.3, we stated that personality is challenging to change, but this is not always true. For example, users may engage in self-development and voluntarily desire to change their personalities. One answer to this issue is to confirm whether or not the user is convinced of the change in personality. However, confirmation might be difficult because the user's declaration cannot be trusted when psychological manipulation techniques of persuasion are being used.

In this study, we proposed a method to verify whether a person is under an APSE attack by examining the consistency of their personality converted sequentially from their behavior logs. However, it is also possible to conduct a personality test once a year, etc., at the time of annual health checkups in the workplace, instead of daily checkups. Just as one manages their weight (whether the weight has not changed from last year, whether they are approaching their ideal weight, etc.), it would be possible to check whether their personality has not changed since last year and whether they are approaching their ideal personality (whether the change is due to self-development).

5   Conclusion

In this study, we define APSE attacks as a form of social engineering in which a target person is subjected to continuous and sophisticated psychological intrusions, leading them to behave spontaneously as the attacker intends. As an effective countermeasure against APSE attacks, we investigated a detection method by determining whether the consistency of human behavioral principles (personality) was compromised.

In a future study, we plan to implement a PoC system using Personality Insights and evaluate the effectiveness of the proposed method through user experiments. The following points are to be confirmed through experiments:

1. Experiments for detecting APSE attacks.
   1-a: Consistency in personality in ordinary situations.
   1-b: Personality changes when subjected to APSE attacks.
2. Comparison of countermeasures in intrusion and manipulation phases.
3. Regarding 1-a, existing studies show that personality does not change abruptly but is consistent [14]. Therefore, we will verify this in a follow-up study. Regarding 1-b, previous studies have shown that mental models are affected by VR [4]. Therefore, we first assume an APSE attack using VR and confirm whether the APSE attacks change personalities.

For 2, we conduct a comparison experiment to determine whether the convincement phase detection method (personality-based detection method) is superior to the intrusion phase detection method and the manipulation phase method (lifestyle authentication method).

Acknowledgments. This study was supported in part by the JST Moonshot-type R&D project JPMJMS2215. We would like to thank Mr. Kazuyasu Shiraishi, an attorney-at-law of TMI Associates for his valuable comment about the frequency of detection in Sect. 4.3.

References

1. Salahdine, F., Kaabouch, N.: Social engineering attacks: a survey. Future Internet **11**(4), 89 (2019)
2. Lopez, J.C., Camargo, J.E.: Social engineering detection using natural language processing and machine learning. In: 2022 5th International Conference on Information and Computer Technologies (ICICT), New York, NY, USA, pp. 177–181 (2022)
3. Kaur, R., Singh, S.: Detecting anomalies in online social networks using graph metrics. In: 2015 Annual IEEE India Conference (INDICON), New Delhi, India, pp. 1–6 (2015)

4. Cheng, K., Tian, J.F., Kohno, T., Roesner, F.: Exploring user reactions and mental models towards perceptual manipulation attacks in mixed reality. In: USENIX Security, vol. 2023 (2023)
5. Cialdini, R. B.: Influence, Port Harcourt A. Michel, vol. 3 (1987)
6. Panarese, P., Baiocchi, A., Colonnese, S.: The extended reality quality riddle: a technological and sociological survey. In: 2023 International Symposium on Image and Signal Processing and Analysis (ISPA), Rome, Italy, pp. 1–6 (2023)
7. Hirano, T., Kanebako, J., Saraiji, M.Y., Peiris, R.L., Minamizawa, K.: Synchronized running: running support system for guide runners by haptic sharing in blind marathons. In: 2019 IEEE World Haptics Conference (WHC), Tokyo, Japan, pp. 25–30 (2019)
8. Li, S., Leider, A., Qiu, M., Gai, K., Liu, M.: Brain-based computer interfaces in virtual reality. In: 2017 IEEE 4th International Conference on Cyber Security and Cloud Computing (CSCloud), New York, NY, USA, pp. 300–305 (2017)
9. Shigetomi Yamaguchi, R., Nakata, T., Kobayashi, R.: Redefine and organize. 4th Authentication Factor, Behavior. In: 2019 Seventh International Symposium on Computing and Networking Workshops (CANDARW), Nagasaki, Japan, pp. 412–415 (2019)
10. Ewen, R.B.: An Introduction to Theories of Personality, Psychology Press, New York (2014)
11. Koichi, K., Nasukawa, T., Kitamura, H.: Personality estimation from Japanese text. In Proceedings of the Workshop on Computational Modeling of People's Opinions, Personality, and Emotions in Social Media, pp. 101–109 (2016)
12. The land blog: Facebook misinformation is bad enough. The Metaverse Will Be Worse. https://www.rand.org/pubs/commentary/2022/08/facebook-misinformation-is-bad-enough-the-metaverse.html. Accessed 25 Nov 2023
13. Benjamin, P., Chapman, L.R., Goldberg.: Act-frequency signatures of the Big Five. Pers. Individ. Diffences **116**(1), 201–205 (2017)
14. Damian, R.I., Spengler, M., Sutu, A., Roberts, B.W.: Sixteen going on sixty-six: a longitudinal study of personality stability and change across 50 years. J. Pers. Soc. Psychol. **117**(3), 674–695 (2019)
15. Bandura, A., Adams, N.E., Beyer, J.: Cognitive processes mediating behavioral change. J. Pers. Soc. Psychol. **35**(3), 125 (1977)
16. Cusack, B., Adedokun, K.: The impact of personality traits on user's susceptibility to social engineering attacks. https://ro.ecu.edu.au/cgi/viewcontent.cgi?article=1228&context=ism
17. Yaser Al-Bustani, A.M., Almutairi, A.K., Alrashed, A., Muzaffar, A.W.: Social engineering via personality psychology - bypassing users based on their personality pattern to raise security awareness. In: 2023 International Conference on IT Innovation and Knowledge Discovery (ITIKD), Manama, Bahrain, pp. 1–8 (2023)
18. Nishikawa, H., et al.: Analysis of malicious email detection using cialdini's principles. In: Proceedings of 2020 Asia Joint Conference on Information Security, pp. 137–142 (2020)
19. Xu, J., Tian, W., Lv, G., Liu, S., Fan, Y.: Prediction of the big five personality traits using static facial images of college students with different academic backgrounds. In: IEEE Access, vol. 9, pp. 76822–76832 (2021)
20. Fleeson, W.: Situation-based contingencies underlying trait-content manifestation in behavior. J. Pers. **75**(4), 825–861 (2007)
21. Chen, R., Mias, G.L., Li-Pook-Than, J., et al.: Personal omics profiling reveals dynamic molecular and medical phenotypes (2012)

Design of Secure Fair Bidding Scheme Based on Threshold Elliptic Curve ElGamal Cryptography

LiPing Shi[✉]

Police Officers College of the Chinese People's Armed Force, Chengdu, China
413234498@qq.com

Abstract. The lowest price winning approach is used in some procurement bidding, which greatly saves the procurement cost, but in reality, there exists the problem of collusion between the purchaser and a bidding supplier, so it is very important to prevent the problem of conspiracy before the bid opening. This paper proposes a bidding scheme based on threshold elliptic curve ElGamal cryptosystem, the bidders jointly generate the public key and use elliptic curve ElGamal to encrypt the bidding price, in the bid opening stage, use the threshold joint decryption. In order to reduce the complexity of encryption and decryption, this paper proposes to group the bidding price numbers, quickly find out the minimum value of the two-digit integers so as to find out the lowest bidder. At the same time, ECDSA digital signature technology is used to achieve the integrity, non-repudiation and public verifiability of the price. By comparing and analyzing the security and performance of this scheme with homomorphic encryption algorithm, it is proved that this scheme improves the security and fairness of the price in the bidding process.

1 Introduction

The lowest bid winning method refers to the evaluation method in which the bidder with the lowest quotation wins the bid. This method is commonly used for projects that require clear cost control, such as engineering construction, procurement, etc. The advantage of using the lowest price bidding method is that it is simple and easy to operate, making the bidding process easier to operate, greatly saving manpower, material resources, and financial resources in the bidding process. The purchaser obtains products or services at the lowest price, which is conducive to saving procurement costs. In addition, the lowest price bidding method can increase competition, encourage suppliers to provide lower prices and better quality, thereby improving procurement efficiency. However, in order to ensure that they can win the bid at the lowest price, some bidders may resort to bribery, bidding manipulation, or colluding with competitors to participate in the bidding. Especially in the case of collusion between the purchaser and a bidding supplier, the purchaser opens the bidding documents of other bidding suppliers before the bid opening and leaks their bidding price information to the bidding supplier, which damages the principles of fairness, impartiality, and openness in the bidding process, harms the interests of other bidders, and may lead to quality and progress issues in the project. Therefore, it is necessary to protect the privacy of bidding prices.

L. Barolli (Ed.): EIDWT 2024, LNDECT 193, pp. 163–172, 2024.
https://doi.org/10.1007/978-3-031-53555-0_16

How to calculate the lowest bidding price while protecting the privacy of bidding prices from all parties. This is a typical issue of data security and privacy protection. In recent years, the rise of privacy computing has been regarded as an important technological means to solve the dilemma of data privacy protection, data security circulation, and data intelligent application. At present, the main ideas for privacy computing solutions are: firstly, based on trusted third-party server computing solutions. The third party is a disinterested party with strong storage and computing power. The scheme generally uses homomorphic encryption technology [1, 2] to construct a computing protocol. Each participating party homomorphic encrypts their private data and hands it over to the third party. The third party performs operations in the ciphertext state and hands the results to the privacy computing requester for decryption to obtain the target computing data. This scheme is mainly used in cloud service application scenarios [3–6]; The second is a secure multi-party computation scheme without a trusted dealer of the key, which allows participants to perform joint computation without exposing the original data [8, 9].

The main security threat of the lowest bid winning method lies in the collusion between the tenderer and bidders. Before the results of the bid opening are available, the tenderer and bidders should not know the prices of other bidders. While after the lowest bidder is determined, the prices are not necessary to be kept secret and can be made public. Therefore, to prevent collusion attacks, the second approach to privacy computing, i.e., a decentralized secure computing scheme, is adopted. This paper proposes a bidding scheme based on the lowest bid method to prevent collusion between any party. Each bidder shall jointly negotiate the encryption public key and first use the EC ElGamal algorithm to encrypt their bid price using the public key. At the bid opening stage, a threshold-based decryption algorithm is used to decrypt the original price only when all bidders participate. When the original price is decrypted, it is also clear who the winning bidder is. Simultaneously encrypting bidding requires bidders to digitally sign the bidding price to prevent tampering and denial.

Section 2 of this paper describes the theoretical basis and algorithms used in the scheme; Sect. 3 designs the overall scheme with specific conventions and protocols to address the security and fairness of the bidding; Sect. 4 provides a comparative analysis of the security and performance of the scheme with homomorphic encryption schemes; and Sect. 5 concludes the paper.

2 Related Theoretical Basis and Algorithm

2.1 Elliptic Curve Cryptography

The Elliptic Curve Cryptosystem (ECC) is a public key cryptosystem based on elliptic curve mathematics., elliptic curves over finite fields are commonly used in cryptography, with the most commonly used being curves defined by the following equation:

$$y^2 \equiv x^3 + ax + b \pmod{p} \, (a, b \in GF(p), 4a^3 + 27b^2 \pmod{p}) \neq 0 \quad (1)$$

Using $E_p(a, b)$ represents the set of points on the elliptic curve defined by curve (1) and the infinite point O, in which a special addition operation is defined, that is, the point addition operation, to calculate the sum of point P and Q on the elliptic curve: Point $R = P + Q$, when $P = Q$, $R = P + P = 2P$, and so on, can calculate nP, which is the multiplication operation on the elliptic curve.

For two points P and Q on an elliptic curve, k is an integer. When P is known, calculating $Q = kP$ is easy, but given P and Q, finding k is very difficult. This is the Elliptic Curve Discrete Logarithm Problem (ECDLP), which is difficult to solve with current computing power. By selecting appropriate elliptic curve parameters a, b, and large prime p, choosing a base point G in $E_p(a, b)$, where k is the private key, and the calculated $Q = kG$ can be used as the public key to achieve encryption and decryption operations. Compared with other public key cryptography systems such as RSA, Diffie Hellman, etc., ECC has the advantages of small computational complexity, short keys, low bandwidth and processor requirements, and higher security performance.

2.2   EC ElGamal

The EC ElGamal cryptosystem [10] is implemented based on elliptic curves in the ElGamal encryption algorithm, described as follows:

Key Generation. Select an elliptic curve $E_p(a, b)$ and its base point G, where n is the order of G. Choose an integer x ($x \in [1, n-1]$) as the private key and calculate $X = xG$, where x is the private key and X is the public key, where base point G 和 and elliptic curve $E_p(a, b)$ as a public parameter.

Encryption. Encode the message m to a point M on $E_p(a, b)$, Calculate the ciphertext of the point M by randomly selecting a positive integer r:

$$E(M) = (c_1, c_2) = (rG, M + rX) \qquad (2)$$

Decryption. Decrypt using private key x, the formula is as follows:

$$M = c_2 - xc_1 \qquad (3)$$

Then the plaintext encoding point M is decoded and finally m is obtained.

2.3   Threshold Decryption

Threshold decryption [11] is an important tool against collusion attacks in secure multi-party computing. In a threshold decryption cryptosystem, n participants jointly generate a public key, and each of them holds a portion of the decryption key. The shared public key can be used directly to encrypt a message during the encryption process, but multiple participants need to work together to decrypt the ciphertext. If it takes at least t individuals to decrypt a message, and less than t individuals will not be able to decrypt the message, this cryptosystem is called (t, n) threshold cryptosystem. In this paper, we need the plain (n, n) threshold cryptosystem to resist the collusion attack by $n - 1$ participants, which can be constructed by using elliptic curve cryptosystem as follows:

Key Generation. Each participant agrees on an elliptic curve $E_p(a, b)$ and a base point G on it, and each participant P_i arbitrarily chooses an integer x_i as the private key, computes $X_i = x_i G$, and jointly generates the public key:

$$X = \sum_{i=1}^{n} X_i = \sum_{i=1}^{n} x_i G \tag{4}$$

Encryption. The message is encoded into points M, and randomly selects a positive integer r, and computes the ciphertext $(c_1, c_2) = (rG, M + rX)$.

Decryption. Each participant P_i computes $x_i c_1$ based on their private key x_i, sends the computation result to other participants, and then all the participants jointly decrypt to get the encoded point M:

$$M = c_2 - \sum_{i=1}^{n} x_i c_1 \tag{5}$$

2.4 Plaintext Encoding Method

With ECC encryption, the plaintext is encoded to a point on the elliptic curve. The plaintext used in this scheme is in the range of integers 0 to 99. Therefore, it is only necessary to map 0 to 99 to a point on the elliptic curve. The steps are as follows [12]:

(1) Choose an auxiliary base parameter k, e.g., $k = 50$ (encryption and decryption parties reach an agreement).
(2) For each m, For $i = 1$ to $k - 1$, let $x = mk + i$ and use the elliptic curve equation to find y. If found, stop. If not found, let $i = i + 1$ and continue searching until it is found. In fact, a point (x, y) can be found, which is the encoding of message m
(3) Decoding. The point (x, y) is decoded as $\frac{x-1}{k}$ (Round down the value)

2.5 Digital Signature Algorithm

To facilitate the establishment of the same common function during the implementation of the scheme, this scheme adopts the Elliptic Curve Digital Signature Algorithm (ECDSA) algorithm based on elliptic curves [13]. The signing process is as follows:

(1) Generate key pairs. Select an elliptic curve $E_p(a, b)$ with a base point G on it, n is the order of G. Choose an integer $d(d \in [1, n - 1])$ as the private key and compute $Q = dG$, where d is the private key and Q is the public key, where the base point G and the elliptic curve $E_p(a, b)$ are used as public parameters.
(2) The data m to be signed is hashed using SHA-1 algorithm $e = H(m)$.
(3) Choose a random number k and compute $P = (x_1, y_1) = kG$, $r = x_1 \bmod n$.
(4) Compute $s = k^{-1}(e + rd) \bmod n$.
(5) The signature results in (r, s).

The process of verifying a signature is:

(1) Receive the signature result (r, s) and the data to be verified m.
(2) Using SHA-1, compute the hash $e = H(m)$ of the data to be verified.

(3) Compute $w = s^{-1} mod n$, $u_1 = ew$, $u_2 = rw$.

(4) Compute the point $X = (x_1, y_1) = u_1 G + u_2 Q$.

(5) If $X = O$, validation fails; otherwise compute $v = x_1 mod n$.

(6) Validation succeeds if and only if $v = r$.

Digital signatures provide the following basic cryptographic services, i.e., message integrity, source trustworthiness, and resistance to repudiation. The role of the ECDSA algorithm in this scheme is to determine the integrity of the bidding price data, to authenticate the source of the price data (i.e., the corresponding bidder), and to ensure that the bidder cannot repudiate the price data and the bidding behavior of the previous bids.

3 Bidding Scheme Design

3.1 Scheme Overview

The participants in the bidding program proposed in this paper are the tenderer, the platform server, and the bidders. The tenderer is responsible for releasing the bidding information on the bidding platform, and does not participate in the price comparison of the bidding process. The platform server is not involved in the calculation and comparison of the bidding price, but is responsible for verifying the signature and controlling the process and progress of the bidding process, and posting the publicly available information of each stage on the bulletin board. The bulletin board information is visible to all users. The interaction of the entities in the scheme is shown in Fig. 1.

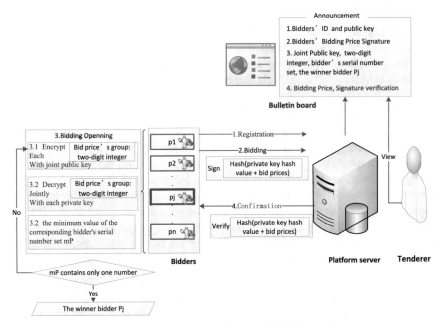

Fig. 1. Workflow diagram of secure fair bidding scheme

The bidding program mainly includes the following steps:

(1) Registration. After the bidder publishes the bidding information on the bidding platform, the bidding platform server initializes the bidding workflow. Bidders who are willing to participate need to hash their identity ID with their own private key and sign it before sending the identity ID and public key to the platform server. The platform server verifies the signature using the bidder's public key. After verification, publish the identity ID and its public key on the bulletin board. Bidders can confirm whether the published identity ID and public key are correct. If they are incorrect, they can appeal to the platform server until it is confirmed that the published identity ID and public key are valid (i.e., their signature is confirmed to be a valid signature).

(2) Bidding. The bidder uses the private key which is used during registration to sign the hash value of the private key hash value plus the bid price, and sends the signature to the server, which publishes its signature. Because the signer did not send the bid price plaintext, so the server is unable to verify their signatures and obtain the bid price plaintext at this stage.

(3) Bid Opening. This stage is a critical and sensitive stage of this plan. Each bidder enters the stage of calculating the lowest bidder in a fair, open, and secure manner according to the instructions on the platform bulletin board. In this stage, the price comparison method based on grouping processing and threshold decryption in chapters 3.2–3.4 is used to calculate the lowest bidder: each bidder encrypts their bid prices using a joint public key, and then jointly decrypts the bid prices of each bidder to select the lowest bidder, which is the winning bidder. Announcement board announces the winning bidder.

(4) Confirmation. Each bidder publicizes its own bid price, the complete bid price and private key hash value sent to the server, the server verifies the signature of each bidder in the bidding stage, and if the verification passes, it indicates that the bidder's publicly disclosed price is the same as with the price at the time of bidding and acknowledges the bidding result. All bids are announced on the bulletin board.

3.2   Quotation Number Grouping Processing

Since ECC calculates points on elliptic curves, when using EC ElGamal for encryption, plaintext numbers need to be encoded and mapped to points on the elliptic curve for encryption. After decryption, decoding is required to obtain the original text. If the bidding price number is large, the computational cost is high. Therefore, in order to improve the comparative efficiency of bidding prices, this scheme groups the bidding price numbers in a unified format, with each group consisting of two digits. The comparison starts from the highest digit group. If there is a unique minimum value in the highest digit group, the comparison ends. If there are two or more identical minimum values, the comparison continues to the next digit group. The unified bidding data format convention is:

Agreement: Temporarily, the highest digit is tens of millions (of course, it can be expanded to billions); Divide into 5 groups; Each group is represented as an integer ranging from 0 to 99; If there is no high digit, use 0 to represent that digit group.

Assuming that the 1st bidder's bid price m_1 for 14905703.87 yuan, with the numerical value is $m_1 = \{m_{11}, m_{12}, m_{13}, m_{14}, m_{15}\} = \{14, 90, 57, 03, 87\}$, the 2nd bidder's bid price of m_2 for 14674902.73 yuan:$m_2 = \{m_{21}, m_{22}, m_{23}, m_{24}, m_{25}\} = \{14,67,49,02,73\}$, and bidder No. 3's bid price m_3 is 152657368.00 yuan:$m_3 = \{m_{31}, m_{32}, m_{33}, m_{34}, m_{35}\} = \{15,26,73,68, 00\}$. The grouping corresponds to the following in a uniform format: "ten-million million, hundred-thousand ten-hundred, thousand hundred, ten one, decile percentile".

$m_1 = \{14,90,57,03,87\}$
$m_2 = \{14,67,49,02,73\}$
$m_3 = \{15,26,73,68,00\}$

Firstly, by decrypting and decoding m_{11}, m_{21}, m_{31}, it can be seen that both m_{11} and m_{21} are the minimum value 14, so then compare m_{12} and m_{22}, then get the lowest bidding price for the bidder's bid price of No. 2, you can declare the bidder No. 2 as the winning bidder.

3.3 Two-Digit Integer Minimum Calculation Protocol Based on Threshold Decryption

Protocol: calculating the minimum value of a two-digit integer with equal status among bidders to resist collusion.

Input: bidders $P_1, P_2, \ldots, P_n$'s respective secret bid prices grouped two-digit integers $m_1, m_2, \ldots, m_{i,\ldots}, m_n$.

Output: Min Value $minM = \min\{m_1, m_2, \ldots, m_{i,\ldots}, m_n\}$; The set of corresponding bidder numbers for the minimum value $mP = \{i\}$ (when $m_i = minM$, $i \in [1, n]$)

Step:

(1) Bidders $P_1, P_2, \ldots, P_n$ agree on an elliptical curve $E_p(a, b)$ and its generator G (requiring G has a sufficiently large order L), as well as encoding auxiliary parameter k, each bidder selects their own private key x_i, $(2 \leq x_i \leq L - 1)$, generate its public key as $X_i = x_i G$. Each bidder publishes their public key through the server, and the server coordinates to send the public key of the bidder P_1 to the next bidder P_2, the bidder P_2 computes $X = X_1 + X_2$. Finally, the bidder P_n generate joint public key X:

$$X = \sum_{i=1}^{n} X_i = \sum_{i=1}^{n} x_i G \tag{6}$$

(2) Each bidder casts a two-digit integer m_i to a point M_i of the elliptic curve $E_p(a, b)$ with the help of the encoding method of embedding the plaintext message into the elliptic curve and the encoding auxiliary parameter k.

(3) Each bidder P_i chooses an arbitrary random number r_i on Z_n^*, encrypts the respective M_i using the joint public key X, and publicly releases the ciphertext C_i:

$$C_i = EM_i = (c_{i1}, c_{i2}) = (r_i G, M_i + r_i X)$$

(4) All bidders jointly decrypt the bidder P_i price ciphertext C_i. Each bidder $P_j (j \in [1, n])$ computes $C_{i1j} = x_j c_{i1}$ according to its own private key x_j and sends

C_{i1j} to the other bidders, then all the bidders jointly decrypt C_i to get the encoding point M_i:

$$M_i = c_{i2} - \sum_{j=1}^{n} C_{i1j} = c_{i2} - \sum_{j=1}^{n} x_j c_{i1} \tag{7}$$

Correctness analysis:

$$c_{i2} - \sum_{j=1}^{n} x_j c_{i1} = M_i + r_i X - \sum_{j=1}^{n} x_j r_i G = M_i + r_i \sum_{j=1}^{n} x_j G - r_i \sum_{j=1}^{n} x_j G = M_i \tag{8}$$

Security Analysis:

Since each bidder P_i jointly generates the public key of the threshold elliptic curve cryptosystem, i.e., $X = \sum_{i=1}^{n} x_i G$, with the private key x_i. Decryption requires the participation of all bidders. As long as one bidder does not participate in the decryption process, the ciphertext cannot be decrypted correctly, and neither the renderer nor the server can participate in the encryption and decryption process, thus resisting collusion attacks.

(5) The server decodes the decryption result to get two-digit integers m_i which is a group of the bidding price of each bidder, and selects the minimum value $minM = \min\{m_1, m_2, \ldots, m_{i,\ldots,}m_n\}$, and if $m_i = minM$, then put the bidder's serial number i into the set mP.

3.4 Calculate the Lowest Bidder

During the bidding stage, each bidder P_i expresses its bid as 5 sets of integers $m_i = \{m_{i1}, m_{i2}, m_{i3}, m_{i4}, m_{i5}\}$ according to the Agreement. During the bid opening stage, all bidders jointly calculate the minimum value of m_{i1} corresponding to the bidder's serial number set mP_1. if there are two or more serial numbers in mP_1, the bidders corresponding to the serial number encrypt and jointly decrypt m_{i2} according to the Protocol again, so on and so forth until the minimum value of the corresponding bidder's serial number set mP_k ($k \in [1, 5]$) has only one serial number, that is, the only minimum value is found, and the serial number in the set mP_k is the serial number of the bidder with the lowest bid.

4 Security and Performance Analysis

4.1 Security Analysis

(1) Resistance to collusion attack: Based on the algorithm of threshold decryption, the decryption at the bid opening stage requires all participants to complete. The tenderer is not a participant of decryption, and cannot obtain any bid price information in advance, so naturally it cannot conspire with any bidder, thus this program can resist collusion attack.

(2) Legitimacy and uniqueness: the bidders are authorized to participate in the bidding only after they have registered and passed the digital signature verification, and each bidder has a unique identity ID, which avoids dishonest bidders from bidding multiple times.

(3) Verifiability and non-controversiality: In the bidding process, the elliptic curve public parameters and public key used for encryption and decryption are made public through a bulletin board. In the bidding stage, bidders sign their own bid prices and publicize them through the platform, and bidders are able to verify relatively easily whether their bid prices have been tampered with. In the confirmation stage, the platform verifies whether the signatures of the bidders in the bidding stage are valid to prevent the bidders from denying their previous bidding prices.

4.2 Efficiency Analysis

The commonly used method for privacy computation is based on homomorphic encryption, here the threshold-based EC ElGamal algorithm of this scheme is compared with homomorphic encryption-based algorithm for minimization [14]. Reference [14] proposes a secure minimax computation protocol based on Paillier homomorphic encryption. The protocol involves two participants, the client and the server. The client encrypts its private input with Paillier homomorphic encryption and sends the ciphertext to the server. The server uses Paillier homomorphic addition and multiplication operations to compare ciphertext and returns the ciphertext of the minimum value to the client. The client decrypts the ciphertext of the minimum value and obtains the plaintext result with the minimum value.

1. Computational complexity analysis: Paillier encryption algorithm requires two modulo exponential operations for encryption, one modulo exponential operation for decryption [15], and homomorphic addition operations such as ciphertext multiplication for ciphertext comparison. EC ElGamal encryption scheme requires two elliptic curve multiplication operations for encryption and one elliptic curve multiplication operation for decryption. In this scheme, all participants encrypt and decrypt the ciphertext of price grouping, the encryption and decryption objects are two-digit integers, the computational complexity is low, at least one group of two-digit integers need to be encrypted and decrypted, at most five groups of two-digit integers need to be encrypted and decrypted. Therefore, the computational complexity of EC Elgamal scheme is lower than that of Paillier addition operation.

2. Protocol complexity analysis. The protocol based homomorphic Paillier encryption algorithm needs to build at least two server roles, one role is responsible for homomorphic computing, one role (usually the initiator) is responsible for homomorphic decryption, and the two server roles need to agree in advance that there is no conspiracy. Therefore, the protocol based homomorphic encryption algorithm cannot effectively solve the problem of the renderer and server conspiracy in the bidding, if you want to solve the problem, you need to join the trusted third-party server. The protocol based on threshold decryption in this scheme only requires the participation of bidders, without the need for a trusted third party to join the algorithm, without the need for renderer to participate in the encryption and decryption, the protocol is simple and effective in preventing the problem of conspiracy.

5 Conclusion

In this paper, through the bidding scheme based on threshold elliptic curve ElGamal encryption and ECDSA digital signature cryptosystem, the security and fairness of the bidding process based on the lowest bid method are realized by effectively preventing the problem of conspiracy from the tenderer and some bidders. Through the analysis of security and performance, the scheme has high operational efficiency and can meet the bidding requirements of the lowest-bid-award method, which has certain application value. The next research direction is to address the safety and fairness issues of multi factor bidding methods other than price.

References

1. Paillier, P.: Public-key cryptosystems based on composite degree residuosity classes. In: Stern, J. (ed.) Advances in Cryptology — EUROCRYPT 1999. EUROCRYPT 1999. LNCS, vol. 1592, pp. 223–238. Springer, Berlin (1999). https://doi.org/10.1007/3-540-48910-X_16
2. Gentry, C.: Fully homomorphic encryption using ideal lattices. In: The Forty-First Annual ACM Symposium on Theory of Computing, pp. 169–178 (2009)
3. Rivest, R.L., Shen, E., Adida, B.: Remote electronic voting with fully homomorphic encryption. In: Financial Cryptography and Data Security, pp. 108–115 (2008)
4. Patel, S., Persiano G.Y.: Private stateful information retrieval. In: 2018 ACM SIGSAC Conference on Computer and Communications Security, pp. 1002–1019 (2018)
5. Wang, C., Wang, Q., Ren, K., Lou, W.: Privacy-preserving public auditing for data storage security in cloud computing. IEEE Trans. Comput. $62(2)$, 362–375 (2016)
6. Ju, Y., Li, H., Huang, X.: Homomorphic encryption-based secure outsourcing for deep learning. IEEE Trans. Dependable Secure Comput. $17(1)$, 120–132 (2019)
7. Lindell, Y., Pinkas, B.: Privacy preserving data mining. J. Cryptol. $15(3)$, 177–206 (2002). https://doi.org/10.1007/3-540-44598-6_3
8. Lindell, Y.: Fast secure multiparty computation for small circuits: Beyond dishonest majority. In: Proceedings of the 2017 ACM SIGSAC Conference on Computer and Communications Security, pp. 1853–1871 (2017)
9. Koblitz, N.: Elliptic curve cryptosystems, Math.Comp. 48, 203–209 (1987)
10. Yang, B.: Foundations of Modern Cryptography. Tsinghua University Press, Beijing (2022)
11. Long, Y., Chen, K., Mao, X.: New constructions of dynamic threshold cryptosystem. J. Shanghai Jiaotong Univ. (Science) 19, 431–435 (2014). https://doi.org/10.1007/s12204-014-1520-8
12. Bhp, P., Chandravathi, D., Roja, P.P.: Encoding and decoding of a message in the implementation of Elliptic Curve Cryptography using Koblitz's method. Int. J. Comput. Sci. Eng. $2(5)$, 1904–1907 (2010)
13. Don, B.J.: Alfred, J.M.: Elliptic curve digital signature algorithm (ECDSA) [Z/OL]. http://www.certicom.com/resources/download/ecdsa.ps
14. Li, M., Sun, X., Wang, W., Liu, J., Wang, J.: Secure minimum computation using paillier homomorphic encryption. IEEE Trans. Inf. Forensics Secur. $14(8)$, 2029–2042 (2019). https://doi.org/10.1109/TIFS.2019.2892158
15. LI, S.-D., Kang, J., Yang, X.-Y., Dou, J.-W., Liu, X.: Secure multiparty characters sorting. Chin. J. Comput. $41(5)$, 1172-P1181 (2018)

Monitoring Power Usage Effectiveness to Detect Cooling Systems Attacks and Failures in Cloud Data Centers

Michele Mastroianni[✉][iD], Massimo Ficco[iD], Francesco Palmieri[iD], and Vincenzo Emanuele Martone

Dipartimento di Informatica, Università degli Studi di Salerno,
Via Giovanni Paolo II, 132 84084 Fisciano, SA, Italy
mmastroianni@unisa.it

Abstract. Energy-related Denial of Service (DoS) attacks have the potential to impact not only the quality or availability of services provided by large-scale data center (DC) infrastructures but also the operating expenses incurred by the organizations managing them, essentially in terms of energy bills. More specifically, the impact on the overall energy usage and, consequently, on the associated expenses, increases with the amount of time required to identify the attack. Therefore, the degradation of the environmental control systems in the buildings/facilities hosting the computing or storage nodes poses an especially insidious threat, which could result in a novel kind of attack to the involved infrastructures. Due to the limited ability to observe events in cyber-physical systems, recognizing these violations is extremely challenging for data center administrators. This paper proposed a new methodology for detecting cooling systems attacks based on continuous Power Usage Effectiveness (PUE) monitoring. This kind of measurement is quite simple to arrange in a data center, and can help to detect, in a limited amount of time, both attacks and failures on a DC cooling system, thus helping the system administration to limit expenses and service outages.

Keywords: Power Usage Effectiveness (PUE) · Energy-related DoS · GreenCloud · cloud simulator

1 Introduction

The problems related to the energy consumed by data centers are assuming an ever increasing importance: in 2018, it was estimated that the electricity consumption of Data Centers accounted for 1% of the global electricity consumption [14], and the growing trend in consumption needs careful monitoring, aimed at limiting the impact on global energy demand. Among the strategies mentioned to counteract this phenomenon, there is the promotion of energy efficiency standards for servers, storage devices, and networking equipment, as well as for benchmarking these devices.

On the other side, there are emerging attacks that leverage the energy consumption of IT devices in various ways [4]. Moreover, attacks may be also directed to cooling systems and other ancillary data center subsystems [15].

© The Author(s), under exclusive license to Springer Nature Switzerland AG 2024
L. Barolli (Ed.): EIDWT 2024, LNDECT 193, pp. 173–184, 2024.
https://doi.org/10.1007/978-3-031-53555-0_17

Regarding attacks directed to cooling systems, in [3] the authors demonstrated that a hostile user can breach the data center's environmental control systems to target a significant computing infrastructure, causing service outage, forcing system administrators to stop service and manually recover the data center.

Accordingly, this paper proposes a methodology to detect the attacks directed to cooling systems. The methodology is based on the continuous measuring of Power Usage Effectiveness (PUE), a widely recognized data center performance indicator of energy effectiveness. To prove the feasibility of this technique, a data center has been simulated using the GreenCloud simulator and the results are shown.

This paper is organized as follows: in the next section are listed the main related works, while in Sect. 3 the methodology proposed is detailed. The simulation scenario and the results are shown in Sect. 5. The last section exposes the conclusion and the possible future work.

2   Related Work

Sophisticated denial-of-service attacks have been used against cloud data centers in recent years to cause substantial resource and energy exhaustion. Energy-related attacks on data centers were described for the first time in [16] and then in [11,20], along with an analysis of the magnitude of the effects that were caused.

Palmieri et al. [17] presented an overview of the various DDoS attacks (e-DDoS) linked to energy exploitation, that can target massive cloud systems and can pose a significant risk to large-scale cloud infrastructures, if designed to harm specific vulnerabilities in the target system's architecture.

Attack patterns can be designed to do the most damage possible while maintaining a low profile (low message/job rate, for example) to avoid detection by defense systems that are meant to prevent brute-force or high-rate attacks. In [5], for instance, the authors described a variety of attacks designed to take advantage of the cloud's persistence, forcing the cloud resource manager to migrate and scale virtual machines severely and waste energy.

Additionally, Park et al. [18] presented a cyberattack in the style of an Advanced Persistent Threat conceived for distributed energy resources in cyber-physical contexts. In order to develop potential attack techniques, they offered a cyber kill chain model and testbed for a distributed energy resource system that was based on the most recent version of MITRE's cyber kill chain model.

Chlela et al. [2] proposed a mitigation strategy for denial-of-service (DoS) attacks that targets the most important distributed energy resource in a microgrid that powers critical loads of data centers and uninterrupted manufacturing processes for dependable power.

At the best of our knowledge, there is no other paper proposing techniques specifically suited for data center ancillary equipment.

3   The Methodology

The main reasons to use the PUE monitoring as an attack indicator is the ease of availability of PUE information at low cost. Indeed, due to the widespread adoption of PUE as a basic energy effectiveness indicator, the PUE metric is easy to compute for most Cloud providers in their data centers [9], and the data needed for this are already available without additional costs. In any case, the PUE measurement could be performed in a relatively simple way. In the following Table 1 the three main level of measurement accuracy are shown [1].

Table 1. PUE measurement point and frequency.

	Level1 - Basic	Level 2 - Intermediate	Level 3 - Advanced
IT Equipment Energy	UPS Outputs	PDU Outputs	IT Equipment Input
Total Facility Energy	Utility Inputs	Utility Inputs	Utility Inputs
Measurement Interval	Monthly/Weekly	Daily/Hourly	Continuous (15 min)

In the first case, the measurement points are the Facility inputs and the UPS outputs, that are very easy to measure.

3.1   Power Usage Effectiveness (PUE)

Power Usage Effectiveness [10, 12, 19] is the most frequently used energy-related performance metric, which was first introduced by The Green Grid. PUE is defined as *"The ratio of total facilities energy to all ICT equipment energy"* [7]:

$$PUE = \frac{E_{TOT}}{E_{IT}}; PUE \geq 1. \tag{1}$$

In formula 1, E_{TOT} represents the total energy supplied to the whole DC and E_{IT} is the sum of energy supplied to servers, switches, routers, etc. While the ideal best value of PUE is 1.0, its value for a typical data center could be around 2.0 when load is around 50% of maximum capacity. When the PUE assumes the value of 2, it indicates that the overall energy used for the entire DC facility is twice as much as the energy used for all IT devices.

The PUE may be represented with a curve in which its values depends on the instantaneous IT load and on the performance and physical configuration of the other non-IT ancillary systems located in the data center (cooling systems, UPS, lights etc.). To show the PUE trend versus the IT energy load, it is possible to use a tool implemented by Scheider Electric, the *Data Center Efficiency and PUE Calculator*[1]

The Fig. 1 shows the tool screenshot. In the left panels it is possible to set the configuration of the cooling system, UPS and the other ancillary system of the

---

[1] [Online]   https://www.se.com/ww/en/work/solutions/system/s1/data-center-and-network-systems/trade-off-tools/data-center-efficiency-and-pue-calculator/.

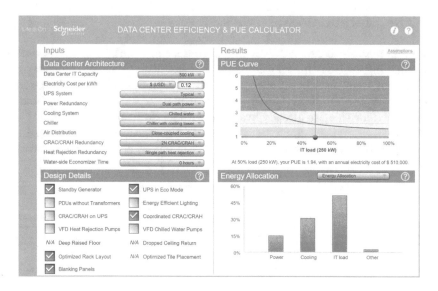

Fig. 1. The Schneider calculator tool.

facility, while in the right panels are shown, respectively, the PUE values (green curve) versus ITload, in terms of percentage of energy related to the maximum available to IT devices (in the specific case, 250 kW out to 500, so 50%), and the percentage of energy consumed by all subsystems.

Regarding the PUE curve, it is necessary to highlight that the vertical green line represent the working point of the system. Note that in the lower part of the horizontal axis (less than 40%) the curve has a steepest slope while in the other part the slope is less steep. With the given settings, the value of PUE is 1.94. If the load increases at 80%, the value of PUE will decrease to 1.65, while if the load decreases to 30% the PUE will increase to 2.47. Modifying the setting of the subsystems, the shape of the PUE curve will change reflecting the new configuration.

3.2   Effect of Energy Attacks on PUE Curve

The energy-related attacks on data centers may be directed to two category of devices: i) IT equipment (servers, routers, switches,...) or ii) service facilities of the DC (cooling systems, fans, UPS,...). In the first scenario, an attack will try to put as much load as possible on the IT infrastructure's components in an effort to prevent them from ever reaching low power consumption states [17] and hence to benefit of their energy-efficient design.

In the second case, also the ancillary equipment of the DC may be subjected to attacks: in today data centers all those devices are connected to the network due to needs of environmental monitoring and/or maintenance activities (often provided by external business organizations). Moreover, those devices are in most cases equipped with low-cost IoT-like boards. In this case, the attacks aim to modify the configuration of the devices; as an example, An attack against the cooling system could attempt to maintain the DC temperature below what is required, or distort the values of environmental parameters, or disable the free-cooling fans. In this situation, the result is a change of the shape of PUE curve.

The effect of both kinds of attacks (if successful) is represented in Fig. 2. A breach to IT infrastructure will not affect the configuration of the cooling system, so the Working Point (WP) will move on the same PUE curve (Curve A); if an IT system overload occurs, the WP moves forward on Curve A (1), lowering the PUE, while if i.e. there was a shutdown of some servers the WP moves backward, increasing the PUE. It will be noted that, although it is possible to detect an attack using this technique, a change in the ITload amount could be caused also by an increase (or decrease) of users' activities. This issue is outside the boundaries of this work and will be examined in further studies.

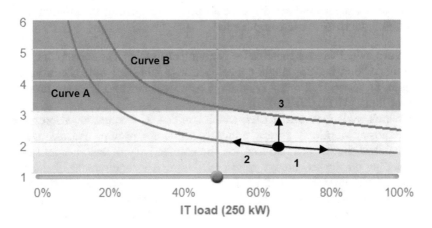

Fig. 2. Working point shift in case of attack to IT facilities (Curve A) and attack to cooling system (Curve B).

The case of a breach occurred on a service facility is different. In this case, the change of the setting reflects in a change of the PUE curve, and the behaviour of the system is represented by *a different PUE curve* (Curve B), and the working point of the system moves to (3) raising the PUE without any change to ITload. In this condition, if the ITload varies, the working point moves on Curve B rather than on Curve A, and it is possible to detect the change of the working condition of the data center. Hence, this technique deserves further investigation.

4   Simulation Environment

4.1   GreeenCloud Simulator

Greencloud [8] is an open source simulator written in C++ and OTcl for academic purposes aimed to explore the energy usage in data centers. It operates by
starting from the network level, hence allowing users to evaluate the performance
of different energy-saving policies and algorithms by accurately estimating the
energy usage patterns of network and server-related components.

GreenCloud is built on the top of the network simulator NS2[2], which was
developed by the Information Sciences Institute (ISI) at University of Southern
California. Despite the fact that NS2 it is no longer supported by its developing
institution, it is in use by a relevant part of the scientific community; in fact, a
recent survey [21] estimated that circa 14% of the scholars still use NS2 in their
scientific work, due to the availability of a number of libraries, components and
the support of a wide range of protocols [6].

The basic GreenCloud architecture is shown in Fig. 3. In the simulator there
are to highlight three basic components:

- **Servers**, which are responsible for carrying out tasks. The server components
  in GreenCloud feature single core nodes with different job scheduling techniques, an associated memory/storage resource size, and a preset processing
  power limit (MIPS/MFLOPS);
- **Network switches and links**, which compose the communication network
  and transfer workloads to any compute server for prompt execution. Different cabling choices are required for the connectivity of switches and servers
  depending on the technology used and the quality requirement of the links;
- **Workloads**, used for modeling of different cloud services and are represented
  as a set of jobs in which each job is divided into a group of tasks, which
  may start to execute independently or in dependence on the results of other
  activities.

It should be noted that, despite its effectiveness in energy monitoring, Green-
Cloud suffers of scalability issues (inherited by NS2), so the simulation time is
quite long, and the amount of memory used is high, which can limit the size
and complexity of the simulated scenarios [13]. Moreover, GreenCloud do not
implement ancillary equipment of the data center, such as cooling systems, UPS,
and so on.

4.2   PUE Calculation and Data Center Design

Since GreenCloud do not implement cooling system objects, it has been chosen
to estimate the values of PUE in function of the ITload by using the Schneider
calculator previously described using the following configuration:

---

[2] http://www.isi.edu/nsnam/ns/.

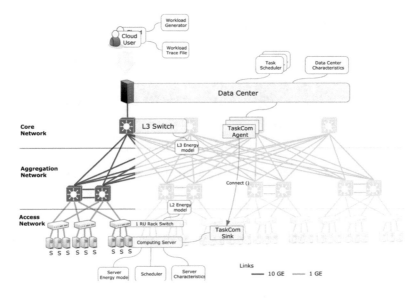

Fig. 3. Basic GreenCloud Architecture (adapted from [8]).

- Maximum power for IT equipment: 500 kW
- Power redundancy: dual path power
- Cooling system: chilled water
- Air distribution: close-coupled cooling
- CRAC/CRAH Redundancy: single path

The Curve A of the Fig. 4 is the PUE curve for the system in regular operating conditions. In the configuration shown it has been supposed that the CRAC/CRAH subsystem is configured with a *COLD* redundancy, so the other CRAC devices are not active, waiting for possible failures. In order to represent the behaviour of the cooling system after a successful attack, it has been supposed that the configuration of the cooling system has been changed by putting the cold spare CRAC/CRAH in ON status. In this case, the PUE curve changes as shown in Fig. 4 (Curve B).

Due to the fact that the GreenCloud simulator do not implement any feature related to the cooling system, the values obtained with the Schneider calculator have been used to perform a curve fitting[3], and the functions obtained for the PUE trends for the two different curves are the following:

$$PUE_A = 725.31 * ITload^{-0.575} \tag{2}$$

$$PUE_B = 1337.4 * ITload^{-0.623} \tag{3}$$

---

[3] In order to perform the curve fitting, it has been used Excel "power" function.

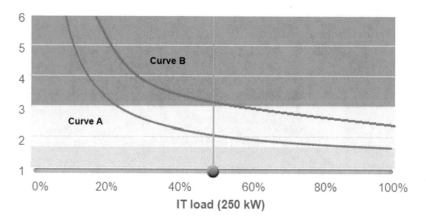

Fig. 4. PUE curves for regular operating conditions (Curve A) and after a successful cooling systems attack (Curve B).

5   The Example Scenario

The simulation scenario is a data center implementation using GreenCloud. The architecture simulated is a basic three-tiered, as shown in Fig. 3. The scenario simulates 144 servers, 6 switches at the access layer, 2 switches at the aggregation layer, and 1 switch at the core layer.

In the three-tier architecture each server is featured with a 4-core CPU, 8 GB RAM, and a 500 GB hard drive. Each CPU core provides a computing power of 150015 MIPS resulting in a total of 600060 MIPS per single server, that consumes 201 W under maximum load and 50% of the maximum (100.5 W) during idle periods. Furthermore, the energy profile of the core and aggregation layer switches is characterized by a power consumption of 1558 W for chassis, 1212 W for line cards, and 27 W for ports, while for the access layer switches, the power consumption amounts to 146W for chassis and 0.42 W for ports. Finally, in the access layer, 1 GE links are employed, while in the aggregation and core layers, 10 GE links are used.

While we have a clear indication of a single server maximum energy consumption, regarding the switches, the documentation of the simulator and the source code did not provide a clear indication of their maximum achievable energy consumption. To overcome this issue, multiple simulations were carried out with all energy-saving mechanisms disabled, ensuring that switches operated at their highest power levels. These simulations demonstrated that core and aggregation layer switches consume 2824W, while access layer switches consume 166W at their maximum load (all ports active).

The simulation has been performed for a ITload varying between 10% and 100% of maximum load. Figure 5 shows the comparison between the PUE values in case of regular and attacked cooling system. It can be easily seen that the difference between the values is bigger when the ITload value is low due to the shape of PUE curve (Fig. 4).

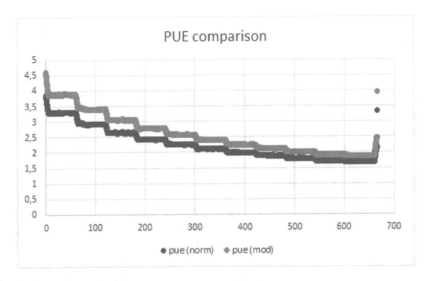

Fig. 5. PUE values comparison vs. simulation time.

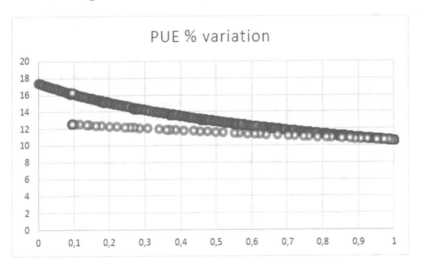

Fig. 6. Percentage of PUE difference vs. ITload%.

Despite this, it is possible to identify the difference between PUE values also in case of an high value of ITload. The following Fig. 6 shows the minimum and maximum difference between PUE values, and the difference have the maximum value of 17.4% and the minimum of 10.5%.

The simulation results show that the approach proposed could be effectively used for the detection of attacks, or more in general, anomalies, affecting cooling system in data centers. It is also to be noted that, if the change of shape of

PUE curve was not caused by an attack (in other words, a false positive), it stills deserve attention by system administrators, as it indicates a failure of the cooling system or other ancillary systems in DC.

6   Conclusions and Future Work

Looking at simulation results it can be concluded that the proposed approach can provide a valid and low cost option in order to detect attacks directed on cooling systems and other ancillary equipment of a data center.

It is also to highlight that the main limitation of this work is due to the use of Schneider calculator to estimate the relationship between the PUE indicator and the energy supplied to IT devices. In fact, only cooling solutions provided by Schneider can be experimented, and the features of the calculator are limited.

Therefore, future works will be focused in two directions. Firstly, it is necessary to experiment with the real devices that will be used to measure energy (watt-meters, etc.) and to implement a tool for measurement to be used in an anomaly detection framework. On the other side, GreenCloud lacks a module for managing the thermal aspects of the Data Center and for PUE calculation, therefore, it would be useful to study and implement such a module for handling these aspects.

Acknowledgment. This work was partially supported by project SERICS (D43C22003050001 - SPK 4 "Operating Systems and Virtualization Security" - SecCo) under the NRRP MUR program funded by the EU - NGEU, and it is part of the research activity realized within the project "Federated Learning for Generative Emulation of Advanced Persistent Threats (FLEGREA)", Bando PRIN 2022. This work was also partially supported by project SERICS (D43C22003050001 - SPK 3 "GEneralized Real-time On-line National Internet MOnitoring" - GERONIMO) under the NRRP MUR program funded by the EU - NGEU, and it is part of the research activity realized within the project PON "Ricerca e Innovazione" 2014-2020, action IV.6 "Contratti di ricerca su tematiche Green", issued by Italian Ministry of University and Research.

References

1. Avelar, V., Azevedo, D., French, A., Power, E.N.: PUE: a comprehensive examination of the metric. White paper **49** (2012)
2. Chlela, M., Mascarella, D., Joós, G., Kassouf, M.: Fallback control for isochronous energy storage systems in autonomous microgrids under denial-of-service cyberattacks. IEEE Trans. Smart Grid **9**(5), 4702–4711 (2018). https://doi.org/10.1109/TSG.2017.2667586
3. Chung, K., Formicola, V., Kalbarczyk, Z.T., Iyer, R.K., Withers, A., Slagell, A.J.: Attacking supercomputers through targeted alteration of environmental control: a data driven case study. In: 2016 IEEE Conference on Communications and Network Security (CNS), pp. 406–410 (2016). https://doi.org/10.1109/CNS.2016.7860528

4. Ficco, M.: Could emerging fraudulent energy consumption attacks make the cloud infrastructure costs unsustainable? Inf. Sci. **476**, 474–490 (2019). https://doi.org/10.1016/j.ins.2018.05.029, https://www.sciencedirect.com/science/article/pii/S0020025518303955

5. Ficco, M., Palmieri, F.: Introducing fraudulent energy consumption in cloud infrastructures: a new generation of denial-of-service attacks. IEEE Syst. J. **11**(2), 460–470 (2017). https://doi.org/10.1109/JSYST.2015.2414822

6. Ghazi, M.U., Khan Khattak, M.A., Shabir, B., Malik, A.W., Sher Ramzan, M.: Emergency message dissemination in vehicular networks: a review. IEEE Access **8**, 38606–38621 (2020). https://doi.org/10.1109/ACCESS.2020.2975110

7. Grid, T.G.: Green grid data center power efficiency metrics: Pue and dcie, the green grid. White Paper **6**, 1–16 (2008)

8. Kliazovich, D., Bouvry, P., Khan, S.U.: Greencloud: a packet-level simulator of energy-aware cloud computing data centers. J. Supercomput. **62**, 1263–1283 (2012)

9. Kurpicz, M., Orgerie, A.C., Sobe, A.: How much does a VM cost? energy-proportional accounting in vm-based environments. In: 2016 24th Euromicro International Conference on Parallel, Distributed, and Network-Based Processing (PDP), pp. 651–658 (2016). https://doi.org/10.1109/PDP.2016.70

10. Li, J., Jurasz, J., Li, H., Tao, W.Q., Duan, Y., Yan, J.: A new indicator for a fair comparison on the energy performance of data centers. Appl. Energy **276**, 115497 (2020). https://doi.org/10.1016/j.apenergy.2020.115497, https://www.sciencedirect.com/science/article/pii/S0306261920310096

11. Lin, J.C., Leu, F.Y., Chen, Y.p.: Analyzing job completion reliability and job energy consumption for a general mapreduce infrastructure. Journal of High Speed Networks **19**, 203–214 (2013). https://doi.org/10.3233/JHS-130473

12. Long, S., Li, Y., Huang, J., Li, Z., Li, Y.: A review of energy efficiency evaluation technologies in cloud data centers. Energy and Buildings **260**, 111848 (2022). https://doi.org/10.1016/j.enbuild.2022.111848, https://www.sciencedirect.com/science/article/pii/S0378778822000196

13. Mansouri, N., Ghafari, R., Zade, B.M.H.: Cloud computing simulators: a comprehensive review. Simul. Model. Pract. Theory **104**, 102144 (2020)

14. Masanet, E., Shehabi, A., Lei, N., Smith, S., Koomey, J.: Recalibrating global data center energy-use estimates. Science **367**(6481), 984–986 (2020)

15. Mastroianni, M., Palmieri, F.: Energy-aware optimization of data centers and cybersecurity issues. In: 2022 IEEE International Conference on Dependable, Autonomic and Secure Computing, International Conference on Pervasive Intelligence and Computing, International Conference on Cloud and Big Data Computing, International Conference on Cyber Science and Technology Congress (DASC/PiCom/CBDCom/CyberSciTech). pp. 1–7 (2022). https://doi.org/10.1109/DASC/PiCom/CBDCom/Cy55231.2022.9927965

16. Palmieri, F., Ricciardi, S., Fiore, U.: Evaluating network-based dos attacks under the energy consumption perspective: new security issues in the coming green ICT area. In: 2011 International Conference on Broadband and Wireless Computing, Communication and Applications, pp. 374–379. IEEE (2011)

17. Palmieri, F., Ricciardi, S., Fiore, U., Ficco, M., Castiglione, A.: Energy-oriented denial of service attacks: an emerging menace for large cloud infrastructures. J. Supercomput. **71**(5), 1620–1641 (2015). https://doi.org/10.1007/s11227-014-1242-6

18. Park, K., et al.: An advanced persistent threat (apt)-style cyberattack testbed for distributed energy resources (der). In: 2021 IEEE Design Methodologies Conference (DMC), pp. 1–5 (2021). https://doi.org/10.1109/DMC51747.2021.9529953

19. Shaikh, A., Uddin, M., Elmagzoub, M.A., Alghamdi, A.: PEMC: power efficiency measurement calculator to compute power efficiency and co2 emissions in cloud data centers. IEEE Access **8**, 195216–195228 (2020). https://doi.org/10.1109/ACCESS.2020.3033791

20. Wu, Z., Xie, M., Wang, H.: On energy security of server systems. IEEE Trans. Dependable Secure Comput. **9**(6), 865–876 (2012). https://doi.org/10.1109/TDSC.2012.70

21. Yugha, R., Chithra, S.: A survey on technologies and security protocols: reference for future generation IoT. J. Netw. Comput. Appl. **169**, 102763 (2020) https://doi.org/10.1016/j.jnca.2020.102763, https://www.sciencedirect.com/science/article/pii/S108480452030237X

A Method for Estimating the Number of Diseases in J-MID Database: Application to CT Report

Koji Sakai[✉], Yu Ohara, Yosuke Maehara, Takeshi Takahashi, and Kei Yamada

Kyoto Prefectural University of Medicine, Kawaramachi Hirokoji Agaru Kajiicho, Kyoto, Japan
{sakai3,y-ohara,ymaehara,kei}@koto.kpu-m.ac.jp

Abstract. Databases containing medical images play a crucial role in advancing healthcare and broadening our understanding of medical information. They serve essential functions, aiding in diagnosis, facilitating medical training, ensuring quality management, contributing to research on the prevalence of diseases and treatment outcomes. A comprehensive grasp of disease nomenclature within these databases offers a significant advantage, enabling deeper insights into various disease types and distribution patterns. Nevertheless, precisely quantifying the myriad of distinct disease names within image databases poses a significant challenge in numerous healthcare institutions. To address this, our study aimed to estimate disease counts by systematically extracting disease names matching registered medical conditions. This was achieved using a created predicates lexicon specific to disease names in image diagnostic reports written in Japanese. Utilizing this predicate lexicon, our study successfully extracted disease-affirming sentences with high accuracy, demonstrating a sensitivity of $87.9 \pm 0.8\%$, specificity of $93.0 \pm 0.6\%$, and accuracy of $90.5 \pm 0.4\%$. These findings highlight the importance of linguistic tools and systematic approaches in managing medical image databases, ultimately enhancing the efficiency of healthcare processes and the quality of research in the field.

1 Introduction

Databases containing medical images play a crucial role in advancing healthcare and broadening our understanding of medical information. These databases aid healthcare professionals by providing diagnostic assistance, facilitating variable trainings, enabling the medical image analysis, ensuring quality management, and supporting research on treatment outcomes based on images. By this contribution, they promote the advancements in the field of medicine. The construction of purpose-specific image databases has already advanced, with a global database established for Alzheimer's disease. Additionally, efforts are underway to establish national databases to address racial specificity.

In Japan, the development of a National Medical Image Database is also underway [1]. The Japan-Medical Image Database (J-MID) is a database established by the Japan Radiological Society (JRS) in 2018 to aggregate and centrally manage medical images taken at domestic healthcare institutions. Through J-MID, the JRS is actively promoting research and development efforts aimed at improving radiological healthcare. This

© The Author(s), under exclusive license to Springer Nature Switzerland AG 2024
L. Barolli (Ed.): EIDWT 2024, LNDECT 193, pp. 185–193, 2024.
https://doi.org/10.1007/978-3-031-53555-0_18

includes not only the utilization of image diagnostic support AI technologies [2] but also advancements in the accuracy and efficiency of image diagnosis, appropriate utilization of imaging devices and systems, image standardization, and radiation exposure management. Currently, there are 10 facilities in Japan participating as data providers and 14 facilities as collaborative research institutions within J-MID. We also actively participate in research and development as J-MID users.

Within the J-MID database, achieving precise determination of the number of disease names and obtaining this information has proven to be elusive. Although the feasibility of structured reporting is being contemplated, practical realization has not been yet achieved [3]. Hence, in the selection of research targets, it becomes necessary to repeatedly verify data counts for specific diseases. This makes it challenging to choose target diseases solely for machine learning study based on data quantity. By using searching query within the database, it is possible to look up words like disease names mentioned in the reports. However, with this query, it is only indicated whether the presence of a disease name is affirmative, and it cannot be discerned whether it carries a positive or negative meaning. While there are techniques available to analyze emotions conveyed in text data and forecasting the presence of positive or negative aspects [4], a challenge arises when disease names are expressed in negative terms, for which appropriate methods are currently lacking [5].

In Fig. 1, we show the conceptual framework, which can be used in many Japanese hospitals where the quantity of disease names linked to images remains unknown. Consequently, the effective utilization of stored image data for machine learning and making decision model poses a significant difficulty. Initially, we identified disease names corresponding to the conditions registered in the medical diagnosis master. Nevertheless, as we solely extracted disease names in the first step, discerning whether they were employed affirmatively or negatively proved challenging. Then, we created a lexicon and ascertained the context in which disease names appeared in the reports, thereby approximating the count of mentioned diseases.

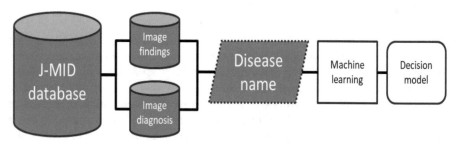

Fig. 1. Conceptual framework.

2   Materials and Methods

2.1   Computed Tomography Image Reports

The objective of this study is to analyze 184,465 computed tomography (CT) image reports archived in the J-MID database as of 2022. We received approval from the Ethics Committee of Kyoto Prefectural University of Medicine (ERB-C-1262-1).

2.2   Data Preprocessing and Computational Environment

The data processing procedures is shown in Fig. 2. By carrying out the cleansing procedure is eliminated the extraneous information and standardizing half-width and full-width characters. We considered 27,164 injuries and illnesses from the disease name master [6]. In cases when the findings and diagnoses contained commas, they were segregated into distinct sentences. By using MeCab [7], we carried out morphological analysis with medical terminology morpheme dictionary from Cabocha [8].

As computational environment, we used a desktop computer operating on Windows 11 with these specifications: Intel® Core™ i9-10900K CPU @3.7 GHz and 64 GB of RAM. We used as graphics card NVIDIA GeForce RTX 3090 and the e analysis was conducted through an in-house program developed with MATLAB R2023a.

2.3   Predicate Extraction

Through morphological analysis, a collection of chunks was obtained. Then, by subsequent dependency analysis was identified those directly related to injury and illness names, establishing a predicate thesaurus. Chunks associated with injury and illness names without apparent dependencies were classified as lacking a predicate.

2.4   Estimation of Number of Diseases

From predicates that directly modify the injury and illness names, we chose those with positive connotations. A radiologist with 8 years of professional experience selected the predicates and the combinations of injury and illness names with positive predicates were considered as the target disease count.

2.5   Extraction Performance Test

We formed two distinct sets considering 184,465 CT diagnostic reports compiled from the J-MID database in 2022. The first set included 1,000 report texts that unequivocally indicated the presence of a disease, encompassing the disease name.

The second set (comprised 1,000 report texts), which did not clearly identify the disease had 250 negative and 250 ambiguous reports. Then, a random subset of 1,000 report texts was further extracted to construct 10 test sets. We considered two methods to accurately identify and validate the presence of disease in each test set: (a) extracting diseases that match disease name master with an affirmative predicate; (b) extracting diseases that match disease name master and lacking a negative predicate.

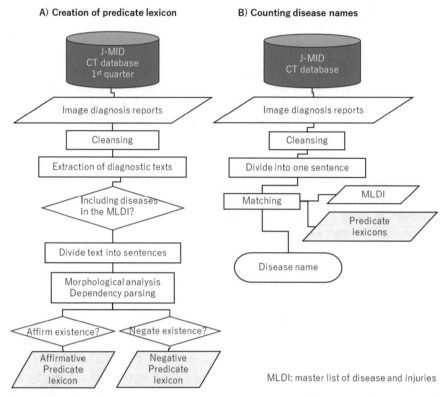

Fig. 2. Data processing procedures: A) Formulation of predicate lexicon through morphological analysis and dependency parsing. B) Enumeration of disease names using MLDI (master list of disease and injuries) and the predicate lexicon.

The sensitivity, specificity, accuracy, and precision were computed using the following formulas:

$$\text{Sensitivity} = TP / (TP + FN),$$
$$\text{Specificity} = TN / (TN + FP),$$
$$\text{Accuracy} = (TP + TN) / (TP + FP + TN + FN),$$
$$\text{Precision} = TP / (TP + FP),$$

where T: true, F: false, P: positive and N: negative.

3 Evaluation Results

3.1 Number of Extracted Disease Names from CT Report

Figure 3 illustrates the quantity of J-MID CT reports spanning from 2018 to 2022. Throughout the five-year duration, a total of 993,286 CT reports were recorded. The surges in reports observed in 2019 and 2020 were deemed to distinctly mirror the influence of the COVID-19 pandemic.

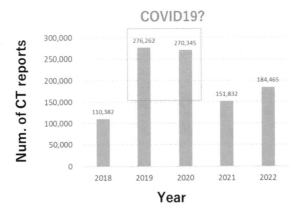

Fig. 3. The number of J-MID CT reports from 2018 to 2022.

3.2 Popular Predicates

In the J-MID CT reports of 2022, 1,209 affirmative predicates and 295 negative predicates were extracted. Table 1 displays the top 10 representative positive and negative predicates with high extraction frequencies (%). For reader comprehension, English translations of these predicates are also supplied. In expression conveying the same meaning, the usage of honorifics such as "-masu" and "-masen" is flexible, making Japanese a linguistically complex language with diverse expressions that very among radiologists.

There are 1,209 distinct types as affirmative predicates. Top 10 affirmative predicates accounted for 89.2% of the total. In contrast, for negative predicates, the top three alone constituted nearly 90%. These results show that affirmative predicates exhibit diversity, while there are only three words as negative predicates. Particularly, expression such as "mitome-masen" (I do not approve), "shiteki-deki-masen" (I cannot point out), and "ari-masen" (there is not) accounted for 84.1% of the negative predicates. Thus, the skillful incorporation of negative predicates could enhance the extraction of disease names.

3.3 Extracted Disease Names

Figure 4(a) and Fig. 4(b) display the prevalent disease names with a substantial number of extractions.

Inflammations were the most frequently extracted by the affirmative predicates with the highest number of 7,485 cases for pneumonia. Pneumonia predominantly manifested as interstitial and bronchopneumonia. The predominant type of cancer was lung cancer, with adenocarcinoma following closely, each yielding over thousand extracted cases, positioning them as promising candidates for machine learning analysis. In contrast to affirmative negative, lymph node diseases were extracted by negative predicates. The lesions, pleural effusion, cancers, ascites, tumors, pneumonia, mass were collectively extracted by affirmative and negative predicates.

Table 1. 10 notable instances of affirmative and negative predicates found in radiological CT reports within J-MID database.

#	Affirmative predicates			
	Japanese	English	Count	%
1	ari	there is / to exist	153,858	36.8
2	utagai-masu	I doubt	50,909	12.2
3	utagau	to doubt	47,089	11.3
4	mitome-masu	I acknowledge	44,556	10.7
5	kangae-masu	to think	23,445	5.6
6	utagai	doubt / suspicion	18,634	4.5
7	tokki-subeki	noteworthy	9,533	2.3
8	utagaware-masu	suspected	9,313	2.2
9	omoware-masu	it seems	8,281	2.0
10	tomonau	accompany with	7,279	1.7
#	**Negative predicates**			
	Japanese	**English**	**Count**	**%**
1	mitome-masen (CC)	I do not approve	176,248	48.0
2	shiteki-deki-masen	I cannot point out	73,622	20.0
3	ari-masen	there is not	58,996	16.1
4	henka-nasi	no change	21,115	5.7
5	mirare-masen (CC)	it is not observed	7,882	2.1
6	izyou-nasi	no abnormality	7,022	1.9
7	mirare-masen	it is not observed	6,397	1.7
8	mitome-masen	I do not approve	5,995	1.6
9	mitome-nai	I do not approve	5,728	1.6
10	shiteki-deki-nai	I cannot point out	4,273	1.2

CC: written by Chinese character.

3.4   Results of Extraction Test

The results of disease name extraction test are shown in Fig. 5. Ten test trials were executed to identify statements confirming the presence of the disease name. The sensitivity averaged 89.2 ± 1.2%, specificity was 59.5 ± 2.0%, accuracy stood at 74.4 ± 1.1%, and the precision was 68.8 ± 1.1%. The average measurement time was 46.22 ± 0.65 s. Also, ten test trials were conducted to identify statements which did not negate the disease name presence. The sensitivity averaged 87.9 ± 0.8%, specificity was 93.0 ± 0.6%, accuracy reached 90.5 ± 0.4%, and the precision was 92.7 ± 0.6%. The average measurement time for these trails was 47.98 ± 0.77 s.

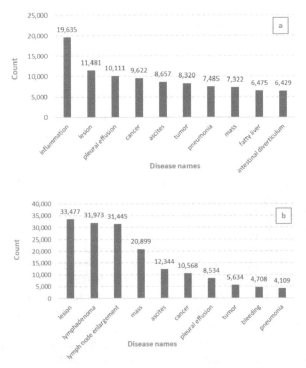

Fig. 4. Representative 10 examples disease names in radiological CT reports at J-MID (2022). a) extracted by affirmative predicates, b) extracted by negative predicates.

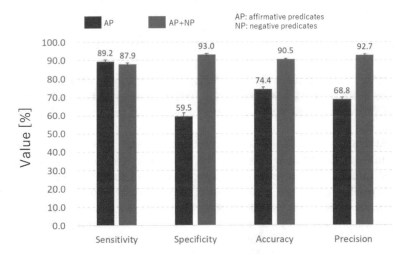

Fig. 5. Results of disease name extraction test.

These findings clearly demonstrated that the approach of avoiding predicates explicitly denying the presence of the disease name is effective in extracting disease names.

4 Discussions

In general, Japanese medical reports have some specific disease names and symptoms. However, the presence or absence of predicates indicating the presence of disease names has not been previously investigated. In this study, we explored a method to estimate the presence of disease names in the database by utilizing affirmative and negative predicates extracted from the radiology report texts by an experienced radiologist.

Among the predicates selected from the diagnostic sentences, there were 1,209 affirmative predicates and 295 negative predicates. Top 10 affirmative predicates represented 89.2% of the total, while the top 3 negative predicates alone contributed to 84.1%. The use of negative predicates indicated the potential for extracting the presence of a pathological condition. The reasons for the incomplete extraction results included cases where there were no corresponding predicates for certain disease names, such as "Postoperative esophageal cancer". We will consider these issues in our future studies.

There are some scenarios where dependency analysis does not perform optimally. In this research, the proposed lexicon underwent validation by a single radiologist. In the future work, we should consider multiple radiologists or some mechanisms for selection of appropriate predicate sets. Within the test dataset, the presence or absence of diseases was assessed on an individual sentence basis. However, original reports consist of multiple sentences, demanding a comprehensive evaluation that takes them into account. Additionally, it is worth exploring the possibility of considering an approach for the selection of images in the image database considering individual sentences.

5 Conclusions

We introduced a methodology for constructing a diagnostic predicate lexicon and estimating the quantity of disease names in the J-MID CT image database. This approach enables the efficient selection of data subsets conducive to machine learning within the J-MID database.

In this research work, the proposed lexicon underwent validation by a single radiologist. In the future work, we should consider multiple radiologists or some mechanisms for selection of appropriate predicate sets. Also, we found that some incomplete extraction results in cases where there were no corresponding predicates for certain disease names, such as "Postoperative esophageal cancer". We will consider this issue in our future work. We also will consider a new approach for selection of images in the image database considering individual sentences. While this study was focused on CT images from various facilities within a single database, in ongoing work we will address variances between facilities.

Acknowledgments. This research received support from Japan Society for the Promotion of Science under Grant Numbers JP21K07683 and JP21K07652.

References

1. J-MID. https://www.juntendo.ac.jp/news/13203.html. Accessed 25 Oct 2023
2. Toda, N., et al.: Deep learning algorithm for fully automated detection of small ($\leq$4 cm) renal cell carcinoma in contrast-enhanced computed. Invest. Radiol.Radiol. **57**, 327–333 (2022)
3. Nobel, J.M., van Geel, K., Robben, S.G.F.: Structured reporting in radiology: a systematic review to explore its potential. Eur. Radiol.Radiol. **32**(4), 2837–2854 (2022)
4. Evaluate sentiment. https://jp.mathworks.com/help/textanalytics/ug/analyze-sentiment-in-text.html. Accessed 25 Oct 2023
5. Japanese Sentiment Polarity Dictionary. http://www.cl.ecei.tohoku.ac.jp/Open_Resources-Japanese_Sentiment_Polarity_Dictionary.html. Accessed 25 Oct 2023
6. Injury and Illness Master, Health Insurance Claims Review & Reimbursement Services, Japan, https://www.ssk.or.jp/seikyushiharai/tensuhyo/kihonmasta/kihonmasta07.html#cms01. Accessed 25 Oct 2023
7. Kudo, T., Yamamoto, K., Matsumoto, Y.: Applying conditional random fields to japanese morphological analysis. In: Proceedings of the 2004 Conference on Empirical Methods in Natural Language Processing (EMNLP-2004), pp. 230–237 (2004)
8. Kubo et al.: http://taku910.github.io/cabocha/. Accessed 25 Oct 2023, 2023

Text Mining with Finite State Automata via Compound Words Ontologies

Alberto Postiglione[✉]

Business Science and Management & Innovation Systems Dep. (DISA-MIS),
University of Salerno, Via San Giovanni Paolo II, 84084 Fisciano, SA, Italy
ap@unisa.it

Abstract. The paper introduces an efficient text mining method using finite automata to extract knowledge domains from textual documents. It focuses on identifying multi-word units within terminological ontologies. Unlike simple words, multi-word units (e.g., *credit card*) possess a monosemic nature and are relatively few and diverse from each other, precisely pinpointing a semantic area. The algorithm, designed to handle challenges posed by even very long multi-word units composed of a variable number of simple words, integrates selected ontologies into a single finite automaton. At runtime, it efficiently recognizes and outputs the knowledge domain associated with each multi-word unit, even when they partially or completely overlap. Benefits of the system include minimal IT maintenance for ontologies, continuous updates without additional computational costs, and no need for software training. The proposed approach demonstrates robust performance on both short and long documents, validated through tests on multiple textual documents, with a specific test outlined in the paper.

1 Introduction

When a search engine retrieves unstructured natural language text documents that are partially incorrect or unrelated to an input query, it is evident that an imperfect linguistic investigation was conducted during the analysis and cataloging of these documents. In general, unstructured natural language texts pose challenges for computers to process due to a wide variety of formatting and standardization issues. These problems arise because natural language texts predominantly originate from humans with different backgrounds and cultures. As a result, these texts inevitably exhibit different *writing styles*, may contain *poorly formed expressions*, or present *spelling errors*. Furthermore, natural language is characterized by a significant presence of *polysemous words*. Finally, human beings often tend to *express emotions and feelings*; they can use *irony*, *metaphors* and other *figures of speech*. All of these features have the effect of hiding information beyond direct readability in the text. Understanding these hidden layers with technology remains a daunting challenge, still largely unsolved, so we must accept semantic compromises, even if they are acceptable.

The largest number of words occurring in a sentence can be of two main types: **simple words**, which are sequences of uninterrupted letters with meaning delimited by spaces, and **multi-word units (MWUs)** [8, 10, 28], composed of two or more words separated by spaces or other diacritical marks. The meaning of multi-word units is, to

L. Barolli (Ed.): EIDWT 2024, LNDECT 193, pp. 194–205, 2024.
https://doi.org/10.1007/978-3-031-53555-0_19

some extent, independent of that of their components. For example, the multi-word unit *mathematical analysis*[1] is *monosemic*, as it precisely identifies a precise semantic area or a well-defined object. In contrast, a simple word (*mathematics* or *analysis*, taken individually) is normally *polysemic*, having various meanings and belonging to many different semantic areas, often very distant from the actual meaning of that word in a specific context[2].

Text analysis and cataloging systems are primarily built upon sophisticated investigations into simple words. However, these systems may not be sufficient to fully capture all the semantics of a text. Established linguistic studies indicate that, in natural language, the majority of cognitive information is conveyed through multi-word units. A single word always requires inclusion in a sentence to be disambiguated (e.g., 'Alice did the *analysis*', but what kind of analysis?). In contrast, a multi-word unit retains its meaning regardless of the sentences and discourses in which it occurs and specifies its semantic content (e.g., 'Alice did the *market analysis*').

Multi-word units are regarded as *'units of fixed meaning'* with specific formal and semantic characteristics: they are not extemporaneous combinations of simple words and should not be exclusively addressed through statistical methods. Solely relying on statistical approaches may result in a semantic mismatch between the genuine, underlying meaning of a text and how it is perceived and categorized.

There is a close relationship between multi-word units and terminology. Specialized field lexicons are primarily composed of multi-word units, but this relationship extends beyond specialized lexical domains and encompasses all semantic areas. In a particular knowledge domain, the terminological multi-word units can be regarded as metadata for the corresponding ontology. In summary, for the effective retrieval of the semantic content in a textual document written in natural language, our initial linguistic theoretical hypothesis emphasizes the importance of focusing on multi-word units.

In this paper, we propose AUTOMETA, a terminological text-mining system based on the efficient use of finite automata and a consistent natural language formalization. This formalization is essentially built on the universal concepts of *'unit of meaning'* and *'multiword unit'*. AUTOMETA identifies multi-word units within natural language texts and associates a knowledge domain with the text. It relies on the correspondence between a digitized text and an ontology composed of multi-word units. We tested a prototype of the system on 100 input documents, interacting with a union of one generic multi-word unit ontology and 22 terminological multi-word unit ontologies, comprising a total of 21,396 entries. The system accurately identified knowledge domains in over 90% of the analyzed texts. AUTOMETA implements algorithms based on theoretical computer science research on string matching via finite automata [3,6,11,18]. Its linguistic roots refer to studies on Lexicon-Grammar (LG) [12,13,16,17,24,26].

---

[1] *"disciplinary sector of mathematics that studies the functions defined on sets of real or complex numbers with real or complex values and, in any case, properties connected with the notion of limit, derivative, differential equation, integral, series"* (Enciclopedia online Treccani, https://www.treccani.it/).

[2] For example, for the word "analysis": *mathematics*, but also *psychology, statistics, medicine, economy, physics, chemistry, optics*, and so on; similarly, for the word "mathematical": *arithmetic, algebra, geometry, physics, statistics, economy, sounds, computer science*, and so on.

This paper is so structured: In Sect. 2, the main concepts of text mining are outlined. In Sect. 3, the features of AUTOMETA are described. In Sect. 4, the performance of AUTOMETA is discussed, including a comparison with other algorithmic methods addressing the same problem. In Sect. 5, an example of AUTOMETA running on real data is provided. In Sect. 6, conclusions are drawn, and suggestions for possible future research are offered.

2   Text Mining

Until recently, the number of textual documents has remained relatively small, allowing effective management through classical algorithmic methods. However, in recent times, there has been a significant surge in the volume of textual data. Factors such as the growth of corporate data and the widespread adoption of the Internet of Things (IoT) [4,5,7,27,33], coupled with the rapid expansion of digital text documents influenced by government policies promoting the dematerialization of paper documents and the increasing ubiquity of social media in the daily lives of billions of people, have all contributed to this phenomenon. This surge has resulted in an unstoppable growth of *unstructured*[3] or *semi-structured*[4] textual documents. As a result, it is widely recognized that a substantial amount of information is now stored in unstructured or semi-structured textual form. In [30,34], it is estimated that at least 80% of real-world data comprises unstructured or semi-structured natural language textual documents, such as books, news, research articles, web pages, emails, social media posts (chat, blog, forum), etc., with only a small percentage being structured and/or numeric data. Similarly, in [31], it is stated that over 85% of current data is unstructured or semi-structured natural language textual data. Regardless of the exact percentage, the volume of natural language textual data generated has now attained the characteristics of Big Data. These data are characterized by their enormity, heterogeneity, poor typing, and continuous generation. The sheer volume of data generated surpasses the capabilities of traditional databases and algorithmic methods, rendering it impossible for these methods to efficiently manipulate, manage, and process such data. This challenge is further compounded by the necessity to process these data swiftly, if not instantaneously. Consequently, there arises a need for robust data management systems capable of handling these large datasets and conducting real-time analytics.

Text Mining [1,2,20,31,32,34] is a comprehensive combination of various technologies that addresses challenges in different fields. Its primary goal is to automatically identify meaningful patterns, reveal new insights, uncover hidden relationships, and, in general, extract meaningful knowledge and information from unstructured or semi-structured text documents in natural language. Text mining is a kind of specialization of data mining [31]: while data mining is the process of finding patterns in structured data, text mining is the process of finding patterns in unstructured or semi-

---

[3] Unstructured textual data refers to text that is not organized in a predefined manner, such as free-form natural language textual documents, emails, social media posts, articles, and so on.

[4] Semi-structured textual data has some organization but lacks sufficient structure to be stored in a relational database, examples of which include XML, JSON, and HTML files.

structured documents. In general, unstructured text is easily manipulated by people, but it is very complex for computer programs to handle.

Text mining is continuously applied in industry, academia, web applications, internet and other fields [22]. Application areas like search engines, customer relationship management system, filter emails, product suggestion analysis, fraud detection, and social media analytics use text mining for opinion mining, feature extraction, sentiment, predictive, and trend analysis. Furthermore, text mining is applied in fields such as: Risk management [9], Predictive maintenance of industrial machines [25], financial domains [21], Healthcare [14], medicine [23], social networks [19], social media [29], education and training [15], and so on.

In [34] a text mining system is categorized into pre-processing, text representation, and four operational phases:

1. Pre-processing
2. Text Representation
3. Dimensionality Reduction
4. Features Extraction
5. Document Classification
6. Evaluation

During the initial phases (Phases 1 and 2), the objective is to transform unstructured text into structured features, aiming to normalize the input text to a standard format. This process involves cleaning the text data by excluding unnecessary characters and words. In [20] (Fig. 1) are reported the interaction among the four operational phases: Feature Extraction, Dimensionality Reduction, Classification and Evaluation.

Fig. 1. Text Mining systems phases (from [20])

3   AUTOMETA

AUTOMETA is an algorithmic-linguistic text mining system based on the efficient use of finite automata and a consistent natural language formalization that identifies terminological terms from an input text to determine, without human reading, the main knowledge domains of the text. The system operates by comparing the input text with terminological multi-word units (metadata) present in some ontologies.

AUTOMETA is configured as stand-alone software, embeddable within websites and portals for online use. The system is suitable for any type of digitized text. AUTOMETA traces its remote ancestry back to "Cataloga" [12, 13], also a finite-automata-based system. However, it differs drastically from Cataloga due to technological, algorithmic, coding, and interface choices.

3.1   Ontologies System

As mentioned above, information in any text is provided mainly by terminological multi-word units. AUTOMETA accesses terminological ontologies in which each entry is assigned an ontological identification consisting of tags (or labels) referring to one or more knowledge domains or semantic fields (metadata) in which the entry is commonly used and has an unambiguous meaning. We define this tagging as a lexical ontology, representing a unique and unambiguous relationship between a multi-word unit in an electronic terminology dictionary and one or more tags indicating the semantic domain(s) to which it belongs. In other words, it indicates the specific domain(s) of knowledge where that multi-word unit has a well-defined meaning. Our lexical ontologies do not serve encyclopedic purposes; rather, they provide the ontology used by a specific domain to unambiguously refer to an established, standardized, and universally shared body of knowledge. Each ontology was created and verified under the supervision of experts in the sector.

3.2   System Functionality

The system consists of a preliminary linguistic phase for the design and development of the IT data structure for the ontologies, and an algorithmic phase.

3.2.1   Preliminary Linguistic Step

In order for AUTOMETA to recognize a particular knowledge domain, the system requires that an ontology of terminological multi-word units is available for that domain. If the ontology is not available, it must be built; this operation must be performed only once and must be carried out in close contact with experts in the sector for which the ontology is being built. At runtime, the system offers the possibility of combining multiple ontologies from different knowledge domains and then integrating them into a single structure.

The preliminary Linguistic step consists of these elementary steps:

Step 1   multi-word units Choice and Normalization: Experts in the sector must identify all the multi-word units of the knowledge domain (metadata), choose the normalization parameters of the text[5], and normalize the data. In other words, they need to transform the set of multi-word units towards a standard format, following the normalization choices determined in the previous step.

Step 2   Association of the Semantic Domain to Each multi-word unit. The system builds the linguistic model that associates one or more semantic domains with each of the messages coming from the previous step.

Step 3   Ontology Construction: A digital data structure is built containing all possible multi-word units, in the form of an ontology in which each line contains a multi-word unit and its semantic domains.

---

[5] For example: the format of its capitalization, which special characters to keep in the text or with which character to replace them, and so on.

3.2.2 Algorithmic Step

The algorithmic process involves a one-time pre-processing phase and a two-stage processing phase for each input text: Matching and Analysis.

Pre-Processing: The system builds a finite automaton data structure, based on some user-selected ontologies.

Matching: The algorithm reads the input text character by character, navigating through the finite automaton (each character of the input text is a state of the FA). It identifies immediately and simultaneously all terminological multi-word units, even if partially or totally overlapped without revisiting previous characters.

Analysis: After reading the entire input text, the algorithm analyzes metadata and classifies the text based on its knowledge domains.

4   AUTOMETA Performances

The system reads the input text navigating a finite automaton. Upon reaching a state corresponding to the end of a multi-word unit, it outputs the associated semantic domain(s). The algorithms are highly performant, fast, and precise, adept at identifying all occurrences of multi-word units, even if partially or totally overlapping.

4.1   Algorithm Performances

Phase 1: Preprocessing

Initially, the user selects ontologies, and the system constructs a finite automaton in the computer's memory based on these choices. This occurs once per session and whenever reference ontologies change. The time complexity of the preprocessing phase (i.e. the total number of elementary algorithmic steps, hence the number of state transitions on the finite automaton) is $O(m)$ [3], where m is linearly proportional to the sum of the lengths of all entries in the selected ontologies (n elements), as given by:

$$m = \sum_{i=1}^{n} length(a_i)$$

Phase 2: Matching

The automaton continuously reads lines of input text, character by character, never stepping back through the input text. The characters guide it through the finite automaton; when it encounters a state corresponding to the end of a multi-word unit, it emits its associated semantic domains. All terminological multi-word units are recognized simultaneously in a single pass (i.e., with only one reading of the input text) even if partially or totally overlapping, however long the multi-word units are. Thus, the algorithm works in linear time with respect to the number of characters in the input text [3]; therefore, its performance does not depend on the size of the ontologies (that is, the set of all possible multi-word units). It also runs on a data structure completely on the computer's main memory.

Phase 3: Analysis

At the end of the Matching phase, i.e. when the input text has been completely read, the system counts the frequency of each of the semantic fields that have emerged and proposes a classification of the input text knowledge domain based on the most frequent ones.

4.2   Running System Requirements and Performances

There are no particular hardware or software requirements, but the architecture on which it runs is standard. To carry out the tests, specifically, we used the hardware/software configuration described below. The hardware used for our tests is a DELL INSPIRON 16 laptop with CPU 11th Gen Intel(R) Core(TM) i7-11800H @ 2.30 GHz 2.30 GHz, 32.0 GB of installed RAM and OS Windows 11 Home, 64-bit, x64, ver. 22H2. Neither a particular video card nor mass memory beyond the standards are required. The technology used to develop this software is "Embarcadero Delphi 11.2". Delphi is a powerful RAD (Rapid Application Development) visual software development tool based on an Object Programming Language. While the program was running, the computer was mainly running system applications.

During the startup phase, the software implements all the chosen ontologies in a single Finite Automaton, which resides completely in RAM. We currently have 1 generic multi-word unit ontology and 22 terminological multi-word unit ontologies containing a total of 21,396 entries, and the time needed to build the finite automaton for the entire set of ontologies ranges from 10 to 12 s, depending on the overhead of the system.

The time to pre-process a text ranges from 0.1 s for the smallest text (103 words and 644 characters) up to 2.86 s for the longest text (556,907 words and 3,449,325 characters).

The processing time for a text varies from 0.1 s for the smallest text up to 1.35 s for the longest text.

The maximum space used by the entire system, including the Finite Automaton (23 ontologies and 21,396 elements), is 709 MB for the longest text.

4.3   Comparison with Other Methods

Current analysis systems for unstructured texts in natural language rely mainly on frequency-based statistical surveys, even highly sophisticated ones, applied to simple words. However, these systems are not sufficient to efficiently recover the full semantic content of a document. For a more precise automatic analysis of a text, it becomes crucial to focus on identifying terminological multi-word units.

The real challenge of this approach lies in the fact that multi-word units are not easy to process, given some peculiarities:

- **A multi-word unit can be very long and is composed of any number of simple words**: multi-word units lack explicit terminators to signal their start and end, unlike white-space or punctuation that separates simple words. Consequently, when reading an input text, it remains uncertain how many consecutive elementary words must be read to form a multi-word unit;
- they can be **partially or totally overlapping** one another;
- **in texts they occur with low frequency** compared to single words; therefore, in an overall statistical analysis, they tend to disappear. Thus, statistics-based parsers may not correctly recognize even very frequent multi-word units as cohesive units of unitary meaning.

Algorithmically, a classical database-like method based on the matching of words against a dictionary is very time-consuming and depends on the size of the dictionary;

the larger it is, the longer you have to wait for the answer. In fact, the identification of a multi-word unit in a text with n words requires

- $O(n^2)$ searches in the ontology, because for each word we have to try the chains with all its successive words;
- each search in an ontology with m items costs $O(logm)$-accesses, presumably to a secondary memory, to an ordered data structure;

thus, the algorithm requires $O(n^2 logm)$ accesses to a sorted data structure for a text with n words. Furthermore, the ontologies must be sorted, and therefore, in case of deletion, insertion, or modification of an entry, they must be reordered. This costs at least $O(logk)$ operations, possibly involving the physical movement of entries from one memory area to another

With our proposal, we overcome the above problems. Our system:

- Works regardless of the length of the multi-word unit and the number of elementary words that make up a multi-word unit.
- Recognizes, without any additional computational effort, all partially or completely overlapping multi-word units.
- The entire ontology is in a finite automaton in central memory, so the access time to data is many orders of magnitude lower than secondary memory, even if it were solid-state memory.
- The processing time, on a text of k characters, requires $O(k)$ state transitions. Therefore, analyzing an input text is performed in linear time with respect to the number of characters in the text. Thus, the dictionary size could be very large without any problem.
- Works for both long and (very) short texts.
- Does not require any initial learning process but is able to recognize the semantic domain of a document without first having to read other documents of the same knowledge domain.
- The ontologies do not need to be sorted, so adding, changing, or deleting one or more entries is done at virtually no computational cost; we simply insert a new word at the end of the ontology or modify or delete a word without touching the others.
- Furthermore, it recognizes multi-word units directly while reading the input text, without providing any kind of feedback.
- involves a "small" pre-processing phase to normalize the formal spelling of words (the multi-word units are few and very different from each other)
- have few flected forms.

All these advantages make our method for the automatic recognition of multi-word units decidedly preferable to classical methods.

In conclusion, some of the desirable characteristics of our proposal are:

1. The system should limit pre-processing and dimensionality reduction as much as possible because they are time-consuming and error-generating tasks, as well as being tied to the specific natural language.
2. The system should be able to auto-correct as many errors as possible in "useful" but ill-written pieces of text.

3. The system should work well on both very small (a few hundred words) and very large input documents.
4. The system should detect as few "pieces of text" as possible, but each of which should carry a lot of semantic information and have few synonymy relationships.
5. The system should work without first having to acquire other similar documents, or having acquired as few as possible.

5   A Sample of AUTOMETA Text Mining

We tested the system on 100 input documents interacting with the union of 1 generic multi-word unit ontology and 22 terminological multi-word unit ontologies containing a total of 21,396 entries. The system correctly recognized the knowledge domains for the vast majority of texts ($> 90\%$) that we analyzed, which seems like a good start, considering that the system is only in the prototypal phase and the ontologies are in mostly incomplete and in need of improvements.

In this paper, we report, by way of example, a test conducted on an Astronomy document[6] of 625 words, 4,231 characters, and 12 paragraphs. We conducted the test first on the entire document and then on four much smaller subsets of it: the second test on the first two-thirds of the document, while the third test on the first third; the fourth test on the central part and the last on the union of the first and last paragraphs of the document. Note that even with very few words (less than 200), the system hits the target. The result of the analysis for the whole document is summarized in Fig. 2, while the summary of the 5 tests is in Table 1.

AUTOMETA (Rel. 0.2)

File name: Esoterre 1.txt

Total Number of Lines in the input text:	12
Total Number of Words in the input text:	647
Total Number of Chars in the input text:	4.270
Total Number of Compound Words in the Ontologies:	21.396
Total Number of Compound Words in the input text:	10
Total Number of Different Comp. Words in the input text:	7

The Text talks about (sorted by freq):

Astronomy, Architecture, Optics, Aircraft

Astronomy	7	70.0%
Optics	1	10.0%
Architecture	1	10.0%
Aircraft	1	10.0%

COMPOUND WORDS (sorted by name)

Compound Words	TIMES	Ontology
anni luce	3	Astronomy
chiave di volta	1	Architecture
corpo celeste	1	Astronomy
immagini dirette	1	Optics
missioni spaziali	1	Aircraft
nana rossa	1	Astronomy
sistema solare	2	Astronomy

Fig. 2. Example of AUTOMETA output

---

[6] 'Two potentially habitable exo-earths discovered 16 light years from us', 'Le Scienze' (Italian edition of 'Scientific American') Dec -15-2022 (https://www.lescienze.it/news/).

Table 1. Summary of the 5 subtests for the example

Paragraphs	Words	Chars	Knowledge Domains
12	647	4,270	Astronomy (7), Architecture (1), Aircraft (1), Optics (1)
10	440	2,884	Astronomy (7), Aircraft (1), Optics (1)
4	168	1,093	Astronomy (6)
7	186	1,220	Astronomy (2)
4	182	1,178	Astronomy (3), Architecture (1)

6   Conclusions and Future Research Perspectives

This paper introduces an efficient algorithmic-linguistic text mining approach that identifies knowledge domains of a text by detecting terminological multi-word units and comparing them with ontologies. Each multi-word unit is linked to a knowledge domain (metadata). The algorithm uses a finite state automaton for all ontologies, emitting knowledge domains upon recognition of multi-word units. After reading the entire input text, the algorithm analyzes the metadata issued to associate one or more knowledge domains with the document. The approach is applicable to both long and short texts.

The algorithm is highly efficient, fast, and precise due to the following reasons:

- The running time is independent of the number of ontology entries, as it reads the input text character by character without revisiting previous characters. Therefore, the execution time is solely dependent on the number of characters in the input text.
- All terminological multi-word units are recognized in a single pass, even if they overlap partially or entirely.
- Modification of each ontology incurs no computational cost, as they are not sorted.
- The system operates on a data structure entirely in RAM.

In conclusion, we state that our approach, focused on semantics-based text mining and information retrieval, is highly suitable for achieving the goals of automatic recognition, and related semantic cataloging, of (semi) unstructured natural language texts.

To attain efficient semantic-based text mining, special consideration should be given to the analysis and swift capture of linguistic changes, primarily arising from the evolution of civilization. Thus, potential future research perspectives could be directed towards:

- **Ontologies Structure Update**: Focus on updating existing ontologies and creating new electronic terminology ontologies for various semantic domains, such as e-government, biomedicine, ecological transition, etc. The flexibility of our approach allows modifications without the need to reorder ontology items.
- **Testing and Validation**: Conduct thorough testing and validation of the existing ontologies by applying them to more extensive corpora.
- **Integration into Websites and Portals**: Explore integrating ontologies and software tools into websites and portals, enabling internet users to test the system and provide feedback on its quality and usability.

- **Automatic Translation of multi-word units**: Explore the addition of a translation feature to our approach, allowing the translation of multi-word units from one language to another. This capability is crucial as translating multi-word units is not always straightforward by translating their individual components, since a multi-word unit is a whole. The rapid translation feature could address errors in technological and scientific translation caused by the lack of reliable terminological bilingual electronic glossaries/dictionaries.

References

1. Aggarwal, C.C., Zhai, C.: Mining Text Data. Springer, Boston (2012)
2. Aggarwal, C.C., Zhai, C.: A survey of text classification algorithms. In: Aggarwal, C.C., Zhai, C., (eds.) Mining Text Data, pp. 163–222. Springer US, Boston (2012). Cited by: 1243; All Open Access, Green Open Access
3. Aho, A.V., Corasick, M.J.: Efficient string matching: an aid to bibliographic search. Commun. ACM **18**(6), 333–340: Cited by: 2250. All Open Access, Bronze Open Access (1975)
4. Al-Fuqaha, A., Guizani, M., Mohammadi, M., Aledhari, M., Ayyash, M.: Internet of things: a survey on enabling technologies, protocols, and applications. IEEE Commun. Surv. Tutorials **17**(4), 2347–2376 (2015). Cited by: 5626. All Open Access, Bronze Open Access
5. Atzori, L., Iera, A., Morabito, G.: The internet of things: A survey. Comput. Networks **54**(15), 2787–2805 (2010). Cited by: 10936
6. Boyer, R.S., Strother Moore, J.: A fast string searching algorithm. Commun. ACM **20**(10), 762–772 (1977). Cited by: 1769. All Open Access, Bronze Open Access
7. Boyes, H., Hallaq, B., Cunningham, J., Watson, T.: The industrial internet of things (iiot): an analysis framework. Comput. Ind. **101**, 1–12 (2018). Cited by: 824. All Open Access, Green Open Access, Hybrid Gold Open Access
8. Calzolari, N., et al.: Towards best practice for multiword expressions in computational lexicons. In: Proceedings of the 3rd International Conference on Language Resources and Evaluation, LREC 2002, pp. 1934–1940. European Language Resources Association (ELRA) (2002). Cited by: 88
9. Chu, C.-Y., Park, K., Kremer, G.E.: A global supply chain risk management framework: an application of text-mining to identify region-specific supply chain risks. Adv. Eng. Inform. **45** (2020). Cited by: 69
10. Constant, M., et al.: Multiword expression processing: a survey. Comput. Linguistics, **43**(4), 837–892 (2017). Cited by: 129. All Open Access, Gold Open Access, Green Open Access
11. Crochemore, M., Hancart, C., Lecroq, T.: Algorithms on strings, vol. 9780521848992. Cambridge University Press (2007). Cited by: 391
12. Elia, A., Monteleone, M., Postiglione, A.: Cataloga: a software for semantic-based terminological data mining. In: 1st International Conference on Data Compression, Communication and Processing, IEEE, Palinuro (SA), June 21-24, pp. 153–156. IEEE Computer Society, June 2011
13. Elia, A., Postiglione, A., Monteleone, M., Monti, J., Guglielmo, D.: Cataloga: a software for semantic and terminological information retrieval. In: ACM International Conference Proceeding Series, pp. 1–9 (2011)
14. Fenza, G., Orciuoli, F., Peduto, A., Postiglione, A.: Healthcare conversational agents: Chatbot for improving patient-reported outcomes. Lecture Notes in Networks and Systems, vol. 661, LNNS, pp. 137–148 (2023)
15. Ferreira-Mello, R., André, M., Pinheiro, A., Costa, E., Romero, C.: Text mining in education. Wiley Interdisciplinary Reviews: Data Mining and Knowledge Discovery 9(6) (2019). Cited by: 102

16. Gross, M.: The construction of electronic dictionaries; [la construction de dictionnaires électroniques]. Annales Des Télécommunications **44**(1-2), 4–19 (1989). Cited by: 21

17. Gross, M.: The use of finite automata in the lexical representation of natural language. In: Gross, M., Perrin, D. (eds.) LITP 1987. LNCS, vol. 377, pp. 34–50. Springer, Heidelberg (1989). https://doi.org/10.1007/3-540-51465-1_3

18. Hakak, S.I., et al.: Exact string matching algorithms: Survey, issues, and future research directions. IEEE Access **7**, 69614–69637 (2019). Cited by: 61. All Open Access, Gold Open Access, Green Open Access (2019)

19. Irfan, R., et al.: Survey on text mining in social networks. Knowl. Eng. Rev. **30**(2), 157–170 (2015). Cited by: 87. All Open Access, Bronze Open Access (2015)

20. Kowsari, K., et al.: Text classification algorithms: a survey. Information (Switzerland), 10(4) (2019). Cited by: 587. All Open Access, Gold Open Access, Green Open Access (2019)

21. Shravan Kumar, B., Ravi, V.: A survey of the applications of text mining in financial domain. Knowl.-Based Syst. **114**, 128–147 (2016). Cited by: 161

22. Liao, S.-H., Chu, P.-H., Hsiao, P.-Y.: Data mining techniques and applications - a decade review from 2000 to 2011. Expert Syst. Appl. **39**(12), 11303–11311 (2012). Cited by: 511

23. Luque, C., Luna, J.M., Luque, M., Ventura, S.: An advanced review on text mining in medicine. Wiley Interdisciplinary Reviews: Data Mining and Knowledge Discovery, 9(3), 2019. Cited by: 40

24. Monteleone, M.: Nooj for artificial intelligence: an anthropic approach. Commun. Comput. Inf. Sci. **1389**, 173–184 (2021). Cited by: 1

25. Nota, G., Postiglione, A., Carvello, R.: Text mining techniques for the management of predictive maintenance. In: Padovano, A., Longo, F., Affenzeller, M. (ed.) 3rd International Conference on Industry 4.0 and Smart Manufacturing, ISM 2021; 19-21 November 2021, vol. 200, pp. 778–792. Elsevier B.V., 2022. Cited by: 8; All Open Access, Gold Open Access

26. Postiglione, A., De Bueriis, G.: On web's contact structure. J. Ambient. Intell. Humaniz. Comput. **10**(7), 2829–2841 (2019)

27. Qadri, Y.A., Nauman, A., Zikria, Y.B., Vasilakos, A.V., Kim, S.W.: The future of healthcare internet of things: a survey of emerging technologies. IEEE Commun. Surv. Tutorials **22**(2), 1121–1167 (2020). Cited by: 449

28. Sag, I.A., Baldwin, T., Bond, F., Copestake, A., Flickinger, D.: Multiword expressions: a pain in the neck for NLP. In: Gelbukh, A. (ed.) CICLing 2002. LNCS, vol. 2276, pp. 1–15. Springer, Heidelberg (2002). https://doi.org/10.1007/3-540-45715-1_1

29. Salloum, S.A., Al-Emran, M., Monem, A.A., Khaled Shaalan, A.: Survey of text mining in social media: Facebook and twitter perspectives. Adv. Sci. Technol. Eng. Syst. **2**(1), 127–133,: Cited by: 151. All Open Access, Gold Open Access (2017)

30. Mukund Tamizharasan, R.S., Shahana, Subathra, P.: Topic modeling-based approach for word prediction using automata. J. Critical Rev. **7**(7), 744–749 (2020). Cited by: 7. All Open Access, Bronze Open Access (2020)

31. Tandel, S.S., Jamadar, A., Dudugu, S.: A survey on text mining techniques. In: 2019 5th International Conference on Advanced Computing and Communication Systems, ICACCS 2019, pp. 1022–1026. Institute of Electrical and Electronics Engineers Inc., 2019. Cited by: 31

32. Usai, A., Pironti, M., Mital, M., Mejri, C.A.: Knowledge discovery out of text data: a systematic review via text mining. J. Knowl. Manage. **22**(7), 1471–1488 (2018). Cited by: 54. All Open Access, Green Open Access (2018)

33. Xu, L.D., He, W., Li, S.: Internet of things in industries: a survey. IEEE Trans. Ind. Inform. **10**(4), 2233–2243 (2014). Cited by: 3715

34. Zong, C., Xia, R., Zhang, J., Mining, T.D., Singapore, S.: All Open Access, Green Open Access (2021). Cited by: 23

Binary Firmware Static Vulnerability Mining Based on Semantic Attributes and Graph Embedding Network

Feng Tian[1], Baojiang Cui[1(✉)], and Chen Chen[2]

[1] Beijing University of Post and Telecommunications, Beijing Haidiandian District West, TuCheng Road 10, Beijing, China
{fourler,cuibj}@bupt.edu.cn

[2] Air Force Engineering University, No. 1, Changle East Road, Baqiao District, Xi'an, Shaanxi, China

Abstract. For static vulnerability detection technology, traditional machine learning vulnerability detection methods mostly use abstract syntax trees as code representations. This will ignore semantic information such as code logical structure and data flow direction, which will ultimately affect the accuracy of vulnerability detection. In response to the above problems, this paper proposes a new attribute program slicing graph (APSG), which uses program slicing to simplify the structure of the graph based on the program dependency graph, and it also retains the semantic attributes of nodes. In addition, this article also uses graph embedding network to extract feature vectors and builds multiple neural network prediction models according to different vulnerability types, and finally achieves function-level vulnerability existence and type prediction for unknown binary files. Experiments have proven that the vulnerability prediction method proposed in this article is more accurate in predicting the existence and type of vulnerabilities than the existing binary vulnerability detection methods.

1 Introduction

According to statistics from the general vulnerability disclosure website CVE [1], the number of included vulnerabilities has increased year by year this year, and IoT-related vulnerabilities have also increased from 15% in 2016 to 37% in 2022. In addition, according to statistics from the BangBang Security IoT Security White Paper, a total of 20,065 security risks were detected for 433 samples, with an average of 46.34 security risks per IoT device. This further highlights the importance of security protection for IoT devices.

Traditional machine learning vulnerability detection methods mostly use source code as the object to generate abstract syntax trees as the basis for research. This method mainly has two flaws. One is that the compiled binary firmware cannot directly parse the abstract syntax tree, making it impossible to perform. The next step of analysis is to take the abstract syntax tree as the research object, which will ignore semantic information such as code logical structure and data flow direction, and will introduce many irrelevant statement fragments, affecting the final semantic accuracy of the vulnerability. In response to the above problems, this paper proposes a static vulnerability detection method for general-architecture IoT binary firmware, which specifically includes the following three aspects of work:

1. In view of the problems that existing binary vulnerability data sets have such problems as unclear vulnerability classification, inconsistent vulnerability function labels, and vague vulnerability description information, this paper selects vulnerability source codes of 16 high-frequency vulnerability types (CWEID) based on the SARD source code data set.
2. This paper proposes a program dependency graph generation and slicing method based on disassembly software and encodes node attributes to achieve automatic generation of attribute slice graphs. This method fully retains the structural semantics and attribute semantics of the original vulnerability.
3. For vulnerability detection scenarios, this paper improves the struc2vec model and proposes a path sampling algorithm for multi-layer directed weighted graphs. Experimental results prove that its effect is better than the existing graph embedding algorithm.

2   Background

This section will focus on explaining the research background and related work on static IoT firmware vulnerability detection, which mainly includes data embedding related technology research and existing static vulnerability detection technology.

Graph embedding network technology embeds nodes or edges in the graph into a low-dimensional vector space and retains the structural information of the original graph as much as possible.

Word2vec is the basis of the graph embedding network algorithm. It is an algorithm proposed by researchers such as Mikilov [2]. It uses semantic windows to capture the semantic context in each sentence and uses sliding context windows to capture the relationship between nodes on sequence data.

The DeepWalk model is a classic model proposed by Perozzi et al. at KDD 2014 [3]. It first uses a random walk to sample graph nodes, converts the local topology of the nodes in the graph into sequence information, and then uses the skip-gram model to learn the sequence information.

Node2vec [4] is a representation model for learning the homogeneity and similarity of graph nodes proposed by Jure Leskovec. It proposes a biased random walk strategy. Compared with the DeepWalk model, Node2vec can learn a more complete node representation when facing complex network structures.

The method of combining deep learning with vulnerability detection provides a new direction in the field of modern vulnerability detection. Li proposed VulDeePecker [5] in 2018, which is a vulnerability detection method based on deep learning. Its core approach is to segment the source code according to semantics, then mark these segmented blocks, and finally model these segmented blocks. Train. Although VulDeePecker focuses on content semantics, it ignores the structural semantics of the program, and its experimental results prove that it has lower false negatives.

μ VulDeePecker [6] is also an improvement on VulDeePecker, which is a multi-type vulnerability detection system. It adds control flow dependencies based on the original content semantics and introduces the concept of code attention for the first time to represent the local features of the program and uses local features to help detect vulnerability types.

3 System Overview

For the current vulnerability detection work of IoT firmware, the main problem is that there is insufficient focus on semantics such as binary context information, data flow information, and control flow information, and there is too many redundant path information in the code. Therefore, such vulnerabilities Detection methods generally suffer from low accuracy and high false alarm rates.

In order to solve the above problems, this paper proposes a semantic attribute-oriented feature extraction method for binary files. The overall framework is shown in Fig. 1, which can be divided into the following stages:

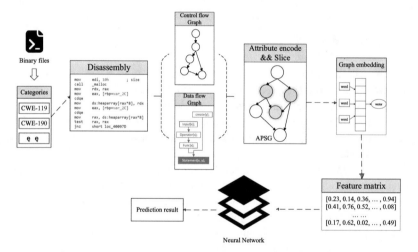

Fig. 1. Architecture of static IoT firmware vulnerability detection.

Program Slicing and Semantic Extraction: After having the abstract data and control flow graph, starting from the focus statement, the program dependency graph is sliced and irrelevant program fragments are eliminated. Finally, semantic attributes are extracted from the remaining dependency graph to obtain an attribute slice graph (Attribute Program Slice Graph) referred to as APSG graph.

Since different types of vulnerabilities have different vulnerability characteristics, a unified feature attribute extraction standard cannot be used. For example, for overflow vulnerabilities, the boundary attributes of the array need to be considered, for vulnerabilities within memory operations, for memory operations, However, the number and sequence attributes of instructions need to be focused on.

Feature Vector Generation: The main purpose of this stage is to encode the vulnerability characteristics in the form of graph representation and convert them into a vector matrix set to facilitate subsequent neural network processing. The module uses the graph embedding model [7] to characterize the APSG graph to explore different levels and nodes in the network graph. Deeper hidden associations and information between them, and finally generate a low-dimensional vector representation set.

Vulnerability Detection: The main work of this stage is to use graph neural network to establish a suitable vulnerability detection model. This model uses the feature vector generated in the previous step as the input of the neural network [8] to learn different types of vulnerability characteristics, and finally generates multiple types of vulnerability detection sub-models, to detect the existence of vulnerabilities in unknown binary software and predict the types of vulnerabilities.

4   System Design

4.1   Preprocessing

This article uses the C++ source program files provided by the Software Assurance Reference Dataset (SARD) [9] as the initial data set. This data set provides classified source program codes, and the functions with vulnerabilities in the program are distinguished by good and bad. It greatly facilitates the preliminary preparation work.

This article mainly uses two processing rules to unify the source code of the program before compilation as much as possible into a coding specification, which mainly includes:

Advanced Syntax Feature Replacement Operation: For modern C++, it provides a lot of advanced feature syntax sugar, such as lambda expressions, template partial specialization, etc. These features will generate some uncommon anonymous classes during compilation, or generate a large amount of generic code, resulting in code expansion, which is reflected in IDA's disassembly code as a large number of difficult-to-understand assembly instructions, or a large number of repeated function codes with different types.

Simplification Operation of Synonymous Statements: This operation is mainly to solve the situation of diverse writing methods mentioned above. Its main operation object is to unify function code fragments with the same semantics, such as merging some complex operations or multi-line operations into a single operation operator [10].

After the source code of the data source is uniformly processed, batch compilation with debugging information is required to obtain the final data set with symbol table information. The purpose of this step is to save the function name information obtained by IDA decompiled.

4.2   Slice Processing Based on Multiple Information Flow Graphs

This paper proposes a program slice graph APSG with attributes, which is a composite data structure that combines the composition theory of control flow graph and data flow graph commonly used in traditional program analysis. In addition, it uses a slicing algorithm to further simplify the vulnerability feature semantics necessary for the vulnerability function. Figure 2 shows the process from program flow graph to program slicing.

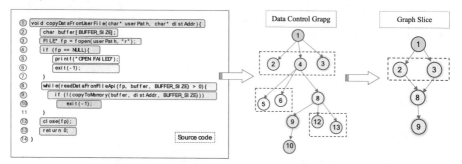

Fig. 2. Demonstrates how to extract control flow graphs and data flow graphs from a piece of real vulnerable code and slice them based on points of interest.

a) Analysis and generation of control flow graphs and data flow graphs

For a control flow graph CFG, it is constructed by IDA Python, which is a script file that can be run in IDA. It can operate on the assembly instructions and data provided in IDA, and can conveniently and directly obtain the address functions displayed in the assembly interface, including function names, functions Address, etc. It also provides rich and powerful API interfaces and documents, including obtaining the segment where the current address is located, the current function attribute set.

The construction of the data flow diagram is relatively simple, and it mainly relies on the back reference function provided by IDA. For a function, the parameters used in it can simply find its definition location and other complete usage paths through back references.

b) Program slicing [11]

Vulnerabilities caused by binary programs are almost always due to incorrect use of system call functions provided by the operating system, or malicious transmission of carefully designed attack loads to system calls, resulting in security vulnerabilities. Therefore, this paper proposes a backward slicing algorithm based on the reachability of interest points.

The interest point is a collection of a series of system calls used in this function, and the interest point is the node corresponding to the program slicing criterion. Starting from the interest point, a backward search traverse is performed along the data dependency edge and control dependency edge, and all the traversed reachable nodes are added to the slice, thereby retaining only statements that may have associated semantics and reducing semantic interference caused by irrelevant code fragments.

4.3 Semantic-Based Feature Extraction

Program semantics is to give each legal program a clear meaning and accurately define the behavior and results of this program execution. For example, for an expression, its meaning is how to evaluate and what kind of value it is. The semantic rules of a program can generally be divided into three broad categories:

Context-Dependent. This rule mainly includes the program logical structure, function objects before and after jumps, transfer based variable attributes, etc., the number of basic block branches in the program, etc.

Types. In computer languages, especially strongly typed languages such as C/C++, types are an important carrier of semantics. Semantic rules about types mainly include type detection, type inference and other attribute calculation operations.

Memory Operation Instructions. For an executable program, it needs to be loaded into memory before it can be run, and during the running process, the program will frequently perform operations such as opening and destroying stacks, applying for reclaimed memory space, etc. Therefore, memory operation execution instructions are also an important part of the description of semantic rules.

4.4 Feature Vector Generation Based on Graph Embedding

This paper proposes a new path sampling algorithm to address this problem. Practice has proven that it is better than Struc2vec [12]. The traditional path sampling algorithm used in this article has better semantic aggregation in the vulnerability model feature transformation scenario.

This section will focus on describing how to use the graph embedding algorithm based on the struc2vec model to convert complex APSG graphs into low-dimensional vector sets.

(1) Definition of distance between vertices

Struc2vec is learned for structural similarity, while the evaluation of structural similarity is mainly based on hierarchical structure measurement. For a graph $G = (V, E)$, V represents the set of all vertices in the graph, and E represents the set of all edges in the graph. Now select two vertices u and v in the graph G, and define $R_k(u)$ to represent the vertex u A set of vertices with a distance of k, and $s(S)$ represents the ordered degree sequence of the vertex set S. Use $g(D_1, D_2)$ as the calculation function of the distance between the ordered degree sequences D_1 and D_2, and let $f_k(u, v)$ represent the loop structure distance of k between vertices u and v. The calculation formula is as follows:

$$f_k(u, v) = f_{k-1}(u, v) + g(s(R_k(u)), s(R_k(v)))$$ (1)

$$k \geq 0 \, and \, |R_k(u)|, |R_k(v)| > 0$$

Considering that $s(R_k(u))$ and $s(R_k(v))$ represent the ordered degree sequence of different vertices, their value range is $[1, n-1]$. . Therefore, for this model, the dynamic time warping algorithm (DTW) is used to calculate the distance between two ordered degree sequences. For sequence a and sequence b, define $d(a, b)$ as the distance calculation function of sequence a and b, then its formula is as shown in formula (2):

$$d(a, b) = \frac{\max(a,b)}{\min(a,b)} - 1 \tag{2}$$

From the above formula, we can see that when two sequences are the same, $d(a, b) = 0$, that is the distance between the sequences is 0.

(2) Constructing a multi-level directed weighted graph

Define the graph as M. Each layer in the graph M contains all nodes in the graph network. For the two vertices u and v in the k_{th} layer, the edge weight calculation formula $w_k(u, v)$ is as shown in the formula (3):

$$w_k(u, v) = e^{-f_k(u,v)}, k = 0, 1, \ldots, k^* \tag{3}$$

According to the above formula, the weight between pairs of nodes in each layer is inversely proportional to their structural similarity. A multi-layer weighted graph with four vertices, an example of the k layer and k-1 layer is shown in Fig. 3:

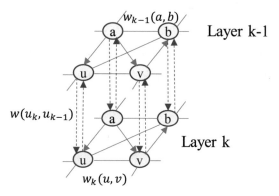

Fig. 3. Layer k and k-1 in a multi-layer directed weighted graph with four vertices.

Similarly, for each vertex in a certain layer, there will be corresponding upper layer or lower layer vertices connected to it. The edge weight $w(u_k, u_{k+1})$ is defined using the following formula (4) (5):Then its random sampling transfer formula is:

$$w(u_k, u_{k+1}) = \log(\Gamma_k(u) + e), k = 0, \ldots, k^* - 1 \tag{4}$$

$$w(u_k, u_{k-1}) = 1, k = 1, \ldots, k^* \tag{5}$$

where $\Gamma_k(u)$ is the number of edges in the k-th layer connected to u whose edge weight is greater than the average edge weight.

(3) Path sampling to obtain vertex sequence

After constructing the multi-layer weighted path graph, the path information needs to be sampled to build complete context information for each node.

For a certain layer L, the directed graph described by it is G, then for any vertex u, assuming its adjacent edge set is E, then the normalization factor for the vertex u is:

$$Z_k(u) = \sum_{\substack{v \in E \\ v \neq u}} e^{-f_k(u,v)}. \tag{6}$$

Then its random sampling transfer formula is:

$$p(u|v) = \begin{cases} \frac{e^{-f_k(u,v)}}{Z_k(u)}, & (u, v) \in E \\ 0, & (u, v) \notin E \end{cases} \tag{7}$$

$p(u|v)$ represents the probability of selecting node v at the next moment starting from point u.

In addition to considering the walking probability of the current layer, you also need to consider whether you need to switch to other layers for sampling every time you sample. Assuming that the walking probability of k in the current layer is p, there is a 1-p probability of switching to layer k-1 and layer $k + 1$. The transition probability formula is as follows:

$$p_k(u_k, u_{k+1}) = \frac{w(u_k, u_{k+1})}{w(u_k, u_{k+1}) + w(u_k, u_{k-1})}, \tag{8}$$

$$p_k(u_k, u_{k-1}) = 1 - p_k(u_k, u_{k+1}), \tag{9}$$

Based on the above probability, a vertex is randomly selected from the graph as the next transfer node at each moment until the path length exceeds the preset path length threshold γ, or there is no successor node.

For a multi-level directed weighted graph M, node set V, edge set E, edge weight calculation formula w, current layer walking probability p, and current layer number k, the path sampling algorithm proposed in this section is described as follows:

Algorithm 1: Entitled Random Walk Sampling

input : V, E, γ, p, k

---
1. $v_0 = random\_choice(V)$
2. $path = \{v_0\}$
3. $t = 0$
3. **for** v_i **in** V:
4. $compute(Z_k(v_i) \parallel p(v_0|v_i) \parallel p_k(u_k, u_{k+1}) \parallel p_k(u_k, u_{k-1}))$
5. **endfor**
6. **while** $[path] < \gamma$:
7. $v_{t+1} = (p + (1 - p) * (p_k(u_k, u_{k+1}) + p_k(u_k, u_{k-1}))) * p(v_0|v_{t+1})$
8. $path.append(v_{t+1})$
9. **endwhile**

---

5   Experiments

5.1   Datasets

This article uses Juliet Test Suite For C/C++ v1.3 in the SARD data set maintained by NIST as the original data set which is a collection of 64,099 source vulnerability file programs, covering 118 types of software weaknesses. This article selects 18 mainstream vulnerability types and conducts secondary classification based on the similarity of vulnerability type characteristics. Table 1 shows the correspondence between different CWE names and vulnerability categories.

Table 1. Classification of secondary vulnerability types with different CWEIDs.

CWEID	Name Of CWE	Vulnerability Category
CWE-119	Improper Restriction of Operations within the Bounds of a Memory Buffer	Buffer overflow
CWE-121	Stack-based Buffer Overflow	
CWE-122	Heap-based Buffer Overflow	
CWE-124	Buffer Underwrite	
CWE-125	Out-of-bounds Read	
CWE-126	Buffer Over-read	
CWE-127	Buffer Under-read	
CWE-190	Integer Overflow or Wraparound	Integer Overflow
CWE-191	Integer Underflow	
CWE-401	Missing Release of Memory after Effective Lifetime	Memory leak
CWE-415	Double Free	User after Free
CWE-416	Use After Free	
CWE-457	Use of Uninitialized Variable	Uninitialized variable
CWE-824	Access of Uninitialized Pointer	
CWE-476	NULL Pointer Dereference	Null Pointer Dereference
CWE-690	Unchecked Return Value to NULL Pointer Dereference	
CWE-561	Dead Code	Dead Code
CWE-563	Unused Variable	

5.2 Experiment Results and Analysis

5.2.1 Vulnerability Feature Extraction

(1) Model evaluation method

The purpose of extracting and transforming semantic attribute features of vulnerabilities is to obtain low-dimensional vector sets. For the same type of vulnerability code, the positions of its semantic feature vectors in the vector space should be close. This article uses the minimum spanning tree length to quantify the vector set distance, and uses the minimum length generated by the word2vec model as the baseline model to calculate the minimum spanning tree compression ratio Z_k under different algorithms. The smaller the compression ratio is, the better the model transformation effect.

(2) Experimental results and analysis

Taking the vulnerable function sub_407B20 existing in CVE-2017-17215 as an example, Fig. 4 shows the distribution of vector sets in the vector space after conversion by different graph embedding models.

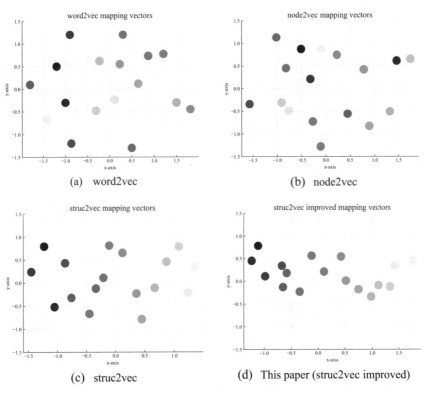

Fig. 4. Comparison of feature vector generation for vulnerability functions using multiple graphs embedding models.

This article uses the prim algorithm to generate a minimum tree for the vector point set. The final results are shown in Fig. 5. It can be seen that the proposed model has the highest degree of semantic aggregation.

5.2.2 Vulnerability Feature Extraction

(1) Model evaluation method

In order to verify the effectiveness of the vulnerability existence and type prediction proposed in this paper, this section uses Precision, Accuracy, and Recall to evaluate this paper.

This research uses the binary-oriented vulnerability detection tool DeepFuzz, as well as the source code-based vulnerability detection tools Flawfinder [13] and VulDeePacker, etc. to conduct controlled experiments.

(2) Experimental results and analysis

After testing all vulnerability data sets in Juliet Test Suite For C/C++ v1.3, we compare the experimental data.

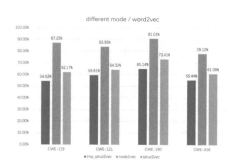
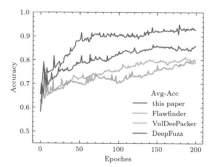

Fig. 5. Compression ratio of spanning trees for different graph embedding networks.

Fig. 6. Comparison of vulnerability type prediction accuracy of different models.

Table 2. Prediction results of proposed model compared with DeepFuzz for different epochs.

Epoch	75			100		
Evaluation	Accuracy	Precision	Recall	Accuracy	Precision	Recall
DeepFuzz	79.45	85.31	69.10	83.19	91.98	70.81
This Paper	87.92	89.22	73.21	93.12	92.10	74.32

From the test data of Table 2, it can be clearly seen that the accuracy, precision and recall rate of the proposed vulnerability prediction model are higher than DeepFuzz. The reason is that we conduct separate model training for different types of vulnerabilities, so the effect is better.

Tables 3 shows the comparison of the proposed model with other vulnerability detection models in the source code scenario. It can be seen that the model in this article has similar effects to μ VulDeePacker. However, considering that μ VulDeePacker parses the source code, the difficulty and sufficiency of semantic extraction are more difficult than binary extraction.

Figure 6 shows the accuracy of multiple models in predicting vulnerability types. It can be seen that the accuracy of this model is 8%–14% higher than other models.

Table 3. Comparison of various indicators for vulnerability detection between this model and other source code-based models.

Epoch	100		
Evaluation	Accuracy	Precision	Recall
Flawfinder	60.31	53.43	38.24
VulDeePacker	74.27	72.11	76.23
μ VulDeePacker	86.28	90.63	81.23
This Paper	87.92	89.22	73.21

6   Conclusions

This paper takes IoT firmware as the research object and proposes a targeted static vulnerability detection method. The experimental results prove that the method proposed in this paper has improved the accuracy of vulnerability existence prediction and vulnerability type prediction to varying degrees compared with existing tools.

In future work, we can consider how to predict vulnerability types from the perspective of multi-type vulnerability feature fusion, and propose better algorithm designs to extract IoT firmware vulnerability features.

References

1. CVE. https://cve.mitre.org/
2. Church, K.W.: Word2Vec. Natural Lang. Eng. **23**(1), 155–162 (2017)
3. Perozzi, B., Al-Rfou, R., Skiena, S.: Deepwalk: online learning of social representations. In: Proceedings of the 20th ACM SIGKDD International Conference on Knowledge Discovery and Data Mining, pp. 701–710 (2014)
4. Grover, A., Leskovec, J.: node2vec: scalable feature learning for networks. In: Proceedings of the 22nd ACM SIGKDD International Conference on Knowledge Discovery and Data Mining, pp. 855–864 (2016)

5. Li, Z., et al.: Vuldeepecker: a deep learning-based system for vulnerability detection. arxiv preprint https://arxiv.org/abs/1801.01681 (2018)
6. Zou, D., Wang, S., Xu, S., Li, Z., Hai, J.: μVulDeePecker: a deep learning-based system for multiclass vulnerability detection. IEEE Trans. Dependable Secure Comput. **18**(5), 2224–2236 (2019)
7. Xu, M.: Understanding graph embedding methods and their applications. SIAM Rev. **63**(4), 825–853 (2021)
8. Abdi, H.: A neural network primer. J. Biological Syst. **2**(03), 247–281 (1994)
9. Fan, J., Li, Y., Wang, S., Nguyen, T.N.: AC/C++ code vulnerability dataset with code changes and CVE summaries. In: Proceedings of the 17th International Conference on Mining Software Repositories, pp. 508–512 (2020)
10. Cheng, W., Hua, X., Sui, Y.: Deepwukong: statically detecting software vulnerabilities using deep graph neural network. ACM Trans. Softw. Eng. Methodol. (TOSEM) **30**(3), 1–33 (2021)
11. Binkley, W., Gallagher, K.B.: Program slicing. Adv. Comput. **43**, 1–50 (1996)
12. Ribeiro, F.R., Saverese, P., Figueiredo, D.R.: struc2vec: learning node representations from structural identity. In: Proceedings of the 23rd ACM SIGKDD International Conference on Knowledge Discovery and Data Mining, pp. 385–394 (2017)
13. Ferschke, O., Iryna, G., Rittberger, M.: FlawFinder: a modular system for predicting quality flaws in Wikipedia. In: CLEF (Online Working Notes/Labs/Workshop), pp. 1–10 (2012)

Probing Cross-lingual Transfer of XLM Multi-language Model

Raffaele Guarasci[(✉)], Stefano Silvestri, and Massimo Esposito

National Research Council of Italy (CNR),
Institute for High Performance Computing and Networking (ICAR),
Napoli, Italy
{raffaele.guarasci,stefano.silvestri,massimo.esposito}@icar.cnr.it

Abstract. This paper investigates the ability of XLM language model to transfer linguistic knowledge cross-lingually, verifying if and to which extent syntactic dependency relationships learnt in a language are maintained in other languages. In detail, a structural probe is developed to analyse the cross-lingual syntactic transfer capability of XLM model and comparison of cross-language syntactic transfer among languages belonging to different families from a typological classification, which are characterised by very different syntactic constructions. The probe aims to reconstruct the dependency parse tree of a sentence in order to representing the input sentences with the contextual embeddings from XLM layers. The results of the experimental assessment improved the previous results obtained using mBERT model.

1 Introduction

In recent years, the Natural Language Processing field has undergone tremendous growth with the rise of Large Language Models (LLMs). In particular, since the release of Transformer-based models [12], ushered in a new generation of approaches that raised the bar of achievable results higher and higher in many NLP tasks, ranging from different types of text classification [3,16,35], information extraction [15,16], sentiment analysis [23], and anaphora and coreference resolution [13,17,25,32]. Given this enormous potential offered by LLMs, the interest of scholars has shifted to their ability to encode linguistic knowledge[22] and transfer it from one language to another. The realm that stands to benefit the most from recent cross-lingual research in NLP is Syntax [1,17–19,24], due the possibility to use Dependency Parse Trees (DPTs) as shared representations to represent the universal relationship. Probe-based approaches [34] have shown that models like BERT embeds in distinct levels the different types of linguistic information [22]. The introduction of large-scale multilingual NLMs has facilitated a wide range of cross-lingual natural language tasks, including Named Entity Recognition, Part of Speech Tagging, Neural Machine Translation, and Text Classification [8,26,30].

L. Barolli (Ed.): EIDWT 2024, LNDECT 193, pp. 219–228, 2024.
https://doi.org/10.1007/978-3-031-53555-0_21

Starting from recent approaches, this work aims to test the possibilities of the application of XLM multilingual model [11] on a parallel corpus, to show the capabilities of transferring syntactic knowledge from one language to another.

The paper is structured as follows. Section 2 provides an overview of the recent related works. The next Sect. 3 describes the research methodology, including details of the structural probe, the LLM used, and the used dataset. Section 4 describes the experimental assessment, and then presents and discusses the results. Finally, Sect. 5 summarises the paper and hints at future developments.

2   Related Work

Interest in syntactic knowledge transfer-although always present in Theoretical Linguistics-only with the advent of NLMs such as BERT has aroused interest from scholars in the NLP field [36]. In [7], attention heads in the BERT architecture were probed for linguistic phenomena, demonstrating that certain heads corresponded well to specific syntactic relations. The findings of [33], which showed that BERT encodes syntax more prominently than semantics. In [14] is illustrated instead that the model can acquire significant syntax knowledge without task-specific training. The authors of [22] have performed several experiments to discover which elements of English language structure are learned by the BERT model. Results have highlighted that BERT different layers capture diverse levels of language complexity. In [21] a structural probing model has been trained to show that learned spaces of language models such as BERT and ELMo are better for reconstructing dependency trees than baselines. The same probing approach is used in [6] to examine how mBERT learns a cross-lingual representation of the syntactic structure.

More recently, some studies have investigated these phenomena from a typological point of view. The authors of [2] have formulated the hypothesis that a mBERT model can embed typological information from the input data. They have verified their hypothesis using different tasks in 40 different languages. In [27], decoding experiments for mBERT across 18 languages have been presented. These experiments aim to test if dependency syntax is reflected in attention patterns. Experiments have been carried out on 18 languages, demonstrating that mBERT can decode dependency trees from attention patterns more accurately than an adjacent-linking baseline.

3   Methodology

This paper presents an analysis of the cross-lingual syntax transfer capability of the XLM model, using three different languages, namely English, Italian, and French, in particular focused on dependency DPTs, adopting the structural probe proposed by [21]. To test this feature, we firstly trained three different probe models using annotated DTP datasets and the XLM pretrained language model, respectively in each one of the considered languages. Then, we cross-tested their capability of transferring the syntax between these languages, and we also investigated a specific syntactic phenomena (subject omissibility), exploiting a parallel dataset annotated with DPTs.

3.1 Neural Language Model

The authors of [8] introduced two innovative learning methodologies for cross-lingual representation applied to Transformer-based architecture, with the purpose of improving the performance obtained by the classic mBERT model [9], which exploited a BERT model [12] pretrained on a huge multilanguage corpus [10] counting more than 100 different languages, with the main purpose of the application to machine translation and cross lingual tasks [26]. In detail, they defined the *Cross-lingual Language Model Pretraining* (XLM), exploiting the Casual Language Modeling (CLM), which is based on an unsupervised method for learning the probability of a word given the previous words in a sentence, and the Translation Language Modeling (TLM), which exploits a supervised learning objective to improve cross-lingual pretraining on parallel multilanguage data. In detail, CLM trains the model to predict the token following a sequence of tokens. This task, such as BERT Masked Language Model (MLM) approach [12], works only on monolingual data, being unsupervised. On the other hand, TLM is defined as the classic next sentence prediction task of the BERT model [12], but it uses a set of concatenated parallel multilingual sentences as input [8] during its training phase. Both CLM/MLM and TLM approaches are used to pretrain the XLM models. The authors made several different XLM pretrained models publicly available[1], using different training corpora (including different languages) and training methods (CLM/MLM or both CLM/MLM and TLM). Compared to the mBERT, it has been demonstrated that XLM allows for the training of a neural language model for many languages while not sacrificing per-language performance, producing state of the art performances in many NLP cross-lingual tasks [8], as well as produces good performances on cross-lingual dynamic word embeddings [9].

In our experiments, we exploited the XLM model pretrained using CLM/MLM approach on a multi-language corpus that includes 17 languages, among which Italian, French and English are available (the languages considered in the experiments presented in this paper). In detail, this model is based on a Transformer architecture with 1280 hidden states, 16 attention heads, 16 layers, GELU activation function, and the total number of parameters equal to 570M. The model is pretrained with a dropout rate of 0.1 and learned positional embedding, using the Adam optimiser. The training corpus has a vocabulary size equal to approximately 200,000 words, and is preprocessed with tokenization and rare-words subwords splitting [28].

---

[1] XLM models are available at https://github.com/facebookresearch/XLM.

3.2   Structural Probe

A structural probe is employed to extract a syntactic structure from a model-learned representation, serving as evidence of a particular phenomenon.

The structural probe is conceptualised as a model M that takes as input the sequence of contextual embeddings $\mathbf{h}1 : n^m$ derived from a neural language model, such as BERT. This sequence represents the vectorized form of the n words $w1 : n^m$ in the m-th sentence, where the sentence length is denoted as n. The probe operates under the hypothesis that the dependency tree structure is embedded if the new space resulting from the linear transformation satisfies the condition that the squared L2 distance between vectors of two words corresponds to the number of edges between those words in the parse tree. Consequently, the structural probe is defined as an inner product on the original space, where squared distances and norms encode syntax trees. This definition involves a family of inner products $\mathbf{h}^T A \mathbf{h}$, parameterised by any symmetric, positive semi-definite matrix A, leveraging the properties of the dot product. The linear transformation B can be specified such that $A = B^T B$, and subsequently, the inner product can be expressed as $(B\mathbf{h})^T(B\mathbf{h})$, which also represents the norm of $\mathbf{h}$ transformed by B. Accordingly, for each sentence, a family of squared distances is defined as:

$$d_B(\mathbf{h}_i^m, \mathbf{h}_j^m)^2 = (B(\mathbf{h}_i^m - \mathbf{h}_j^m))^T(B(\mathbf{h}_i^m - \mathbf{h}_j^m)) \tag{1}$$

Here, i, j denote the indices of words in sentence m.

The coefficients of the matrix B serve as trainable parameters tasked with predicting the tree distance between all words in each sentence of the training set, leveraging gradient descent:

$$\min_B \sum_m \frac{1}{|s^m|^2} \sum_{i,j} |d_{TM}(w_i^m, w_j^m) - d_B(\mathbf{h}_i^m, \mathbf{h}_j^m)^2| \tag{2}$$

Here, $|s^m|$ represents the length of the m-th sentence, and the normalisation by $|s^m|^2$ (the square of the length of the m-th sentence) is necessary because each sentence has $|s^m|^2$ word pairs.

This structural probe defines a distance metric by exhibiting symmetric and non-negative properties. Simultaneously, it tests the existence of an inner product in the representation space, where the squared distance encodes syntax tree distance. This allows the model not only to discern relationships between individual words but also to capture the proximity of each word to every other word in the syntax tree. For a more detailed implementation of the probe, please refer to [21], producing the requisite parse-tree-like representation in the output.

4   Experiments

Several syntactic probes exploiting each one of the 12 layers of the XLM model have been trained in each considered language, obtaining in total 36 different trained models. Therefore, each model has been assessed on the sentences of a parallel multi-language test set, which contains the same sentences in English, French and Italian annotated with the corresponding DPTs, investigating the cross-language capability of XLM and which layer of the model learns syntactic constructs.

In next Sect. 4.3, each one of these experiments is named with the abbreviation of the language used to train the probe (respectively EN, IT, FR) and the language used to test the probe. Therefore, a model trained in English and tested in Italian is defined as *EN-IT*. Then, a final experiment leverages the best performing models, corresponding to probes that uses the embeddings extracted from a specific layer of the XLM model, to test the subject omissibility phenomenon, running these probes a parallel dataset formed by sentences where this syntactic construct appears in the Italian version.

4.1   Dataset

Three datasets annotated with the Universal Dependencies (UD) standard (CoNLL-U format), respectively in English, Italian and French, have been used to train the structural probe models. In the case of English language, *the Universal Dependencies English Web Treebank (EN-EWT)*[2] [29] has been adopted. In the case of the Italian language, *Italian ISDT Treebank (IT-ISDT)*[3] [4,31] has been used. For the French language, two datasets have been merged: *UD-French-GSD (FR-GSD)*[4] [20] and *UD-French-Sequoia (FR-SEQ)*[5] [5]. These datasets have a comparable number of sentences and words, counting approximately 12,000 sentences and 250,000 words each one. The *Parallel Universal Dependencies* (PUD)[6] corpus has been used to test the trained models. PUD is a set of parallel treebanks created for the CoNLL 2017 shared task on Multilingual Parsing from Raw Text to Universal Dependencies, and contains parallel sentences manually translated and annotated with the DPTs in English, Italian and French. This dataset contains 1,000 parallel sentences. Furthermore, 100 sentences where the subject is omitted in the Italian version have been selected from PUD, to the end of investigating the capability of XLM to transfer the considered syntactic phenomena between different languages.

4.2   Metric

The performances of the models have been evaluated using *Sentence-based UAS* (*sUAS*) metrics. This score is calculated from the **Unlabeled Attachment**

---

[2] https://github.com/UniversalDependencies/UD_English-EWT.
[3] https://github.com/UniversalDependencies/UD_Italian-ISDT.
[4] https://github.com/UniversalDependencies/UD_French-GSD.
[5] https://github.com/UniversalDependencies/UD_French-Sequoia.
[6] https://universaldependencies.org/treebanks/en_pud/index.html.

Score (*UAS*) metric, by micro-averaging the UAS on each sentence of the dataset. In detail, sUAS is defined as:

$$sUAS = \frac{1}{m} \sum_{i=1}^{m} \frac{PEdges_i}{Edges_i} \tag{3}$$

where m is the total number of sentences in the dataset, $PEdges_i$ is the number of correctly predicted edges for the *ith* sentence and $Edges_i$ is the total number of true edges of the *ith* sentence.

4.3   Results and Discussion

Table 1 summarises the *sUAS* metric obtained in the cross language experiments performed. In detail, a probe has been trained leveraging a different XLM layer to represent the input text, using respectively the English, the Italian and French training set. Then, each probe has been tested on the whole parallel PUD test set, considering each time a different language, obtaining the monolingual (EN-EN, IT-IT, FR-FR) and cross-lingual (EN-IT, EN-FR, IT-EN, IT-FR, FR-EN, FR-IT) experiments reported in Table 1. Analysing the results, it is possible to confirm that the layers of the XLM model where the syntactic information is embedded are the central upper ones experiments, as previously observed in monolingual English BERT-Base [21,22] and in cross language mBERT model [18]. In detail, the best results are obtained at layers 9 and 10 of XLM, depending on the specific monolingual and cross-lingual test.

The obtained results prove the ability of the XLM model to embed the sentence structures is not affected by a specific language, demonstrating its cross-lingual syntax transfer capability. As expected, the cross lingual tests have shown a slight performance drop compared to monolingual ones, as well as the cross language transfer between French and English is more effective, compared to Italian, transfer, in accordance with the Theoretical Linguistics.

The additional experiments, devoted to investigate the syntactic transfer of the omissibility of subject from Italian to other languages, allowed us to further investigate the behaviour of XLM. As explained above, we tested the best performing probes (trained using the embeddings from the layers 9 and 10 of the XLM model) on a subset of parallel sentences where, in the case of the Italian version, the subject is omitted. As reported in Table 2, the results obtained with the probes trained in Italian and tested in English and French are comparable with the ones obtained in the corresponding monolingual experiments, providing further evidences that XLM has embedded during the multilingual pretraining the knowledge of the subject agreement, although the languages used to train the probe allows do not allow the pro-drop phenomenon. It is worth noting that in the case of the French test set, the probes respectively trained in Italian and in French have produced almost similar results, probably also thanks to the high level of similarity among these two languages.

Table 1. Results for each English probe respectively tested on PUD English, Italian and French test sets.

XLM layer	sUAS EN-EN	sUAS EN-IT	sUAS EN-FR	sUAS IT-IT	sUAS IT-EN	sUAS IT-FR	sUAS FR-FR	sUAS FR-EN	sUAS FR-IT
0	0.5280	0.4267	0.4244	0.5191	0.4166	0.4345	0.5198	0.4170	0.4651
1	0.5991	0.5216	0.5174	0.5211	0.4434	0.4473	0.5901	0.4563	0.5516
2	0.6212	0.5327	0.5742	0.6202	0.5331	0.5466	0.6311	0.5655	0.5923
3	0.6433	0.5715	0.5982	0.6333	0.5571	0.5782	0.6516	0.5963	0.6113
4	0.7211	0.6957	0.6988	0.7231	0.6518	0.6727	0.7199	0.6784	0.7002
5	0.7288	0.7117	0.7001	0.7285	0.6663	0.6818	0.7301	0.6811	0.7121
6	0.7344	0.7201	0.7116	0.7315	0.6826	0.6997	0.7399	0.6913	0.7255
7	0.7707	0.7327	0.7452	0.7682	0.7061	0.7116	0.7613	0.7122	0.7454
8	0.7997	0.7443	0.7528	0.7917	0.7241	0.7334	0.7771	0.7223	0.7642
9	**0.8389**	**0.7643**	**0.7581**	0.8212	**0.7471**	**0.7421**	**0.8381**	0.7309	0.7997
10	0.8321	0.7600	0.7555	**0.8224**	0.7411	0.7382	0.8322	**0.7353**	**0.8011**
11	0.8267	0.7597	0.7488	0.8181	0.7333	0.7280	0.8087	0.7300	0.7887
12	0.8153	0.7466	0.7312	0.8012	0.7166	0.7176	0.7880	0.7367	0.7711
13	0.7951	0.7125	0.7258	0.7764	0.6972	0.6885	0.7313	0.7001	0.7265
14	0.7322	0.7077	0.7111	0.7285	0.6782	0.6887	0.7283	0.6901	0.7125
15	0.6824	0.6755	0.6554	0.6933	0.6414	0.6400	0.6797	0.6437	0.6528

Table 2. Results for the best performing EN probes (layer 9 and 10) respectively tested in English, Italian and French on a subset of sentences where the subject is omitted in Italian

XLM layer	sUAS EN-EN	sUAS EN-IT	sUAS EN-FR	sUAS IT-IT	sUAS IT-EN	sUAS IT-FR	sUAS FR-FR	sUAS FR-EN	sUAS FR-IT
9	**0.8345**	**0.7414**	**0, 7388**	**0.8499**	**0.7243**	**0.7817**	**0.8401**	**0.7096**	**0.7841**
10	0.8311	0.7366	0, 7311	0.8222	0.7188	0.7745	0.8086	0.7001	0.7698

5 Conclusions and Future Work

In this work, an experiment aiming to validate the transfer capability of the XLM multilingual model on the parallel PUD corpus has been described. Results confirm that the XLM model's proficiency in embedding sentence structures remains unaffected by specific languages, underscoring its cross-lingual syntax transfer capability. As anticipated, the cross-lingual tests indicate a marginal performance decrease compared to monolingual assessments. Additionally, the cross-language transfer between French and English is observed to be more effective than the transfer involving Italian, aligning with principles in Theoretical Linguistics. For future work, new approaches will be tested that promise to exceed the current performance achievable with NLMs. In particular, novel perspectives on analysis have opened up with the emergence of Quantum NLP, or that sub-branch of NLP that uses methods derived from quantum theory to increase performance [18].

Acknowledgements. This work is supported by European Union - NextGenerationEU - National Recovery and Resilience Plan (Piano Nazionale di Ripresa e Resilienza, PNRR) - Project: "SoBigData.it - Strengthening the Italian RI for Social Mining and Big Data Analytics" - Prot. IR0000013 - Avviso n. 3264 del 28/12/2021

References

1. Arslan, T.P., Eryiğit, G.: Incorporating dropped pronouns into coreference resolution: the case for Turkish. In: Proceedings of the 17th Conference of the European Chapter of the Association for Computational Linguistics: Student Research Workshop, pp. 14–25 (2023)
2. Bjerva, J., Augenstein, I.: Does typological blinding impede cross-lingual sharing? In: Proceedings of the 16th Conference of the European Chapter of the Association for Computational Linguistics: Main Volume, pp. 480–486. Association for Computational Linguistics, Online (2021)
3. Bonetti, F., Leonardelli, E., Trotta, D., Guarasci, R., Tonelli, S.: Work hard, play hard: collecting acceptability annotations through a 3D game. In: Proceedings of the Thirteenth Language Resources and Evaluation Conference, LREC 2022, pp. 1740–1750. ELRA, Marseille, France (2022)
4. Bosco, C., Montemagni, S., Simi, M.: Converting Italian treebanks: towards an Italian Stanford dependency treebank. In: Proceedings of the 7th Linguistic Annotation Workshop and Interoperability with Discourse, pp. 61–69. ACL, Sofia, Bulgaria (2013)
5. Candito, M., et al.: Deep syntax annotation of the Sequoia French Treebank. In: Calzolari, N., et al. (eds.) Proceedings of the Ninth International Conference on Language Resources and Evaluation, LREC 2014, pp. 2298–2305. European Language Resources Association (ELRA), Reykjavik, Iceland (2014)
6. Chi, E.A., Hewitt, J., Manning, C.D.: Finding universal grammatical relations in multilingual BERT. In: Proceedings of the 58th Annual Meeting of the Association for Computational Linguistics, pp. 5564–5577. Association for Computational Linguistics, Online (2020). https://doi.org/10.18653/v1/2020.acl-main.493
7. Clark, K., Khandelwal, U., Levy, O., Manning, C.D.: What does BERT look at? An analysis of BERT's attention. In: Proceedings of the 2019 ACL Workshop BlackboxNLP: analyzing and Interpreting Neural Networks for NLP, pp. 276–286. ACL, Florence, Italy (2019). https://doi.org/10.18653/v1/W19-4828
8. Conneau, A., et al.: Unsupervised cross-lingual representation learning at scale. In: Proceedings of the 58th Annual Meeting of the Association for Computational Linguistics, pp. 8440–8451. ACL, Online (2020)
9. Conneau, A., Lample, G.: Cross-lingual language model pretraining. In: Advances in Neural Information Processing Systems 32: Annual Conference on Neural Information Processing Systems 2019. NeurIPS 2019, pp. 7057–7067. Vancouver, BC, Canada (2019)
10. Conneau, A., Rinott, R., Lample, G., Williams, A., Bowman, S.R., Schwenk, H., Stoyanov, V.: XNLI: evaluating cross-lingual sentence representations. In: Proceedings of the 2018 Conference on Empirical Methods in Natural Language Processing, Brussels, Belgium, pp. 2475–2485. ACL (2018). https://doi.org/10.18653/v1/d18-1269
11. Conneau, A., Wu, S., Li, H., Zettlemoyer, L., Stoyanov, V.: Emerging cross-lingual structure in pretrained language models. In: Proceedings of the 58th Annual Meeting of the Association for Computational Linguistics, pp. 6022–6034. Association

for Computational Linguistics, Online (2020). https://doi.org/10.18653/v1/2020.acl-main.536

12. Devlin, J., Chang, M.W., Lee, K., Toutanova, K.: BERT: pre-training of deep bidirectional transformers for language understanding. In: Proceedings of the 2019 Conference of the North American Chapter of the Association for Computational Linguistics: Human Language Technologies, Volume 1 (Long and Short Papers), pp. 4171–4186. ACL, Minneapolis, Minnesota (2019). https://doi.org/10.18653/v1/N19-1423

13. Gargiulo, F., et al.: An electra-based model for neural coreference resolution. IEEE Access **10**, 75144–75157 (2022). https://doi.org/10.1109/ACCESS.2022.3189956

14. Goldberg, Y.: Assessing BERT's syntactic abilities. CoRR abs/1901.05287 (2019)

15. Guarasci, R., Damiano, E., Minutolo, A., Esposito, M.: When lexicon-grammar meets open information extraction: a computational experiment for Italian sentences. In: Proceedings of the Sixth Italian Conference on Computational Linguistics CLIC-IT, vol. 2481. CEUR-WS.org, Bari, Italy (2019)

16. Guarasci, R., Damiano, E., Minutolo, A., Esposito, M., De Pietro, G.: Lexicon-grammar based open information extraction from natural language sentences in Italian. Expert Syst. Appl. **143**, 112954 (2020). https://doi.org/10.1016/j.eswa.2019.112954

17. Guarasci, R., Minutolo, A., Damiano, E., De Pietro, G., Fujita, H., Esposito, M.: ELECTRA for neural coreference resolution in Italian. IEEE Access **9**, 115,643–115,654 (2021). https://doi.org/10.1109/ACCESS.2021.3105278

18. Guarasci, R., Silvestri, S., De Pietro, G., Fujita, H., Esposito, M.: BERT syntactic transfer: a computational experiment on Italian, French and English languages. Comput. Speech Lang. **71** (2022). https://doi.org/10.1016/j.csl.2021.101261

19. Guarasci, R., Silvestri, S., De Pietro, G., Fujita, H., Esposito, M.: Assessing BERT'S ability to learn Italian syntax: a study on null-subject and agreement phenomena. J. Ambient. Intell. Humaniz. Comput. **14**(1), 289–303 (2023)

20. Guillaume, B., de Marneffe, M.C., Perrier, G.: Conversion et améliorations de corpus du Français annotés en Universal Dependencies. Traitement Automatique des Langues **60**(2), 71–95 (2019)

21. Hewitt, J., Manning, C.D.: A structural probe for finding syntax in word representations. In: Proceedings of the 2019 Conference of the North American Chapter of the Association for Computational Linguistics: Human Language Technologies, Volume 1 (Long and Short Papers), pp. 4129–4138. ACL, Minneapolis, Minnesota (2019). https://doi.org/10.18653/v1/N19-1419

22. Jawahar, G., Sagot, B., Seddah, D.: What does BERT learn about the structure of language? In: Proceedings of the 57th Annual Meeting of the Association for Computational Linguistics, pp. 3651–3657. ACL, Florence, Italy (2019). https://doi.org/10.18653/v1/P19-1356

23. Li, W., Zhu, L., Shi, Y., Guo, K., Cambria, E.: User reviews: sentiment analysis using lexicon integrated two-channel CNN-LSTM family models. Appl. Soft Comput. **94**, 106435 (2020). https://doi.org/10.1016/j.asoc.2020.106435

24. Linzen, T., Baroni, M.: Syntactic structure from deep learning. Annu. Rev. Linguist. **7**, 195–212 (2021). https://doi.org/10.1146/annurev-linguistics-032020-051035

25. Minutolo, A., Guarasci, R., Damiano, E., De Pietro, G., Fujita, H., Esposito, M.: A multi-level methodology for the automated translation of a coreference resolution dataset: an application to the Italian language. Neural Comput. Appl. **34**(24), 22,493–22,518 (2022). https://doi.org/10.1007/s00521-022-07641-3

26. Pires, T., Schlinger, E., Garrette, D.: How multilingual is multilingual BERT? In: Proceedings of the 57th Annual Meeting of the Association for Computational Linguistics, pp. 4996–5001. ACL, Florence, Italy (2019). https://doi.org/10.18653/v1/P19-1493

27. Ravishankar, V., Kulmizev, A., Abdou, M., Søgaard, A., Nivre, J.: Attention can reflect syntactic structure (if you let it). In: Proceedings of the 16th Conference of the European Chapter of the Association for Computational Linguistics: Main Volume, pp. 3031–3045. Association for Computational Linguistics, Online (2021)

28. Sennrich, R., Haddow, B., Birch, A.: Neural machine translation of rare words with subword units. In: Erk, K., Smith, N.A. (eds.) Proceedings of the 54th Annual Meeting of the Association for Computational Linguistics (Volume 1: Long Papers), pp. 1715–1725. Association for Computational Linguistics, Berlin, Germany (2016). https://doi.org/10.18653/v1/P16-1162

29. Silveira, N., et al.: A gold standard dependency corpus for English. In: Proceedings of the Ninth International Conference on Language Resources and Evaluation (LREC'14), pp. 2897–2904. ELRA, Reykjavik, Iceland (2014)

30. Silvestri, S., Gargiulo, F., Ciampi, M., De Pietro, G.: Exploit multilingual language model at scale for ICD-10 clinical text classification. In: ISCC 2020, pp. 1–7. IEEE, Rennes, France (2020). https://doi.org/10.1109/ISCC50000.2020.9219640

31. Simi, M., Bosco, C., Montemagni, S.: Less is more? Towards a reduced inventory of categories for training a parser for the Italian Stanford dependencies. In: Proceedings of the Ninth International Conference on Language Resources and Evaluation (LREC'14), pp. 83–90. ELRA, Reykjavik, Iceland (2014)

32. Sukthanker, R., Poria, S., Cambria, E., Thirunavukarasu, R.: Anaphora and coreference resolution: a review. Inf. Fusion **59**, 139–162 (2020). https://doi.org/10.1016/j.inffus.2020.01.010

33. Tenney, I., Das, D., Pavlick, E.: BERT rediscovers the classical NLP pipeline. In: Proceedings of the 57th Annual Meeting of the Association for Computational Linguistics, pp. 4593–4601. ACL, Florence, Italy (2019). https://doi.org/10.18653/v1/P19-1452

34. Tenney, I., et al.: What do you learn from context? probing for sentence structure in contextualized word representations. In: 7th International Conference on Learning Representations, ICLR 2019. New Orleans, LA, USA (2019)

35. Trotta, D., Guarasci, R., Leonardelli, E., Tonelli, S.: Monolingual and cross-lingual acceptability judgments with the Italian CoLA corpus. In: M.F. Moens, X. Huang, L. Specia, S.W.t. Yih (eds.) Findings of the Association for Computational Linguistics: EMNLP 2021, pp. 2929–2940. Association for Computational Linguistics, Punta Cana, Dominican Republic (2021). https://doi.org/10.18653/v1/2021.findings-emnlp.250

36. Warstadt, A., Singh, A., Bowman, S.R.: Neural network acceptability judgments. Trans. Assoc. Comput. Linguist. **7**, 625–641 (2019). https://doi.org/10.1162/tac_a_00290

Rev Gadget: A Java Deserialization Gadget Chains Discover Tool Based on Reverse Semantics and Taint Analysis

Yifan Luo and Baojiang Cui[✉]

Beijing University of Posts and Telecommunications, Beijing, China
{luexp,cuibj}@bupt.edu.cn

Abstract. Java is a widely utilized object-oriented programming language known for its powerful cross-platform features. The object serialization mechanism in Java enables the persistence and propagation of objects over the network. However, this capability also introduces deserialization vulnerabilities. Java deserialization vulnerabilities arise when an application has a deserialization entry point. Attackers can exploit this by constructing malicious serialized data, leading to the invocation of dangerous methods and resulting in issues such as command execution and information leakage. The malicious serialized data typically consists of a series of method gadget chains, commonly referred to as 'gadget chains.'To mitigate deserialization vulnerabilities, it is crucial to discover and understand the gadget chains within a program. This article introduces a tool named Rev Gadget, designed for discovering Java deserialization gadget chains. The tool employs reverse semantics and taint analysis for this purpose. The process begins with static analysis using CodeQL to identify entry and dangerous methods. Subsequently, reverse semantics are defined, and taint analysis is utilized to uncover potential deserialization gadget chains. Comparative experiments conducted using Gadget Inspector demonstrate that Rev Gadget exhibits superior detection rates, accuracy, and overall performance.

1 Introduction

Java is a widely-used object-oriented programming language renowned for its powerful cross-platform features.

Its object serialization mechanism enables the persistent storage or transmission of objects over the network in the form of byte streams, facilitating data exchange between various applications and systems. Nevertheless, this powerful feature introduces a significant security concern in Java-deserialization vulnerabilities.

The core of the Java deserialization vulnerability lies in scenarios where an application receives serialized data from an untrusted source and attempts to deserialize it. In such situations, an attacker can construct malicious serialized data containing harmful code, exploiting the deserialization process. This vulnerability opens the door to potential remote code execution (RCE) attacks,

L. Barolli (Ed.): EIDWT 2024, LNDECT 193, pp. 229–240, 2024.
https://doi.org/10.1007/978-3-031-53555-0_22

authentication control bypass, sensitive data theft, denial-of-service attacks, and other malicious activities, as illustrated in Fig. 1.

Deserialization vulnerabilities, though not new, have garnered significant attention in recent years, particularly due to their discovery in numerous large enterprise and open-source applications, such as Apache Struts, Spring Framework, WebLogic, and WebSphere (e.g., CVE-2016-4437 [1]). The impact of deserialization vulnerabilities is evident, affecting systems using versions preceding Shiro 1.2.4, as demonstrated by CVE-2016-4437. The exposure of these vulnerabilities has spurred increased research efforts and a demand for solutions to address Java deserialization vulnerabilities.

Researchers, security experts, and developers have diligently sought ways to mitigate these vulnerabilities and safeguard Java applications from potential attacks. Strategies include implementing deserialization whitelists, utilizing secure deserialization libraries, adhering to the principle of least privilege, and enhancing the overall security of the Java platform itself.

Furthermore, the academic and security communities have engaged in discussions regarding the intricacies of Java deserialization vulnerabilities, potential attack scenarios, vulnerability classification and analysis, as well as the development of new attack and defense techniques. The identification of deserialization vulnerabilities necessitates the discovery of gadget chains. The automation of gadget chians discovering has become a focal point for an increasing number of researchers. These studies contribute to a deeper understanding of the issue and aid in enhancing the security of Java applications.

Within this context, this article aims to explore the intricacies of Java deserialization vulnerabilities and propose a dadget chains discovering method based on reverse semantics and taint analysis. It is intended to furnish researchers, security experts, and developers with valuable insights on effectively addressing these risks, ensuring the security and reliability of Java applications.

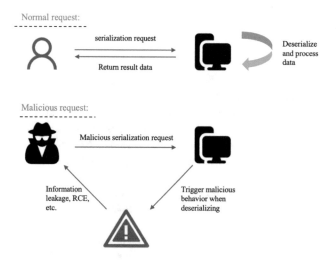

Fig. 1. Java Deserialization Vulnerability Attack

2   Related Work

Enhancing the efficiency and accuracy of the automated discovery of gadget chains in Java deserialization vulnerabilities has consistently posed a significant challenge for researchers in this field.

V. Benjamin Livshits [8] et al. proposed a static analysis technology to detect many recently discovered application vulnerabilities, such as SQL injection, cross-site scripting attacks and HTTP splitting attacks. These vulnerabilities originate from unchecked input and are widely considered the most common source of security vulnerabilities in web applications. A static analysis method based on scalable and accurate pointer analysis is proposed. In their system, user-supplied vulnerability specifications are automatically translated into static analyzers. Their approach is able to find all vulnerabilities matching the specification in statically analyzed code.

Shawn Rasheed [11] et al. proposed a hybrid method that combines static analysis with fuzzing to detect serialization vulnerabilities. The novelty of this approach is that it uses heap abstraction to guide fuzz testing of serialization vulnerabilities in Java libraries. This produces results quickly and efficiently, and automatically validates static analysis reports. Sicong Cao [2] et al. proposed a new gadget chain discovering method, GCMiner, which captures explicit and implicit method calls to identify more gadget chains, and adopts a coverage-oriented object generation method to generate effective injections. Object used for fuzz testing.

Xingchen Chen [4] and others proposed a Java Gadget detection framework, which uses the Soot framework to perform static analysis to generate graph data on Neo4j, and then performs dynamic analysis on Neo4j to mine potential Java deserialization gadget chains. Sicong Cao [3] et al. proposed a novel hybrid solution called ODDFUZZ to efficiently discover Java deserialization vulnerabilities. First, ODDFUZZ performs lightweight static taint analysis to identify candidate gadget chains that may lead to deserialization vulnerabilities. In this step, ODDFUZZ tries to find all candidate chains and avoid false positive results. ODDFUZZ then performs directed gray-box fuzzing (DGF) to explore these candidate chains and generate PoC test cases to improve accuracy. Specifically, ODDFUZZ applies a structure-aware seed generation method to ensure test case validity and adopts a novel hybrid feedback and stepping strategy to guide directed fuzz testing.

2.1   Taint Analysis

Taint analysis is a static analysis technology that performs security analysis on software without running the program to detect whether there are software vulnerabilities, such as information leakage, SQL injection, remote method execution, memory overflow, etc.

Taint analysis technology first needs to analyze the control flow graph, program dependency graph, etc. of the software. Soot [12] is a Java static analysis tool that analyzes Java byte data and converts it into Jimple. Jimple is an intermediate language that stores typical three-address code data and is used to describe the grammatical structure of the source code. Control flow graphs can be easily generated by analyzing Jimple. After Tabby [13] used Soot to analyze the target program to generate Jimple, he further analyzed it on Jimple to obtain the control flow graph, then used Neo4j to store the graph information, and finally used the query function provided by Neo4j to analyze the software vulnerabilities. Neo4j [9] is a graph database that can conveniently store graph data and use query syntax to obtain target nodes. CodeQL [5] is a multi-language static analysis tool. It can directly generate the control flow graph, program dependency graph, etc. of the program by analyzing the grammatical structure of the source code. It also provides query syntax and can directly perform target data on the graph. Query without relying on a third-party graph database.

Taint analysis can be abstracted into the form of a triplet [sources, sinks, sanitizers], where source is the taint source, which represents the direct introduction of untrusted data or confidential data into the system; sink is the taint collection point, which represents the direct generation of Security-sensitive operations (violating data integrity) or leaking private data to the outside world (violating data confidentiality); sanitizer means harmless processing, which means that data transmission is no longer harmful to the information security of the software system through data encryption or removal of harmful operations. cause harm. Taint analysis is to analyze whether the data introduced by the taint source in the program can be directly propagated to the taint collection point without harmless processing. If not, it means that the system is information flow safe; otherwise, it means that the system has security issues such as privacy data leakage or dangerous data operations.

2.2 Fuzz

Fuzz testing is a dynamic analysis technique that aims to pass a large amount of random, abnormal, invalid or illegal input data to the input of the program under test to observe how the program behaves in response to these abnormal inputs.

In fuzz testing, the input of the test program is often called a seed. The generation, mutation and screening of seeds are the focus of fuzz testing research. AFL [7] is a fuzz testing framework that is often used for testing C language programs. It uses coverage as a criterion for screening seeds. When the coverage rate increases, it indicates that it is a high-quality seed and can be selected into the seed pool for the next test. Iterate. AFL also defines several mutation behaviors, such as bit flipping, shuffling, deletion, etc. Syzkaller [6] is a fuzz testing framework for the Linux kernel. It generates a running program based on the system call template, and then starts the virtual machine to run the target program to detect whether the system will trigger a crash.

In the field of Java, Padhye, R. [10] proposed a Java fuzz testing platform JQF based on coverage feedback. It is based on Junit's QuickCheck to write a test method. The method contains some parameters, and then in the method body, Call the method to be tested, etc., and verify the parameters and results in the code. JQF will detect the byte code of the program, generate seeds and run them to obtain coverage to guide subsequent generation. The generation algorithm here is extensible, so researchers can define their own fuzz algorithms. ODDFuzz proposes a fuzz testing method for Java deserialization vulnerabilities. After obtaining the potential gadget chains through taint analysis, it uses the attribute graph to generate structured seeds, and then provides guidance based on the seed distance and Gadget coverage of the seeds. Subsequent generation of mutations.

3   Gadget Chains Discovering Algorithm

The gadget chains is the calling information of a series of methods in the program that can trigger dangerous methods. Once the gadget chains of the current program is known, dangerous operations such as remote command execution can be realized through deserialization entry.

The conventional gadget chains discovering method starts from the entry method and looks for paths that can trigger dangerous methods. This article defines a reverse semantics, starting from the dangerous method, reversely searching the calling path, and through taint analysis, verifying whether the data of the calling chain can be delivered. The steps included in this article's gadget chains discovering algorithm are roughly as follows:

1. Define method predicate semantics, analyze the program to obtain entry methods and dangerous methods.
2. Define reverse semantics and analyze the program to obtain potential gadget chains.
3. Use taint analysis to analyze whether the entry parameters of the gadget chains are controllable.

The detailed method of each step will be expanded in detail below.

3.1   Method Predicate Analysis

The entry method and dangerous method in the deserialization gadget chains have certain universality. For example, the entry method is a serializable method, and the dangerous method is a method that can execute commands. By collecting and analyzing existing gadget chains, a list of entry and exit methods and dangerous methods is summarized.

Table 1 records the entry method that is often used as a deserialization gadget chains. It is usually a method of a serializable class. During deserialization, this method will be called and can be used.

Table 1. Source methods

Class/Interface	Method
Any	hashcode()
Any	equals()
java.util.Map	get()
java.util.Map	put()
java.io.Serializable	readObject()
java.io.Serializable	readObjectNoData()
java.io.Serializable	readResolve()
java.util.Comparator	compare()
java.io.Externalizable	readExternal()
java.io.ObjectInputValidation	validateObject()
java.lang.reflect.InvocationHandler	Invoke()
javassist.util.proxy.MethodHandler	Invoke()
groovy.lang.MetaClass	invokeMethod()
groovy.lang.MetaClass	invokeConstructor()
groovy.lang.MetaClass	invokeStaticMethod()

Table 2. Sink methods

Class/Interface	Method
java.lang.Runtime	exec()
java.lang.reflect.Method	invoke()
java.net.URL	openStream()
java.net.URLConnection	connect()
java.lang.ProcessBuilder	ProcessBuilder()
java.nio.file.Files	newInputStream()
java.nio.file.Files	newOutputStream()
java.nio.file.Files	newBufferedReader()
java.nio.file.Files	newBufferedWriter()
java.io.FileInputStream	FileInputStream()
java.io.FileOutputStream	FileOutputStream()
javax.script.ScriptEngine	Eval()

Table 2 records dangerous methods that are often used as deserialization gadget chains, which are usually methods that perform dangerous behaviors. Dangerous operations such as command execution, file reading, remote class loading, etc. When this method is called, it may cause problems such as RCE and information leakage.

Whether a program has a gadget chains, you first need to determine whether it has an entry method or a dangerous method. If it does not exist, there is no need to perform subsequent analysis. CodeQL is a static analysis tool that provides a predicate definition function that can be used to filter methods in the program to obtain the target method. This section analyzes the characteristics of entry methods and dangerous methods and defines method predicate semantics to determine whether there are entry methods and dangerous methods in the program.

The definition method of method predicate semantics is qualified by the package name, class or interface name and method name. First, obtain information about all methods and classes of the program through CodeQL static analysis, and then compare whether the method names in the method and predicate definitions are the same. If they are the same, obtain all the classes and interfaces inherited by the method class, and then use these classes and interfaces with the predicate semantics The package name in the definition is compared with the class or interface name. If the comparison is successful, it is the method defined by the predicate semantics.

Specifically, if you want to obtain all readObject() methods that inherit the java.io.Serializable interface class, you only need to extract the java.io package name, Serializable class name, and readObject method name, and then define the method through the method predicate semantic algorithm Predicate semantics, you can get the target method list.

Through the entry method and dangerous method information collected in Table 1 and Table 2, and then defined by the method predicate semantics, method predicate analysis can be performed, as shown in Fig. 2. For all method nodes, after method predicate analysis, the Source node and Sink node in the method node can be obtained, such as the green and red parts in Fig. 2, so that it can be judged whether subsequent analysis is required.

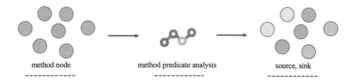

method node   method predicate analysis   source, sink

Fig. 2. Method predicate analysis

3.2 Reverse Semantic Analysis

Reverse semantics refers to the reverse call semantics of a method. The gadget chains is composed of a series of sequential method calls, such as A→B→C, that is, A calls B, and B calls C. Reverse semantics is the reverse process of the calling sequence. For example, we define the reverse semantics of C to refer to the method that can finally be called to C through 0 or more intermediate methods, that is, the A and B methods in the gadget chains shown. In the discovering gadget chains, we obtain the method call node graph by defining the reverse semantics of dangerous methods, and filter out the entry method nodes through the entry method predicate definition in the graph, and finally obtain the calls from the entry method to the dangerous method in the graph path.

Using the powerful static analysis function of CodeQL, we implemented the definition of reverse semantics. Figure 3 is the overall process of reverse semantic analysis. Specifically, first define the method node A that directly calls the dangerous method, then define the method node B that calls the method node A, and so on, recursively define subsequent method nodes. Finally, all nodes that directly or indirectly call dangerous method nodes can be obtained, as shown in the reverse semantic part of Fig. 3, where red nodes are dangerous method nodes and blue nodes are nodes that directly or indirectly call dangerous methods.

After obtaining the call node graph, use the entry method predicate semantics to obtain the entry method node from the graph. When there is an entry method node in the graph, it indicates that there is a calling chain from the entry method to the dangerous method. As shown in the entry method predicate in Fig. 3, the green node is the found entry method node. After finding the entry method node, you can find a calling link from the entry method node to the dangerous method node.

The gadget chains graph records the potential gadget chains, which methods are used to finally call the dangerous method step by step, but it cannot be determined whether the parameters of the entry method can be controllably passed to the dangerous method, so further taint analysis is needed. Filter the gadget chains.

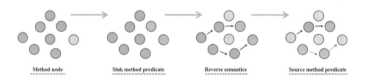

Fig. 3. Reverse semantic analysis

3.3 Taint Analysis Screening

Taint analysis is used to determine whether tainted data can be transferred from the tainted source to the target source. In the previous step, we obtained the gadget chains graph. The taint source is the member variable of the class in the gadget chains, and the target source is the parameter of the dangerous method. Using taint analysis, you can determine whether the gadget chains is available and further filter the results, thereby improving the accuracy of detection.

In taint analysis, the most critical thing is the tainted data flow rules. Predicates are used in CodeQL to define the tainted data flow rules. The rules are shown in Table 3.

Table 3. Tainted data transfer rules

Type	Stmt
Simple assignment	a = b
Create new variable	a = new statement
Class attribute assignment	a.f = b
Class attribute loading	a = b.f
Static class attribute assignment	Class.field = b
Static class attribute loading	a = Class.field
Array assignment	a[i] = b
Array loading	a = b[i]
Forced conversion	a = (T) b
Assignment function call	a = b.func(c)
Function call	b.func(c)
Function return	return stmt

After defining the taint analysis rules, perform taint analysis on the gadget chains mined by reverse semantics, as shown in Fig. 4. First, analyze the gadget chains information obtained by reverse semantic discovering, and obtain the member variables of all classes in the gadget chains. Then use taint analysis to perform taint data on the taint source data in turn to determine whether the taint data can be transferred to the target source. If not, continue to analyze the next taint source. When all taint sources cannot be delivered to the target source, it indicates that the gadget chains is unavailable.

Eventually, we will be able to obtain an exploitable gadget chains in the target program, which can be exploited by the deserialization vulnerability.

Fig. 4. Reverse semantic analysis

4   Experiment

For the Java basic library: Apache-commons-collection 4.0, a comparative experiment was conducted using the tools Rev Gadget and Gadget Inspector designed in this paper. A host using a 3.2 GHz Intel 6-core i7 processor and 16 GB of memory was selected for the experiment. The selected operating system was Ubuntu20.04.

Table 4. Total number of gadget

Tool	Number
Gadget Inspector	4
Rev Gadget	31

Table 4 records the total number of gadget chains discovered by the two tools respectively. Gadget Inspector found 4 gadget chains, and Rev Gadget found 31 gadget chains. Experiments have proven that Rev Gadget can discover more gadget chains than Gadget Inspector. This is because Rev Gadget uses reverse semantics to analyze gadget chains and can more comprehensively dig out the call links from entry methods to dangerous methods. Therefore, A higher total number of gadget chains is detected.

For each gadget chains information, a corresponding malicious serialized object is constructed, and then the serialized binary object is serialized through the deserialization interface to analyze whether the gadget chains can be exploited. Taking the chain from Flat3Map::readObject() to InvokerTransformer::transform() as an example, we construct a link that calls get() of the DefaultedMap object in FlatMap, and then calls the transform of InvokerTransformer to verify whether

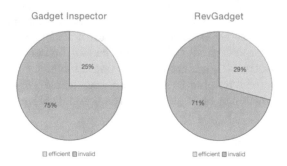

Fig. 5. Detect Accuracy

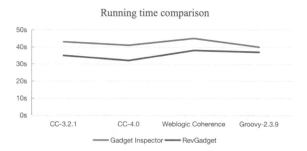

Fig. 6. Running time comparison

the chain can be exploited. The experiment ultimately found that only one of the four gadget chains discovered by Gadget Inspector could be exploited. Of the 31 gadget chains discovered by Rev Gadget, 9 are exploitable. The accuracy is calculated using Publication 1, and Fig. 5 is obtained. Experiments have proven that Rev Gadget has higher accuracy than Gadget Inspector.

$$P = (M/N) * 100\%, \tag{1}$$

In addition, time-consuming analysis was conducted for CC-3.2.1, cc-4.0, weblogic coherence, and groovy-2.3.9 respectively, and Fig. 6 was obtained. Experiments show that RevGadget has a higher detection speed than Gadget Inspector.

5   Conclusion

In this article, we propose a new method for discovering Java deserialization gadget chains.

We first use CodeQL to conduct static analysis, obtain entry method and dangerous method information, define reverse semantics, and finally use taint analysis to obtain potential gadget chains. And compared with the Gadget Inspector tool, experiments show that the tool Rev Inspector proposed in this article has higher detection rate, accuracy and detection rate. After defining the reverse semantics in this article, it is screened through taint analysis, and Fuzz technology can be used to improve the accuracy of screening. This article uses manual analysis when verifying the gadget chains. Later, the logic of manual analysis can be abstracted into a model to implement automated verification analysis.

References

1. CVE-2016-4437.   http://cve.mitre.org/cgi-bin/cvename.cgi?name=CVE-2016-4437
2. Cao, S., et al.: Improving java deserialization gadget chain mining via overriding-guided object generation (2023). arXiv preprint arXiv:2303.07593
3. Cao, S., et al.: ODDFUZZ: Discovering Java Deserialization Vulnerabilities via Structure-Aware Directed Greybox Fuzzing (2023). arXiv preprint arXiv:2304.04233
4. Chen, X., Wang, B., Jin, Z., Feng, Y., Li, X., Feng, X., Liu, Q.: Tabby: automated gadget chain detection for java deserialization vulnerabilities. In: 2023 53rd Annual IEEE/IFIP International Conference on Dependable Systems and Networks (DSN), pp. 179–192. IEEE, June 2023
5. GitHub. (2023). CodeQL. GitHub. Retrieved November 14, 2023. https://github.com/github/codeql
6. Google. syzkaller. GitHub. Retrieved November 13, 2023 (2023). https://github.com/google/syzkaller
7. Google. American Fuzzy Lop. GitHub. Retrieved Month Day, Year (2021). https://github.com/google/AFL

8. Livshits, V.B., Lam, M.S.: Finding security vulnerabilities in Java applications with static analysis. In: USENIX Security Symposium, vol. 14, pp. 18, August 2005

9. Neo4j: Neo4j Graph Database. Retrieved November 14, 2023, (2023). https://neo4j. com/

10. Padhye, R., Lemieux, C., Sen, K.: JQF: coverage-guided property-based testing in Java. In Proceedings of the 28th ACM SIGSOFT International Symposium on Software Testing and Analysis, pp. 398–401, July 2019

11. Rasheed, S., Dietrich, J.: A hybrid analysis to detect Java serialisation vulnerabilities. In: Proceedings of the 35th IEEE/ACM International Conference on Automated Software Engineering, pp. 1209–1213, December 2020

12. Soot-oss. Soot: A Java Optimization Framework. GitHub (2023). https://github. com/soot-oss/soot

13. wh1t3p1g. Tabby: A Cat-Inspired Terminal App with Advanced Features. GitHub (2023). https://github.com/wh1t3p1g/tabby

A Method for Calculating Degree of Deviation of Posts on Social Media for Public Opinion in Flaming Prediction

Yusuke Yoshida[1(✉)] and Kosuke Takano[2]

[1] Course of Information and Computer Sciences, Graduate School of Engineering,
Kanagawa Institute of Technology, Atsugi, Japan
s2285011@cco.kanagawa-it.ac.jp
[2] Department of Information and Computer Sciences, Faculty of Information,
Kanagawa Institute of Technology, Atsugi, Japan
takano@ic.kanagawa-it.ac.jp

Abstract. This study presents a method to calculate the degree of deviation of posts on social media from a set of opinions from diverse viewpoints. It's important to prevent the post on social media from getting flamed; however, even if the post does not contain malice intent, it is very difficult to predict whether the post blows up, since it may occur due to the context-dependent reasons such that the post is not accepted by public opinion. The feature of proposed method is to calculate the degree of deviation based on the magnitude of conflict between the opinions from various viewpoints, which is constructed as external resources. This allows our method to judge how moderate the opinion will be accepted by public opinion before posting it. In this study, by the experiments using a pseudo dataset of public opinion created using a Large Language Model, we evaluate the feasibility of the proposed method.

1 Introduction

While the spread of social media [1–3] has increased opportunities for individuals to share information, the phenomenon of flames over opinions and comments posted on social media has become a social problem. Risks caused by flaming varies from loss of social trust leading a business withdrawal to personal attack such as leaks of personal information. Against this background, it's an important issue to prevent the post on social media from getting flamed and various flaming detection methods have been proposed focusing on specific flaming [4–6]; however, even if the post does not contain malice intent, it is very difficult to predict whether the post blows up or not, since it may occur due to the context-dependent reasons such that the post is not accepted by public opinion.

It is deemed that one of the factors causing such unintended flaming is that our individual values and perceptions vary due to the diversity of opinions among people, and as the result, there is a possibility that divergences in thoughts and perceptions between posted opinion and public opinion occur. Therefore, it is significantly needed to develop a method to determine the degree of conflicts arising from public opinion consisting of various opinions, even if opinions in the posted opinions seem to have no problems.

In this study, we propose a method to calculate the degree of deviation of posts on social media from a set of opinions from diverse viewpoints. The feature of proposed method is to calculate the degree of deviation based on the magnitude of conflict between the opinions from various viewpoints, which is constructed as external resources. This allows our method to judge how moderate the opinion will be accepted by public opinion before posting it, so that it is expected to reduce the risk of flaming.

In this study, by the experiments using a pseudo dataset of public opinion for flaming estimation created using a Large Language Model (LLM), we evaluate the feasibility of the proposed method. The detail of pseudo dataset of public opinion is introduced in Sect. 4.

2   Related Work

There have been many methods for the flaming detection. Some studies focus on number of posting, discriminatory and defamatory expression, date and time of posting, person of posting [4, 5, 7–10].

Osone et al. proposed a method for detecting the flaming status and quantifying the scale of flames basing on the time trend of the number of comments appeared on the Web [4]. Sakai et al. proposed a method to use YouTube comments for the flaming detection and judge the flaming status with prediction models trained by SVM (Support Vector Machine), decision trees, and MLP (Multi Layer Perceptron) [5]. Kawakami et al. classified posts into flaming posts and non-flaming posts and applied Word2Vec [6] trained with words appeared in the flaming posts to determine whether a target post gets flamed [7].

Yoshida et al. proposed a method for judging offensive sentences using fine-tuned model trained with a set of offensive sentences collected manually [8]. Takahashi et al. proposed a method for the flaming prediction by calculating positive and negative polarity based on a sentiment dictionary created based on word dependency and emoticons appeared in each post [9]. Segawa et al. analyzed characteristics of offensive user and classified attack types based on the characteristics of attackers and attackers' relationships [10]. Iwasaki et al. proposed a method for judging if flaming is arising or not by calculating a ratio of positive to negative responses to timeline in posting [11].

However, these conventional studies cannot solve the issue to predict the possibility of flaming before posting opinions even if malice intent, discriminatory and defamatory expression are not included in them.

3   Proposed Method

Figure 1 shows a flaming estimation model that focuses on four factors causing flaming: (1) degree of deviation, (2) expressions, (3) topicality, and (4) speed of propagation of topics. In Fig. 1, "degree of deviation" represents the degree of deviation between posted opinion and public opinion, "expression" represents expressions that cause flames due to including discriminatory and slanderous expressions, "topicality" represents the impact of the topic in society, and "speed of propagation" represents the ease with which the postings spreads on each social media. In this study, we especially focus on "degree of deviation" described in Sect. 2 and propose a method to calculate the degree of deviation of posts on social media from a set of opinions from diverse viewpoints.

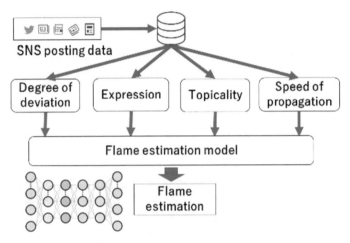

Fig. 1. Flame estimation model for several factors

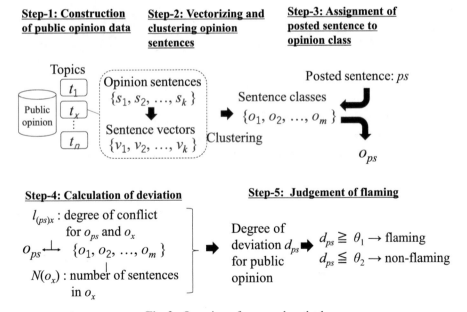

Fig. 2. Overview of proposed method

Figure 2 shows the overview of the proposed method. Our method executes in the following Steps 1 to 5.

Step-1: Construction of an opinion data set representing public opinion with various viewpoints

In Step-1, we construct an opinion data set representing public opinion with various viewpoints using the schema as described in Sect. 3. We assume that public opinion is formed from several topics T:

$$T = \{t_1, t_2, ..., t_n\} \tag{1}$$

where, each t_x consists of a set of opinion classes O that comes from various viewpoints for the topic t_x. In addition, each opinion class o_y has a set of opinion sentences S consisting of opinion sentences:

$$O = \{o_1, o_2, ..., o_m\} \tag{2}$$

$$S = \{s_1, s_2, ..., s_j\} \tag{3}$$

Step-2: Vectorizing and clustering opinion sentences in the opinion data set

In Step-2, opinion sentences are vectorized into sentence vectors V using an embedded representation for term or sentence learned by BERT (Bidirectional Encoder Representations from Transformers) [16], and so on.

$$V = \{v_1, v_2, ..., v_j\} \tag{4}$$

Sentence vectors are classified into m-groups using clustering method such as k-means [14] with cosine measure. After clustering, the classified m-groups form opinion classes O defined in Step-1. The magnitude of conflict between each opinion class is represented as L:

$$L = \{l_{12}, l_{13}, l_{23}, l_{24}, ..., l_{pq}\} \tag{5}$$

where, l_{pq} is the magnitude of conflict between opinion classes o_p and o_q.

Step-3: Assignment of posted sentence to opinion class

In order to assign a posted sentence ps to the corresponding opinion class o_p, similarity between posted sentence and each sentence s in the cluster in the opinion data set is calculated using a measure such as cosine similarity.

$$Similarity(ps, s) = (ps \cdot s)/|ps||s| \tag{6}$$

For example, using k-nearest neighbor, the opinion class o_{ps} for the posted sentence ps can be predicted according to the number of sentences in the opinion class that are similar to the posted sentence ps.

$$Assignment(ps, O, k) = \text{argmax}\left(N_{ps}^k(O_1), N_{ps}^k(O_2), ..., N_{ps}^k(O_m)\right) \tag{7}$$

where, $N_{ps}^k(O_m)$ means the number of the sentences belonging to the opinion class O_m and similar to the posted sentence ps within top-k.

Step-4: Calculation of deviation

Using the magnitude of conflict L among opinion classes in the target opinion O_{target}, deviation d_{ps} for the O_{target} $(= O)$ is calculated. Suppose the number of opinions in o_x be $N(o_x)$, the deviation d_{ps} of ps for the O_{target} is calculated by weighing the number of opinions in o_x with the conflict degree between o_{ps} and o_x and summing them as follows:

$$d_{ps} = \sum_{x=1}^{m} l_{(ps)x} N(o_x)/m \qquad (8)$$

Step-5: Judgement of flaming

If d_{ps} exceeds a threshold θ_1, the posted sentence ps is predicted that there is a risk for flaming on social media, and if d_{ps} is less than a threshold θ_2, predicted that the posted sentence ps has a less risk.

$$Faming\left(ps, d_{ps}\right) = \begin{cases} 1 \text{ if } d_{ps} \geq \theta_1 \\ 0 \text{ if } d_{ps} \leq \theta_2 \end{cases} \qquad (9)$$

where, 1 and 0 mean the flaming risk is high and low, respectively.

4 Pseudo Dataset of Public Opinion

We define the pseudo dataset of public opinion that consists of a set of sentences for a specified topic in public opinion. This dataset includes classification information for sentences, date and time posted, and degree of conflict between sentences (Table 1).

The magnitude of conflict is defined as an n-step integer for each pair of sentences. For example, the higher the likelihood of conflict, the higher the conflict level. The degree of magnitude is used to calculate the degree of deviation of opinions. The degree of deviation can be calculated using the 3-sigma rule [12]. Since opinion sentences are posted on social media in chronological order, date and time information is assigned to all opinions. Tables 2 and 3 show the schema of the proposed pseudo dataset. As shown in Table 4, the degree of conflict can be assigned to the opinions belonging to each classification based on the degree of conflict of the pairs of classification information.

There are several methods for creating pseudo datasets, such as acquiring data posted on social media. However, in many cases, there are restrictions on the acquisition of posted data in the APIs provided by social media, and posts that have been flamed are often deleted. On the other hand, a Large Language Model (LLM) such as GPT-3 (Generative Pre-trained Transformer 3) [13] has made remarkable progress and succeeded in generating human-like sentences. For this reason, this study challenges to generate a pseudo dataset of public opinion using LLM. The evaluation results on the generation of pseudo dataset by LLM are described in Sect. 5.

Table 1. Pseudo dataset of public opinion

Item	Description
Opinion sentences	A set of pseudo-opinion sentences posted on social media
Class of opinion sentences	Opinion class to which each sentence belongs
Date and time	Date and time opinion sentence was posted on
Degree of conflict	Degree to which opinion sentences conflict with each other

Table 2. Schema of opinion

{(ID, INT), (Opinion, TEXT), (Classification Information, INT), (Time-series, TIME STAMP)}

Table 3. Schema of degree of conflict

{(Classification Information A, INT), (Classification Information B, INT), (Degree of Conflict, INT)}

Table 4. Assigning degrees of conflict to pairs of opinion data

Sentence A	Sentence B	Degree of conflict
[Original Japanese sentence] Nannno maebure mo naku, konna koto ga okoru nante, hontōni kowai desu [English translation] I am really scared that something like this could happen without warning	[Original Japanese sentence] Jishin ga okitara, jibun ya kazoku wo mamoru tame ni, jishin ga okita toki no taisho ho wo shitte okou [English translation] Know what to do in the event of an earthquake to protect yourself and your family	1

5 Experiments

5.1 Experiment 1

We confirm the feasibility of building a pseudo dataset of public opinion using a Large Language Model (LLM).

5.1.1 Experimental Environment

We used the GPT-3 model "text-davinci-003" to collect sentences and generated 1,048 sentences for a topic "Earthquake and disaster area". The pre-trained BERT model "bert-based-japanese-whole-word-masking" was used to generate a sentence vector from the

1,048 sentences. For the clustering of the generated sentences, the elbow method [15] was used to determine the number of clusters.

5.1.2 Experimental Method

The properness and diversity of generated sentences for the specified topic are confirmed.

5.1.3 Results

Table 5 shows an example of generated sentences. Figure 3 shows the results of the elbow method. Table 6 shows the descriptions of each cluster and the number of sentences in each cluster when the number of categories is divided into 10 based on the results of the elbow method (Fig. 3). The average similarity of all the clusters ranged from 0.78 to 0.92. The similarity of the sentences in each cluster was calculated on a cosine scale.

From the results in Tables 5, we can confirm that LLM can generate good quality of sentences for the topic "Earthquake and disaster area". In addition, the result of Table 6 shows that LLM can also generate a set of opinions on the topic from a variety of viewpoints. From these results, we confirm the feasibility of building a pseudo dataset of public opinion using LLM.

Table 5. Example of generated sentences

ID	Sentences
43	[Original Japanese sentence] Jishin de nokosareta hakai no kazukazu wo mite kokoro ga itamimasu [English translation] It pains me to see the destruction left behind by the earthquake
109	[Original Japanese sentence] Jishin to tsunami niyoru hakai wa shinjirarenai hodo desu. Ie ya kaisha no ato wo miruto kokoro ga itamimasu [English translation] The destruction caused by the earthquake and tsunami is incredible. My heart aches when I see the remains of houses and businesses

5.2 Experiment 2

We confirm the feasibility of the proposed method by showing that our method can calculate the degree of deviation based on the magnitude of conflict between the opinions from various viewpoints and judge if posted opinions have the flaming risk.

5.2.1 Experimental Environment

We use the same dataset as in Experiment 1. In addition, 50 flaming sentences and 50 non-flaming sentences, totally 100 sentences, are used. For the vectorization, we used TFIDF-based method. Thresholds θ_1 and θ_2 in Step-5 of Sect. 4 are set to 60% and 40%, respectively.

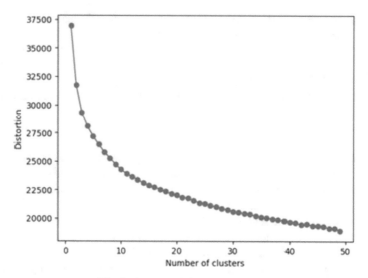

Fig. 3. Results of the elbow method

Table 6. Description of classification information

ID	Description of classification	Number
0	Worried about disaster area and victims	175
1	Impressions of fear	177
2	Damage of earthquake was beyond imagination	79
3	Sympathy for the victims	80
4	Use exclamation mark	114
5	The extent of damage from the earthquake	141
6	Prayers for Survivors	85
7	heart-wrenching feeling	62
8	Comments to facilitate confirmation	86
9	The need to be prepared and united	111

5.2.2 Experimental Method

For the flaming sentence, if the degree of deviation exceeds thresholds θ_1, we regard the prediction result is correct. Meanwhile, for the non-flaming sentence, if the degree of deviation is less than thresholds θ_2, we regard the prediction result is correct. For the evaluation, we calculate the correctness, precision, and recall using the following equations.

$\text{Correctness} = (N_{fl} + N_{non\text{-}fl}) / N_{all}.$
$\text{Precision} = NA_{fl} / NSJ_{fl}.$
$\text{Recall} = NA_{fl} / NS_{fl}.$

where,

N_{all}: number of all sentences.

NA_{fl}: number of correct answers for flaming sentences.

$NA_{non\text{-}fl}$: number of correct answers for non-flaming sentences.

NSJ_{fl}: number of sentences judged to be correct.

NS_{fl}: number of correct sentences for flaming sentences.

5.2.3 Results

For 50 flaming sentences and 50 non-flaming, the average rates of deviation degree are 65.15% and 31.89%, respectively. The number of correct answers for flaming sentences and non-flaming sentences are 36 and 44, respectively. In addition, correctness, precision, and recall are 80.0%, 72.0%, and 85.7%.

These results shows that our method can calculate the degree of deviation based on the magnitude of conflict between the opinions from various viewpoints and judge if posted opinions have the flaming risk or not.

6 Conclusions

In this study, we proposed a method to calculate the degree of deviation of posts on social media from a set of opinions from diverse viewpoints. By the experiments using a pseudo dataset of public opinion created by a Large Language Model, we evaluated the feasibility of the proposed method by showing that our method can calculate the degree of deviation based on the magnitude of conflict between the opinions from various viewpoints and judge whether posted opinions have the flaming risk.

In our future work, we plan to increase the scale of opinion sentences and make the pseudo dataset available to the public. In addition, we will apply this dataset to construct neural models with a transformer architecture for estimating the likelihood of flaming to verify the effectiveness of the neural models for the flaming prediction. Applying fine-tuned large language models to the flaming prediction is also within our future plans.

References

1. X. https://twitter.com/. Visited 22 Nov 2023
2. Facebook. https://www.facebook.com/. Visited 22 Nov 2023
3. mixi. https://mixi.jp/. Visited 22 Nov 2023
4. Tadashi, O., Koji, F.: Verification of an algorithm for detecting the negative reactions based on viral media. Inf. Syst. Soc. Japan (2016)
5. Sakai, Y., Takeuchi, K., Ito, E.: A study on flame video detection using comments. IPSJ SIG Tech. Rep., 1–5 (2021)
6. Mikolov, T., Chen, K., Corrado, G., Dean, J.: Efficient estimation of word representations in vector space. arXiv preprint arXiv:1301.3781v3 (2013)
7. Kawakami, M., Iyatomi, H.: Development of a flaming detection system for Twitter. In: Proceedings of the Fuzzy System Symposium, pp. 705–708 (2016)

8. Yoshida, M., Matsumoto, K, Yoshida, M.: Correction of inappropriate sentences using toxic sentence judging. Forum Inf. Technol., 265–266 (2023)
9. Takahashi, N., Higaki, Y.: Flaming detection and analysis using emotion analysis on Twitter. In: Information Processing Society of Japan, Institute of Electronics, Information and Communication Engineers, pp. 135–140 (2017)
10. Segawa, T., Asatani, K., Sakata, I.: User-focused analysis and characterization of cyberbullying. In: The 35th Annual Conference of the Japanese Society for Artificial Intelligence (2021)
11. Iwasaki, Y., Orihara, R., Sei, Y., Nakagawa, Y., Tahara, Y., Ohsuga, A.: Analysis of flaming and its applications in CGM. Trans. Jpn. Soc. Artif. Intell. **30**(1), 152–160 (2015)
12. Kazmier, L.J.: Schaum's Outline of Business Statistics. Schaum Pub. Co. (2003)
13. Brown, T.B., et al.: Language models are few-shot learners. In: Proceedings of the 34th International Conference on Neural Information Processing Systems, pp. 1877–1901 (2020)
14. MacQueen, J.: Some methods for classification and analysis of multivariate observations. In: Proceedings of the Fifth Berkeley Symposium on Mathematical Statistics and Probability, vol. 1, pp. 281–297 (1967)
15. Thorndike, R.L.: Who belongs in the family? Psychometrika **18**(4), 267–276 (1953)
16. Devlin, J., Chang, M.-W., Lee, K., Toutanova, K.: BERT: pre-training of deep bidirectional transformers for language understanding (2018). arXiv preprint. arXiv:1810.04805

Systematic Review of Eye-Tracking Studies

Alba Haveriku[✉], Hakik Paci, Nelda Kote, and Elinda Kajo Meçe

Faculty of Information Technology, Polytechnic University of Tirana, Tirana, Albania
{alba.haveriku,hpaci,nkote,ekajo}@fti.edu.al

Abstract. Numerous studies have illustrated the connection between eye movements during reading and text understanding. Incorporating eye-tracking data into text processing models has proved to enhance the quality of results in automatic text processing. This systematic literature review evaluates the influence of eye tracking data in text processing models. We have investigated articles published between the period 2015–2023, selecting the most relevant papers based on two main aspects: (a) eye-tracking data integrated in natural language processing models, in subfields such as sentiment analysis, text summarization, part of speech tagging and name entity recognition, (b) multilingual eye-tracking the study and comparison of cross-lingual similarities and differences.

1 Introduction

The eye-tracking concept is associated with the detection of eye movements, providing useful information of human visual attention. A considerable number of studies demonstrate how the movement of eyes during reading reflects cognitive processes and text understanding. Eye tracking is linked with the measurement of focus points, fixations, saccades, and pupil detection.

Eye-tracking and text processing are closely related and the integration of these two fields can generate improved results in automatic text processing models. The usage of eye-tracking devices can help in the measurement of two reading aspects: temporal and spatial. The temporal aspect is closely linked with how much time a person is focused on a specific word while reading a text. Meanwhile, the spatial aspect is connected to the location on the screen where readers focus on.

A native reader moves eyes from one word to another, focusing in the most important words that represent the meaning of the text. Some of the words do not even have direct fixation of the eyes, while others may have a first, second or even third fixation, determining their key role in the sentence. Some of the main features which can be collected during eye tracking experiments are fixation time, points of fixation, saccades, scan path, gaze duration and total viewing time.

There exist a variety of devices which can gather the data from eye movements, from simple infrared cameras included in mobile devices to more advanced equipment which comes with a higher cost and better accuracy. The different devices vary depending on the level of sampling, the accuracy with which they capture temporal records of one or both eyes. Nowadays, to conduct proper and accurate studies, eye tracking devices with high

L. Barolli (Ed.): EIDWT 2024, LNDECT 193, pp. 251–260, 2024.
https://doi.org/10.1007/978-3-031-53555-0_24

costs are used, and the experiments are mostly associated with laboratory environments. The aim is to make tracking eye movements and other cognitive data more accessible.

Eye tracking data has been proved to help in a variety of fields from psychology to linguistics, marketing, healthcare, human-computer interaction etc. Meanwhile, some of the main subfields of natural language processing which can be leveraged by using an additional layer of eye-tracking data are the evaluation of machine learning models' results, parameters and attention functions in neural networks, the improvement of automatic part of speech (PoS) tagging, named entity recognition (NER) and sentiment analysis.

Eye tracking data can serve as an effective inductive bias in natural language processing models, leading to enhanced performance. Additionally, this data can be leveraged to assess the outcomes generated by neural network models, thereby enhancing their interpretability.

The structure of this paper is as following: Section 2 presents the research methodology, specifically the two main research questions, the collection and selection of the papers to include in this review; Sect. 3 presents the analysis and results from analyzing the two main research questions; and Sect. 4 present the conclusions achieved and possible future work.

2   Research Methodology

The main databases chosen to collect relevant papers for this systematic literature review are: Springer, ACM Digital Library, ACL Anthology, ScienceDirect, IEEE Xplore, arXiv and Google Scholar. The methodology used to conduct this study is based on Kitchenham et al. [1].

This work is based on two main research questions, each of which is linked with a specific contribution:

First Research Question (RQ1).   Can the usage of eye-tracking data provide improvements in the existing models for automatic text processing?

Studying the relevant literature and answering this question will provide a list of the main papers which present the usage of eye-tracking in different fields of NLP (Contribution 1).

Second Research Question (RQ2).   Is there a necessity to build multilingual corpuses for eye-tracking purposes?

Studying the relevant literature and answering this question will provide a list of the main existing corpuses (in one or more than one language) and the main papers which emphasize the advantages and disadvantages of multilingual scenarios (Contribution 2).

The main steps followed to identify the main papers that are related to each of the above research questions are:

1. Searching in digital databases.

   - Digital databases: ScienceDirect, ACM Digital Library, IEEE Xplore, ACL Anthology, Springer, Google Scholar.

- Search terms: eye-tracking, gaze estimation, low-cost eye-tracking, natural language processing.
- Publication year 2015–2023.

2. Review of the collected papers. The papers collected from searching the digital databases are organized in an Excel file, where the following information was saved:

- List of authors
- Title of the papers
- Year of publication
- Digital Object Identifier (DOI)
- Abstract

3. Determining inclusive and exclusive criteria. After reading the abstract of the collected studies, based on the inclusive and exclusive criteria defined below, it was possible to select the papers which are going to be analyzed.

Inclusive criteria:

- The study describes the combination of eye-tracking and natural language processing.
- The study has been published in journals or conference proceedings.
- The study must be part of the primary research group.
- The year in which the study has been published is between 2015–2023.
- The language in which the study has been published is English.

Exclusive criteria:

- The study is not related to the field of computer science.
- The study does not describe methods that are related to eye-tracking and natural language processing.

4. The final step is to access the full paper which remains after following the above criteria.

In this systematic review, the initial number of papers collected was 260 from which 56 papers were identified as more important to our research purposes and 23 were referred as relevant for the two research questions presented in this work. Future work will include the addition of new research questions were all the identified papers will be analyzed.

The work of Kennedy et al. [2] has been included even though it is not published between 2015–2023, due to its importance in the field. In Table 1 are shown the total number of collected and selected papers.

Table 1. The number of selected research works.

Digital Database	Initial papers	Selected papers
Science Direct	150	25
ACM Digital Library	50	10
IEEE Xplore	35	7
Other (ACL Anthology, Google Scholar, Springer)	25	14
Total	**260**	**56**

3   Analysis and Results

3.1   RQ1: Can the Usage of Eye-Tracking Data Provide Improvements in the Existing Model for Automatic Text Processing?

A series of papers highlight the usage of eye-tracking techniques in problems related to natural language processing, such as PoS tagging, text summarization, entity recognition etc. Table 2 provides a summarization of the relevant works based on the subfield of NLP they discuss and the accuracy achieved when eye-tracking data is included in the model.

Table 2. Eye-tracking and NLP.

Authors & Year	NLP Subfield	Corpus	Type of Model	Accuracy
Barrett & Søgaard, 2015 [3]	Syntactic categories	250 sentences from: Wall Street Journal articles, Wall Street Journal headlines, emails, weblogs, and Twitter	Supervised Logistic Regression	80.7–83.7%
Barrett, Bingel, Keller, & Søgaard, 2016 [4]	PoS Tagging	Dundee Treebank; Penn Treebank PoS annotation	Unsupervised model; Data used during training and testing time	79.77–81.00%

(continued)

Table 2. (*continued*)

Authors & Year	NLP Subfield	Corpus	Type of Model	Accuracy
Mishra, Kanojia, Nagar, Dey, & Bhattacharyya, 2017 [5]	Sentiment Analysis; Sarcasm Detection	**Dataset 1:** 994 texts: 383 positive, 611 negative, 350 sarcastic **Dataset 2:** 1059 texts: positive, negative and objective	Support Vector Machine; Naïve Bayes; Multi-layers Neural Network	Improvement of 3.7% in the first dataset and -9.3% in F-score in the second dataset
Rohanian, Taslimipoor, Yaneva, & Ha, 2017 [6]	Predicting multiword expressions	GECO corpus	Conditional Random Fields (CRF); Data used during training and testing time	Accuracy: F1 score 70.05 for the first word and 54.0 for the last word in the expression
Barret, Bingel, Hollenstein, Rei & Søgaard [7]	Sentiment analysis; grammatical error detection; detection of abusive language	Dundee corpus; ZuCo corpus	Bidirectional LSTM	Reduction of mean error from baseline models of 4.5%
Strzyz, Vilares, & Gómez-Rodríguez, 2019 [8]	Dependency parser	**Parallel data:** Dundee corpus, Dundee treebank **Disjoint data:** Penn treebank (*parsing*) and Dundee Treebank (*eye-movement*)	BiLSTM model; Data used during training time	Parallel setup: 85.36- 85.61 Disjoint data: 93.98–94.12
Klerke & Plank, 2019 [9]	Syntactic Tagging	Dundee Corpus; Ghent Eye-Tracking Corpus (GECO)	BiLSTM model	Dundee: 95.25- 95.48 GECO: 95.25–95.41
Hollenstein & Zhang, 2019 [10]	NER	Dundee Corpus; GECO corpus; ZuCo corpus	BiLSTM model; Gaze feature in an additional embedding layer	Precision 86.92–89.04

(continued)

Table 2. (*continued*)

Authors & Year	NLP Subfield	Corpus	Type of Model	Accuracy
Chen, Mao, Liu, Zhang, & Ma, 2022 [11]	Sentiment Analysis	1224 microblogging; Emotions: positive, neutral, negative (408 for each); Chinese language	Hierarchical Attention Networks; Reinforcement learning	2.13% improvement in F1-sore
Bautista & Naval, 2020 [12]	Predicting gaze features for each word	ZuCo Corpus Provo Corpus UCL Corpus GECO	Bi-LSTM model	Improvement of 2.67%

Analysing the data presented in Table 2 we notice that in the last years, there has been an increased research work to leverage the usage of eye tracking data in estimating and optimizing human reading behaviour. In papers, such as [3, 4, 8, 9] the results achieved from the authors show that gaze features can provide improved performance in PoS taggers and dependency parsers, even on small amount of data. Strzyz et al. [8] also proposes that their method can be even used to test the usage of previously gathered gaze features to improve annotated treebanks without annotated gaze.

Meanwhile, the benefits of eye-tracking data in the field of sentiment analysis are demonstrated in [5, 7, 11]. Mishra et al. [5] use two datasets which are publicly available, for two tasks: sarcasm detection and sentiment annotation. They enhanced these datasets with gaze features and implemented three different classifiers: SVM, Naïve Bayes and multi layered neural network. The best performing classifier resulted SVM, and the improvement of the F-score when adding gaze features was 6.1% in the best case. The authors point to the necessity of using inexpensive mobile devices to make the usage of eye-tracking data more practical and increase the number of participants in the experiments.

Barret et al. [7] uses public corpora, and evaluated their architecture in three sub-fields of NLP: detecting errors in grammatic, detecting abusive language and sentiment analysis. The authors demonstrated that human attention information provides an inductive bias which is useful in recurrent network architectures. Meanwhile Chen et al. [11] have built a model by adding human behaviour which surpasses classical models with an improvement over 2.13% in F1-score.

Other meaningful improvements are shown in the work of Rohanian et al. [6], where the authors present a study for the automatic identification of multi-word expressions based on the previously known gaze features. Hollenstein & Zhang [10] studied the usage of eye-tracking data for NER (named entity recognition). They achieved two meaningful results:

1. Gaze features on word level, eliminate the necessity to include eye-tracking information at testing;
2. NER models augmented with gaze data outperform classic models.

3.2 RQ2: Is There a Necessity to Build Multilingual Corpuses for Eye-Tracking Purposes?

Several studies have been carried on for languages such as English and German, but there is a lack of studies in low-resource languages. In reading research, it is important to study which aspects exist in different written languages. That is why it is important to see the influence of reading processes and eye movements.

This section is going to focus on the main existing corpuses and the main papers that determine the necessity of multilingual corpuses and research work, presented in Table 3.

Table 3. One and multilingual corpuses.

Corpus	Participants	Languages
Raymond et al. corpus [13]	40 participants 97 trials	English, Spanish, Chinese, Hindi, Russian, Arabic, Japanese, Kazakh, Urdu, Vietnamese
Dundee Corpus [2]	10 participants	English, French
Ghent Eye-Tracking Corpus (GECO) [14]	33 participants	English, Dutch
ZuCo Corpus [15]	12 participants	English
Multilingual Eye Movement Corpus (MECO) [16]	580 participants	Dutch, English, Estonian, Finnish, German, Greek, Hebrew, Italian, Korean, Norwegian, Russian, Spanish and Turkish
Russian Sentence Corpus [17]	96 participants	Russian
RatrOS Corpus [18]	393 participants	Brazilian Portuguese
The Provo Corpus [19]	84 participants	English
WebQAmGaze [20]	332 participants	English, Spanish and German
CopCo (Copenhagen Corpus) [21]	22 participants	Danish
RaCCooNS (Radbound Coregistration Corpus of Narrative Sentences) [22]	37 participants	Dutch

Hollenstein et al. [15] created ZuCo (Zurich Cognitive Language Processing Corpus) where eye tracking information is combined with electroencephalography (EEG). The dataset contains EEG and eye tracking data of 21.629 words and 154.173 fixations. The combination of EEG and eye tracking data provides a good resource for future work and for training machine learning models for tasks such as sentiment analysis, relation extraction and entity recognition. Another corpus is RastrOS, developed by Leal et al. [18]. This corpus is composed of eye-tracking data collected from university students in Brazilian Portuguese.

Slegelman et al. [16] investigate 13 different languages to study the difference of reading behavior amongst different languages. The authors created MECO as a multilingual eye movement corpus which can help in distinguishing eye tracking in various languages. The main finding of their study is that the skipping rate has a considerate difference between the languages considered, a result which the authors link with the different word length distribution across languages. Possible future work that authors mention is the investigation of the effects of frequency and word length, also proving that the skipping and fixation probability are the two main features that change in cross linguistic contexts.

The corpus created by Raymond et al. [13] is an open access reading dataset which includes several underrepresented languages in eye-tracking research. The authors mention two main reasons why eye tracking corpuses do not have a large diversity of languages: economic factors (the high cost of devices and software) and eye trackers designed mostly for Latin languages. Future work includes the collection of data for new languages: Ukrainian, Telugu, indigenous languages in Africa.

Barret et al. [23] study the correlation between gaze and PoS (Part of Speech) in two languages: English and French. The authors demonstrate that English gaze data can be used to assist NLP models in French language, demonstrating that gaze based PoS tagging models can be generalized from one language to the other. The experiments were conducted with portions of data from the Dundee eye tracking corpus.

Hollenstein et al. [24] analyze the performance of multilingual pre trained models to predict reading behavior. In their research they consider languages such as Dutch, German, English and Russian. In this work BERT and XLM models are used to predict the features of eye tracking.

The results show that BERT multilingual gives a better generalization across different languages than XLM models. Meanwhile, XLM models provide a better performance when fine-tuned on a smaller amount of eye tracking data.

4   Conclusions and Future Work

Studying the effects of using eye-tracking data to leverage natural language processing models has led to optimistic results. Various authors in their research work have demonstrated that adding eye movement data in training or testing time has improved the performance of their machine learning or neural network models. Subfields of NLP such as sentiment analysis, text summarization, PoS tagging, and name entity recognition have achieved improvements due to involving gaze information.

Meanwhile, one of the main on-going issues is to expand eye-tracking data in more languages, especially in low-resource languages. Having more available data in different languages can lead to future work in the comparison of eye-tracking models in different languages and prove which features seem to remain similar and which are the main differences between human reading behavior in different types of languages.

From the papers studied, we demonstrated that it is possible to utilize eye-tracking data to achieve better results in NLP tasks. Several corpuses and libraries listed in this review article can become of help to future research works done in this field.

Possible future work includes the extension of the current research questions, to study and compare low and high-cost eye-tracking alternatives.

References

1. Kitchenham, B., Charters, S.M.: Guidelines for performing systematic literature reviews in software engineering. Technical report, EBSE Technical Report EBSE-2007-01 (2007)
2. Kennedy, A., Hill, R., Pynte, J.: The dundee corpus. In: Proceedings of the 12th European Conference on Eye Movement (2003)
3. Barrett, M., Søgaard, A.: Reading behavior predicts syntactic categories. In: Proceedings of the 19th Conference on Computational Language Learning, pp. 345–349. Association for Computational Linguistics, Beijing, China (2015). https://doi.org/10.18653/v1/k15-1038
4. Barrett, M., Bingel, J., Keller, F., Søgaard, A.: Weakly supervised part-of-speech tagging using eye-tracking data. In: Proceedings of the 54th Annual Meeting of the Association for Computational Linguistics (Volume 2: Short Papers), pp. 579–584. Association for Computational Linguistics, Berlin, Germany (2016)
5. Mishra, A., Kanojia, D., Nagar, S., Dey, K., Bhattacharyya, P.: Leveraging cognitive features for sentiment analysis. In: The SIGNLL Conference on Computational Natural Language Learning (CoNLL 2016) (2017). https://arxiv.org/pdf/1701.05581.pdf
6. Rohanian, O., Taslimipoor, S., Yaneva, V., Ha, L.A.: Using gaze data to predict multiword expressions. In: Proceedings of the International Conference Recent Advances in Natural Language Processing, RANLP 2017, pp. 601–609. INCOMA Ltd., Varna, Bulgaria (2017). https://doi.org/10.26615/978-954-452-049-6_078
7. Barret, M., Bingel, J., Hollenstein, N., Rei, M., Søgaard, A.: Sequence classification with human attention. In: Proceedings of the 22nd Conference on Computational Natural Language Learning (CoNLL 2018), pp. 302–312. Association for Computational Linguistics, Brussels, Belgium (2018). https://doi.org/10.18653/v1/K18-1030
8. Strzyz, M., Vilares, D., Gómez-Rodríguez, C.: Towards making a dependency parser see. Comput. Lang. (2019). https://arxiv.org/pdf/1909.01053v1.pdf
9. Klerke, S., Plank, B.: At a glance: the impact of gaze aggregation views on syntactic tagging. In: Proceedings of the Beyond Vision and LANguage: inTEgrating Real-world kNowledge (LANTERN), pp. 51–61. Association for Computational Linguistics, Hong Kong, China (2019). https://doi.org/10.18653/v1/D19-6408
10. Hollenstein, N., Zhang, C.: Entity recognition at first sight: improving NER with eye movement information. In: Proceedings of the 2019 Conference of the North American Chapter of the Association for Computational Linguistics: Human Language Technologies, Volume 1 (Long and Short Papers), pp. 1–10. Association for Computational Linguistics, Minneapolis, Minnesota (2019). https://doi.org/10.18653/v1/N19-1001
11. Chen, X., Mao, J., Liu, Y., Zhang, M., Ma, S.: Investigating human reading behavior during sentiment judgment. Int. J. Mach. Learn. Cybern., 2283–2296 (2022). https://doi.org/10.1007/s13042-022-01523-9
12. Bautista, L.G., Naval, P.J.: Towards learning to read like humans. In: 12th International Conference on Computational Collective Intelligence, vol. 12496, pp. 779–791. Springer, Cham (2020). https://doi.org/10.1007/978-3-030-63007-2_61

13. Raymond, O., Moldagali, Y., Madi, N.A.: A dataset of underrepresented languages in eye tracking research. In: Proceedings of the 2023 Symposium on Eye Tracking Research and Applications (ETRA 2023), pp. 1–2 (2023). Association for Computing Machinery. https://doi.org/10.1145/3588015.3590128

14. Cop, U., Dirix, N., Drieghe, D., Duyck, W.: Presenting GECO: an eyetracking corpus of monolingual and bilingual sentence reading. Behav. Res. Methods 49, 602–615 (2017). https://doi.org/10.3758/s13428-016-0734-0

15. Hollenstein, N., Rotsztejn, J., Troendle, M., Pedroni, A., Zhang, C., Langer, N.: Data descriptor: ZuCo, a simultaneous EEG and eye-tracking resource for natural sentence reading. Sci. Data 5 (2018). https://doi.org/10.1038/sdata.2018.291

16. Slegelman, N., et al.: Expanding horizons of cross-linguistic research on reading: the multilingual eye-movement corpus (MECO). Behav. Res. Methods, 2843–2863 (2022). https://doi.org/10.3758/s13428-021-01772-6

17. Laurinavichyute, A.K., Sekerina, I.A., Alexeeva, S., Bagfasaryan, K., Kliegl, R.: Russian sentence corpus: benchmark measures of eye movements in reading in Russian. Behav. Res. Methods 51, 1161–1178 (2018). https://doi.org/10.3758/s13428-018-1051-6

18. Leal, S.E., Lukasova, K., Carthery-Goulart, M.T., Aluisio, S.M.: RastrOS project: natural language processing contributions to the development of an eye tracking corpus with predictability norms for Brazilian Portuguese. Lang. Resour. Eval., 1333–1372 (2022). https://doi.org/10.1007/s10579-022-09609-0

19. Luke, S.G., Christianson, K.: The Provo corpus: a large eye-tracking corpus with predictability norms. Behav. Res. Methods 50 (2017). https://doi.org/10.3758/s13428-017-0908-4

20. Ribeiro, T., Brandl, S., Søgaard, A., Hollenstein, N.: WebQAmGaze: a multilingual webcam eye-tracking-while-reading dataset. Comput. Lang. (cs.CL) (2023). https://arxiv.org/pdf/2303.17876v2.pdf

21. Hollenstein, N., Barrett, M., Bjornsdottir, M.: The copenhagen corpus of eye tracking recordings from natural. In: Proceedings of the 13th Conference on Language Resources and Evaluation (LREC 2022), pp. 1712–1720. Marseille (2022). https://aclanthology.org/2022.lrec-1.182.pdf

22. Frank, S.L., Aumeistere, A.: An eye tracking with EEG coregistration corpus of narrative sentences. Lang. Resour. Eval. (2023). https://doi.org/10.1007/s10579-023-09684-x

23. Barret, M., Keller, F., Sogaard, A.: Cross-lingual transfer of correlations between parts of speech and gaze features. In: Proceedings of COLING 2016, the 26th International Conference on Computational Linguistics: Technical Papers, pp. 1330–1339. The COLING 2016 Organizing Committee, Osaka, Japan (2016). https://aclanthology.org/C16-1126

24. Hollestein, N., Pirovano, F., Zhang, C., Beinborn, L., Jäger, L.: Multilingual language models predict human reading behavior (2021). https://doi.org/10.48550/arXiv.2104.05433

TPLUG: An Efficient Framework Through Token Pair Linking and Undirected Graph for Chinese Event Extraction

Ting Hu[1(✉)], Fang Deng[1], and Zhiqiang Chu[2]

[1] Beijing University of Posts and Telecommunications, Beijing, China
{tinagioro,dengfang}@bupt.edu.cn
[2] Information Center, State Administration for Market Regulation, Beijing, China
chuzhiqiang@samr.gov.cn

Abstract. Chinese Event extraction (CEE) from text has attracted increasing attention in recent years but remains challenging. Some EE models dealing with overlap and nesting employ multi-stage classification, potentially causing misinformation spread. Alternatively, EE often focuses on triggers but ignores triggerless scenarios. Meanwhile, multiple same-type events in a sentence can't be distinguished by the trigger word alone. Therefore, we design a CEE model TPLUG to solve the above problems. To unify scenarios with and without trigger words, we treat the trigger word as an argument role, combine event type and argument role in pairs into multiple classes, and transform argument extraction into a scoring task of token pairs under different classes. Furthermore, an undirected complete graph is employed to depict relationships among different arguments, facilitating event partitioning by searching complete subgraphs. Experiments on DuEE1.0 datasets demonstrate the proposed approach effectively addresses these challenges and yields competitive results.

1 Introduction

Chinese Event extraction (CEE) is an important and challenging task in natural language understanding research [1–4]. EE facilitates the retrieval of event-related information and enables the analysis of human behavior, thus fostering applications such as information retrieval [5], recommendation systems, intelligent question-answering, knowledge graph construction [6, 7], and other related domains [8–10].

This task aims to detect specified events in Chinese text, referred to as Event Detection, and extract event arguments based on their associated argument roles, known as Event Argument Extraction. CEE frequently encounters the issue of roles overlap where An argument plays different roles in multiple events as exemplified in Fig. 1(b). Typical pipeline extraction models [11–13] break EE into multi-stage classification, like the CasEE model by Sheng et al. [14], which depends on trigger word extraction to handle role overlap. However, inaccuracies in trigger word recognition can impact argument recognition accuracy, especially in Chinese event extraction, where trigger word misidentification is frequent. Figure 1(a) illustrates the entity nesting problem where "Yuncheng"

© The Author(s), under exclusive license to Springer Nature Switzerland AG 2024
L. Barolli (Ed.): EIDWT 2024, LNDECT 193, pp. 261–272, 2024.
https://doi.org/10.1007/978-3-031-53555-0_25

not only acts as the "place" role, but also exists in the argument "Yuncheng Industrial Expo" in the argument role "activity name". CRF-based models [15–17] describe event extraction as a sequence labeling task, but cannot address entity nesting. Cao et al. [18] proposed a new tagging scheme that casts event extraction as a word-word relation recognition task to solve overlapped and nested EE, but relies on the annotation of trigger words. Furthermore, the existing models do not consider the partition problem of multiple same-type events like Fig. 1(c).

Fig. 1. Examples of three main problems that EE faces, including entity nesting (a), overlapping roles (b), multiple same-type events (c).

We propose a grid labeling scheme named Efficient-Attention-Tagging based on TPLinker [19] to solve the problem of entity nesting and role overlap. We combine event types and argument roles in pairs into multiple classes, and transform the argument extraction task into a scoring task of entity head to entity tail (EH to ET) under different classes based on Efficient-Attention-Tagging. As shown in Fig. 2, the token pairs "Yun" and "Hui" are scored as 1 under the class "(closing, activity name)" and this does not affect the score of 1 for the token pairs "Yun" and "Cheng" under the class "(closing, place)". Therefore, Efficient-Attention-Tagging is not affected by the issue of nested entities. It is obvious that different event types and argument roles will not affect the scoring of Efficient-Attention-Tagging even if the same argument exists. Therefore, our scoring model skillfully solves the role overlap problem.

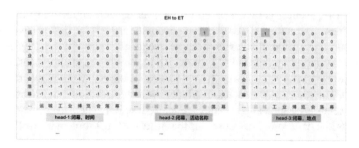

Fig. 2. An example to illustrate our tagging scheme.

If there are multiple identical events in a sentence, it is impossible to distinguish which event an argument belongs to by only using the above module. Therefore, we propose a multi-event partition module based on undirected graph. We assume that the arguments of the same event are related to each other and convert the acquisition of this relation into a token pairs scoring task, as shown Fig. 3(a). In addition to the argument entity corresponding to the imprisonment event type, Efficient-Attention-Tagging also labels the head of argument s to the head of argument o (SH to OH) and the tail of argument s to the tail of argument s (ST to OT). When two arguments can be connected from head to head and from tail to tail, they can be connected by an edge in an undirected graph like Fig. 3(b). When the arguments are connected to each other, they belong to the same event, so that we can solve multiple same-type event problems.

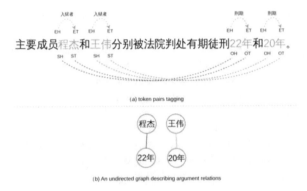

Fig. 3. An example to illustrate event partitioning module.

Based on above modules, we propose TPLUG Chinese event extraction model. First, it adopts Bert [20] as an encoder to get contextualized word representations and adopts Positional Encoding [21] to get relative position information, then use Efficient-Attention-Tagging to score token pairs, then use EH to ET to decode argument entities, SH to OH and ST to OT to decode the relations between argument entities under a certain event type. The undirected graph is used to partition the events, and finally the arguments and the corresponding argument roles under each event are obtained.

We evaluate TPLUG on DuEE1.0 which is the largest Chinese event extraction dataset to date, and conduct extensive experiments and analyses. Our contributions can be summarized as follows:

- We design a new tagging scheme named Efficient-Attention-Tagging to score token pairs to solve the problem of entity nesting and role overlap.
- We design Event partitioning module which relies on Efficient-Attention-Tagging to get scores of argument relations and based on undirected graph to partition the events.
- We propose a Chinese event extraction model TPLUG based on token pair linking and undirected graph, which can be applied to as many scenarios as possible, including no trigger words and multiple same-type events in a sentence.

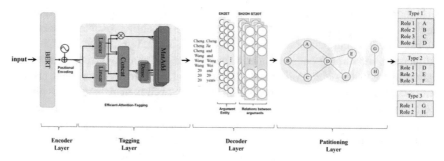

Fig. 4. The architecture of our TPLUG.

2 Methodology

The architecture of our model is illustrated in Fig. 4, which mainly consists of four layers. First, in the encoder layer the widely-used pre-trained language model Bert [20] is used as the encoder to yield contextualized word representations from the input sentences and Positional Encoding is used to get relative position information of word representations. Afterwards, in the tagging layer Efficient-Attention-Tagging is employed to get scores of different types of token-pairs. Next, in the decoder layer we use EH to ET to obtain the argument information under different combination classes, and SH to OH and ST to OT to obtain the relationship between these arguments. Last partitioning layer use an undirected graph to complete the event partitioning.

2.1 Encoder Layer

Bert. Since BERT has achieved significant performance gains on several NLP benchmark tasks, we use Bert as the text encoder. Bert considers both left and right context for each word, allowing it to capture a more comprehensive understanding of the context in which words appear. Give an input sentence $X = \{x_1, x_2, ...x_n\}$, we convert each token x_i into word pieces and then feed them into Bert module. After the Bert calculation, each sentential word may involve vectorial representations of several pieces. After encoding, the vector sequence $H = \{h_1, h_2, ..., h_n\}$. Here, n denotes the sequence length.

Positional Encoding. We obtain the relative position information using Position embedding matrix from Vaswani et al. [21]. The position embedding matrix is denoted as $P \in \mathbb{R}^{n \times d_{model}}$. Here, d_{model} denotes the word embedding dimension.

$$P_{(pos,2i)} = \sin(\frac{pos}{1000^{2i/d_{model}}}), \tag{1}$$

$$P_{(pos,2i+1)} = \cos(\frac{pos}{1000^{2i/d_{model}}}). \tag{2}$$

2.2 Tagging Layer

In order to unify the scenarios with and without trigger words, the trigger words are regarded as argument roles, and the event types and argument roles are cross-combined to obtain multiple classes, and argument extraction is converted into a scoring task of token pairs under different classes.

Specifically, let the input X of length n be encoded by Bert and position to obtain a vector sequence $H = \{h_1, h_2, ..., h_n\}$. We apply a linear transformation to H

$$Q_i = W_q h_i, \tag{3}$$

$$K_i = W_k h_i. \tag{4}$$

Efficient-Attention-Tagging. Efficient-Attention-Tagging decomposes argument extraction into two main steps including entity extraction and entity classification. In the entity extraction stage, we mainly extract the entity fragments with specific semantic meaning from the text. However, in the entity classification stage, the specific type of each entity needs to be determined. Therefore, we can view the entity extraction step as an entity extraction task with a single type, so we only need a scoring matrix to evaluate and identify the entity fragments. For the entity classification step, we utilize the feature concatenation method combined with the Dense layer to build the classification model. Combine these two steps to form a new scoring function

$$s_\gamma(i, j) = \omega_\gamma^T[h_i; h_j] + (W_q h_i)^T(W_k h_j). \tag{5}$$

$s_\gamma(i, j)$ as a continuous segment from i to j is the score of an entity of type. The parameters of the entity extraction part are shared by all entity types, so for each new entity type, we only need to add new parameters to the corresponding $\omega_\gamma \in \mathbb{R}^{2D}$, that is, the number of parameters for each new entity type is only increased by $2D$. To further reduce the number of parameters, using $[Q_i; K_i]$ instead of h_i, then

$$s_\gamma(i, j) = \omega_\gamma^T[Q_i; K_i; Q_j; K_i] + Q_i^T K_j. \tag{6}$$

At this point $\omega_\gamma \in \mathbb{R}^{4D}$, therefore each new entity type increases the number of parameters by $4D$. Because $D \ll d_{model}$ Eq. (6) has far fewer parameters than Eq. (5), which is the final scoring model Efficient-Attention-tagging.

2.3 Decoder Layer

EH to EH. After tagging by the Efficient-Attention-tagging layer, in the decoding stage, all the segments satisfying $s_\gamma(i, j)$ are regarded as the entity output of class γ, that is, the corresponding argument entities of different (event type, argument role) are obtained.

We consider that the arguments of the same event are related, and the acquisition of this connection is transformed into argument relation extraction. Argument relation extraction is defined as the extraction of a triple (s, p, o), and the expansion can be considered as the extraction of a quintuple (s_h, s_t, p, o_h, o_t), where s_h and s_t are the head and tail positions of an argument s, while o_h, o_t are the head and tail positions of an argument, respectively. We present the hypothetical model

$$S(s_h, s_t, p, o_h, o_t) = S(s_h, s_t) + S(o_h, o_t) + S(s_h, o_h|p) + S(s_t, o_t|p), \quad (7)$$

where $S(s_h, s_t)$, $S(o_h, o_t)$ is used to identify the entities corresponding to arguments s and o. It is equivalent to the argument extraction task with two entity types, so it can be completed with an EH to EH. $S(s_h, o_h|p)$ represents a matching with the head feature of s argument and o argument as their own representation, and considering the possibility of entity nesting, it is necessary to use $S(s_t, o_t|p)$ to match the tail of the entity again. We use **SH to OH** to decode $S(s_h, o_h|p)$ and **ST to OT** to decode $S(s_t, o_t|p)$.

2.4 Partitioning Layer

After obtaining the argument relations, we use undirected graphs to describe the connection of arguments between events. We treat each argument as a node on the graph and join any two arguments in the same event into adjacent nodes to form a complete graph. If two arguments never occur in the same event, there is no edge between their corresponding nodes, that is, they are not adjacent. Thus, event partitioning translates into a search for complete subgraphs on the graph. We propose recursive search algorithms for event partitioning on undirected graphs.

(1) First, enumerate all pairs of nodes in the graph. If all pairs of nodes are adjacent, then the graph is complete and is returned directly. If there are non-adjacent node pairs, execute step 2.
(2) For each pair of non-adjacent nodes, find out all the node sets adjacent to them to form a subgraph, and then execute step (1) for each subgraph set.

When performing a search, we only need to focus on nodes of the same event type. Typically, the number of arguments for the same event type is in the single digits. Therefore, although the above algorithms may seem complex, they are actually fast.

3 Experiments Settings

3.1 Dataset

We conduct experiments on a Chinese event extraction benchmark DuEE1.0 [22] which is the largest Chinese event extraction dataset to date. The dataset has a total of 11958 training samples and a total of 1498 validation data, and as shown in Fig. 5 there is a large amount of entity nesting and role overlapping data as shown in the figure. In addition, there are 129 cases with multiple events of the same type in the training set and 35 cases in the validation set.

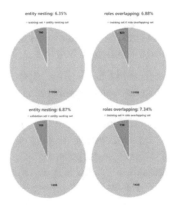

Fig. 5. Entity nesting and role overlap of DuEE1.0

3.2 Evaluation Metrics

We use the evaluation metrics proposed in the Language and Intelligent Technology Competition held by Baidu in 2021. The output event argument results are matched with the manually labeled event argument results, and word-level matching F1 is used as the evaluation metric. The matching process is case insensitive. If the argument has more than one statement, the highest value of F1 among multiple matches is taken. F1 values are calculated as follows:

$$F1 = 2\frac{P \times R}{P + R}, \tag{2}$$

where P represents the ratio of the sum of predicted argument scores to the number of all predicted arguments, and R represents the ratio of the sum of predicted argument scores to the number of all manually annotated arguments.

3.3 Implementation Details

We employ the Chinese Bert-base model for DuEE1.0 and set the hidden layer dimension to 128, batch size to 32, learning rate to 0.001, dropout to 0.5, and epochs to 200. All the hyper-parameters are tuned on the development set.

3.4 Baselines

Sequence Labeling Methods for Flat EE. Within the domain of Event Extraction (EE), a collection of methods transforms the task into a sequence labeling endeavor by assigning unique labels to each individual token. **BERT-Softmax** method leverages BERT to extract vital feature representations necessary for identifying triggers and arguments. Additionally, it incorporates an extra layer of Conditional Random Field (CRF) built on Bert [15] to enhance the overall model performance. **BERT-CRF** is designed to capture intricate label dependencies [16]. **BERT-CRF-joint** extends the BIO tagging scheme by introducing joint labels (B/I/O-type-role) inspired by joint entity and relation extraction [17]. However, these approaches, while effective, encounter challenges in handling overlapping issues, mainly due to conflicts in labeling.

Methods for Overlapped and Nested EE. PLMEE [3] tackles the challenge of parameter overlap by systematically isolating role-specific parameters through a sequential process that relies on triggers anticipated by the trigger extractor. **MQAEE** [23] employed a reading comprehension approach to formulate question templates for multi-stage event extraction. Initially, it anticipated overlapping triggers based on their respective types, followed by the prediction of overlapping arguments in accordance with the identified trigger types. **CasEE** [14] sequentially conducts type, trigger, and argument extraction, extracting overlapping targets based on preceding predictions. The model collectively learns all subtasks in an integrated fashion. **OneEE** [18] achieves concurrent identification of relationships among trigger words or arguments within a single stage, employing parallel grid labeling to enhance the efficiency of event extraction. Currently, the OneEE model demonstrates superior performance across various datasets.

4   Experimental Results

4.1   Main Results

Table 1. Comparison experiments between TPLUG model and other models.

	model	P	R	F1
Flat EE	BERT-Softmax	72.2	69.8	71.0
	BERT-CRF	74.6	70.1	72.3
	BERT-CRF-joint	77.4	73.3	75.3
Ovlp. & Nest.EE	PLMEE	79.5	77.9	78.7
	MQAEE	80.2	79.3	79.7
	CasEE	83.4	81.5	82.4
	OneEE	83.9	82.3	**83.1**
	TPLUG	82.7	**83.0**	82.8

Table 1 shows the result of all methods on the overlapped EE dataset DuEE1.0. We can observe that,

1) Our model achieves competitive experimental results.
2) Our model also performs well when it does not rely on trigger words. Compared with BERT-CRF-joint, the precision is improved by 5.3%, the recall rate is improved by 9.7%, and the overall F1 value is 7.5% higher. This is because the sequence labeling method can only solve the flat EE problem and cannot solve the overlapping and nesting problem, and TPLUG is effective in the overlapping and nesting problem, so the experimental results are better than the sequence labeling method.

3) Compared with the EE model for the overlapping nesting problem, our model still performs well, with the best accuracy, and the F1 value is only lower than the OneEE model. It proves that the model can still effectively complete the argument extraction task without trigger words.

4.2 Comparison on Speed and Parameter Number

As Table 2 shows the number of parameters of CasEE and OneEE is about half that of PLMEE, while the number of parameters of TPLUG is even less, only 80.2M, which proves that Tagging can greatly reduce the parameters required by the model. In addition, the inference speed of TPLUG is 5 times faster than PLMEE and 1.2 times faster than OneEE, which also proves the advantage of the model, supporting parallel inference. In summary, TPLUG achieves better performance and inference speed by using fewer parameters.

Table 2. Parameter number and inference speed comparisons on DuEE1.0. All models are tested with batch size 1, The ratio denotes the multiple of the speed increase with regard to PLMEE.

Model	stage	#Param.	Speed(sent/s)	Ratio
PLMEE	Two	227.8M	20.2	×1
CasEE	**Three**	145.3M	67.3	×3.3
OneEE	One	133.5M	84.1	×4.2
TPLUG	One	80.2M	102.3	×5.1

4.3 Effects of the Modules

To verify the effectiveness of each component, we conduct ablation studies on DuEE1.0 dataset, as shown in Table 3. Firstly, after removing the relative position encoding component, the F1 score decreases by 5.9%, which indicates that the relative position information component can capture more position information of the sequence and significantly improve the performance of the event extraction task. Removing the event partitioning component, we observe a slight performance degradation, probably because there are only 1% instances of multiple identical events in a sentence on the DuEE1.0 training dataset and only 35 instances on the validation dataset, so the effect is not significant. If we increase the number of identical events in the dataset, the effect will be more obvious.

In addition, we also explore the impact of changing the dimension of Bert's hidden layer on the DuEE1.0 data event extraction task. According to Fig. 6, it can be seen that in the DuEE1.0 dataset, the hidden layer dimension achieves a good recognition effect in both 128 and 256 dimensions. Compared with 32 dimensions, the precision, recall and F1 value of 128 dimensions are improved by about 11.3%, 8.8% and 10.5%, respectively. There is a slight drop when reaching 512 dimensions. When the dimension of the hidden layer is low, TPLUG cannot fully learn the features in the text, which affects the effect of recognition. However, if the dimension is set too high, it may lead to over-fitting phenomenon, resulting in the decline of recognition effect.

Table 3. Ablation studies

	P	R	F1	
TPLUG	82.7	83.0	82.8	
w/o positional encoding	77.3	76.5	76.9	
w/o partitioning layer	81.2	82.8	82.0	

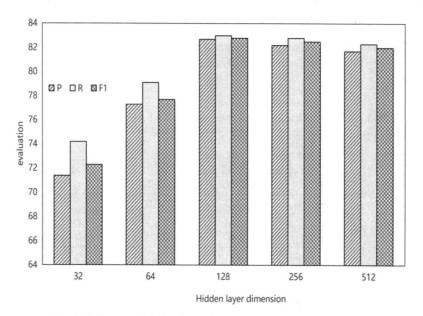

Fig. 6. Influence of hidden layer dimension on event extraction task.

4.4 Effects of Dataset Size

In order to explore the performance of TPLUG on different data sizes, we conduct experiments on training set sizes of 100, 500, 1000, 3000, 5000 and 10,000, and the experimental results are shown in Fig. 7. When the training data is less, the difference between precision and recall is larger, and the F1 value is smaller. As the amount of data increases, the f1 value gradually increases and finally approaches 82 or so. This indicates that TPLUG is more dependent on the data and less effective in the case of small samples. The reason is that TPLUG is complete enough to adapt to enough scenarios, and if the prior information is insufficient, the learning difficulty will increase.

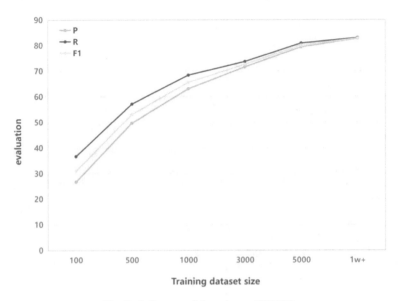

Fig. 7. Influence of data size on TPLUG.

5 Conclusion

In this paper, we propose an EE model based on token pair linking and undirected graph for Chinese event extraction. The model can solve the problems of entity nesting, role overlapping and multiple event division of the same type. We propose Efficient-attention-tagging, an Efficient token pair scoring model, which decomposes argument extraction into two main steps including entity extraction and entity classification. In addition, We propose an undirected graph-based multi-event partitioning model, which captures the relationship between arguments based on Efficient-Attention-Tagging, constructs an undirected graph to describe the relationship between event arguments, and searches for complete subgraphs in the undirected graph to complete event partitioning. Experimental results show that the proposed model achieves competitive results on the Chinese dataset DuEE1.0, and is faster than the SOTA model. Through ablation studies, we find that event partitioning is effective. In future work, we will explore how to improve the performance of event extraction models on small sample datasets.

References

1. Doddington, G.R., Mitchell, A., Przybocki, M.A., Ramshaw, L.A., Strassel, S.M., Weischedel, R.M.: The automatic content extraction (ACE) program - tasks, data, and evaluation. In: LREC (2004)
2. Chen, Y.B., Xu, L.H., Liu, K., et al.: Event extraction via dynamic multi-pooling convolutional neural networks. In: Proceedings of the 53rd Annual Meeting of the Association for Computational Linguistics and the 7th International Joint Conference on Natural Language Processing of the Asian Federation of Natural Language Processing, pp. 167–176 (2015)

3. Yang, S., Feng, D., Qiao, L., et al.: Exploring pre-trained language models for event extraction and generation. In: Proceedings of the 57th Annual Meeting of the Association for Computational Linguistics, pp. 5284–5294 (2019)
4. Lu, Y., Liu, Q., Dai, D., et al.: Unified structure generation for universal information extraction. arXiv preprint arXiv:2203.12277 (2022)
5. Zhang, W., Zhao, X., Zhao, L., Yin, D., Yang, G.H.: DRL4IR: 2nd workshop on deep reinforcement learning for information retrieval. In: ACM SIGIR (2021)
6. Wu, X., Wu, J., Fu, X., Li, J., Zhou, P., Jiang, X.: Automatic knowledge graph construction: a report on the 2019 ICDM/ICBK contest. In: ICDM (2019)
7. Bosselut, A., Bras, R.L., Choi, Y.: Dynamic neuro-symbolic knowledge graph construction for zero-shot commonsense question answering. In: AAAI (2021)
8. Su, X., et al.: A comprehensive survey on community detection with deep learning, CoRR (2021)
9. Liu, F.: Deep learning for community detection: progress, challenges and opportunities. In: IJCAI (2020)
10. Ma, X., Wu, J., Xue, S., Yang, J., Sheng, Q.Z., Xiong, H.: A comprehensive survey on graph anomaly detection with deep learning. CoRR (2021)
11. Chen, Y., Xu, L., Liu, K., Zeng, D., Zhao, J.: Event extraction via dynamic multi-pooling convolutional neural networks. In: ACL (2015)
12. Subburathinam, A.: Cross-lingual structure transfer for relation and event extraction. In: EMNLP-IJCNLP (2019)
13. Cheng, Q., Fu, Y., Huang, J., et al.: Event detection based on the label attention mechanism. Int. J. Mach. Learn. Cybern. 14(2), 633–641 (2023)
14. Sheng, J., Guo, S., Yu, B., et al.: CasEE: a joint learning framework with cascade decoding for overlapping event extraction. arXiv preprint arXiv:2107.01583 (2021)
15. Ramrakhiyani, N., Hingmire, S., Patil, S., et al.: Extracting events from industrial incident reports. In: Proceedings of the 4th Workshop on Challenges and Applications of Automated Extraction of Socio-political Events from Text (CASE 2021), pp. 58–67 (2021)
16. Rui-Fang, H.E., Shao-Yang, D.: Joint Chinese event extraction based multi-task learning. J. Softw. (2019)
17. Zeng, Y.: Scale up event extraction learning via automatic training data generation. In: Proceedings of the 32nd AAAI Conference on Artificial Intelligence and the 30th Innovative Applications of Artificial Intelligence Conference and the 8th AAAI Symposium on Educational Advances in Artificial Intelligence, vol. 742. AAAI, New Orleans (2018). https://doi.org/10.5555/3504035.3504777
18. Cao, H., Li, J., Su, F., et al.: OneEE: a one-stage framework for fast overlapping and nested event extraction. arXiv preprint arXiv:2209.02693 (2022)
19. Wang, Y., Yu, B., Zhang, Y., et al.: TPLinker: single-stage joint extraction of entities and relations through token pair linking. arXiv preprint arXiv:2010.13415 (2020)
20. Devlin, J., Chang, M.-W., Lee, K., Toutanova, K.: BERT: pre-training of deep bidirectional transformers for language understanding. In: Proceedings of the North American Chapter of the Association for Computational Linguistics, pp. 4171–4186 (2019)
21. Vaswani, A., Shazeer, N., Parmar, N., et al.: Attention is all you need. In: Advances in Neural Information Processing Systems, vol. 30 (2017)
22. Li, X., et al.: DuEE: a large-scale dataset for Chinese event extraction in real-world scenarios. In: Zhu, X., Zhang, M., Hong, Y., He, R. (eds.) Natural Language Processing and Chinese Computing. LNCS (LNAI), vol. 12431, pp. 534–545. Springer, Cham (2020). https://doi.org/10.1007/978-3-030-60457-8_44
23. Li, F., et al.: Event extraction as multi-turn question answering. In: Proceedings of the 2020 Conference on Empirical Methods in Natural Language Processing: Findings, pp. 829–838 (2020)

Usability and Effectiveness of a Developed IoT-Based Smart Beverage Dispenser

Dennis R. dela Cruz[✉], Dyan G. Rodriguez, and Hazel A. Caparas

Bulacan State University, Malolos City, Philippines
{dennis.delacruz,dyan.rodriguez,hazel.caparas}@bulsu.edu.ph

Abstract. PET packaging accounted for 44.7% of single-serve beverage packaging and 12% of solid waste globally in 2021. In the Philippines, to regulate plastic bags and other single-use plastics, a few towns and provinces have implemented legislation to support SDG 12, responsible consumption and production. In this era of smart technologies, one remarkable application that has gained substantial attention is the IoT-based Smart Beverage Dispenser. This paper explores this novel technology to evaluate its usability and effectiveness in redefining the beverage consumption experience. Usability testing, user surveys, and performance evaluations are all included in the methodology to give readers a comprehensive picture of the operation of the application software and the machine. After extensive testing and study of usability, the system was discovered to demonstrate a high degree of usability. The system's user-friendly interface design, transparent feedback mechanisms, and simple controls and signals made this excellent usability possible.

1 Introduction

The convenience of easy access to drinks or beverages is attributed to plastics. Plastic is still the most efficient packaging material, both in terms of convenience, cost, and time taken to manufacture - the perfect solution for beverage companies. Also, they are lighter and more space-efficient than other materials, so more can be packed and stored [1]. Today, the world is experiencing a plastic waste crisis since many single-use plastic products are more likely to end up in our seas than various reusable options. Aside from the bottle, other waste like corks, labels, foils, and the package came in. Eventually, these substances will seep into the surroundings and persist for a minimum of 400 years until they degrade into microplastics. Despite the undeniable advantages that plastics provide to the beverage sector, these companies have yet to abandon plastic packaging in the near future [2].

In order to combat plastic pollution and marine litter and expedite the shift towards a circular plastics economy, the European Union (EU) is implementing measures. These actions include single-use plastics, plastic packaging, microplastics, and soon-to-be-included bio-based, biodegradable, and compostable plastics [3]. Meanwhile, India has taken a significant step to address the escalating pollution issue by imposing a ban on single-use plastics, which include items like straws and cigarette packets. This ban addresses the mounting waste problem on the country's streets [4]. Countries like the

L. Barolli (Ed.): EIDWT 2024, LNDECT 193, pp. 273–284, 2024.
https://doi.org/10.1007/978-3-031-53555-0_26

Philippines have implemented several policies and initiatives to combat plastic pollution, including the Ecological Solid Waste Management Act of 2000 or the Republic Act (RA) 9003 [5].

Aside from recycling, the beverage industry can stop billions of plastic bottles from polluting the earth by switching from single-use, throwaway bottles to refillable ones. Today, Oceana, the leading global advocacy organization committed exclusively to preserving our oceans, published a report revealing that the beverage sector has the potential to reduce marine plastic pollution by 4.5 billion to 7.6 billion bottles annually, which accounts for a significant 22% decrease. It can be achieved by a mere 10% increase in the market share of soft drinks and water sold in refillable bottles as a substitute for single-use PET bottles discarded after use. Additionally, the report notes that recent life cycle analysis studies conducted in Germany and Chile have revealed that refillable bottles have a lower carbon footprint than single-use plastic bottles. The findings indicate that refillable PET bottles, in particular, offer substantial environmental benefits. In the report, Dr. Wilts highlights that these analyses demonstrate refillable's potential to conserve up to 40% of raw materials and reduce greenhouse gas emissions by 50% [6]. While no longer popular in the United States, refillable systems account for more than 30% of beverages sold in major markets, including Germany, Mexico, the Philippines, and Indonesia [6].

The Internet of Things (IoT) can assist beverage companies in transitioning or improving the use of refillable options from single-use plastic bottles. By enabling smart tracking and management of refillable containers, IoT sensors can be embedded in these containers to monitor their usage, track refilling patterns, and optimize distribution and collection processes, ensuring efficient utilization and reducing waste. This paper explores this novel technology to evaluate its usability and effectiveness in redefining the beverage consumption experience.

Interesting problems about usability and effectiveness are brought up by incorporating IoT technology into beverage dispensers. Can people choose their preferred beverage with ease using the device's interface? Does IoT functionality improve the effectiveness and user experience of preparing beverages? These queries highlight how crucial it is to conduct a thorough analysis to gauge the real-world applications of this cutting-edge technology.

With the development of an IoT-based smart beverage dispenser, this research aims to close this gap in the literature by extensively evaluating its usability and efficiency. The proponents used a holistic approach to analyze the user experience, drawing ideas from well-established frameworks for usability evaluation and human-computer interaction.

2   Purpose of the Paper

Usability testing, user surveys, and performance evaluations are all included in the methodology of the article to give readers a comprehensive picture of the operation of the application software and the machine. By doing this, the proponents hope to clarify the following significant aspects:

Usability: How user-friendly is the interface of the Smart Beverage Dispenser? Can individuals of varying technical backgrounds effortlessly interact with the device to select and dispense beverages?

Effectiveness: To what extent does the IoT integration enhance the overall effectiveness and efficiency of the beverage dispensing process? Does it align with user expectations for customization and convenience?

User Satisfaction: What are users' sentiments regarding their experience with the IoT-Based Smart Beverage Dispenser? To what extent does the device fulfill their preferences and expectations?

IoT Integration: How seamlessly does the device integrate into IoT ecosystems, and how effectively does it leverage IoT capabilities to redefine the beverage preparation experience?

The results of this study have implications for the optimization of smart beverage dispensers as well as for a more general discussion of IoT-powered kitchen equipment. We strive to provide insightful analysis that resonates with academics, designers, and users navigating the expanding landscape of smart, connected devices by analyzing usability and efficacy.

3 System Overview

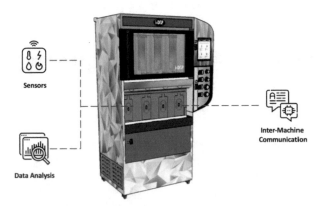

Fig. 1. Smart beverage dispenser framework

As shown in Fig. 1, Smart Dispenser is an innovative beverage machine solution. Creating this system aims to monitor the volume of beverages dispensed by a customer in a given location and even learn the customer behavior. It is essential to analyze user behavior on how much, which kind, and where the beverage intake happens. With behavior analysis, the product manager can gain more insights to optimize product deployment and placement for more effective marketing. As the sensor will continuously catch the real-time data for beverage consumption and dispenser status, through data analytics, the system could easily predict how much product is likely to be refilled in a specific period and location to avoid lost sales due to out-of-stock conditions.

The IoT-Based Real-Time Control and Monitoring System for Smart Beverage Dispensing with Data Analytics will be guided by the system architecture illustrated in Fig. 2 to realize the project objectives.

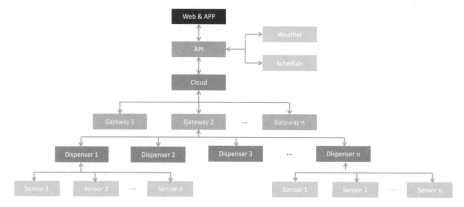

Fig. 2. Systems architecture of the intelligent water dispenser

Additionally, a set of protocols can be deployed to enable system communication. Historical data will be examined to detect anomalies through the dataset. The idea is to establish proactive problem management to reduce service and maintenance costs by deploying alerts when it senses conditions that typically precede problems and repair problems before equipment stops to prevent lost sales.

Fig. 3. Sample website interface

Figure 2 shows the system architecture of the proposed research project and was adapted from [7]. By using the Internet of Things to carry out Cyber-Physical Systems, the smart dispenser can control and monitor this usage and connect it to humans. Several sensors will be attached to the dispenser and connected to the gateway. The connection of several smart dispensers in a given area, like a building, complex, or station, can be handled by a single gateway. Afterward, all the data will be sent to the cloud, which is then connected to the Application Programming Interface (API). This API handles all platforms that will be designed and developed, including the website, shown in Fig. 3, and the mobile application.

4   Methodology

Fig. 4. Usability testing and user experience methodological framework

The methodological framework described in Fig. 4 is a structured approach frequently used in usability testing and user experience (UX) research to assess the usability and user satisfaction of products, interfaces, or systems. It consists of a pre-task survey, task performance assessment, and post-task survey. The pre-task survey is the first stage of the usability testing procedure and assists in collecting crucial data about the participants before they begin tasks. The usability testing focuses on task performance evaluation, which examines how users interact with the Smart Beverage Dispenser while completing their orders. This stage offers clear insights into problems with usability, user effectiveness, and efficiency. Participants can reflect on how they performed on the task and offer comments on the whole experience in the post-task survey, including how well it was used, whether they were satisfied with it, and any improvements they would want to see.

4.1   Participants

Twenty-seven people between the ages of 17 and 50 who represented a variety of genders and educational levels took part in this study. With varied degrees of tech-savvy, the sample included working professionals and students. All participants gave their informed agreement to participate after being recruited via the snowball method. The purpose of the survey was to learn more about this demographic's preferences and the app's usability and user experience.

4.2   Context of Use

The Usability Testing Task involves participants purchasing their preferred drink using an IoT-Based Smart Beverage Dispenser. During this task, respondents perform a series of steps to evaluate the usability of each stage. The study's participants were divided into two groups: the first group (called the controlled group shown in Fig. 5 (a) and (b)) received written instructions on how to use the application. In contrast, the second group (the uncontrolled group) was free to use it without help.

The tasks include (1) initiating an order, (2) selecting a flavor, (3) choosing the size of the drink, (4) selecting the payment method, (5) logging into the account to pay, (6) selecting account options, (7) placing the cup for dispensing, and (8) getting the drink.

The time participants spend on each task will be evaluated to determine which steps contribute to task completion. This analysis aims to identify potential difficulties or uncertainties encountered during product interaction. The study employed usability metrics, including efficiency (task completion time), effectiveness (task success rate), consistency, error tolerance, learnability, user satisfaction, and memorability.

Fig. 5. Actual testing of a respondent under the controlled group

When developing an app, consistency refers to how uniformly and predictably certain features (such as buttons, navigation menus, language, and visual design) behave throughout the program [8]. Thanks to consistency, users can move seamlessly from one section of the software to another. The program's error tolerance measures how well it manages user errors, such as incorrect input or misunderstanding. It shows how effectively the program can avoid, find, and fix mistakes while minimizing user inconvenience and annoyance [9]. Learnability gauges how quickly users can grasp concepts and master the app's interface. It focuses on how new users feel during their first interaction with the system and how simple it is for them to complete simple tasks [10]. Users' level of enjoyment, pleasantness, and emotional appeal for the application is referred to as its likeability [11]. Aesthetics, visual design, and general user satisfaction are all included. After a time of inactivity, users' capacity to recall how to operate an application is measured by its memorability. It gauges how well an app can encourage re-engagement and leave a lasting impression [9]. The post-task survey phase measures all the identified usability metrics, which generally cover various aspects such as user satisfaction, perceived ease of use, and feedback on system usability. These measures help assess user perceptions, identify issues, and gather valuable insights to improve the tested product or system's overall usability and user experience.

4.3 Design of Usability Test

Different methods are employed based on the research objectives, various types of data are considered, and active involvement of end-users occurs. This process facilitates evaluating the ease of use and identifying areas for improvement for the IoT-Based Smart Beverage Dispenser. Firstly, categorized by the evaluation purpose, the methodology adopts a Summative Evaluation approach, typically executed at the product development's culmination to validate overall quality and determine areas requiring improvement. Secondly, grounded in data type considerations, the approach incorporates quantitative and qualitative methods to ensure a holistic usability assessment. Lastly, centered on participant involvement, the usability testing involves a User-based Field Study Style of Evaluation. This enables respondents to engage with the product within the actual environment where the product is normally used, facilitating a real-world evaluation of system performance. The evaluation protocol utilizes protocol analysis techniques, comprising conventional paper and pencil methods and video recording, to capture detailed user interactions. Additionally, Observational Techniques are deployed, specifically Query Techniques using a survey questionnaire, to assess learnability, memorability, error tolerance, consistency, and overall likability of the system, thus furnishing a comprehensive usability assessment.

4.4 Data Analysis

For the pre-task survey, descriptive statistics were used to summarize and present the demographic information and participants' background characteristics collected.

The mean completion time for each task and other calculated and provided descriptive statistics for task performance indicators were compiled. Inferential statistics were also applied to determine whether the two participant groups' task performance differed statistically significantly. It was tested using the means of two samples (t-test). Task-level data were examined to find specific usability problems or patterns. Participants' mistakes that were classified as being of the most prevalent sorts were recorded.

The usability and satisfaction questions from the post-task survey were addressed using descriptive statistics. For Likert-scale items, calculated means and medians were compiled. Like the task performance phase, inferential statistics (t-test) were also applied to determine if there are statistically significant differences in user satisfaction or perceived usability between controlled and uncontrolled groups. Regression analysis (Analysis of Variance) also helped predict the significance of usability metric predictors and the categorical predictors (controlled and uncontrolled group) to the performance time. Open-ended questions underwent qualitative analysis to find recurrent participant concerns. The overall usability scores were computed to examine how these scores differ between these two groups.

5   Discussion of Results

5.1 Pre-task Survey

Twenty-seven people between the ages of 17 and 50 who represented a variety of genders and educational levels took part in this study. With varied degrees of tech-savvy, the sample included working professionals and students. Twenty-nine percent were 21 years old. There were 58% male and 42% female respondents. Students comprised 84% of the respondents, while the rest were employees.

5.2 Task Performance

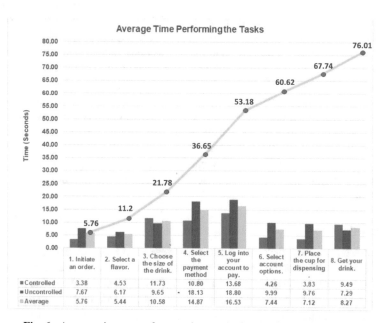

Fig. 6. Average time to perform various tasks for the testing procedure.

During the usability testing session, the task performance phase was completed. Controlled participants received written instructions and were coached through the activities; uncontrolled participants received neither written instructions nor a guide. A thorough time analysis was conducted using video recordings of participant performance. The comparison of each task's performance times for the two groups is shown in Fig. 6.

According to Fig. 6, except for tasks 3 (choosing the size of the drinks) and 8 (obtaining the drinks), the uncontrolled group took longer to complete each job.

The means of two samples test (t-test) was used to determine if there were significant differences in the task times of each group. The average times for tasks 4, 6, 7, and 8, and the overall time for the two groups differed significantly with p-values less than 0.05. About all tasks but task 8, the uncontrolled group took longer on average.

Figure 6 also shows the average time in which tasks significantly contribute to the overall completion time. Tasks 5 (Log into your account to pay), 4 (Select the payment method), and 3 (Choose the amount of the drink) together account for 55.2% of the overall completion time, according to the chart.

5.3 Post-Task Survey

After completing the tasks, participants were given the post-task survey immediately while their memories of the experience were still fresh. An online poll and interviews were used to collect the data. Consistency, error toleration, learnability, likeability, and memorability were used to cluster questions accordingly.

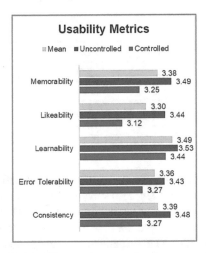

Fig. 7. Usability metric means for the two groups.

The uncontrolled group exhibited greater means for all usability indicators than the controlled group, as shown in Fig. 7. Except for likeability, there were no significant differences in the means of the two groups (see Table 1). The findings show that the opinions of the uncontrolled and controlled groups on the device's usability and user experience are similar or comparable, except for likeability.

Disregarding the sample groups, all the usability metrics scored an average of more than 3, which indicates that users approved of all the usability metrics, according to the survey findings addressing the user experience and usability of the device (see Table 2). The study found that learnability has the greatest mean, followed by consistency and memorability, error tolerance, and likeability.

Table 1. Summary of the Significance of User Experience and Usability

Usability Metrics	Significantly Different?	Higher mean?
Consistency	No	Equal
Error Tolerability	No	Equal
Learnability	No	Equal
Likeability	Yes	Controlled
Memorability	No	Equal

Relationships between the total completion time, demographic profiles, total completion time, and usability metrics between groups were explored using regression analyses. The demographic profiles (age, gender, and occupation) do not significantly affect the completion time.

Table 2. Summary of the User Experience and Usability

Metrics	Mean	Median	Interpretation
Consistency	3.3852	3.6000	Agree
Error Tolerability	3.3611	3.5000	Agree
Learnability	3.4938	3.6667	Agree
Likeability	3.2988	3.3333	Agree
Memorability	3.3827	3.3333	Agree

Additionally, learnability and group significantly impacted the overall completion time, with p-values of 0.03 and 0.024, respectively. Greater completion time was seen in the uncontrolled group. Additionally, learnability and overall completion time produced opposing results. The overall completion time lowers as learnability rises.

The participants were allowed to respond to open-ended questions with further comments, suggestions, and experience-based insights. According to the users, the most negative aspects of the system were the issue with the internet connectivity, the response of the machine sensor, frequent errors due to the buggy system, and taking more time to complete the order due to the interface of the application software. On the other hand, the most positive aspects of the system were its being user-friendly, cashless, helpful, and reliable. Users also recommended improving the app's user interface, security, and payment system interface and making the system more user-friendly, especially for the elderly and those not technology savvy.

6   Summary, Conclusions and Recommendations

The authors of this study set out to conduct a thorough assessment of the usability and performance of the IoT-Based Smart Beverage Dispenser, a technological advancement that has the potential to revolutionize how people prepare and consume beverages—understanding how this IoT-based equipment matches user expectations and how it improves the whole experience of preparing and consuming beverages served as the foundation for our research.

The IoT-Based Smart Beverage Dispenser was discovered to demonstrate a high degree of usability after extensive testing and study of usability. Participants were able to navigate the user interface (UI), personalize it to their preferences in beverages, and efficiently operate the device via IoT connectivity. The system's user-friendly interface design, transparent feedback mechanisms, and simple controls and signals made this excellent usability possible.

In terms of improving beverage preparation efficiency, IoT technology changed the game. Users were grateful for the real-time monitoring and control options, which allowed them to customize their beverages precisely and conveniently. These IoT-driven capabilities considerably streamlined the entire process. Users' satisfaction ratings and post-task questionnaires both showed high levels of participant satisfaction. Users commended the IoT-Based Smart Beverage Dispenser for being likable, practical, and helpful for everyday routines. High results on standardized usability tests served as additional evidence of this contentment.

The memorability of the device was a noteworthy finding. Participants could quickly resume using the device even after a break from use, demonstrating how well it works to leave a lasting impression and encourage re-adoption.

The proponents give the following recommendations based on the findings for future improvement and optimization of the IoT-Based Smart Beverage Dispenser and comparable IoT-driven culinary technology. A continuing usability test is advised to find and fix minor usability problems that might develop over time. To promote continual improvement, users should provide regular feedback.

Additionally, it is advised to consider enhancing IoT features to enable more complex customization choices, such as adding more beverage choices, automating reminders or notifications, and working with virtual assistants. To build a comprehensive beverage ecosystem that easily connects with the IoT-Based Smart Beverage Dispenser, consider market expansion initiatives, including relationships with beverage providers.

The IoT-Based Smart Beverage Dispenser is one example of an IoT-based smart device with a bright future. This technology has the potential to completely change how people make and enjoy beverages in their daily lives because of the convergence of its usability, efficiency, and user satisfaction. Manufacturers can keep innovating and providing outstanding IoT-driven culinary experiences by following our advice and staying aware of user preferences.

Acknowledgments. We sincerely thank the Department of Science and Technology (DOST) for their support through the Collaborative Research and Development to Leverage Philippine Economy (CRADLE) Program, serving as the Funding Agency of the IoT-Based Smart Beverage Dispenser. Our heartfelt thanks also extend to the Philippine Council for Industry, Energy, and Emerging Technology Research and Development (PCIEERD) for their continual guidance in project development, acting as the Monitoring Agency. We would like to acknowledge Purenectar Company's steadfast support, which has significantly contributed to advancing scientific knowledge and economic growth in the Philippines through this project. Finally, our deep appreciation goes to the i-DRIP project team, including Engr. Reagan Galvez, Engr. Aldrin Bernardo and all project staff for their tireless dedication and effort in developing the project.

References

1. XL Plastics: 8 Advantages of Using Plastic Packaging Bags for Food Packaging. XL Plastics (2018). https://www.xlplastics.com/8-advantages-using-plastic-packaging-bags-food-packaging/. Accessed 25 May 2023

2. Weina, K.: Beverage packaging: history of plastic packaging and our zero-plastic way out. Evergreen Labs, 2 August 2021. https://evergreenlabs.org/beverage-packaging-history-of-plastic-packaging-and-our-zero-plastic-way-out/. Accessed 15 July 2022
3. European Commission: Single-use plastics. European Commission. https://environment.ec.europa.eu/topics/plastics/single-use-plastics_en. Accessed 15 July 2022
4. World Economic Forum: This country has imposed a ban on single-use plastic to tackle pollution. World Economic Forum, 6 July 2022. https://www.weforum.org/agenda/2022/07/india-ban-policy-single-use-plastic-pollution/. Accessed 15 July 2022
5. One Planet: Ecological Solid Waste Management Act of 2000 (RA 9003). One Planet, 19 February 2021). https://www.oneplanetnetwork.org/knowledge-centre/policies/ecological-solid-waste-management-act-2000-ra-9003#:~:text=The%20law%20provides%20for%20a,acts%20prohibited%2C%20and%20providing%20penalties. Accessed 12 June 2023
6. Oceana: Oceana report: soft drink industry can stop billions of plastic bottles from polluting the ocean by switching from single-use, throwaway bottles to refillables. Oceana, 27 January 2020. https://oceana.org/press-releases/oceana-report-soft-drink-industry-can-stop-billions-plastic-bottles/. Accessed 15 July 2022
7. Yonathan, S.-Y.C., Cheng, R.-G., Dewebharata, A.: A novel design of intelligent energy efficient drinking water dispensing systems. Transdisc. Eng. Compl. Socio-tech. Syst. (2019)
8. Novák, J.Š, Masner, J., Vaněk, J., Šimek, P., Hennyeyová, K.: User experience and usability in agriculture-selected aspects for design systems. Agris Online Papers Econ. Inform. 11(4), 75–83 (2019)
9. Kureerung, P., Ramingwong, L., Ramingwong, S., Cosh, K., Eiamkanitchat, N.: A framework for designing usability: usability redesign of a mobile government application. Information (Switzerland) 13(10) (2022)
10. Nathoo, A., Bekaroo, G., Gangabissoon, T., Santokhee, A.: Using tangible user interfaces for teaching concepts of internet of things: usability and learning effectiveness. Interact. Technol. Smart Educ. 17(2), 133–158 (2020)
11. Gonçalves, R., Rocha, T., Martins, J., Branco, F., Au-Yong-Oliveira, M.: Evaluation of e-commerce websites accessibility and usability: an e-commerce platform analysis with the inclusion of blind users. Univers. Access Inf. Soc. 17(3), 567–583 (2018)

An Algorithm to Change the TBFC Model to Reduce the Energy Consumption of the IoT

Dilawaer Duolikun[1]([✉]), Tomoya Enokido[2], and Makoto Takizawa[1]

[1] RCCMS, Hosei University, Tokyo, Japan
dilewerdolkun@gmail.com, makoto.takizawa@computer.org
[2] Faculty of Business Administration, Rissho University, Tokyo, Japan
eno@ris.ac.jp

Abstract. The TBFC (Tree-Based Fog Computing) model is proposed to energy-efficiently realize the FC model. Here, fog nodes are tree-structured and support application processes for sensor data. A target node is a most congested one in a TBFC tree. In our approach to reducing the total energy consumption of the IoT, a TBFC tree is changed by applying am operation *op*, i.e. *migrate*, *replicate*, or *expand* on a target node f. Processes of the node f migrate to child nodes. The node f and each child node are host and guest nodes of the processes, respectively. In addition, the node f is replicated to multiple sibling replicas and is expanded, i.e. new child nodes are created. In addition to the target node f, newly created nodes and guest nodes of processes migrating are affected by applying an operation since the energy to be consumed by the affected nodes also changes. First, we newly propose formulas to estimate the total energy consumption of all the affected nodes. Then, an algorithm is proposed to decide on an operation *op* on a target node f by using the estimation formulas. In the evaluation, we show the energy consumption of a CC (Cloud Computing) model can be reduced to 0.3 to 0.5% by changing to a TBFC tree of fog nodes in the proposed algorithm.

Keywords: Green computing · IoT · Fog computing (FC) model · Energy consumption · TBFC (Tree-Based FC) model

1 Introduction

In the FC (Fog Computing) model of the IoT [3], sensor data from devices is first processed by application processes installed on fog nodes and processed data is delivered to servers from fog nodes. In the TBFC (Tree-Based FC) model [15,20,22–25], fog nodes are tree-structured where a leaf node is named a device node which is abstraction of a set of devices and a root node stands for a cloud of servers. Each fog node processes input data units (DUs) received from the child nodes and sends a processed DU to the parent node. The CC (Cloud Computing) model [1,2] is considered to be a TBFC model composed of a pair of a root node,

© The Author(s), under exclusive license to Springer Nature Switzerland AG 2024
L. Barolli (Ed.): EIDWT 2024, LNDECT 193, pp. 285–296, 2024.
https://doi.org/10.1007/978-3-031-53555-0_27

i.e. clouds of servers and a device node, i.e. a set of devices. The FTTBFC (Fault-Tolerant TBFC) [25–27], DTBFC (Dynamic TBFC) [20,27], FTBFC (Flexible TBFC) [17,18], and NFC (Network FC) [28] models are so far proposed to make the TBFC model more flexible and reliable.

A *target node* is a most congested one in a TBFC tree. To reduce the total energy consumption of a TBFC tree of nodes, we consider three operations; *migrate*, *replicate*, and *expand* on a target node in this paper. Application processes on the target node migrate to every child node in the migrate operation. Here, the target node and the child nodes are *host* and *guest* nodes of the processes migrating. Replicas of the target node are created as the sibling nodes in the replicate operation. New child nodes of the target node are created and processes migrate from the target node to the created nodes in the expand operation. Here, the target node, nodes created, and guest nodes of migrating processes are *affected* by the operation. In this paper, we newly propose estimation formulas to give the total energy consumption of all the affected nodes. An operation, i.e. migrate, replicate, or expand is selected for a target node to change the tree, where the total energy consumption of the affected nodes is smallest by using the estimation formulas. Then, the tree is changed by applying the operation on the target node. In the evaluation, we show the CC model can be changed to a more energy-efficient TBFC tree of fog nodes in the proposed algorithm, whose energy consumption is only 0.3 to 0.5% of the CC model.

In Sect. 2, we briefly present the TBFC model. In Sect. 3, we discuss the energy consumption of a fog node. In Sect. 4, a homogeneous TBFC model is discussed. The estimation formulas of the energy consumption are proposed in Sect. 5. In Sect. 6, we propose the algorithm to change the TBFC tree. The proposed algorithm is evaluated in Sect. 7.

2   The TBFC Model

In the FC (Fog Computing) model, some application processes are installed on fog nodes, by which sensor data from devices is processed. Processed sensor data is finally delivered to servers. In the TBFC (Tree-Based FC) model [15,20,22–25], fog nodes are tree-structured. A *root* node denotes a cloud of servers. A leaf node named a *device* node shows a set of devices. *Fog* nodes are intermediate nodes to interconnect the root node and device nodes. An *edge* fog node has one child device node. A DU (Data Unit) is a unit of data exchanged among nodes. Each node processes data units (DUs) received from the child nodes by its application processes and sends a processed DU to the parent node.

Each node is specified in a form "f_I" where I is a sequence $i_1 i_2 \ldots i_{h-1} i_h$ ($h \geq 1$) of numbers. A root node is denoted as "f". Each node f_I has b_I (≥ 0) child nodes and one parent node. This means, $f_{i_1 \ldots i_{h-1}}$ is a parent node and f_{Ii} ($= f_{i_1 i_2 \ldots i_{h-1} i_h i}$) is a child node of a node f_I ($i = 1, \ldots, b_I$). An edge node f_I has one child device node d_I which is abstracted from a set of nd_I (≥ 1) devices. In this paper, we assume each device sends a same size of sensor data. Hence, the size of an output DU sent by a device node d_I is proportional to the number nd_I of devices. Let nd be the total number of devices in the IoT.

In this paper, an application AP of sensor data is assumed to be realized by a sequence $\langle p_1, \ldots, p_m \rangle$ $(m \geq 1)$ of application processes. First, the *tail* process p_m takes an input DU of sensor data from devices. Each process p_h calculates an output DU od_h on an input DU od_{h+1} from p_{h+1} and forwards od_h to p_{h-1} $(h = m - 1, \ldots, 2)$. Lastly, the *top* process p_1 gets an output DU od_1 which includes actions on actuators of devices and delivers od_1 to the devices via fog nodes. A subsequence P_I $(= \langle p_{tp_I}, p_{tp_I+1}, \ldots, p_{tl_I} \rangle$, $1 \leq tp_I \leq tl_I \leq m)$ of m_I $(= tl_I - tp_I + 1)$ processes of AP is installed on a node f_I. Let P be a process subsequence $\langle p_t, p_{t+1}, \ldots, p_l \rangle$ of AP of a root node f. In a path $\langle f, f_{i_1}, f_{i_1 i_2}, \ldots, f_{i_1 i_2 \ldots i_h} \rangle$ from a root node f to each edge node $f_{i_1 i_2 \ldots i_h}$, a concatenation of the process subsequences P, P_{i_1}, $P_{i_1 i_2}$, $\ldots$, $P_{i_1 i_2 \ldots i_h}$ is AP. Here, no pair of different subsequences $P_{i_1 \ldots i_k}$ and $P_{i_1 \ldots i_l}$ $(i_k \neq i_l)$ have a common process. On the other hand, every pair subsequences P_{Ii} and P_{Ij} have at least one common process, i.e. top process. Thus, each common process p_h is replicated on child nodes $f_{I1}, \ldots, f_{Ib_I}$, i.e. p_h receives smaller input DUs in each child node f_{Ii} and can be executed in parallel with the other child nodes.

Let ID_I be a set $\{id_{I1}, \ldots, id_{Ib_I}\}$ of input DUs of a node f_I, where id_{Ii} is an input DU from each child node f_{Ii}.

3   An Energy Consumption Model

Each fog node f_I follows the SPC (Simple Power Consumption) model [4–12, 16, 21]. This means, f_I consumes the maximum power XP_I [W] if some application process is executed and the minimum power MP_I $(< XP_I)$ [W] otherwise. Each fog node f_I also consumes the power RP_I $(= rp_I \cdot XP_I)$ and SP_I $(= sp_I \cdot XP_I)$ [W] to receive and send a DU where $rp_I \leq 1$ and $sp_I \leq 1$.

A node f_I supports a process subsequence $P_I = \langle p_{tp_I}, p_{tp_I+1}, \ldots, p_{tl_I} \rangle$ of an application $AP = \langle p_1, p_2, \ldots, p_m \rangle$ where $1 \leq tp_I \leq tl_I \leq m$. In this paper, the amount $PCA_h(s)$ of the computation to be done by each process p_h to process an input DU of size s is assumed to be $a_{1h}s^2 + a_{2h}s + a_{3h}$ [bit] where a_{1h}, a_{2h}, and a_{3h} are constants. The *computation rates* CR_I and CR [bps] show how many bits of an input DU a node f_I and a root f process for one second, respectively. The size of an output DU du is $\rho_h s$ where ρ_h is the *output ratio* of p_h. Let PB_I be XP_I/CR_I [W sec (J)/bit] which is the energy to be consumed by a node f_I to execute one bit of an input DU. The execution time $PT_{I,h}(s)$ [sec] is $PCA_h(s)/CR_I$. Here, f_I consumes the energy $PE_{I,h}(s) = PCA_h(s) \cdot PB_I$ [J].

The total computation amount $FCA_I(s, tl_I, tp_I)$ [bit] of a node f_I to execute m_I $(= tl_I - tp_I + 1)$ processes $p_{tl_I}, p_{tl_I-1}, \ldots, p_{tp_I}$ and obtain an output DU od_I for input DUs ID_I of size s is $\Sigma_{h=tl_I}^{tp_I} PCA_h(\rho_{tl_I}\rho_{tl_I-1}\ldots\rho_{tl_I-h}s)$ where $\rho_0 = 1$. The total execution time $FCT_I(s, tl_I, tp_I)$ [sec] to process input DUs of size s is $\Sigma_{h=tl_I}^{tp_I} PT_h(\rho_{tl_I}\rho_{tl_I-1}\ldots\rho_{tl_I-h}s)$. Here, the node f_I consumes the total energy $FCE_I(s, tl_I, tp_I) = \Sigma_{h=tl_I}^{tp_I} PE_h(\rho_{tl_I}\rho_{tl_I-1}\ldots\rho_{tl_I-h}s)$ [W sec (J)]. FCT_I and FCE_I are given as follows:

$$FCT_I(s, l, t) = FCA_I(s, l, t)/CR_I. \tag{1}$$

$$FCE_I(s,l,t) = FCA_I(s,l,t) \; PB_I. \tag{2}$$

The node output ratio π_I of f_I is $\rho_{tl_I}\rho_{tl_I-1}\cdots\rho_{tp_I+1}\rho_{tp_I}$. The size of the output DU od_I is $\pi_I s$.

Let $RA_I(s)$ and $SA_I(s)$ [bit] be the communication amounts of a node f_I to receive and send a DU of size s, $rc_{1I}\cdot s+rc_{2I}$ and $sc_{1I}\cdot s+sc_{2I}$, respectively, where $rc_{1I}, rc_{2I}, sc_{1I}$, and sc_{2I} are constants. It takes time $RT_I(s)$ ($= RA_I(s)/CR_I$) and $ST_I(s)$ ($= SA_I(s)/CR_I$) [sec] for the node f_I to receive and send a DU of size s, respectively. Let RPB_I and SPB_I be the energy $rp_I\cdot PB_I$ and $sp_I\cdot PB_I$ [W sec/bit] for receiving and sending a DU, respectively. A node f_I consumes the energy $RVE_I(s) = RA_I(s)\cdot RPB_I$ and $SE_I(s) = SA_I(s)\cdot SPB_I$ [J] to receive and send a DU of size s [bit], respectively.

A node f_I totally consumes the energy $FE_I(s, tl_I, tp_I) = RA_I(s)\cdot SPB_I + FCA_I(s, tl_I, tp_I)\cdot PB_I + SA_I(\pi_I s)\cdot SPB_I= (RA_I(s)\cdot rp_I + CE_I(s, tl_I, tp_I) + SA_I(\pi_I\cdot s)\cdot sp_I)\cdot PB_I$ [J] for an input DU of size s. The total execution time $FT_I(s, tl_I, tp_I)$ [sec] is ($RA_I(s) + FCA_I(s, tl_I, tp_I) + SA_I(\pi_I\cdot s)$)$/CR_I$.

Each node f_I is initially in *idle* state. On receipt of input DUs ID_I of size s from every child node, f_I transits to *receiving* state and the communication residue RR_I [bit] is $RA_I(s)$. For each second, RR_I is decremented by CR_I. If $RR_I \leq 0$, f_I transits to *executing* state and the computation residue RC_I [bit] is the computation amount $FCA_I(s)$. For each second, RR_I is decremented by CR_I if *state = executing*. If $RR_I \leq 0$, f_I transits to an *sending* state and the communication residue $RS_I = SA_I(\pi_I s)$. For each second, RS_I is decremented by CR_I if *state = sending*. If $RS_I \leq 0$, f_I transits to the *idle* state.

4   A Homogeneous Estimation Model

We consider a homogeneous TBFC (HTBFC) model for an application $AP = \langle p_1, \ldots, p_m \rangle$. Here, each fog node f_I supports m_I ($= tl_I - tp_I + 1$) processes $p_{tl_I}, p_{tl_I-1}, \ldots, p_{tp_I}$ in AP and satisfies the following properties:

1. The computation rate CR_I is FCR ($\leq CR$).
2. $rc_{1I} = rc_1, rc_{2I} = rc_2, sc_{1I} = sc_1$, and $sc_{2I} = sc_2$ of the communication amount RA_I and SA_I.

That is, every fog node f_I is realized by a same type of a computer like Raspberry Pi [29] while a root node f is realized by a server which is faster than a fog node f_I.

We make the following assumptions on each application process p_h in an application AP:

1. The computation amount $PCA_h(s)$ is $a_1 s^2 + a_2 s + a_3$. That is, $a_{1h} = a_1, a_{2h} = a_2$, and $a_{3h} = a_3$.
2. The output ratio ρ_h is ρ.

The output ratio π_I of a node f_I is ρ^{m_I}. A node f_I gets an output DU od_I of size $\pi_I s$ from input DUs of size s.

Suppose a node f_I receives input DUs ID_I of size s. Here, the total computation amount $FCA_I(s, tl_I, tp_I)$ [bit] of the node f_I to get an output DU od_I from the input DUs ID_I is $\Sigma_{h=tl_I}^{tp_I} PCA_h(\rho^{tl_I-h}s)$:

$$FCA_I(s, tl, tp) = \begin{cases} a_1 \dfrac{1 - \rho^{2u}}{1 - \rho^2} s^2 + a_2 \dfrac{1 - \rho^u}{1 - \rho} s + a_3 u \\ \qquad \text{where } u = tl - tp + 1 \text{ and } tl \geq tp. \end{cases} \tag{3}$$

A node f_I consumes the energy $RVE_I(s)$ and $SDE_I(s)$ [J] to receive and send a DU of size s, respectively:

$$RVE_I(s) = RA_I(s) \; RPB_I = (rc_1 s + rc_2) rp_I PB_I. \tag{4}$$

$$SDE_I(s) = SA_I(s) \; SPB_I. = (sc_1 s + sc_2) sp_I PB_I \tag{5}$$

Suppose input DUs $ID_I = \{id_{I1}, \ldots, id_{Ib_I}\}$ of total size s [bit] from b_I child nodes $f_{I1}, \ldots, f_{I,b_I}$ are received by a node f_I. Let f_C denote a child node f_{Ii} and f_P stand for the parent node of f_I. The node f_I receives a DU of size $\frac{s}{b_I}$ from each child node f_C. Each child node f_C consumes the energy $SDE_C(\frac{s}{b_I})$ to send a DU of size $\frac{s}{b_I}$ to f_I and then f_I consumes the energy $RVE_I(\frac{s}{b_I})$ to receive the DU from each child node f_C. The node f_I consumes the energy $FCE_I(s, tl_I, tp_I)$ to process ID_I and obtains an output DU od_I of size $\pi_I s$. The node f_I consumes the energy $SDE_I(\pi_I s)$ to send the output DU od_I to the parent node f_P. The energy $RVE_P(\pi_I s)$ is consumed by the parent node f_P to receive the DU od_I. Thus, the energy $NFE_I(s, tl_I, tp_I)$ is consumed by f_I, every child node f_C to send a DU to f_I, and the parent node f_P to receive a DU from f_I [Fig. 1]:

$$NFE_I(s, tl_I, tp_I) = \begin{cases} SDE_C(\frac{s}{b_I}) \cdot b_I + \\ RVE_I(\frac{s}{b_I}) \cdot b_I + FCE_I(s, tl_I, tp_I) + SDE_I(\pi_I s) + \\ RVE_P(\pi_I s). \end{cases} \tag{6}$$

5   Formulas to Estimate Energy Consumption

We consider a homogeneous TBFC tree T for an application $AP = \langle p_1, \ldots, p_m \rangle$. Suppose a process subsequence $P_I = \langle p_{tp_I}, p_{tp_I+1}, \ldots, p_{tl_I} \rangle$ is installed on a node f_I. Suppose the node f_I receives input DUs ID_I of total size s [bit] from the b_I child nodes. Each child node f_C sends an input DU of size $\frac{s}{b_I}$ [bit] to f_I.

We consider the following three operations on a target node f_I to change the tree T [17–19]. Here, n is the number of replica nodes to be created and k shows the number of processes to migrate from f_I.

1. Migration: $MG(f_I, k)$.
2. Replicate: $\langle f_{N_1}, \ldots, f_{N_{n-1}} \rangle = RP(f_I, n)$.
3. Expand: $\langle f_{N_1}, \ldots, f_{N_n} \rangle \;\; = EP(f_I, n, k)$.

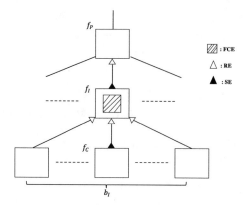

Fig. 1. Energy consumption of f_I and affected nodes.

A node whose energy consumption changes by applying an operation op $\in\{MG, RP, EP\}$ on a target node f_I is *affected* by the operation op. We have to reduce the energy consumption of not only the target node f_I but also all the affected nodes. By using the estimation formulas, the total energy consumption of the target node f_I and the affected nodes is estimated.

In the *migrate* operation $MG(f_I, k)$, k ($< m_I$) processes $p_{tl_I}, \ldots, p_{tl_I-k+1}$ from a tail process p_{tl_I} migrate from a target node f_I to every child node f_C through the live migration of virtual machines [11–15]. Thus, every child node f_C is affected. In each child node f_C, the k processes $p_{tl_I}, \ldots, p_{tl_I-k+1}$ migrating from f_I are additionally executed for a DU of size $\frac{s}{b_I}$ passed from the top process p_{tp_C}. Hence, each child node f_C consumes the energy $FCE_C(\frac{s}{b_I}, tl_I, tl_I - k + 1)$ to execute the k processes migrating from f_I. Since the node f_I receives input DUs ID_I of total size $\rho^k s$ ($= \frac{\rho^k s}{b_I} b_I$) from the b_I child nodes and the $(m_I - k)$ processes $p_{tl_I-k}, p_{tl_I-k-1}, \ldots, p_{tp_I}$ are executed for ID_I, f_I consumes the energy $FCE_I(\rho^k s, tl_I - k, tp_I)$ to execute the processes. Here, the energy $ME_I(s, tl_I, tp_I, k)$ is consumed by f_I and affected nodes:

$$ME_I(s, tl_I, tp_I, k) = \begin{cases} [FCE_C(\frac{s}{b_I}, tl_I, tl_I - k + 1) + SDE_C(\frac{\rho^k s}{b_I})]b_I + \\ RVE_I(\frac{\rho^k s}{b_I})b_I + FCE_I(\rho^k s, tl_I - k, tp_I) + SDE_I(\pi_I s) + \\ RVE_P(\pi_I s). \end{cases} \tag{7}$$

In the *replicate* operation $RP(f_I, n)$ ($n \geq 2$), $(n - 1)$ sibling replica nodes $f_{N_1}, \ldots, f_{N_{n-1}}$ of the node f_I are created. In result, the parent node f_P has totally n child replica nodes f_I and $f_{N_1}, \ldots, f_{N_{n-1}}$. Then, the $\frac{b_I}{n}$ nodes of the nodes $f_{I1}, \ldots, f_{Ib_I}$ are reconnected to each replica node as the child nodes. Here, $b_I \geq n$. If f_I is an edge node, f_I has a child device node d_I which sends a DU of size s to f_I. A device node d_{N_i} is created as a child node of each newly created node f_{N_i}. Each of the device nodes $d_I, d_{N_1}, \ldots, d_{N_{n-1}}$ sends a DU of size $\frac{s}{n}$ [bit] to the parent edge node after replicating f_I. Each of the n sibling replica nodes receives an input DU of size $\frac{s}{n}$ [bit]. Hence, each of the n sibling nodes consumes the energy $FCE_I(\frac{s}{n}, tl_I, tp_I)$ to process input DUs. The parent node

f_P, each replica node f_N, and each child node f_C are affected by the operation $RP(f_I, n)$. The energy $RE_I(s, tl_I, tp_I, n)$ is consumed by the target node f_I and every affected node:

$$RE_I(s, tl_I, tp_I, n) = \begin{cases} SDE_C(\frac{s}{b_I})b_I + \\ (RVE_I(\frac{s}{b_I})\frac{b_I}{n} + FCE_I(\frac{s}{n}, tl_I, tp_I) + SDE_I(\frac{\pi_I s}{n}))n + \\ RVE_P(\frac{\pi_I s}{n})n. \end{cases} \quad (8)$$

In the *expand* operation $EP(f_I, n, k)$, n (≥ 1) child replica nodes $f_{N_1}, \ldots, f_{N_n}$ of a target node f_I are created and k $(\leq m_I)$ processes $p_{tl_I}, p_{tl_I-1}, \ldots, p_{tl_I-k+1}$ migrate to each child replica node. Let f_N denote a created node f_{Ni}. Each node f_N has $\frac{b_I}{n}$ child nodes of the child nodes $f_{I1}, \ldots, f_{Ib_I}$ for $b_I \geq n$. The node f_I supports $(m_I - k)$ processes $p_{tl_I-k}, \ldots, p_{tp_I}$ while each node f_N additionally supports k processes $p_{tl_I}, p_{tl_I-1}, \ldots, p_{tl_I-k+1}$ migrating from f_I. Since each node f_N receives an input DU of size $\frac{s}{n}$, f_N consumes the energy $FCE_N(\frac{s}{n}, tl_I, tl_I - k + 1)$. Since f_I receives input DUs ID_I of size $\rho^k s$ from the n child nodes, f_I consumes the energy $FCE_I(\rho^k s, tl_I - k, tp_I)$ to execute the processes $p_{tl_I-k}, p_{tl_I-k-1}, \ldots, p_{tp_I}$. The target node f_I and the affected nodes f_N totally consume the energy $EE_I(s, tl_I, tp_I, n, k)$ after expanding f_I:

$$EE_I(s, tl_I, tp_I, n, k) = \begin{cases} SDE_C(\frac{s}{b_I}) \cdot b_I + \\ (RVE_I(\frac{s}{b_I})\frac{b_I}{n} + FCE_N(\frac{s}{n}, tl_I, tl_I - k + 1) + \\ SDE_N(\frac{\rho^k s}{n}))n + \\ RVE_I(\frac{\rho^k s}{n}) \cdot n + FCE_I(\rho^k s, tl_I - k, tp_I) + SDE_I(\pi_I s) + \\ RVE_P(\pi_I s). \end{cases} \quad (9)$$

6 An Algorithm to Change a Tree

A node f_I supports a subsequence $\langle p_{tp_I}, p_{tp_I+1}, \ldots, p_{tl_I} \rangle$ $(1 \leq tp_I \leq tl_I \leq m)$ of m_I $(= tl_I - tp_I + 1)$ processes of an application $AP = \langle p_1, \ldots, p_m \rangle$. There are b_I (≥ 0) child nodes $f_{I1}, \ldots, f_{I,b_I}$ of a node f_I. Let f_C and f_P denote a child node f_{Ii} and a parent node of f_I as presented in the preceding subsection. Let f_N denote a node f_{N_i} newly created in the operations $RP(f_I, n)$ and $EP(f_I, n, k)$.

A target node f_I to apply an operation op $(\in \{MG, RP, EP\})$ to change a tree T is selected in the tree T, whose receipt queue is the longest. Let MT_T be an ordered set of target nodes in a tree T, where target nodes are ordered in the descending order of levels in the tree T. This means, a target node at lower level precedes a target node at higher level. Let $f_I = dequeue(MT_T)$ be a function to take a top node f_I in the ordered set MT_T.

First, a target node f_I is taken by $f_I = dequeue(MT_T)$. Then, a tuple $\langle op, n, k \rangle$, where $op \in \{MG, RP, EP\}$, n is the number of nodes to be newly created, and k is the number of processes to migrate from f_I, are given by the function $\langle op, n, k \rangle = select\_operation(f_I, s)$ where s is the size of input DUs of the node f_I. Here, let $ME_I(s, tl_I, tp_I, km)$, $RE_I(s, tl_I, tp_I, ns)$, and $EE_I(s, tl_I, tp_I, ne, ke)$ be the smallest energy consumption of the target node

f_I and the affected nodes for MG, RP, and EP operations, respectively. For example, $ME_I(s, tl_I, tp_I, km) \leq ME_I(s, l_I, tp_I, kk)$ for every $kk = 1, \ldots, m_I - 1$. If $ME_I(s, tl_I, tp_I, km)$, $RE_I(s, tl_I, tp_I, ns)$, and $EE_I(s, tl_I, tp_I, ne, ke)$ for the operations $MG(f_I, km)$, $RP(f_I, ns, ks)$, and $EP(f_I, ne, ke)$, respectively, is smaller than $NE_I(s)$, f_I can be changed by the operations since the energy consumption is reduced. That is, $op = MG$ and $k = km$ if $ME_I(s, tl_I, tp_I, km)$ is the smallest, $op = RP$ and $n = ns$ if $RE_I(s, tl_I, tp_I, ns)$ is the smallest, $op = EP, n = ne$, and $k = ke$ if $EE_I(s, tl_I, tp_I, ne, ke)$ is the smallest. Then, the operation op whose energy consumption is the smallest in $ME_I(s, tl_I, tp_I, km)$, $RE_I(s, tl_I, tp_I, ns)$, and $EE_I(s, tl_I, tp_I, ne, ke)$ is taken, The operation $op(f_I, n, k)$ is applied on the target node f_I to really change the tree.

[Algorithm to select an operation]
$select\_operation(f_I, s)$ {
$E = NE_I(s, tl_I, tp_I);$ $k = n = 0;$ $op = 0:$
for $kk = 1, \ldots, m_I - 1$
$E_{MG} = ME_I(s, tl_I, tp_I, kk);$ /* Migrate */
if $E_{MG} < E$, $\{op = MG; \; E = E_{MG}; \; n = 0; \; k = kk;\}$
for end;
for $nn = 1, \ldots, b_I$
$E_{RP} = RE_I(s, tl_I, tp_I, nn);$
if $E_{RP} < E$, $\{op = RP; \; E = E_{RP}; \; n = nn; \; k = 0; \};$ /* Replicate */
for $kk = 1, \ldots, np$
$E_{EP} = EE_I(s, tl_I, tp_I, nn, nk);$
if $E_{EP} < E$, $\{op = EP; \; E = E_{EP}; \; n = nn; \; k = kk; \};$ /* Expand */
for end;
for end;
return $(\langle op, n, k \rangle);$ };

The following procedure is periodically applied on a tree T. First, each target node f_I in the tree T is collected in the ordered set MT. For every target node f_I in MT, an operation $op \in \{MG, RP, EP\}$ is selected by the function $select\_operation$, where the energy consumption of all the affected nodes can be mostly reduced. If op is found, the tree T is changed by applying the operation op on the target node f_I. By the function $execute(T, op, n, k)$, the tree T is changed.

[Procedure to change a tree]
MT = ordered set of target nodes in a tree T;
while $(MT \neq \phi)$ {
$f_I = dequeue(MT);$ s = size of input DUs of f_I;
$\langle op, n, k \rangle = select\_operation(f_I, s);$
if op is found, $T = execute(T, op, n, k);$ /* tree is changed */
while end;

7   Evaluation

An application AP is a sequence of $m(\geq 1)$ processes $p_1, \ldots, p_m$. In the simulation, we assume the output ratio ρ of each process p_h is 0.5 and for the constants of the computation amount $PCA_h(s) = a_1 s^2 + a_2 s + a_3$, $a_1 = 0.1$, $a_2 = 1$, and $a_3 = 10$. For the communication amount $RA_I(s)$ and $SA_I(s)$ [bit] for communication, $r_1 = s_1 = 0.01$ and $r_2 = s_2 = 0.02$. The computation rate CR is 800 [Kbps], maximum power XP is 301.3 [W], and minimum power $MP = 126.1$ [W] for a root node, i.e. server f. $PB = 0.37$ [J/Kbit]. The computation rate CR_I is 148 [Kbs], maximum power $XP_I = 3.7$ [W], and $se_I = 0.68$ and $re_I = 0.73$ for each fog node f_I [22,24]. $PB_I = 0.025$ [J/Kbit]. In this evaluation, there are totally 1,000 devices, i.e. $nd = 1,000$ and each device sends a DU of sensor data every 10 [sec]. The size of each DU is 8 [Kbit]. Hence, all the devices send totally 8,000 [Kbit] to edge nodes every 10 [sec].

In the replicate RP operation on a target node f_I, if f_I has b_I (≥ 1) child nodes, at most $b_I - 1$ replica sibling nodes of f_I are created. In the expand EP operation, at most b_I nodes are created as child nodes of f_I. Suppose a target node f_I is an edge node and its child device node d_I includes nd_I devices. In the RP and EP operations, if $nd_I \leq 8$, at most only seven replica nodes $(n = 7)$ and eight child nodes $(n = 8)$ are created, respectively. Otherwise, at most $nd_I - 1$ and nd_I nodes are created, respectively. Suppose n edge nodes are created for a target edge node f_I where the child device node d_I has nd_I devices in the RP and EP operations. Here, a device node is created for each newly created edge node and is composed of $\frac{nd}{n}$ devices in the RP and EP operations.

Initially, we consider a tree T denoting a CC model which is composed of one root node f and one device node d which shows a collection of nd $(= 1,000)$ devices. Every 100 [sec], every node is checked and a collection MT_T of target nods is found. If a target node f_I is found, i.e. $MT_T \neq \phi$, an operation op to change the tree T is found by the function $select\_operation$. If such an operation op can be reduced is found, T is changed by applying the operation op on the target node f_I. In the simulation, the procedure is applied until the tree T cannot be changed. For each number m of processes $(m = 10, 20, 30)$, the change procedure is applied on the tree T every 100 [sec]. For each ith application of the procedure, the total energy consumption TE_i of a tree T is obtained. TE_1 shows the total energy consumption of the initial tree, i.e. CC model.

Figure 2 shows the ratios of TE_i/TE_1 for application number i. For $m = 30$ and $m = 20$, the procedure terminates at $i = 10$ and for $m = 10$, $i = 11$. A CC model of the IoT can be changed to a TBFC tree whose energy consumption is only 0.3 to 0.5 [%] of the CC model in the proposed algorithm. For $m = 10$, the changed tree includes 247 nodes and the height is six. For $m = 20$, the changed tree includes 227 nodes and the height is five. For $m = 30$, the changed tree includes 199 nodes and the height is five.

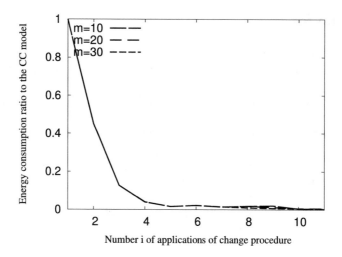

Fig. 2. Ratio of total energy consumption

8   Concluding Remarks

In this paper, we discuss how to reduce the energy consumption of the TBFC model of the IoT where fog nodes are tree-structured. Here, the TBFC tree is changed by making processes of a node migrate to child nodes, replicating a node, and expanding a node, i.e. new child nodes are created. By changing a target node, the energy consumption of the target node, parent node, child nodes, and nodes newly created is also changed. The nodes are affected by changing the target node. In this paper, we newly proposed the formulas to estimate the total energy consumption of not only the target node but also the other affected nodes. By using the estimation formulas, if the estimated energy consumption of all the affected nodes is smaller, the tree is changed. In the evaluation, starting from the CC model, the TBFC tree is obtained by the proposed algorithm. We showed the tree whose total energy consumption is smaller than the CC model can be obtained by replication and expansion of a most congested node and migration of processes. The total energy consumption of the tree obtained by the proposed algorithm is only 0.3 to 0.5 [%] of the CC model.

Acknowledgment. This work is supported by Japan Society for the Promotion of Science (JSPS) KAKENHI Grant Number 22K12018.

References

1. Dayarathna, M., Wen, Y., Fan, R.: Data center energy consumption modeling: a survey. IEEE Commun. Surv. Tutor. **18**(1), 732–787 (2016)
2. Qian, L., Luo, Z., Du, Y., Guo, L.: Cloud computing: an overview. In: Proceedings of the 1st International Conference on Cloud Computing, pp. 626–631 (2009)

3. Rahmani, A. M., Liljeberg, P., Preden, J.-S., Jantsch, A.: Fog Computing in the Internet of Things, 1st esn, 172 p. Springer, Cham (2018). https://doi.org/10.1007/978-3-319-57639-8

4. Enokido, T., Aikebaier, A., Takizawa, M.: Process allocation algorithms for saving power consumption in peer-to-peer systems. IEEE Trans. Ind. Electron. **58**(6), 2097–2105 (2011)

5. Enokido, T., Aikebaier, A., Takizawa, M.: A model for reducing power consumption in peer-to-peer systems. IEEE Syst. J. **4**(2), 221–229 (2010)

6. Enokido, T., Aikebaier, A., Takizawa, M.: An extended simple power consumption model for selecting a server to perform computation type processes in digital ecosystems. IEEE Trans. Ind. Inform. **10**(2), 1627–1636 (2014)

7. Enokido, T., Takizawa, M.: Integrated power consumption model for distributed systems. IEEE Trans. Ind. Electron. **60**(2), 824–836 (2013)

8. Kataoka, H., Duolikun, D., Enokido, T., Takizawa, M.: Energy-efficient virtualisation of threads in a server cluster. In: Proceedings of the 10th International Conference on Broadband and Wireless Computing, Communication and Applications (BWCCA-2015), pp. 288–295 (2015)

9. Kataoka, H., Duolikun, D., Sawada, A., Enokido, T., Takizawa, M.: Energy-aware server selection algorithms in a scalable cluster. In: Proceedings of the 30th International Conference on Advanced Information Networking and Applications, pp. 565–572 (2016)

10. Kataoka, H., Nakamura, S., Duolikun, D., Enokido, T., Takizawa, M.: Multi-level power consumption model and energy-aware server selection algorithm. Int. J. Grid Util. Comput. **8**(3), 201–210 (2017)

11. Duolikun, D., Enokido, T., Takizawa, M.: Energy-efficient dynamic clusters of servers, In: Proceedings of the 8th International Conference on Broadband and Wireless Computing, Communication and Applications, pp. 253–260 (2013)

12. Duolikun, D., Enokido, T., Takizawa, M.: Static and dynamic group migration algorithms of virtual machines to reduce energy consumption of a server cluster. Trans. Comput. Collect. Intell. **XXXIII**, 144–166 (2019)

13. Duolikun, D., Enokido, T., Takizawa, M.: Simple algorithms for selecting an energy-efficient server in a cluster of servers. Int. J. Commun. Netw. Distrib. Syst. **21**(1), 1–25 (2018), pp. 145–155 (2019)

14. Duolikun, D., Enokido, T., Barolli, L., Takizawa, M.: A monotonically increasing (MI) algorithm to estimate energy consumption and execution time of processes on a server. In: Proceedings of the 24th International Conference on Network-based Information Systems, pp. 1–12 (2021)

15. Duolikun, D., Nakamura, S., Enokido, T., Takizawa, M. : Energy-consumption evaluation of the tree-based fog computing (TBFC) model. In: Proceedings of the 22nd International Conference on Broadband and Wireless Computing, Communication and Applications, pp. 66–77 (2022)

16. Duolikun, D., Enokido, Takizawa, M.: Energy-efficient multi-version concurrency control (EEMVCC) for object-based systems. In: Proceedings of the 25th International Conference on Network-Based Information Systems (NBiS-2022), Sanda-Shi, Japan, pp. 13–24 (2022)

17. Duolikun, D., Enokido, T., Barolli, L., Takizawa, M.: A flexible fog computing (FTBFC) model to reduce energy consumption of the IoT. In: Proceedings of the 10th International Conference on Emerging Internet, Data and Web Technologies, pp. 256–262 (2022)

18. Duolikun, D., Enokido, Takizawa, M.: An energy-aware algorithm for changing tree structure and process migration in the flexible tree-based fog computing model. In: Proceedings of the 37th International Conference on Advanced Information Networking and Applications, pp. 268–278 (2023)

19. Duolikun, D., Enokido, Takizawa, M.: An energy-aware dynamic algorithm for the FTBFC model of the IoT. In: Proceedings of the 17th International Conference on Complex, Intelligent and Software Intensive Systems (CISIS-2023), Toronto, ON, Canada, pp. 38–47 (2023)

20. Mukae, K., Saito, T., Nakamura, S., Enokido, T., Takizawa, M.: Design and implementing of the dynamic tree-based fog computing (DTBFC) model to realize the energy-efficient IoT. In: Proceedings of the 9th International Conference on Emerging Internet, Data and Web Technologies, pp. 71–81 (2021)

21. Inoue, T., Aikebaier, A., Enokido, T., Takizawa, M.: Algorithms for selecting energy-efficient storage servers in storage and computation oriented applications. In: Proceedings of IEEE the 26th International Conference on Advanced Information Networking and Applications, (AINA-2016), pp. 920–927 (2016)

22. Oma, R., Nakamura, S., Duolikun, D., Enokido, T., Takizawa, M.: An energy-efficient model for fog computing in the Internet of Things (IoT). Internet Things **1–2**, 14–26 (2018)

23. Oma, R., Nakamura, S., Enokido, T., Takizawa, M.: A tree-based model of energy-efficient fog computing systems in IoT. In: Proceedings of the 12th International Confernce on Complex, Intelligent, and Software Intensive Systems, pp. 991–1001 (2018)

24. Oma, R., Nakamura, S., Duolikun, D., Enokido, T., Takizawa, M.: Evaluation of an energy-efficient tree-based model of fog computing. In: Proceedings of the 21st International Conference on Network-based Information Systems, pp. 99–109 (2018)

25. Oma, R., Nakamura, S., Duolikun, D., Enokido, T., Takizawa, M.: A fault-tolerant tree-based fog computing model. Int. J. Web Grid Serv. **15**(3), 219–239 (2019)

26. Oma, R., Nakamura, S., Duolikun, D., Enokido, T., Takizawa, M.: Energy-efficient recovery algorithm in the fault-tolerant tree-based fog computing (FTBFC) model. In: Proceedings of the 33rd International Conference on Advanced Information Networking and Applications (AINA 2019), pp. 132–143 (2019)

27. Oma, R., Nakamura, S., Enokido, T., Takizawa, M.: A dynamic tree-based fog computing (DTBFC) model for the energy-efficient IoT. In: Proceedings of the 8th International Conference on Emerging Internet, Data and Web Technologies, pp. 24–34 (2020)

28. Guo, Y., Saito, T., Oma, R., Nakamura, S., Enokido, T., Takizawa, M.: Distributed approach to fog computing with auction method. In: Proceedings of the 34th International Conference on Advanced Information Networking and Applications, pp. 268–275 (2020)

29. Raspberry pi 3 model b (2016). https://www.raspberrypi.org/products/raspberry-pi-3-model-b

An Efficient Cross-Contract Vulnerability Detection Model Integrating Machine Learning and Fuzz Testing

Huipeng Liu[1], Baojiang Cui[1(✉)], Jie Xu[1], and Lihua Niu[2]

[1] School of Cyberspace Security, Beijing University of Posts and Telecommunications, Beijing, China
{lhp6666,cuibj,cheer1107}@bupt.edu.cn
[2] School of Computer Science, Beijing University of Posts and Telecommunications, Beijing, China
nlh@bupt.edu.cn

Abstract. Due to the lengthy processing time associated with traditional fuzz testing methods for identifying vulnerable contracts, we have integrated machine learning with improved fuzz testing tools. We employ decision tree, EEC, and random forest models for filtering vulnerable contracts, complemented by an enhanced fuzz testing tool. Starting with the opcode features of smart contracts, we utilize the N-Gram algorithm to extract multidimensional features, assess feature importance using the TF-IWF algorithm, and construct a comprehensive 2312-dimensional feature matrix for training machine learning models to filter cross-contract vulnerabilities. Experimental results demonstrate the robust performance of the model in cross-contract vulnerability detection, successfully identifying 15 new instances of cross-contract reentrancy vulnerabilities with a precision of 93.6% and an F1-Score of 87.5%. A comparison with advanced fuzz testing tools confirms the superiority of our model in cross-contract vulnerability detection.

Keywords: Smart Contract · Fuzz · Cross-contract Vulnerability · Machine Learning

1 Introduction

Since 2009, the decentralized blockchain [1] technology and industry have attracted a large number of investors and practitioners, and have achieved considerable development, and compared with the traditional network, it has the characteristics of immutable, decentralized, traceable and so on. With the development of blockchain, smart contracts have been widely applied in various fields including infrastructure, business retail, gaming, as well as social media and communication. However, this widespread adoption has also given rise to numerous challenges and issues [2–4].

Although existing vulnerability detection tools have identified numerous vulnerabilities, one crucial category, namely cross-contract vulnerabilities, has largely been overlooked to date. Cross-contract vulnerabilities are exploitable flaws that manifest

L. Barolli (Ed.): EIDWT 2024, LNDECT 193, pp. 297–306, 2024.
https://doi.org/10.1007/978-3-031-53555-0_28

only when two or more interacting contracts are present. For instance, the phishing contract shown in Fig. 1 occurs only when three contracts interact in a specific order. Other fuzz testing tools such as ContractFuzzer [5] and sFuzz [6], limited to analyzing only two contracts at a time, failed to discover this vulnerability.

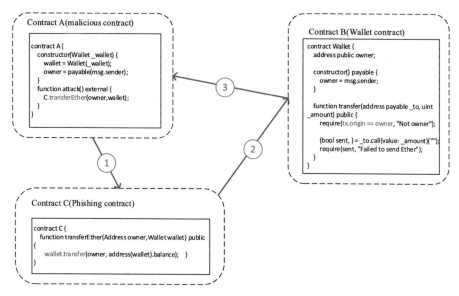

Fig. 1. Cross-contract phishing vulnerability contract.

Therefore, security testing, particularly for cross-contract attacks, is crucial before deployment. We propose a novel machine learning-based cross-contract fuzz testing model. The contributions of the proposed model are listed below.

1. Combining machine learning models with traditional fuzz testing enhances detection accuracy by analyzing and filtering contracts. Fuzz testing rapidly generates test cases, comprehensively detecting cross-contract vulnerabilities in smart contracts.
2. Introducing a machine learning model based on opcode features of smart contracts, utilizing N-Gram and TF-IWF algorithms to convert opcodes into vectors. Eight statistical features are extracted to enrich the vectors, ensuring the quality and diversity of training data.
3. Experimental results demonstrate that our machine learning-based cross-contract fuzz testing tool outperforms other existing tools, offering higher efficiency and accuracy in cross-contract vulnerability detection.

This paper is structured into five chapters. In the introduction, we address the security challenges posed by smart contract vulnerabilities in recent years and outline our contributions to mitigating cross-contract vulnerabilities. The chapter on related work provides an overview of existing research on traditional fuzz testing, machine learning, and the amalgamation of both approaches. The model's structure and principles are then presented, detailing the vulnerability handling process, the principles of opcode vectorization, and the introduction of 11 types of opcode. Moving on to the results and discussion chapter, we elaborate on the experimental process and analyze the results, drawing comparisons with other available smart contract detection models. Finally, the conclusion chapter summarizes the strengths, weaknesses, and outlines future research directions.

2 Related Work

Fuzz testing for smart contract vulnerability detection involves generating diverse test data randomly. Jiang et al. [5] introduced ContractFuzzer in 2018, which statically analyzes the contract's ABI, extracts function signatures, and uses adaptive strategies for fuzz testing. Tai et al. [6] proposed SFuzz in 2020, an adaptive strategy-based fuzz testing tool that monitors the testing process, optimizes test seed generation, and enhances code coverage and efficiency in testing smart contracts.

Machine learning in smart contract vulnerability detection uses historical data and computational power to mimic human learning. In 2018, Tann et al. [7] introduced LSTM networks for sequential learning of smart contract weaknesses. This method maps bytecode sequences to vector sequences and trains a binary classification model using MAIAN-detected smart contracts. Liu et al. [8], also in 2018, proposed S-gram, a technique combining N-gram language modeling and lightweight static semantic labeling. S-gram uses an N-gram training engine to train a language model based on character sequences with semantic metadata markers, enabling contract vulnerability detection.

In the pursuit of enhanced vulnerability detection and to address the limitations of singular analysis techniques, combination techniques (integrating various program analysis methods for smart contract vulnerability detection) have garnered increasing attention from industry professionals since 2019. Y et al. [9] introduced xFuzz, a cross-contract vulnerability detection tool that combines machine learning with fuzz testing. This tool initially employs a machine learning model to select contracts prone to vulnerabilities, assesses the suspicion scores of functions within contracts, prioritizes them, and ultimately utilizes an improved fuzz testing approach to detect relevant cross-contract vulnerabilities.

3 Model's Structure and Principles

3.1 Model's Structure

Our proposed machine learning-based cross-contract fuzz testing tool conducts smart contract vulnerability detection as follows. Figure 2 illustrates the overall vulnerability detection process.

(1) We utilized a web scraper script to extract smart contracts from the real-world Ethereum blockchain (https://etherscan.io/), resulting in 148,500 contracts for our initial test dataset. To ensure representativity, we compared the bytecode of contracts and removed duplicate samples with identical bytecode. The deduplicated dataset serves as the sample set for machine learning.

(2) We employed three Solidity smart contract vulnerability detection tools, namely Slither [10], Securify [11], and Mythril [12], to perform vulnerability analysis on the sample set. Vulnerability labels for integer overflow, reentrancy, and Tx-origin were added to the contract samples.

(3) Following the definition of bytecode by Jay et al. [13], we categorized the bytecode into 11 classes. Based on the smart contract sample set, we obtained the Solidity compiler version used for each contract. We compiled the contracts using the Solidity compiler to generate the corresponding bytecode files. We simplified and classified the bytecode to obtain the smart contract bytecode dataset.

(4) We applied the N-Gram algorithm to process the obtained bytecode dataset, segmenting the bytecode fragments in smart contract bytecode files and generating multidimensional N-Gram features. The TF-IWF algorithm was used to evaluate the importance of different bytecode fragment features, calculating TF-IWF values for bigram features. Subsequently, based on the frequency of different types of bytecode in smart contract bytecode files, we constructed some statistical features. Finally, we merged the 2304-dimensional bigram features with the 8-dimensional statistical features, transforming the smart contract bytecode dataset into a 2312-dimensional feature matrix for machine learning model training.

(5) Using the feature matrix as input, we selected Support Vector Machine, Random Forest, Decision Tree, K-Nearest Neighbors, Logistic Regression, Gradient Boosting, and EEC models for training. We chose the Decision Tree model, EEC model, and Random Forest model with the best classification performance as our machine learning models for smart contract vulnerability classification.

(6) Using the trained machine learning models, we filtered the test set, obtaining 384 contracts for the next fuzz testing iteration as dataset 1. Dataset 1 was input into the fuzz testing model and compared with Slither and sFuzz as reference points, compiling precision, recall, and F1-score for integer overflow and underflow, reentrancy, and Tx-origin vulnerability detection. Finally, analyze the results and draw conclusions.

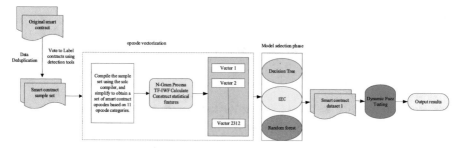

Fig. 2. Vulnerability detection processing.

3.2 The Principle of Model

TF-IWF (Term Frequency-Inverse Word Frequency) [14]. The TF-IWF algorithm is employed to assess the significance of words in the article. A higher TF-IWF value indicates greater importance of the word in the article. Therefore, by computing the TF-IWF values of individual words throughout the article, key terms can be identified.

TF represents term frequency, and its calculation formula is:

$$TF = \frac{count(t)}{M} \tag{1}$$

$count(t)$ represents the number of occurrences of a specific bigram t in the entire contract, and M represents the total number of bigrams in contract. IWF represents the inverse document frequency, and its calculation formula is:

$$IWF = \log\left(\frac{W_c}{W_{c,t}}\right) \tag{2}$$

Here, W_c represents the frequency of all bigrams in the smart contract corpus, and $W_{c,t}$ represents the frequency of the bigram named t in the corpus. Therefore, the TF-IWF value of the bigram t in the document is calculated as:

$$TF - IWF = TF \times IWF \tag{3}$$

Statistical Features. Construct some statistical features based on the usage frequency of different types of opcodes in smart contract opcode files, which can better reflect the behavioral characteristics of smart contract vulnerabilities.

The article selects 7 different types of opcode instructions, calculates their usage frequencies in smart contracts, and constructs a new 8-dimensional feature vector. These features include: 1) the number of arithmetic instructions, 2) the number of logical instructions, 3) the number of call instructions, 4) the number of storage instructions, 5) the number of constant instructions, 6) the number of jump instructions, 7) the number of termination instructions, and 8) the total number of all instructions.

The 11 Types of Opcode. The 11 types of opcodes are shown in Table 1. For each type of opcode, we show the description and specific opcodes.

Table 1. 11 types of opcodes.

Opcode Type	Description	Specific Opcodes
Arithmetic Instructions	Perform arithmetic operations on operands stored on the stack	ADD, MUL, SUB, DIV
Control Instructions	Save operands onto the Ethereum Virtual Machine (EVM) stack	PUSH, DUP, SWAP, LOG

(continued)

Table 1. (*continued*)

Opcode Type	Description	Specific Opcodes
Logical Instructions	Perform logical operations on operands stored on the stack	AND, OR, XOR, NOT
Comparison Instructions	Set and check bit flags	LT, GT, SLT, SGT, EQ, ISZERO
Call Instructions	Facilitate interaction between smart contract accounts through message calls	CALL, CALLCODE, DELEGATECALL, STATICCALL
Storage Instructions	Perform read and write operations in memory and storage	MLOAD, MSTORE, SLOAD, SSTORE
Address Instructions	Access information about smart contract account addresses	ADDRESS, BALANCE, ORIGIN, CALLER
Constant Instructions	Retrieve parameters related to the current block or other block information	BLOCKHASH, COINBASE, TIMESTAMP, NUMBER, DIFFICULTY, GASLIMIT
Jump Instructions	Facilitate jumps within smart contract functions	JUMPDEST, JUMP, JUMPI
Halt Instructions	Terminate smart contract execution	RETURN, STOP, REVERT, INVALID, SELFDESTRUCT

4 Results and Discussion

4.1 Evaluation Metrics

In this paper, a comparison is made using Precision, Recall, and the harmonic mean of the two, F1-Score, to determine the performance of the model. In comparison with other vulnerability detection models, we should focus on performance under similar datasets and evaluation conditions. If the proposed cross-contract vulnerability detection model demonstrates superior performance on these key metrics, such as having a higher F1 score, it can be considered to have a competitive advantage in vulnerability detection.

During the experimentation of cross-contract dynamic fuzz testing, the experimental environment was set up on a local machine running Ubuntu 20.04.1 LTS, with 4GB of allocated memory, a 4-core processor, and a 50GB hard disk. The machine is equipped with an AMD Ryzen 7 5800H with Radeon Graphics CPU and has 16GB of RAM.

4.2 Results and Analysis

The opcode vectorization method was employed to generate a dataset consisting of 38,327 feature matrices for testing contracts. From this dataset, 80% of the data was randomly selected as the training set, while the remaining 20% was used as the testing set. The training involved the selection of SVM, random forest, decision tree, K-nearest neighbor algorithm, logistic regression model, gradient boosting model, and EEC model. The training results are shown in Fig. 3.

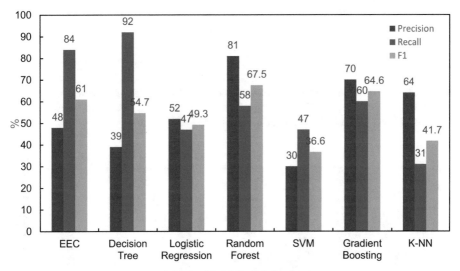

Fig. 3. Model Comparison.

In the binary classification model based on the vectorized dataset of smart contract opcodes, the recall rate of the decision tree is significantly higher than that of other models, with an average recall rate of 92%. The EEC model follows closely, with an average recall rate of 84%. The random forest model achieves a remarkably high average accuracy of 81%, surpassing other models. Therefore, we choose the decision tree model, EEC model, and random forest model as our machine learning models for smart contract vulnerability classification. It is important to note that the goal of training the model is to assist in filtering benign contracts, retaining malicious contracts and suspicious contracts for further screening in the next round of fuzz testing.

Using the test set, 128 contracts with Integer overflow and underflow vulnerabilities, 174 contracts with reentrancy vulnerabilities, and 82 contracts with Tx.origin vulnerabilities were selected as Dataset 1. Initially, Dataset 1 was employed as input, and the machine learning-based cross-contract fuzz testing tool proposed in this paper was compared with the static analysis tool Slither and the fuzz testing tool sFuzz. The results are shown in Table 2.

In Fig. 4, we show comparison of precision, recall, and F1 scores for cross-contract vulnerability detection among three vulnerability detection tools in detecting integer overflow and underflow, Reentrancy, and Tx-origin vulnerabilities.

In Table 3, we show the number of vulnerability detections. Comparing Slither and sFuzz, in terms of the accuracy of detecting Integer Overflow and Underflow Vulnerabilities, Slither does not support the detection of Integer Overflow and Underflow Vulnerabilities, resulting in an accuracy of 0. On the other hand, sFuzz detected a total of 49 Integer Overflow and Underflow Vulnerabilities, but only 47 of them were true positive

Table 2. Comparison of Test Results for Dataset 1

	IOU			Reentrancy			Tx-origin		
	P%	R%	F1%	P%	R%	F1%	P%	R%	F1%
Slither	0	0	0	66.3	74.5	70.1	50	33.3	39.7
SFuzz	95.9	97.9	96.8	81.2	24.2	37.2	0	0	0
This	100	100	100	93.6	82.2	87.5	100	100	100

Note: IOU stands for Integer Overflow and Underflow Vulnerability, P% represents accuracy percentage, R% represents recall percentage, F1 represents F1-score percentage, "This" refers to the cross-contract vulnerability detection tool proposed in this paper

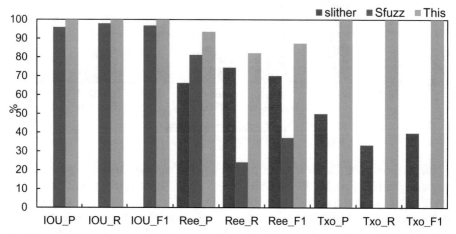

Fig. 4. Comparison of precision, recall, and F1 scores for cross-contract vulnerability detection among three tools. Here, xxx_P represents the precision of detecting a certain vulnerability by the corresponding vulnerability detection tool, xxx_R represents the recall of detecting a certain vulnerability by the corresponding vulnerability detection tool, and xxx_F1 represents the F1 score of detecting a certain vulnerability by the corresponding vulnerability detection tool.

contracts. In contrast, the proposed cross-contract vulnerability detection tool achieved 100% accuracy, detecting a total of 48 true positive contracts. The lower accuracy of sFuzz may be attributed to constraints related to detection time and the mutation speed in covering new path seeds.

In the context of Reentrancy Vulnerability detection, Slither identified 119 Reentrancy Vulnerability contracts, with 79 being true positive contracts, resulting in an accuracy of 66.3%, recall of 74.5%, and a comprehensive F1-score of 70.1%. Among the true positive contracts, 9 had cross-contract Reentrancy Vulnerabilities. On the other hand, sFuzz detected 32 Reentrancy Vulnerability contracts, but only 26 of them were true positive contracts, yielding an accuracy of 81.2%, recall of 24.2%, and a comprehensive F1-score of 37.2%. Among the true positive contracts, 4 had cross-contract

Table 3. Number of vulnerability detections.

	IOU			Reentrancy			Tx-origin		
	Total	True	Cross	Total	True	Cross	Total	True	Cross
Slither	0	0	0	119	79	9	2	1	0
SFuzz	49	47	0	32	26	4	0	0	0
This	48	48	0	94	88	17	3	3	1

Note: IOU stands for Integer Overflow and Underflow Vulnerability, "Total" represents the total number of detected vulnerabilities by the tool, "True" represents the number of true positive vulnerabilities among them, and "Cross" represents the number of true positive vulnerabilities in contracts that have cross-contract vulnerabilities, "This" refers to the cross-contract vulnerability detection tool proposed in this paper

Reentrancy Vulnerabilities. In contrast, the proposed cross-contract vulnerability detection tool achieved an accuracy of 93.6%, detecting a total of 94 Reentrancy Vulnerability contracts, with 88 being true positive contracts, and a comprehensive F1-score of 87.5%. Additionally, the proposed tool detected 17 cross-contract Reentrancy Vulnerabilities, with 15 being true positive cross-contract Reentrancy Vulnerabilities.

Regarding Tx-origin Vulnerability detection, Slither identified 2 Tx-origin Vulnerability contracts, with 1 being a true positive contract, resulting in an accuracy of 50%, recall of 33.3%, and a comprehensive F1-score of 39.7%. sFuzz did not detect any Tx-origin Vulnerability contracts, yielding a comprehensive F1-score of 0. In contrast, the proposed cross-contract vulnerability detection tool achieved 100% accuracy, detecting a total of 3 Tx-origin Vulnerability contracts, all of which were true positive contracts, and a comprehensive F1-score of 100%. Additionally, the proposed tool detected 1 cross-contract vulnerability in the positive contracts.

Overall, our cross-contract vulnerability detection tool outperforms others in various vulnerability types.

5   Conclusions

In this paper, we introduce an innovative cross-contract vulnerability detection tool that effectively reduces the number of contracts for cross-contract vulnerability detection using machine learning algorithms. Subsequently, we employ an enhanced dynamic fuzz testing tool for cross-contract vulnerability detection and comprehensively supplement common vulnerabilities. Experimental results demonstrate the outstanding performance of our proposed cross-contract vulnerability detection tool in terms of accuracy.

The limitation of our study is that the model only detects specific three types of cross-contract vulnerabilities. In the future, we will consider expanding the detection to more types of cross-contract vulnerabilities. Our goal is to develop a feasible and efficient model with broader applicability.

Acknowledgement. This work is supported by national key research and development program (No.2018YFB1800702).

References

1. He, P., Yu, G., Zhang, Y.-F., Bao, Y.-B.: Survey on blockchain technology and its application prospect. Comput. Sci. **44**(4), 1–7 (2017). https://doi.org/10.11896/j.issn.1002-137X.2017.04.001
2. Siegel, D.: Understanding the DAO attack. https://www.coindesk.com/understanding-dao-hack-journalists (2023)
3. BlockCAT. On the Parity multi-sig wallet attack (2017). https://medium.com/blockcat/on-the-parity-multi-sig-wallet-attack-83fb5e7f4b8c
4. Pretrov, S.: Another Parity wallet hack explained (2017). https://medium.com/@Pr0Ger/another-parity-wallet-hack-explained-847ca46a2e1c
5. Jiang, B., Liu, Y., Chan, W.K.: ContractFuzzer: fuzzing smart contracts for vulnerability detection. In: Proceedings of the 33rd ACM/IEEE International Conference on Automated Software Engineering, pp. 259–269. Montpellier: IEEE (2018)
6. Nguyen, T.D., Pham, L.H., Sun, J., Lin, Y., Minh, Q.T.: SFuzz: an efficient adaptive fuzzer for solidity smart contracts. In: Proceedings of the ACM/IEEE 42nd International Conference on Software Engineering (ICSE '20). Association for Computing Machinery, New York, NY, USA, pp. 778–788 (2020). https://doi.org/10.1145/3377811.3380334
7. Tann, W.J.W., Han, X.J., Gupta, S.S., et al.: Towards safer smart contracts: a sequence learning approach to detecting security threats. arXiv preprint arXiv:1811.06632 (2018)
8. Liu, H., Liu, C., Zhao, W., Jiang, Y., Sun, J.: S-GRAM: towards semantic-aware security auditing for ethereum smart contracts. In: 2018 33rd IEEE/ACM International Conference on Automated Software Engineering (ASE), Montpellier, France, pp. 814–819 (2018)
9. Xue, Y., et al.: xFuzz: machine learning guided cross-contract fuzzing. IEEE Trans. Depend. Secure Comput. (2022) https://doi.org/10.1109/TDSC.2022.3182373
10. Feist, J., Greico, G., Groce, A.: Slither: a static analysis framework for smart contracts. In: Proceedings of the 2nd International Workshop on Emerging Trends in Software Engineering for Blockchain (WETSEB '19). IEEE Press, pp. 8–15 (2019). https://doi.org/10.1109/WETSEB.2019.00008
11. Tsankov, P., Dan, A., Drachsler-Cohen, D., Gervais, A., Bünzli, F., Vechev, M.: Securify: practical security analysis of smart contracts. In: Proceedings of the 2018 ACM SIGSAC Conference on Computer and Communications Security (CCS '18). Association for Computing Machinery, New York, NY, USA, 67–82 (2018). https://doi.org/10.1145/3243734.3243780
12. Chaidos, P., Kiayias, A.: Mithril: stake-based threshold multi signatures. Cryptology ePrint Archive (2021)
13. Jay, L., Omar, B.H., Dan, G.: crytic/evm-opcodes. [EB/OL] (2018). https://github.com/crytic/evm-opcodes
14. Wang, X., Yang, L., Wang, D., Zhen, L.: Improved TF-IDF keyword extraction algorithm. Comput. Sci. Appl. **3**, 64–68 (2013)

XRLFuzz: Fuzzing Binaries Guided by Format Information Based on Deep Reinforcement Learning

Rui Liu[1], Baojiang Cui[1(✉)], Chen Chen[2], and Jinxin Ma[3]

[1] Beijing University of Posts and Telecommunications, Beijing, China
{lrzzrr0,cuibj}@bupt.edu.cn
[2] Air Force Engineering University, Xi'an, China
[3] China Information Technology Security Evaluation Center, Beijing, China

Abstract. Fuzzing is a popular and effective automatic vulnerability mining method. More and more fuzzing techniques are starting to integrate reinforcement learning. But traditional fuzzing based on reinforcement learning is blind in sample mutation due to the lack of format information and efficient mutation algorithms. As a result, it is challenging to achieve higher coverage and the number of effective fuzzing is limited, leading to low utilization. To mitigate these problems, this paper proposes a new fuzzer named XRLFuzz, a format information guided fuzzing based on deep reinforcement learning for binaries. We use dynamic instrumentation techniques to provide runtime information, And then we use these information to perform format division and extract keywords based on the invalid mutation reuse algorithm. Format information is incorporated into the action dimension of deep reinforcement learning to guide the selection of mutation strategy. The experimental results show that the format division technology and keyword extraction technology both improve the efficiency of fuzzing, and XRLFuzz achieves code coverage of 106% to 276% of AFL.

1 Introduction

With the development of information technology, program vulnerabilities have become more complex, fuzzing [1] has become an important method for automatic software vulnerability mining. Fuzzing involves continuously feeding mutated data into the target program, aiming to trigger potential vulnerabilities. Based on whether it relies on runtime information of the program. Fuzzing can be categorized into gray-box, black-box, and white-box. Gray-box fuzzing depends on program instrumentation to provide coverage information for guiding the mutation. However, many programs nowadays enforce strong format constraints on their inputs. Gray-box fuzzing relys on program coverage feedback and random mutation strategy can generate numerous invalid samples that do not satisfy the format constraints, making it difficult to achieve higher coverage. Some fuzzers introduce format constraints to improve efficiency [9], but this method cannot fuzz closed-source binaries with unknown formats. In addition,

© The Author(s), under exclusive license to Springer Nature Switzerland AG 2024
L. Barolli (Ed.): EIDWT 2024, LNDECT 193, pp. 307–317, 2024.
https://doi.org/10.1007/978-3-031-53555-0_29

most traditional fuzzers (AFL [2], AFLSmart [9], AFL++ [3], Syzkaller [14], etc.) primarily focus on perceiving changes in coverage. In this process, a mutation that does not produce new path during fuzzing is considered invalid and will be discarded, wasting the effort spent on that mutation. These ineffective mutations constitute a significant portion of the entire fuzzing process.

To mitigate the lack of format information and the low utilization of mutations when fuzzing closed-source programs with unknown formats. We propose XRLFuzz that utilizes dynamic instrumentation techniques to provide runtime coverage and instruction flow information for closed-source programs. With runtime information, we propose a lightweight format division technique based on coverage variation and a keyword extraction technique based on reusing ineffective mutations. We then apply the format information and keyword list obtained from these two techniques to the mutation strategy based on deep reinforcement learning algorithm. This limits the exploration of machine learning to the effective sample space, and increases the utilization rate of mutations.

Our contributions in this paper are as follows.

- We propose a lightweight unknown format division algorithm based on coverage variation.
- We propose a keyword extraction algorithm based on reusing ineffective mutations.
- We implement XRLFuzz, a fuzzer based on deep reinforcement learning and the two aforementioned techniques, which outperforms AFL on the fuzzer-test-suite with a coverage increase of 106% to 276%.

2   Related Work

After years of development, researchers have proposed grey box fuzzing based on coverage, fuzzing based on format constraints, and fuzzing based on machine learning in different improvement directions.

2.1   Coverage-Based Grey Box Fuzzing

Gray-box fuzzing utilizes instrumentation technology to obtain runtime information and guides the mutation based on this information. American Fuzzy Lop (AFL) [2] is the most famous CGF (Coverage-guided Fuzzing) gray-box fuzzing tool, providing mature instrumentation capabilities for open-source code. Many tools have improved on AFL: AFLFAST [8] emphasizes the exploration of low-frequency paths. It utilizes a Markov model for energy distribution. AFL++ [3] is currently a community-maintained version of AFL. It introduces more fuzzing ideas and techniques: Input-To-State, MOpt and Pilot mutation methods.

2.2   Format-Constraint-Based Fuzzing

Format-constrained fuzzing requires describing the format constraints of the target programs based on a specified grammar. The grammar includes specifying

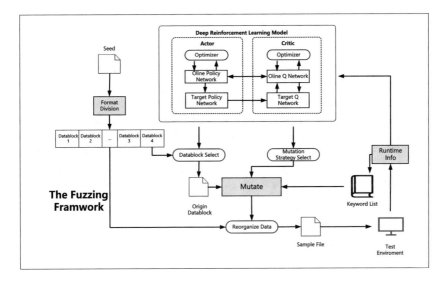

Fig. 1. Fuzzing framwork.

the format division, and telling which input parts need to be mutated and how to mutate them. According to these constraints, the input that meets the constraints is generated. For example, Peach [4] generates highly structured mutated files based on predefined Pit rule files. Each Pit file contains specifications applicable to specific targets, such as the data structures of target intake, the ways in which data flows into and out of the tested application, etc. AFLSMART [9] incorporates format constraints into AFL. It uses Peach Pit file as the input model for formats and introduces higher-order structural mutations. This allows AFLSMART to leverage both the input format constraints provided by Peach and the coverage feedback from the program.

2.3 Machine-Learning-Based Fuzzing

In recent years, machine learning has developed rapidly and has been applied to various stages of fuzzing, including seed file generation, sample filtering, and mutation strategy selection, etc. In the seed file generation stage, SmartSeed [10] utilizes Generic Adversary Networks (GAN) to generate new binary samples from valid sample data found by AFL. Montage [11] uses Long Short-Term Memory (LSTM) to generate JavaScript files for browser-based fuzzy test. In the sample filtering stage, FuzzGuard [12] employs convolutional eural network(CNN) to predict the reachability of mutated samples, filtering out low-quality samples and integrates with AFLGO [13] to improve the efficiency of fuzzing. In the mutation strategy selection stage, Fuzzy Gym [7] combines libFuzzer [16] and Double Deep Q Networks(DDQN) [6] through Remote Procedure Call(RPC).

3   Technology Program

The design of the overall system of XRLFuzz is shown in Fig. 1. The framwork can be divided into three parts: format division, deep reinforcement learning model, keyword extraction. The format division part divides input samples into data blocks. Then, the deep reinforcement learning model outputs the data block that needs to be mutated and the mutation strategy during each mutation step. The mutated data block and original blocks will be eventually reorganized into a new sample, which will be put into the target program. The coverage information of the target program is fed back to the deep reinforcement learning model. Ineffective mutations are reused through the keyword extraction part and applied in subsequent mutation rounds.

3.1   Lightweight Format Division Technique Based on Coverage Variation

Most programs nowadays have formatted inputs, where the input data is often divided into blocks, with each block is responsible for one or more functionalities. For example, the PNG files are typically divided into the Header and Data parts. The Header part contains the magic header, file format version information, and basic properties of the PNG file, while the Data part stores the actual image pixel data. When mutations occur in the header, there is a higher possibility of triggering different paths. Therefore, to improve the efficiency of fuzzing, it is beneficial to concentrate mutations more on the Header part.

To fuzz programs that use open formats, we can directly utilize predefined format rules, such as Peach [4], to guide the mutation algorithm. However, this approach is ineffective for closed-source programs that use private formats. To mitigate this issue, as shown in Fig. 2, this paper proposes a lightweight format divison technique based on coverage variation. It can rapidly divide the private format into data blocks. Providing format information for subsequent mutation processes.

Our format divison technique is based on the observation that disrupting different fields in a sample will have various impacts on runtime information, particularly coverage. We can obtain it using dynamic instrumentation technique.

```
0    bool isFotiFile(const char* filename) {
1        FILE* file = fopen(filename, "rb");
2        ...
3        unsigned char FotiHeader[4];
4        fread(FotiHeader, sizeof(unsigned char), 4, file);
5        fclose(file);
6        // Check if the first three bytes are the Foti identifier.
7        if (FotiHeader[0] == 0x89 && FotiHeader [1] == 70
8        && FotiHeader [2] == 111 && FotiHeader [3] == 116) {
9            return true;
10       } else {
11           return false;
```

```
12      }
13  }
14  int main() {
15      const char* filename = "input.fot";
16      if (isFotiFile(filename)) {
17          //todo normal operations
18          ...
19      } else {
20          printf("The file is not in Foti format. \n");
21          exit(0);
22      }
23      return 0;
24  }
```

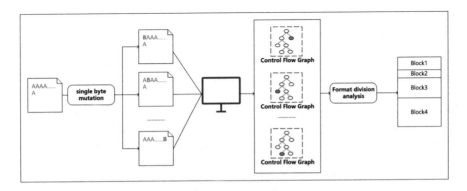

Fig. 2. Format division.

For example, in the given code segment, at line 19, the function isFotiFile is called to check if the current input file is in the Foti format. If it is a Foti format file, the program proceeds with normal data processing logic. Otherwise, it prompts the user that the input file does not meet the format requirements. The main logic of isFotiFile is as follows: at line 4,5,6, it reads the contents of the input file, and at line 9, it checks if the first four bytes of the content match the Foti Magic Header. If this condition is not met, the function returns false.

When our input sample is a valid file conforming to the Foti format, the program goes through the isFotiFile check and then processes file content normally. This leads to a coverage rate (rate1). Considering the impact of single-byte mutations on the first four bytes of the input sample, regardless of which byte is mutated, it results in a mutated sample that fails the isFotiFile check, causing the program to terminate prematurely. As a result, we collect a coverage rate (rate2) that is less than rate1.However, applying single-byte mutations to bytes other than the four bytes constituting the Magic Header does not cause the program to terminate early due to the isFotiFile check. Therefore, the coverage variation observed from mutating these four bytes exhibits approximate similarity compared to mutating other parts of the input. This similarity in coverage variation arises because bytes belonging to the same funtional part often affect nearby or identical basic blocks in the program's control flow graph.

Based on the observations mentioned above, our format division technique performs single-byte mutations on each byte of the initial input sample and collects the coverage of each mutated sample. Then, based on the grouping criterion of generating similar or identical rate, we divide the input sample into several data blocks.

3.2   Keyword Extraction Technique Based on Ineffective Mutation Reuse

Traditional fuzzing algorithms are coverage-driven, aiming to achieve higher coverage and discover more crashes. In this process, a mutation that does not generate new paths is considered ineffective and often discarded, wasting the resources consumed by that mutation. These ineffective mutations constitute a significant proportion of the entire fuzzing. However, from an information acquisition perspective, even though ineffective mutations do not improve the coverage, they still provide valuable runtime information. For example, by adjusting the instrumentation granularity from basic blocks to cmp instructions, we can obtain insights into the impact of input data on instruction operands. Based on this observation, we propose a keyword extraction technique based on the reuse of ineffective mutations. Specifically, our algorithm is based on the following observations:

The input byte stream at the location of a keyword typically maps to one of the operands of a comparison (cmp) instruction. Additionally, the keyword itself is expected to appear as a fixed value in the other operand of the cmp instruction. The algorithm extracts the different information generated by the same cmp instruction during two separate executions. It checks if one operand is a fixed value while the other operand is the value resulting from two consecutive single-byte mutations. If this condition is met, the fixed value is inferred as the keyword and added to the keyword list for future mutation.

In Fig. 3, when the byte stream [0x10, 0x50, 0x4E, 0x47] is input to the program, during the execution of instruction at address 0x400005, al has the value 0x10. After applying a single-byte mutation to the byte stream, resulting in the new byte stream [0x20, 0x50, 0x4E, 0x47], and inputting it into the program again, al becomes 0x20 at address 0x400005. By comparing the operands of the two cmp instructions, we can extract the value 0x89 and add it to the keyword list.

```
0   ; Load pngHeader[0] into the AL register.
1   mov al, byte ptr [pngHeader]
2   ; Compare the value in the AL register with 0x89.
3   cmp al, 0x89
4   ; If they are not equal, jump to the else_label
5   jne else_label
6   jmp normal_label                  ; Skip the else block.
```

```
7
8   else_label:
9   //... exit
10
11  normal_label:
12  //do normal operations
```

We applied the keyword extraction algorithm described above to the fuzzing process of XRLFuzz. By doing so, even ineffective mutations that do not generate new paths can still provide keywords information. The extracted keywords form a keyword list that is synchronized and applied to the mutation algorithm mentioned earlier (Algorithm 10: ReplaceKeyWord) to enhance the efficiency of breaking through certain fields during the fuzzing process.

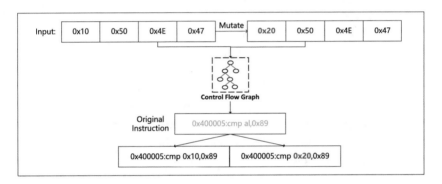

Fig. 3. The mapping of keywords to cmp instruction

3.3   Deep Reinforcement Learning

Traditional fuzzing employs randomly selected mutation algorithms, resulting in a lack of directionality in the mutations. Therefore, we introduce reinforcement learning algorithms to assist us in selecting mutation algorithms by utilizing the runtime information from instrumented grey-box fuzz testing to calculate rewards. Since the objective of reinforcement learning is to find the optimal policy that maximizes rewards, we conduct iterative learning during the training phase of reinforcement learning. Eventually, we obtain a model that seeks to maximize rewards, helping us find the best fuzzing strategy.

The reinforcement learning action space is divided into four dimensions: mutation location, mutation intensity, mutation algorithm, and data blocks after format division. The mutation strategy of XRLFuzz is a series of functions. The input parameters of these functions are the other dimensions in the action space. The output is the mutated byte stream, which is combined with original data blocks to create a new sample. We extended Z. Zhang's work [17] and introduced a new algorithm ReplaceKeyWord in Table 1. The ReplaceKeyWord is based on keywords list which is derived from the keyword extraction technique based on ineffective mutation reuse.

The deep reinforcement learning algorithm used in this paper is DDPG [15] (Deep Deterministic Policy Gradient), which effectively balances the exploration of the state space and the efficiency of action space computation. The work in Z. Zhang [17] has proved the superiority of deep deterministic policy gradient (DDPG) model in fuzzing, so we choose DDPG as the deep reinforcement learning model.

Table 1. Mutate Functions

Mutation Algorithm	Functions
EraseBytes	Delete some bytes
InsertByte	Insert a byte
InsertRepeatedBytes	Insert duplicate bytes
ChangeByte	Change a byte
ChangeBit	Flip a random bit
ShuffleBytes	Select bytes to change the order
ChangeASCIIInteger	Mutate a random integer
ChangeBinaryInteger	Mutate several random bytes
CopyPart	Copy data to another place
ReplaceKeyWord	Replace specific bytes with a keyword

4 Experiments and Evaluation

4.1 Programs and Environment

To evaluate the high effectiveness of XRLFuzz, we selected eight programs from Fuzzer-test-suite [5] for testing. The selected programs includes multiple input

Table 2. Programs

Program	Format	Version
libpng	png	1.2.56
libjpeg	jpg	2017
guetzli	jpg	2017.03.30
libarchieve	zip	2017.01.04
libxml2	xml	v2.9.2
harfbuzz	otf/ttf	1.3.2
freetype2	otf/ttf	2017
json	text	2017.02.12

formats, such as PNG, TTF, JPG, etc. The detailed information of the program is shown in the following Table 2. The testing environment for the experiment was: CPU: Intel Xeon Gold 6258R 2.7GHz, GPU: Nvidia Quadro RTX4000 8GB, RAM: 60GB, OSS: Ubuntu22.04LTS.

4.2 Results

We used the program's edge coverage as the core metric, and firstly compared the impact of enabling or disabling the format division technology and keyword extraction technology on fuzzing: (1) neither is enabled (none); (2) only format division technology is enabled (format); (3) only keyword extraction technology is enabled (only key); (4) both are enabled (all). For each comparative experiment, we used the same program and tested for 12 h, collecting edge coverage at intervals of 30 min. The experimental results are shown in Fig. 4.

From the experimental results, enabling either the format division technology or the keyword extraction technology resulted in higher coverage compared to

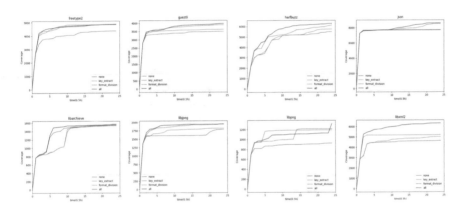

Fig. 4. Coverage contrast between techniques.

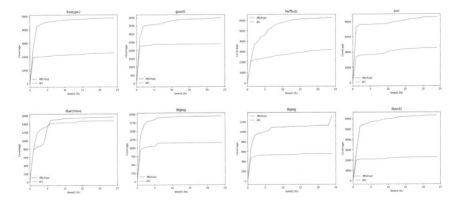

Fig. 5. Coverage contrast between XRLFuzz and AFL.

when neither was enabled. When both were enabled, the effect was the best. When testing target programs such as libxml2 and json, the keyword extraction technology performed better than the format division technology, while for other programs the opposite was true, due to their inputs are text files, more keywords are needed rather than the Header parts.

Finally, we conducted comparative testing between AFL and XRLFuzz. Each set of comparative experiments tested the same program for 12 h, collecting edge coverage and mutation rounds in 30-min increments. During execution, each test program had the same input samples and test commands. As shown in Fig. 5, the XRLFuzz has a clear advantage about 106% to 276% over AFL in terms of the number of covered edges.

Then we calculate seed quality by edge coverage / number of mutations to evaluate the efficiency of mutation. The calculated seed quality is as follows in the Table 3. XRLFuzz has improved the seed quality compared to AFL by a factor of 4.6 to 1225 times. It indicates that XRLFuzz is capable of exploring a greater program state space with fewer high-quality seeds.

Table 3. Seed Quality

Program	XRLFuzz	AFL
freetype	0.040993	0.000102
gueztli	0.046399	0.000513
harfbuzz	0.029860	0.000103
json	0.146161	0.000176
libarchieve	0.007724	0.000051
libjpeg	0.014256	0.003083
libpng	0.002516	0.000012
libxml2	0.123732	0.000101

5   Conclusions

In order to alleviate the shortcomings of traditional Fuzzing on closed-source programs with private formats, we proposed a deep reinforcement learning-based fuzzy testing technique for unknown closed-source binary testing guided by format constraints, and implement a system called XRLFuzz. The use of format divison technology based on coverage variation and keyword extraction algorithm based on ineffective mutations provides format information for the action strategy of reinforcement learning. Experiments on fuzz-test-suite showed that both techniques improve the efficiency of reinforcement learning-based fuzzing, and achieve higher coverage and higher round utilization compared to AFL. In the future, reinforcement learning should be further combined with fuzzing and developed.

References

1. Miller, B.P., Fredriksen, L., So, B.: An empirical study of the reliability of UNIX utilities. Commun. ACM **33**, 32–44 (1990)
2. Zalewski, M.: American Fuzzy Lop (2019). https://github.com/Google/AFL. Accessed 10 Sep 2019
3. Fioraldi, A., Maier, D., Eißfeldt, H., et al.: AFL++: combining incremental steps of fuzzing research. In: 14th USENIX Workshop on Offensive Technologies (WOOT 20) (2020)
4. Eddington, M.: Peach fuzzing platform. http://community.peachfuzzer.com/WhatIsPeach.html
5. https://github.com/google/fuzzer-test-suite
6. Hasselt, H.: Double Q-learning. Adv. Neural. Inf. Process. Syst. **23**, 2613–2621 (2010)
7. A competitive framework for fuzzing and learning. arXiv preprint arXiv:1807.07490 (2018)
8. Böhme, M., Pham, V.T., Roychoudhury, A.: Coverage-based Greybox fuzzing as Markov chain. IEEE Trans. Softw. Eng. **45**, 489–506 (2017)
9. Pham, V., Boehme, M., Santosa, A.E., Caciulescu, A.R., Roychoudhury, A.: Smart Greybox fuzzing. IEEE Trans. Softw. Eng. (2019). https://doi.org/10.1109/TSE.2019.2941681
10. Lyu, C., Ji, S., Li, Y., et al.: SmartSeed: smart seed generation for efficient fuzzing (2018)
11. Son, S., Lee, S., Han, H., et al.: Montage: a neural network language model-guided Javascript fuzzer. In: 20th USENIX Security Symposium (USENIX Security 2020) (2020)
12. Zong, P., Lv, T., Wang, D., et al.: FuzzGuard: filtering out unreachable inputs in directed grey-box fuzzing through deep learning. In: 29th USENIX Security Symposium (USENIX Security 20), pp. 2255–2269 (2020)
13. Böhme, M., Pham, V.T., Nguyen, M.D., et al.: Directed Greybox fuzzing. In: Proceedings of the. ACM SIGSAC Conference on Computer and Communications Security, pp. 2329–2344 (2017)
14. Vykov, D.: Syzkaller. https://github.com/google/syzkaller (2016)
15. Lillicrap, T.P., Hunt, J.J., et al.: Continuous control with deep reinforcement learning. arXiv preprint arXiv:1509.02971 (2015)
16. Serebryany, K.: LibFuzzer-a library for coverage-guided fuzz testing. LLVM project (2015)
17. Zhang, Z., Cui, B., Chen, C.: Reinforcement learning-based fuzzing technology. In: Barolli, L., Poniszewska-Maranda, A., Park, H. (eds.) IMIS 2020. AISC, vol. 1195, pp. 244–253. Springer, Cham (2021). https://doi.org/10.1007/978-3-030-50399-4_24

Implementation of a Fuzzy-Based Testbed for Selection of Radio Access Technologies in 5G Wireless Networks and Its Performance Evaluation

Phudit Ampririt[1]($\boxtimes$), Shunya Higashi[1], Paboth Kraikritayakul[1], Ermioni Qafzezi[2], Keita Matsuo[2], and Leonard Barolli[2]

[1] Graduate School of Engineering, Fukuoka Institute of Technology, 3-30-1 Wajiro-Higashi, Higashi-Ku,, Fukuoka 811-0295, Japan
{bd21201,mgm23108,s23y1004}@bene.fit.ac.jp
[2] Department of Information and Communication Engineering, Fukuoka Institute of Technology, 3-30-1 Wajiro-Higashi, Higashi-Ku,, Fukuoka 811-0295, Japan
qafzezi@bene.fit.ac.jp,{qafzezi,kt-matsuo,barolli}@fit.ac.jp

Abstract. The 5th Generation (5G) wireless networks mark the transition towards Ultra-Dense Heterogeneous Networks (UDHetNets), characterized by densely packed network and a diverse range of networks serving user devices. This evolution brings the challenge of efficiently connecting to optimal Radio Access Technology (RAT). This is a complicated task because there are involved many parameters, which makes it an NP-Hard problem. In order to deal with this issue, in this paper we introduce a Fuzzy-based system and implement a testbed for RAT selection in 5G networks. We consider three input parameters: Coverage (CV), User Priority (UP) and Spectral Efficiency (SE). The output parameter of the decision-making process is Radio Access Technology Decision Value (RDV). The testbed results demonstrate that by increasing CV, UP, and SE values the RDV is increased.

1 Introduction

The 5th Generation (5G) wireless networks will deal with the rapid and unprecedented increase of connected user devices, which exhibit different traffic patterns that threat to overwhelm the Internet and deteriorate the Quality of Service (QoS). The 5G networks will provide substantial improvements in reliability, throughput, latency, and mobility [1].

In order to enhance the performance of 5G Radio Access Technologies (RATs), multiple Base Stations (BSs) utilize a mix of heterogeneous RATs ranging from GSM to Wi-Fi. These technologies offer different levels of radio coverage, from macrocells to picocells, with differentiated transmission power levels, which can provide users with the best possible Quality of Experience (QoE), energy efficiency, redundancy, and reliability [2,3].

© The Author(s), under exclusive license to Springer Nature Switzerland AG 2024
L. Barolli (Ed.): EIDWT 2024, LNDECT 193, pp. 318–331, 2024.
https://doi.org/10.1007/978-3-031-53555-0_30

The 5G networks can support wide spectrum of service and applications, which are pivotal to various aspects of life such as entertainment and transportation. This will be achieved by addressing three key application scenarios. The enhanced Mobile Broadband (eMBB) is designed to provide seamless QoE and superior accessibility to essential services and multimedia content. The massive Machine Type Communication (mMTC) aims to support a vast number of connected devices while extending battery life. Lastly, Ultra-Reliable & Low Latency Communications (URLLC) is intended to facilitate real-time operations with significantly reduced latency and improved reliability, which are critical in applications such as transport security, remote medical procedures, and industrial automation processes [4–6].

Recent research on 5G wireless networks are focused on the design of systems that are well-suited to the enhanced capabilities and requirements of these networks. A notable trend is the incorporation of Network Function Virtualization (NFV) and Software-Defined Networking (SDN), which facilitate flexible and efficient management of technological and administrative networks, including those with extensive computational resources [7,8]. To address the latency issues, innovative mobile handover strategies that leverage SDN are being explored, aiming to significantly reduce processing delays [9–11]. Also, Fuzzy Logic (FL) is used to enhanche the QoS by integrating it into SDN controllers to refine network management and adapt dynamically to changing network conditions and requirements.

In our previous work [12–16], we presented some Fuzzy-based systems for Call Admission Control (CAC) and Handover in 5G Wireless Networks. In this research work, we design and implement an intelligent system and a testbed for RAT selection in 5G wireless networks using FL. For the implementation, we consider three parameters: Coverage (CV), User Priority (UP), and Spectral Efficiency (SE). The output parameter is Radio Access Technology Decision Value (RDV).

The rest of the paper is organized as follows. We introduce SDN in Sect. 2. We present 5G Network Slicing in Sect. 3. We describe the proposed Fuzzy-based system and testbed implementation in Sect. 4. We discuss testbed results in Sect. 5. Finally, conclusions and future work are presented in Sect. 6.

2   Software-Defined Networking (SDN)

The SDN is widely regarded as a revolutionary approach for network management, primarily due to its ability to make networks more programmable and virtualizable. This is achieved by a key architectural feature of SDN: the separation of the network's control plane from its data plane. This separation allows a flexible and dynamic network management, enabling administrators to adapt network behavior via software controls without needing to physically alter the network hardware.

The structure of an SDN is depicted in Fig. 1 which illustrates three distinct layers of SDN architecture.

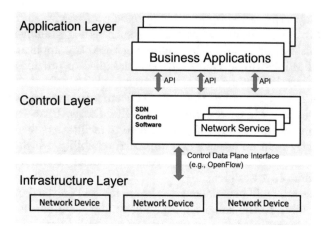

Fig. 1. Structure of SDN.

The Application Layer plays a crucial role in collecting data from the controller to create a comprehensive view of the network, essential for informed decision-making. This layer works in tandem with the Northbound Interfaces, which open up a wide array of network programming possibilities. These interfaces facilitate communication between the Application Layer and the Control Layer, allowing the application to issue specific instructions and relay necessary data to the Control Layer. Based on the application requirements, the Control Layer, guided by the controller, orchestrates the optimal SDN configuration, ensuring that the network delivers the requisite service quality and security.

The Control Layer is responsible for managing the Data Plane. It communicates various types of rules and policies to the Infrastructure Layer through the Southbound Interfaces, following directives received from the Application Layer. These Southbound Interfaces are pivotal protocols that enable the controller to define rules for the forwarding plane, ensuring effective connectivity and coordination between the Control Plane and the Data Plane.

Finally, the Infrastructure Layer is composed of the network physical forwarding devices, such as load balancers, switches, and routers. This layer acts on the instructions received from the SDN controller. By streamlining the flow of information and commands across these layers, the system ensures a seamless, efficient and secure network operation, tailored to the dynamic requirements of 5G and other advanced networking environments.

The SDN components can be effectively managed and utilized through a centralized control plane, offering efficient resource control and adaptation in response to traffic congestion scenarios. SDN enables streamlined data forwarding across various wireless technologies, enhancing mobility management capabilities [17, 18].

3   5G Network Slicing

The Network Slicing (NS) is an advanced networking paradigm that enables the creation of multiple virtual networks, each with its own specific capabilities and characteristics tailored to meet dedicated user-defined purposes. These virtual networks or "slices" are architecturally distinct but operate concurrently on the same underlying physical network infrastructure. This approach allows a higher level of performance and customization than traditional network structures. With NS, each network slice can have its own independent management and control functions, which can be dynamically allocated and scaled to fit diverse user needs. Moreover, NS enhances the overall network reliability and security because each slice operates independently, ensuring that the operations or issues within one slice do not adversely affect any other slice. This logical separation into slices allows an efficient and secure deployment of different and specialized network services [19–22].

The Next Generation Mobile Networks (NGMN) introduced the concept of 5G NS, which is structured into three principal layers: Service Instance Layer, Network Slice Instance Layer, and Resource Layer as shown in Fig. 2.

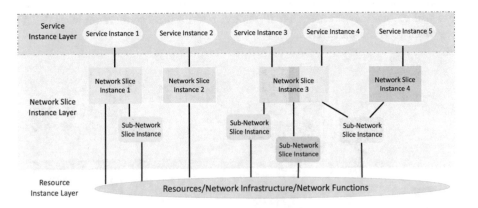

Fig. 2. Concept of 5G NS.

In the layered architecture of NS, the Service Instance Layer encapsulates distinct services that are either directed towards end-users or offered commercially by various sectors such as application providers or mobile network operators. Each service instance in this layer signifies a unique service. In the Network Slice Instance Layer, a NS instance comprises a configured set of network resources that are meticulously tailored to fulfill the particular demands of a given service instance. This layer can contain multiple sub-network instances which may operate in isolation or have shared resources.

Finally, the deployment of these network slices is facilitated by the Resource Layer, which is a composite of both tangible physical resources and abstract logical resources. This layer forms the foundational bedrock upon which network slices are instantiated and managed, ensuring that each service instance is supported by a robust and dedicated slice of the network infrastructure [23,24].

4   Fuzzy-Based System and Implemented Testbed

The overview of our proposed system is shown in Fig. 3. The SDN controller issues commands to evolved Base Stations (eBS), enabling data communication and transfer to User Equipment (UE). Each eBS in the system is designed to handle multiple network slices, each tailored for specific services and applications.

The proposed Fuzzy-based RAT Selection System (FRSS) will be integrated into the SDN controller. This controller will oversee the operations of eBS and other RATs, gathering extensive data regarding the network traffic conditions. The SDN controller serves as an intermediary module, facilitating communication between the RATs' eBS and the core network. For instance, if a User Equipment (UE) is connected to a Wireless LAN (WLAN) but experiences suboptimal QoS, the SDN controller will assess the network data from other RATs and will determine whether the UE should maintain its connection with WLAN or switch to a different RAT.

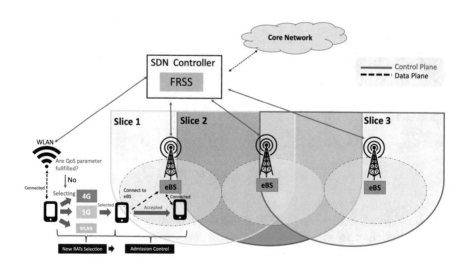

Fig. 3. Proposed system overview.

The FRSS structure is illustrated in Fig. 4. It considers three input parameters: Coverage (CV), User Priority (UP), and Spectral Efficiency (SE), and the output parameter is the RAT Decision Value (RDV).

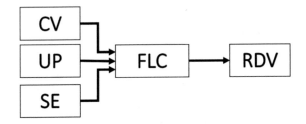

Fig. 4. Proposed system structure.

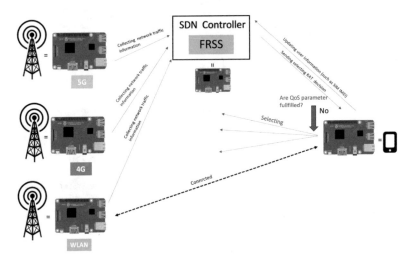

Fig. 5. Testbed structure.

In order to assess the performance of the proposed system, we implemented a testbed shown in Fig. 5, by using a Raspberry Pi Model B (Pi4B) as the SDN controller. We consider another Pi4B as User Equipment (UE), which interacts with three additional Pi4Bs acting as eBSs for different RATs: 5G, 4G, and Wi-Fi, respectively. Each of these Pi4B are equipped with corresponding hats to establish connectivity. This setup facilitates the collection of key network metrics such as RSRP (Reference Signal Received Power), RSSI (Received Signal Strength Indicator), and Data Rate. Additionally, the SDN controller receives user-specific information such as SIM ID and IMEI from the UE Pi4B. Using these data, the FRSS can select the most appropriate RAT, optimizing network performance and user experience based on real-time data and conditions.

The devices used in the testbed are shown in Fig. 6. The WiFi information is obtained through a connection to the router. The 4G data are collected via a 4G antenna and a 4G wireless module attached to the Raspberry Pi. Meanwhile, the 5G information is gathered through Network Simulator 3 (NS-3) [25, 26].

In the following section, we will describe input and output parameters of the FRSS associated with testbed setup.

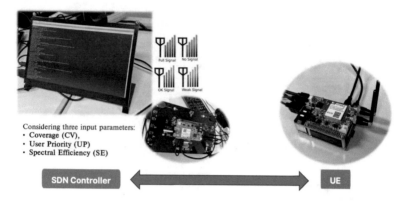

Fig. 6. Devices used for testbed implementation.

Coverage (CV): The CV parameter is a critical factor in assessing the effective range of RAT cell. The CV is determined by considering both the signal strength and RAT-specific signal quality indicators. This parameter varies across different RATs, reflecting their unique characteristics and operational environments.

- **5G** utilizes a wide spectrum that includes high, mid, and low frequency bands. The coverage range of 5G cells varies significantly depending on the used frequency band. Higher frequencies offer greater capacity but shorter range, while lower frequencies provide broader coverage.
- **4G** offers a good coverage and range, especially in low frequency bands. They provide reliable connectivity over relatively larger areas.
- **Wi-Fi** provides coverage within a limited area. It is well-suited for indoor use such as homes, offices, public spaces and offers high-speed wireless connectivity.

In our testbed setup, the CV is determined by RAT signal strength. For 5G and 4G networks, we measure the signal strength using the RSRP metric, while for Wi-Fi networks we employ the RSSI value.

User Priority (UP): The UP is a crucial parameter in our system that reflects the relative importance of users or devices. A higher UP means that the system finds a RAT that can offer better QoS.

In our testbed, the UP value is determined by considering factors such as the specific applications used by the user and the current data pricing models.

Spectral Efficiency (SE): The SE measures the efficiency of spectrum use, calculated as the number of bits transmitted per second per Hertz (bps/Hz). It indicates the data transmission rate possible within a given bandwidth, showcasing the effectiveness of spectrum utilization. SE is calculated using the following equation:

$$SE = \frac{ADR(bps)}{AB(Hz)},$$

where ADR represents Achievable Data Rate, which depends on factors like the modulation technique, bits per symbol, and so on. The AB stands for the Available Bandwidth. The computation of SE considers factors such as channel conditions such as Signal-to-Noise Ratio (SNR), modulation, coding schemes, and other elements pertinent to the communication system being analyzed.

Radio Access Technology Decision Value (RDV): The RDV is the output value, which provides a quantitative measure for selecting the most appropriate RAT.

Table 1. Parameter and their term sets.

Parameters	Term Sets
Coverage (CV)	Small (Sl), Intermediate (Im), Big (Bg)
User Priority (UP)	Low (Lw), Medium (Mu) , High (Hi)
Spectral Efficiency (SE)	Low (Lo), Medium (Me), High (Hg)
Radio Access Technology Decision Value (RDV)	RDV1, RDV2, RDV3, RDV4, RDV5, RDV6, RDV7

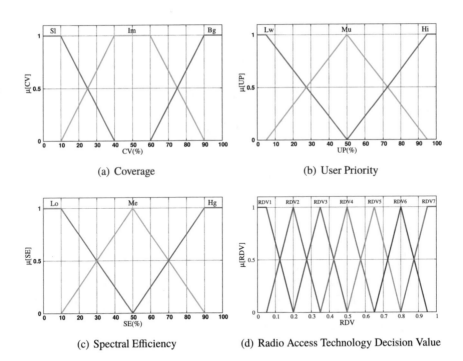

(a) Coverage

(b) User Priority

(c) Spectral Efficiency

(d) Radio Access Technology Decision Value

Fig. 7. Membership functions.

Table 1 shows the parameters utilized in our system along with their respective term sets. The membership functions, which are critical for determining the degree to which specific conditions apply, are shown in Fig. 7. In our system, we

Table 2. FRB.

Rule	CV	UP	SE	RDV
1	Sl	Lw	Lo	RDV1
2	Sl	Lw	Me	RDV2
3	Sl	Lw	Hg	RDV3
4	Sl	Mu	Lo	RDV2
5	Sl	Mu	Me	RDV3
6	Sl	Mu	Hg	RDV4
7	Sl	Hi	Lo	RDV3
8	Sl	Hi	Me	RDV4
9	Sl	Hi	Hg	RDV5
10	Im	Lw	Lo	RDV2
11	Im	Lw	Me	RDV3
12	Im	Lw	Hg	RDV4
13	Im	Mu	Lo	RDV3
14	Im	Mu	Me	RDV4
15	Im	Mu	Hg	RDV5
16	Im	Hi	Lo	RDV4
17	Im	Hi	Me	RDV5
18	Im	Hi	Hg	RDV6
19	Bg	Lw	Lo	RDV3
20	Bg	Lw	Me	RDV4
21	Bg	Lw	Hg	RDV5
22	Bg	Mu	Lo	RDV4
23	Bg	Mu	Me	RDV5
24	Bg	Mu	Hg	RDV6
25	Bg	Hi	Lo	RDV5
26	Bg	Hi	Me	RDV6
27	Bg	Hi	Hg	RDV7

use triangular and trapezoidal membership functions, because these shapes are particularly well-suited for real-time operations [27–30].

The Fuzzy Rule Base (FRB), which consists of 27 rules is shown in Table 2. Each rule follows a standard format: IF "Condition" THEN "Control Action". For example, Rule 22 can be interpretated as follows: "IF CV is Big, UP is Medium and SE is Low, THEN RDV is RDV4".

5   Testbed Results

A snapshot of testbed operation is shown in Fig. 8. It shows a user initiating a data collection sequence for RAT selection, with the system first evaluating Wi-Fi, indicating a strong signal (−46 dBm) and good coverage (CV of 0.73), followed by a good download and upload speeds. Upon switching to a cellular connection, the 4G network data reveals good SIM signal strength (-73 dBm) but lower download and upload speeds compared to Wi-Fi. The system calculates CV, SE, and UP, with user priority set at 0.44. The output of the system for Wi-Fi (Fuzzy Result: 0.66) is better than 4G (Fuzzy Result: 0.34), leading to the final decision to connect to Wi-Fi network.

The testbed results for different CV values are shown in Fig. 9, Fig. 10 and Fig. 11. They show the relationship between RDV and SE for different levels of UP. The averaged lines in red, green and blue colors represent different UP levels (10%, 50%, 90%). Each dot represents a data point from a test case for each technology (WiFi, 4G and 5G).

In Fig. 9 the CV value is 10%. We see that by increasing SE and UP values, the RDV is increased. Specifically, when SE increases from 10% to 50%, there is a 16% increase in RDV, and a further 18% increase when SE increases from 50% to 90%. This trend suggests that for users with high priority, the RAT with greater transmission bandwidth (higher SE value) will have a higher RDV. When UP is decreased from 90% to 50%, and further from 50% to 10%, the RDV is decreased by 13% and 6%, respectively, when SE is 90%. This result indicates that when a user has a low priority, the RAT selection mechanism will allocate resources that might result in lower QoS, suggesting that higher priority users are favored in the allocation of higher QoS network resources.

When CV value is increased from 10% to 50% in Fig. 10, the RDV increases by 6% for users with a 90% UP and a SE value of 50%. This suggests that better coverage values contributes to a higher possibility of selecting a RAT, especially for high-priority users. By increasing CV to 90% in Fig. 11, we see that RDV is increased compared with results in Fig. 9 and Fig. 10.

```
<------------------Collecting info.: [1] ----------------------->
################### WiFi network information [1] ##################
WiFi signal strengths:-46 dBm
CV_wifi (check) = 0.733333
# Ping: 25.5 ms
# Download Rate: 64.53 Mbit/s
# Upload Rate: 92.33 Mbit/s
Switched to cellular connection.
Connection successfully activated (D-Bus active path: /org/freedesktop/NetworkManager/ActiveConnection/35)
################### 4G network information [1] ##################
SIM signal strength: Excellent (-73 dBm)
CV_4G (check) = 0.522222
# Ping: 78.2 ms
# Download Rate: 6.6 Mbit/s
# Upload Rate: 5.6 Mbit/s
Time taken by simulation: 117766359 microseconds
Connection 'ppsim' successfully deactivated (D-Bus active path: /org/freedesktop/NetworkManager/ActiveConnection/35)
Switched to WiFi connection.
#### Calculate and preparing CV, SE, UP [1] #####
CV_wifi = 0.733333
CV_4G = 0.522222
User priority (UP) : 0.44
SE_wifi = 0.78 <--- DL=64.53, UL=92.33
SE_4G = 0.06 <--- DL=6.64, UL=5.60
Fuzzy Result (WiFi): 0.660553
Fuzzy Result (4G): 0.344137
The decsion result is WiFi
```

Fig. 8. A snapshot of testbed operation.

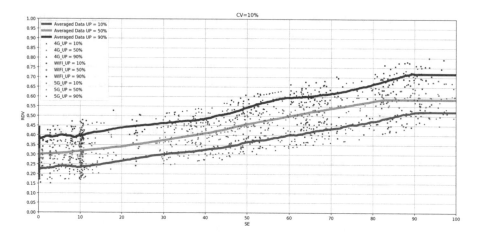

Fig. 9. Testbed results for CV=10%.

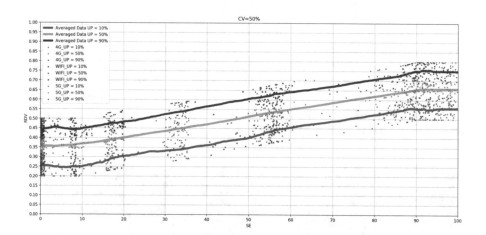

Fig. 10. Testbed results for CV=50%.

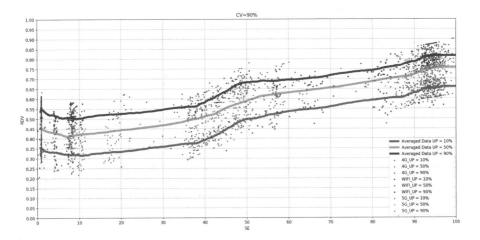

Fig. 11. Testbed results for CV=90%.

6 Conclusions and Future Work

In this paper, we presented the implementation of a FRSS and a testbed for selection of RATs in 5G wireless networks. We consided three input parameters: CV, UP and SE., which are used to decide RDV. We carried out some experiments using the testbed. The experimental results show that the RDV is increased by increasing CV, UP, and SE values.

In our future research, we plan to conduct additional experiments using the testbed by considering more input parameters. Also, we will compare the experimental results with simulation results.

References

1. Navarro-Ortiz, J., Romero-Diaz, P., Sendra, S., Ameigeiras, P., Ramos-Munoz, J.J., Lopez-Soler, J.M.: A survey on 5G usage scenarios and traffic models. IEEE Commun. Surv. Tutorials **22**(2), 905–929 (2020). https://doi.org/10.1109/COMST.2020.2971781
2. Pham, Q.V., et al.: A survey of multi-access edge computing in 5G and beyond: fundamentals, technology integration, and state-of-the-art. IEEE Access **8**, 116, 974–117, 017 (2020). https://doi.org/10.1109/ACCESS.2020.3001277
3. Orsino, A., Araniti, G., Molinaro, A., Iera, A.: Effective rat selection approach for 5G dense wireless networks. In: 2015 IEEE 81st Vehicular Technology Conference (VTC Spring), pp. 1–5 (2015). https://doi.org/10.1109/VTCSpring.2015.7145798
4. Akpakwu, G.A., Silva, B.J., Hancke, G.P., Abu-Mahfouz, A.M.: A survey on 5G networks for the internet of things: communication technologies and challenges. IEEE Access **6**, 3619–3647 (2018)
5. Palmieri, F.: A reliability and latency-aware routing framework for 5G transport infrastructures. Comput. Netw. **179**(9), 107365 (2020). https://doi.org/10.1016/j.comnet.2020.107365

6. Kamil, I.A., Ogundoyin, S.O.: Lightweight privacy-preserving power injection and communication over vehicular networks and 5G smart grid slice with provable security. Internet of Things **8**(100116), 100–116 (2019). https://doi.org/10.1016/j.iot.2019.100116

7. Hossain, E., Hasan, M.: 5G cellular: key enabling technologies and research challenges. IEEE Instrum. Measur. Mag. **18**(3), 11–21 (2015). https://doi.org/10.1109/MIM.2015.7108393

8. Vagionas, C., et al.: End-to-end real-time service provisioning over a SDN-controllable analog mmWave fiber-wireless 5G X-haul network. J. Lightwave Technol. 1–10 (2023). https://doi.org/10.1109/JLT.2023.3234365

9. Yao, D., Su, X., Liu, B., Zeng, J.: A mobile handover mechanism based on fuzzy logic and MPTCP protocol under SDN architecture. In: 18th International Symposium on Communications and Information Technologies (ISCIT-2018), pp. 141–146 (2018). https://doi.org/10.1109/ISCIT.2018.8587956

10. Lee, J., Yoo, Y.: Handover cell selection using user mobility information in a 5G SDN-based network. In: 2017 Ninth International Conference on Ubiquitous and Future Networks (ICUFN-2017), pp. 697–702 (2017). https://doi.org/10.1109/ICUFN.2017.7993880

11. Moravejosharieh, A., Ahmadi, K., Ahmad, S.: A fuzzy logic approach to increase quality of service in software defined networking. In: 2018 International Conference on Advances in Computing, Communication Control and Networking (ICACCCN-2018), pp. 68–73 (2018). https://doi.org/10.1109/ICACCCN.2018.8748678

12. Ampririt, P., Qafzezi, E., Bylykbashi, K., Ikeda, M., Matsuo, K., Barolli, L.: IFACS-Q3S-a new admission control system for 5G wireless networks based on fuzzy logic and its performance evaluation. Int. J. Distrib. Syst. Technol. (IJDST) **13**(1), 1–25 (2022)

13. Ampririt, P., Qafzezi, E., Bylykbashi, K., Ikeda, M., Matsuo, K., Barolli, L.: A fuzzy-based system for handover in 5G wireless networks considering network slicing constraints. In: Computational Intelligence in Security for Information Systems Conference, pp. 180–189. Springer, Cham (2022). https://doi.org/10.1007/978-3-031-08812-4_18

14. Ampririt, P., Qafzezi, E., Bylykbashi, K., Ikeda, M., Matsuo, K., Barolli, L.: A fuzzy-based system for handover in 5G wireless networks considering different network slicing constraints: effects of slice reliability parameter on handover decision. In: International Conference on Broadband and Wireless Computing, Communication and Applications, pp. 27–37. Springer, Cham (2022). https://doi.org/10.1007/978-3-031-20029-8_3

15. Ampririt, P., Ohara, S., Qafzezi, E., Ikeda, M., Matsuo, K., Barolli, L.: An integrated fuzzy-based admission control system (IFACS) for 5G wireless networks: its implementation and performance evaluation. Internet of Things **13**, 100, 351 (2021). https://doi.org/10.1016/j.iot.2020.100351

16. Ampririt, P., Qafzezi, E., Bylykbashi, K., Ikeda, M., Matsuo, K., Barolli, L.: Application of fuzzy logic for slice QoS in 5G networks: a comparison study of two fuzzy-based schemes for admission control. Int. J. Mobile Comput. Multimedia Commun. (IJMCMC) **12**(2), 18–35 (2021)

17. Li, L.E., Mao, Z.M., Rexford, J.: Toward software-defined cellular networks. In: 2012 European Workshop on Software Defined Networking, pp. 7–12 (2012). https://doi.org/10.1109/EWSDN.2012.28

18. Mousa, M., Bahaa-Eldin, A.M., Sobh, M.: Software defined networking concepts and challenges. In: 2016 11th International Conference on Computer Engineering & Systems (ICCES-2016), pp. 79–90. IEEE (2016)

19. An, N., Kim, Y., Park, J., Kwon, D.H., Lim, H.: Slice management for quality of service differentiation in wireless network slicing. Sensors **19**, 2745 (2019). https://doi.org/10.3390/s19122745
20. Jiang, M., Condoluci, M., Mahmoodi, T.: Network slicing management & prioritization in 5G mobile systems. In: European Wireless 2016; 22th European Wireless Conference, pp. 1–6. VDE (2016)
21. Chen, J., et al.: Realizing dynamic network slice resource management based on SDN networks. In: 2019 International Conference on Intelligent Computing and its Emerging Applications (ICEA), pp. 120–125 (2019)
22. Li, X., et al.: Network slicing for 5G: challenges and opportunities. IEEE Internet Comput. **21**(5), 20–27 (2017)
23. Afolabi, I., Taleb, T., Samdanis, K., Ksentini, A., Flinck, H.: Network slicing and softwarization: a survey on principles, enabling technologies, and solutions. IEEE Commun. Surv. Tutorials **20**(3), 2429–2453 (2018). https://doi.org/10.1109/COMST.2018.2815638
24. Alliance, N.: Description of network slicing concept. NGMN 5G P **1**(1), 7 (2016). https://ngmn.org/wp-content/uploads/160113_NGMN_Network_Slicing_v1_0.pdf
25. Patriciello, N., Lagen, S., Bojovic, B., Giupponi, L.: An E2E simulator for 5G NR networks. Simulation Modelling Practice and Theory **96**, 101, 933 (2019)
26. Koutlia, K., Bojovic, B., Ali, Z., Lagén, S.: Calibration of the 5G-LENA system level simulator in 3GPP reference scenarios. Simul. Modelling Pract. Theory **119**, 102, 580 (2022)
27. Norp, T.: 5G requirements and key performance indicators. J. ICT Standard. **6**(1), 15–30 (2018)
28. Parvez, I., Rahmati, A., Guvenc, I., Sarwat, A.I., Dai, H.: A survey on low latency towards 5G: ran, core network and caching solutions. IEEE Commun. Surv. Tutorials **20**(4), 3098–3130 (2018)
29. Kim, Y., Park, J., Kwon, D., Lim, H.: Buffer management of virtualized network slices for quality-of-service satisfaction. In: 2018 IEEE Conference on Network Function Virtualization and Software Defined Networks (NFV-SDN-2018), pp. 1–4 (2018)
30. Barolli, L., Koyama, A., Yamada, T., Yokoyama, S.: An integrated CAC and routing strategy for high-speed large-scale networks using cooperative agents. IPSJ J. **42**(2), 222–233 (2001)

A Fuzzy-Based System for Assessment of Recognition Error in VANETs Considering Inattentive Driving as New Parameter

Ermioni Qafzezi[1]([⊠]), Kevin Bylykbashi[1], Shunya Higashi[2], Phudit Ampririt[2], Keita Matsuo[1], and Leonard Barolli[1]

[1] Department of Information and Communication Engineering, Fukuoka Institute of Technology (FIT), 3-30-1 Wajiro-Higashi, Higashi-Ku, Fukuoka 811-0295, Japan
`qafzezi@bene.fit.ac.jp`, {`kt-matsuo,barolli`}`@fit.ac.jp`
[2] Graduate School of Engineering, Fukuoka Institute of Technology (FIT), 3-30-1 Wajiro-Higashi, Higashi-Ku, Fukuoka 811-0295, Japan
{`mgm23108,bd21201`}`@bene.fit.ac.jp`

Abstract. This paper introduces an innovative system that uses Fuzzy Logic (FL) to detect recognition errors in Vehicular Ad Hoc Networks (VANETs). The system considers various input parameters to assess recognition errors in VANETs. In our previous work, we considered three parameters: Internal and External Distraction, Driver's Inattention, and Inadequate Surveillance. In this paper, we consider Inattentive Driving as a new parameter. By employing FL, we model complex relationships and uncertainties in different situations. This approach transforms multifaceted input data into actionable insights regarding recognition error likelihood. We consider both driver-related and external factors, which affect the recognition accuracy. The simulation results show that the proposed system effectively captures different driver attention situations and environmental conditions. The proposed system can enhance VANET performance by enabling more adaptive and responsive vehicular communication in dynamic road environments.

1 Introduction

Car accidents remain a pressing global issue, incurring substantial human and economic costs. Driver actions are a central contributor to these accidents, as reported by the National Highway Traffic Safety Administration, attributing 94% of accidents to poor decision-making by drivers. The role of drivers in shaping road safety outcomes is crucial, emphasizing the need to explore factors related to drivers for effective accident prevention. Analyzing the causes behind driver actions can facilitate precise interventions to decrease accident rates and mitigate their societal impact.

Recognition errors represent a significant portion of driver actions influencing road accidents. These errors manifest when drivers inaccurately perceive

L. Barolli (Ed.): EIDWT 2024, LNDECT 193, pp. 332–342, 2024.
https://doi.org/10.1007/978-3-031-53555-0_31

and interpret their surroundings, leading to misguided decisions and maneuvers. Various factors contribute to recognition errors, including Internal and External distractions, Driver Inattention, Inadequate Surveillance, and the emerging concern of Inattentive Driving. Internal distractions, involving cognitive load and multitasking, divert attention from the road, while external distractions, such as electronic devices and roadside attractions, increase the associated risks. Driver inattention, often associated with fatigue or preoccupation, can result in delayed reactions during critical situations. Inadequate surveillance, where drivers neglect to thoroughly scan their environment, may lead to collisions with unexpected obstacles. Understanding the intricate dynamics among these factors is crucial for developing targeted interventions to decrease the occurrence of recognition errors and subsequently reduce accident rates.

In our previous work, we considered internal factors, such as recognition errors, decision errors, and performance errors, as the key parameters of driver mistakes. In [12], we proposed a Fuzzy Logic (FL)-based system and explored the role of recognition errors as a significant driver-related cause of car accidents. In our previous research, for recognition errors, we considered three parameters: Internal and External Distraction, Driver's Inattention, and Inadequate Surveillance. In this paper, we consider Inattentive Driving as a new parameter. By analyzing the complex relationship between different parameters, we aim to contribute to the understanding of accident causation and prevention. Utilizing the FL-based recognition error system in Vehicular Ad Hoc Networks (VANETs) offers the potential for real-time detection and mitigation of these errors, thereby enhancing road safety.

The paper's organization is as follows. Section 2 provides an overview of VANETs, tracing their evolution from traditional systems to cloud-fog-edge SDN-VANETs. In Sect. 3, we discuss how emerging technologies contribute to overcoming the limitations of VANETs and fostering a more responsive and agile vehicular communication infrastructure. Section 4 is dedicated to introducing the FL-based system we propose and outlining its implementation. The simulation results are deliberated in Sect. 5. Lastly, Sect. 6 concludes the paper and offers insights into potential future work.

2   Overview of VANETs

Vehicular Ad Hoc Networks (VANETs) represent a specialized form of mobile ad hoc networks (MANETs) designed to facilitate communication among vehicles and roadside infrastructure. In VANETs, vehicles equipped with wireless communication devices form a dynamic network, enabling real-time information exchange to enhance road safety, traffic efficiency, and overall transportation systems [3,4,11,14].

VANETs offer a diverse range of applications that significantly improve the current state of transportation. One prominent application is the provision of intelligent traffic management systems, where vehicles share information about traffic conditions, accidents, and road hazards, enabling dynamic route optimization for improved traffic flow. Cooperative Collision Warning Systems utilize

VANETs to exchange data on vehicle speeds and positions, reducing the risk of accidents. Additionally, VANETs contribute to the development of efficient emergency response systems by enabling swift communication between vehicles and emergency services, thereby enhancing overall road safety and transportation efficiency.

Despite their potential benefits, VANETs face various challenges, including network scalability, security concerns, and the need for efficient data dissemination protocols. Emerging technologies hold promise in addressing these issues. The integration of 5G technology enhances the communication capabilities of VANETs, providing higher data rates, lower latency, and improved network reliability. Software-Defined Networking (SDN) introduces a more flexible and programmable network infrastructure, enabling efficient management of VANET resources and addressing scalability challenges. These technological advancements play a crucial role in overcoming the limitations of VANETs, paving the way for their widespread implementation and ensuring the continued evolution of smart transportation systems.

3   Network Technological Advancement in VANETs

One of the key focal points in the development of VANETs is the enhancement of communication technologies. The integration of 5G networks has revolutionized the landscape by offering higher data rates, reduced latency, and improved reliability. These advancements not only facilitate real-time data exchange among vehicles but also lay the foundation for emerging applications such as cooperative collision warning systems and intelligent traffic management.

Addressing the scalability and flexibility challenges inherent in VANETs, Software-Defined Networking (SDN) emerges as a transformative solution. By introducing a programmable and centralized network management approach, SDN enables efficient resource allocation, dynamic routing, and adaptability to varying network conditions. This not only mitigates the scalability challenges traditionally associated with VANETs but also provides a means to optimize routing, reduce latency, and enhance overall network responsiveness. By decoupling the control and data planes, SDN empowers VANETs with a more agile and adaptable infrastructure, laying the groundwork for seamless integration into the broader framework of intelligent transportation systems.

The integration of Network Slicing, coupled with the convergence of cloud, fog, and edge computing paradigms, marks another significant stride in augmenting the capabilities of VANETs. Network Slicing enables the virtual partitioning of the network, allowing for tailored configurations to meet diverse application requirements within VANETs. This granularity in network customization enhances the efficiency of resource utilization and ensures optimal performance for specific services [1,5,13]. Furthermore, the integration of cloud-fog-edge computing architectures brings computational capabilities closer to the network's edge, reducing latency and enhancing real-time decision-making in VANETs. This integration not only optimizes data processing and storage but also fosters a distributed computing environment that aligns seamlessly with the dynamic nature of vehicular communication. In concert, these advancements usher in a new era for VANETs, where the network infrastructure is not only agile and responsive but also attuned to the diverse and evolving needs of intelligent transportation systems.

4 Proposed Fuzzy-Based System

We introduce a new system, called the Fuzzy System for Assessment of Recognition Errors (FS-ARE), which is a novel and advanced tool in the field of intelligent driving support systems. Our main focus is to make driving safer, and FS-ARE offers a unique approach for assessing recognition errors in real-time by considering a wide range of factors. We examine things like distractions inside and outside the vehicle, driver inattention, and how well the driver is paying attention to their surroundings. FS-ARE's goal is to provide an accurate and flexible evaluation of how a driver is behaving and what potential risks they might face.

Table 1. FS-ARE parameters and their term sets.

Parameters	Term Sets
Internal and External Distraction (IED)	Low (Lo), Medium (Me), High (Hi)
Driver Inattention (DIA)	Low (L), Medium (M), High (H)
Inadequate Surveillance (IS)	Small (GS), Medium (MS), Big (PS)
Inattentive Driving (ID)	Attentive (At), Inattentive (Ia), Very Inattentive (Vin)
Recognition Error (RE)	RE1, RE2, RE3, RE4, RE5 RE6, RE7, RE8, RE9

We explain the input parameters in following.

Internal and External Distraction (IED): Internal distractions involve thoughts, emotions, or actions that draw the driver's focus away from the primary task of driving. On the contrary, external distractions are external stimuli within the surroundings that vie for the driver's attention. Examples of such distractions comprise conversing with passengers, operating a smartphone, tuning the radio, or eating while driving. Both internal and external distractions result in recognition errors by overloading the driver's cognitive capacity, shifting attention away from the road, and hindering the driver's capability to identify and process pertinent information, like road signs, pedestrians, or other vehicles.

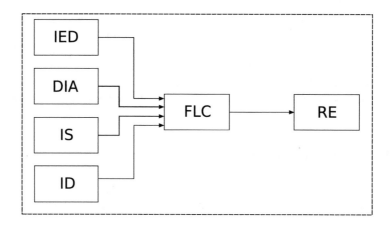

Fig. 1. Proposed system structure.

Driver Inattention (DIA): Driver inattention denotes a deficiency in attentiveness and involvement in the act of driving. It can be triggered by various factors, including weariness, drowsiness, absent-mindedness, or divided focus. Inattention diminishes a driver's ability to perceive and identify vital information within the driving environment, giving rise to recognition errors. For instance, a drowsy driver might overlook significant road signs or neglect abrupt alterations in traffic conditions, which can be a source of recognition errors.

Inadequate Surveillance (IS): Inadequate surveillance transpires when a driver neglects the requisite monitoring of the road and its surroundings. This lapse can arise from overconfidence, complacency, or distraction. When drivers fall short in diligently scrutinizing their environment, they risk overlooking crucial cues, failing to discern potential dangers, or misinterpreting traffic scenarios. Inadequate surveillance contributes to inaccuracies in recognizing road conditions, traffic signals, and unforeseen incidents, thereby elevating the chances of encountering recognition errors.

Inattentive Driving (ID): sometimes referred to as distracted driving, encompasses a broader concept than just the momentary lapse of attention or driver inattention. It involves any behavior where a driver's focus, concentration, and attention are diverted from the primary task of operating the vehicle safely. Inattentive driving can take many forms, including but not limited to: texting or using a smartphone while driving, talking on a cellphone without a hands-free device, eating or drinking while behind the wheel, adjusting the radio, GPS, or other in-car entertainment systems, conversing with passengers and so on.

FS-ARE harnesses the capabilities of Fuzzy Logic (FL) to simulate the intricate and uncertain connections among the input parameters and recognition errors. FL's proficiency in managing imprecise and ambiguous data renders it an apt choice for capturing the subtle interactions among the variables that influ-

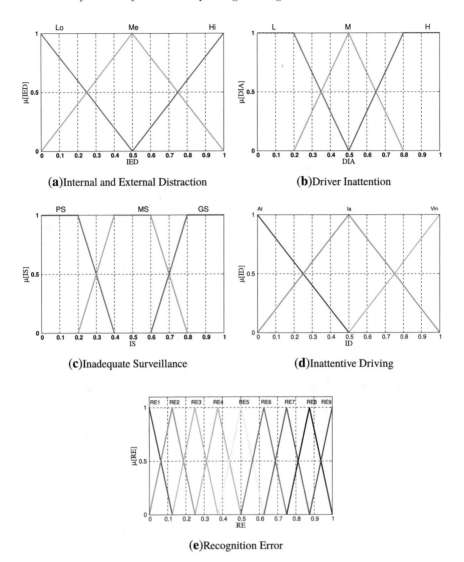

(a) Internal and External Distraction

(b) Driver Inattention

(c) Inadequate Surveillance

(d) Inattentive Driving

(e) Recognition Error

Fig. 2. Membership functions of FS-ARE.

ence driver performance [2,6–10,15–17]. Through the integration of linguistic variables and a rule-based inference mechanism, the system translates unprocessed sensor data and contextual information into meaningful insights regarding the driver's level of attentiveness and the probability of recognition errors.

The design of the proposed system is depicted in Fig. 1. Details regarding the term sets for the input and output parameters are shown in Table 1. The membership functions for the input and output parameters are visually represented in Fig. 2, while the Fuzzy Rule Base (FRB) is comprehensively outlined in Table 2.

Table 2. FRB of FS-ARE.

No	IED	DIA	IS	ID	RE	No	IED	DIA	IS	ID	RE	No	IED	DIA	IS	ID	RE
1	Lo	L	PS	At	RE2	28	Me	L	PS	At	RE3	55	Hi	L	PS	At	RE4
2	Lo	L	PS	Ia	RE2	29	Me	L	PS	Ia	RE4	56	Hi	L	PS	Ia	RE6
3	Lo	L	PS	Vin	RE4	30	Me	L	PS	Vin	RE6	57	Hi	L	PS	Vin	RE7
4	Lo	L	MS	At	RE1	31	Me	L	MS	At	RE2	58	Hi	L	MS	At	RE3
5	Lo	L	MS	Ia	RE2	32	Me	L	MS	Ia	RE3	59	Hi	L	MS	Ia	RE5
6	Lo	L	MS	Vin	RE3	33	Me	L	MS	Vin	RE5	60	Hi	L	MS	Vin	RE6
7	Lo	L	GS	At	RE1	34	Me	L	GS	At	RE1	61	Hi	L	GS	At	RE2
8	Lo	L	GS	Ia	RE1	35	Me	L	GS	Ia	RE2	62	Hi	L	GS	Ia	RE4
9	Lo	L	GS	Vin	RE2	36	Me	L	GS	Vin	RE4	63	Hi	L	GS	Vin	RE5
10	Lo	M	PS	At	RE4	37	Me	M	PS	At	RE5	64	Hi	M	PS	At	RE7
11	Lo	M	PS	Ia	RE5	38	Me	M	PS	Ia	RE7	65	Hi	M	PS	Ia	RE8
12	Lo	M	PS	Vin	RE7	39	Me	M	PS	Vin	RE8	66	Hi	M	PS	Vin	RE9
13	Lo	M	MS	At	RE3	40	Me	M	MS	At	RE4	67	Hi	M	MS	At	RE6
14	Lo	M	MS	Ia	RE4	41	Me	M	MS	Ia	RE6	68	Hi	M	MS	Ia	RE7
15	Lo	M	MS	Vin	RE5	42	Me	M	MS	Vin	RE7	69	Hi	M	MS	Vin	RE8
16	Lo	M	GS	At	RE2	43	Me	M	GS	At	RE3	70	Hi	M	GS	At	RE5
17	Lo	M	GS	Ia	RE3	44	Me	M	GS	Ia	RE5	71	Hi	M	GS	Ia	RE6
18	Lo	M	GS	Vin	RE4	45	Me	M	GS	Vin	RE6	72	Hi	M	GS	Vin	RE7
19	Lo	H	PS	At	RE5	46	Me	H	PS	At	RE7	73	Hi	H	PS	At	RE8
20	Lo	H	PS	Ia	RE7	47	Me	H	PS	Ia	RE8	74	Hi	H	PS	Ia	RE9
21	Lo	H	PS	Vin	RE8	48	Me	H	PS	Vin	RE9	75	Hi	H	PS	Vin	RE9
22	Lo	H	MS	At	RE4	49	Me	H	MS	At	RE6	76	Hi	H	MS	At	RE7
23	Lo	H	MS	Ia	RE5	50	Me	H	MS	Ia	RE7	77	Hi	H	MS	Ia	RE8
24	Lo	H	MS	Vin	RE7	51	Me	H	MS	Vin	RE8	78	Hi	H	MS	Vin	RE9
25	Lo	H	GS	At	RE3	52	Me	H	GS	At	RE5	79	Hi	H	GS	At	RE6
26	Lo	H	GS	Ia	RE4	53	Me	H	GS	Ia	RE6	80	Hi	H	GS	Ia	RE7
27	Lo	H	GS	Vin	RE6	54	Me	H	GS	Vin	RE7	81	Hi	H	GS	Vin	RE8

5   Simulation Results

We carried out simulations employing FuzzyC, covering nine unique scenarios. You can observe the results in Fig. 3(a)–Fig. 5(c). These figures depict the outcomes for low, medium, and high levels of Internal and External Distraction and Driver's Inattention, respectively. They provide insights into the connection between Recognition Error and Inattentive Driving, considering varying levels of Inadequate Surveillance, while keeping Internal and External Distraction (IED) and Driver's Inattention (DIA) values constant.

In the simulation results for low Internal and External Distraction (IED) errors, as demonstrated in Fig. 5, we observe three distinct scenarios with low, medium, and high levels of Driver's Inattention (DIA) as shown in Fig. 3(a), Fig. 3(b), and Fig. 3(c), respectively. Notably, scenarios with low DIA values yield the lowest Recognition Error (RE) values when compared to the other scenarios. This outcome suggests that drivers exhibit a good level of focus and attention, and although there is some distraction, it does not significantly compromise

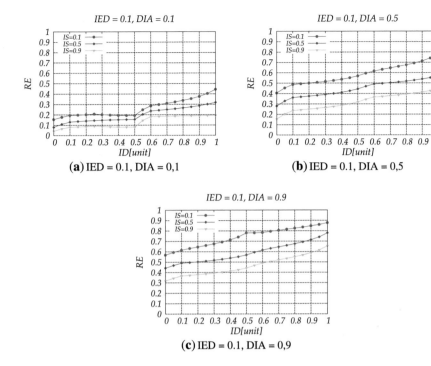

Fig. 3. Simulation results for FS-ARE.

their ability to perceive and react to the driving environment. This results in a manageable rate of recognition errors. Consequently, the likelihood of accidents resulting from recognition errors is significantly reduced in such scenarios. In contrast, medium DIA values lead to an increase in RE values. This can be attributed to the combination of low levels of Internal and External Distraction (IED) and a moderate level of Driver's Inattention (DIA), which collectively contribute to this rise in values. Additionally, we can observe a gradual increase in output values for significant Inadequate Surveillance. For scenarios characterized by high DIA values and low IED values, we notice a deterioration in the error rates across all levels of Inattentive Driving (ID).

When DIA is low, and Internal and IED is at a medium level, recognition errors still remain relatively low as shown in Fig. 4(a). In scenarios with medium and high DIA for medium IED (see Fig. 4(b) and Fig. 4(c) respectively), the balance between distraction and inattention becomes more delicate. Here, the driver is moderately inattentive, and there is a moderate level of distraction. The driver's divided attention and somewhat reduced focus can result in delayed reactions and misinterpretations of critical information on the road. This is shown in high RE values in Fig. 4(b) and Fig. 4(c), despite IS and ID values.

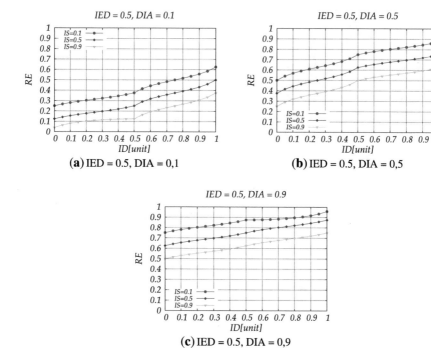

Fig. 4. Simulation results for FS-ARE.

In scenarios where DIA is low, and IED is high (large IED), recognition errors are likely to be elevated (see Fig. 5(a)). We can see that moderate and large ID values affects output error significantly. The moderate inattention, combined with high distraction shown in Fig. 5(b), creates a scenario where the driver's divided attention and reduced focus significantly compromise their ability to make accurate judgments on the road. In scenarios characterized by high DIA and large IED, recognition errors are most pronounced (see Fig. 5(c)). This combination can lead to a substantial increase in recognition errors, as the driver's ability to perceive and respond to the driving environment is severely impaired. This can be seen as RE values are higher than 0.6.

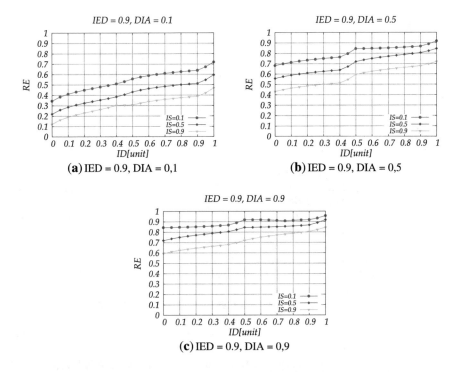

Fig. 5. Simulation results for FS-ARE.

6   Conclusions

Within the domain of road safety, where driver errors remain a significant factor in traffic incidents, VANETs offer a promising avenue for reducing these risks and enhancing overall safety on the roads.

This research showcases the outcomes of the FS-ARE system, which operates in real-time to anticipate and avert recognition errors, thus contributing to accident prevention.

Through the exploration of various scenarios involving the interplay of four key parameters: IED, DIA, IS, and ID, we have gleaned valuable insights into the complex relationships among these variables. It becomes evident that as DIA, IED, IS, and ID values increase, RE values also tend to worsen. This highlights the critical significance of implementing strategic distraction management strategies, including practices like minimizing device usage and promoting attentive driving, to foster safer road conditions.

In the future work, we will consider other parameters which affect the recognition errors.

References

1. Al-Heety, O.S., Zakaria, Z., Ismail, M., Shakir, M.M., Alani, S., Alsariera, H.: A comprehensive survey: benefits, services, recent works, challenges, security, and use cases for SDN-VANET. IEEE Access **8**, 91028–91047 (2020)
2. Bojadziev, G., Bojadziev, M., Zadeh, L.A.: Fuzzy Logic for Business, Finance, and Management, Advances in Fuzzy Systems - Applications and Theory, vol. 12. World Scientific (1997)
3. Hartenstein, H., Laberteaux, K.P. (eds.): VANET: Vehicular Applications and Inter-networking Technologies. Intelligent Transportation Systems. Wiley (2010). https://doi.org/10.1002/9780470740637
4. Hartenstein, H., Laberteaux, L.: A tutorial survey on vehicular ad hoc networks. IEEE Commun. Mag. **46**(6), 164–171 (2008)
5. Hussein, A., Elhajj, I.H., Chehab, A., Kayssi, A.: SDN VANETs in 5G: an architecture for resilient security services. In: 2017 Fourth International Conference on Software Defined Systems (SDS), pp. 67–74 (2017)
6. Kandel, A.: Fuzzy Expert Systems. CRC Press Inc., Boca Raton (1992)
7. Klir, G.J., Folger, T.A.: Fuzzy Sets, Uncertainty, and Information. Prentice Hall, Upper Saddle River (1988)
8. Klir, G.J., Yuan, B.: Fuzzy Sets and Fuzzy Logic - Theory And Applications. Prentice Hall, Upper Saddle River (1995)
9. McNeill, F.M., Thro, E.: Fuzzy Logic: A Practical Approach. Academic Press Professional Inc., San Diego (1994)
10. Munakata, T., Jani, Y.: Fuzzy systems: an overview. Commun. ACM **37**(3), 69–77 (1994)
11. Peixoto, M.L.M., et al.: FogJam: a fog service for detecting traffic congestion in a continuous data stream VANET. Ad Hoc Netw. **140**, 103,046 (2023). https://doi.org/10.1016/j.adhoc.2022.103046
12. Qafzezi, E., Bylykbashi, K., Higashi, S., Ampririt, P., Matsuo, K., Barolli, L.: A fuzzy-based error driving system for improving driving performance in VANETs. In: Barolli, L. (eds.) Complex, Intelligent and Software Intensive Systems, vol. 176, pp. 1–9. Springer, Cham (2023). https://doi.org/10.1007/978-3-031-35734-3_16
13. Sadio, O., Ngom, I., Lishou, C.: SDN architecture for intelligent vehicular sensors networks. In: 2018 UKSim-AMSS 20th International Conference on Computer Modelling and Simulation (UKSim), pp. 139–144 (2018)
14. Schünemann, B., Massow, K., Radusch, I.: Realistic simulation of vehicular communication and vehicle-2-x applications. In: Proceedings of the 1st International Conference on Simulation Tools and Techniques for Communications, Networks and Systems & Workshops, SimuTools 2008, Marseille, France, 3–7 March 2008, p. 62. ICST/ACM (2008). https://doi.org/10.4108/ICST.SIMUTOOLS2008.2949
15. Zadeh, L.A., Kacprzyk, J.: Fuzzy Logic for the Management of Uncertainty. Wiley, New York (1992)
16. Zadeh, L.A., Klir, G.J., Yuan, B.: Fuzzy Sets, Fuzzy Logic, and Fuzzy Systems - Selected Papers by Lotfi A Zadeh, Advances in Fuzzy Systems - Applications and Theory, vol. 6. World Scientific (1996). https://doi.org/10.1142/2895
17. Zimmermann, H.J.: Fuzzy control. In: Fuzzy Set Theory and Its Applications, pp. 203–240. Springer, Cham (1996)

Performance Evaluation of BLX-α Crossover Method for Different Instances of WMNs Considering FC-RDVM Router Replacement Method and Subway Distribution of Mesh Clients

Admir Barolli[1], Shinji Sakamoto[2], Leonard Barolli[3($\boxtimes$)], and Makoto Takizawa[4]

[1] Department of Information Technology, Aleksander Moisiu University of Durres,
L.1, Rruga e Currilave, Durres, Albania
admirbarolli@uamd.edu.al

[2] Department of Information and Computer Science, Kanazawa Institute of Technology,
7-1 Ohgigaoka, Nonoichi, Ishikawa 921-8501, Japan
shinji.sakamoto@ieee.org

[3] Department of Information and Communication Engineering, Fukuoka Institute
of Technology, 3-30-1 Wajiro-Higashi, Higashi-Ku, Fukuoka 811-0295, Japan
barolli@fit.ac.jp

[4] Department of Advanced Sciences, Faculty of Science and Engineering, Hosei University,
3-7-2, Kajino-machi, Koganei-shi, Tokyo 184-8584, Japan
makoto.takizawa@computer.org

Abstract. In this research work, we present a simulation system called WMN-PSOHCDGA for optimization of node placement in Wireless Mesh Networks (WMNs). We implement BLX-α crossover method in our simulation system and carry out simulations for different instances (scales) of WMNs considering Fast Convergence Rational Decrement of Vmax Method (FC-RDVM) and Subway distribution of mesh clients. The simulation results indicate that for both scales of WMNs, all mesh routers are connected and all number of clients are covered. However, the load balancing for middle scale WMNs is better than small scale WMNs.

1 Introduction

There are many complex optimization problems in our everyday life and many research and scientific studies. Therefore are needed high-performance search algorithms to solve these complex optimization problems, especially those that are either NP-hard or NP-complete, has been an interesting research topic for years. In this paper, we consider the optimization problem for Wireless Mesh Networks (WMNs). There are many optimization issues in WSNs, which have showing their usefulness for the efficient design of WMNs. These problems are related with the optimization of the network connectivity, user coverage and stability, which are very important for optimized network performance. Most of optimization problems in WMNs deal with computing optimal

L. Barolli (Ed.): EIDWT 2024, LNDECT 193, pp. 343–352, 2024.
https://doi.org/10.1007/978-3-031-53555-0_32

placement of mesh router nodes. However, node placement problems are known to be complex therefore heuristics methods are used to solve such problems.

The WMNs are very popular because they have many applications. They can be used as last miles networks and for edge computing. However, in the design process of WMNS are needed different parameters, which makes the problem NP-Hard. In order to deal with this problem, the position of mesh routers should be optimized in order to achieve good network connectivity, client coverage and load balancing. For the optimization, we consider three parameters: Size of Giant Component (SGC), Number of Covered Mesh Clients (NCMC) and Number of Covered Mesh Clients per Router (NCMCpR).

There are many research works, which deal with the mesh node placement in WMNs [2,4–6,12,13] and some intelligent algorithms and systems are proposed [1,3, 7,8]. In [9,10], the authors consider simple intelligent algorithms and implement intelligent systems for optimization of mesh routers in WMNs.

In this paper, by considering PSO (Particle Swarm Optimization), HC (Hill Climbing) and DGA (Distributed Genetic Algorithm), we implemented WMN-PSOHCDGA hybrid intelligent simulation system. Also, we implemented BLX-α crossover method and Fast Convergence Rational Decrement of Vmax Method (FC-RDVM) in WMN-PSOHCDGA system. We compare the simulation results for small and middle scale WMNs. The simulation results indicate that all mesh routers are connected for both scales of WMNs, so the size of giant component is maximized. Also, all mesh clients are covered. However, the load balancing for middle scale WMNs is better than small scale WMNs.

The paper is organized as follows. In Sect. 2, we introduce intelligent algorithms. In Sect. 3, we present the implemented hybrid simulation system. The simulation results are given in Sect. 4. Finally, we give conclusions and future work in Sect. 5.

2   Intelligent Algorithms: PSO, HC and DGA

2.1   PSO Algorithm

The swarm-based algorithms are a powerful family of optimization techniques, which are inspired by the collective behavior of animals. In PSO, for the optimization problem a set of candidate solutions is considered as a swarm of particles flowing through the parameter space defining trajectories driven by their own and neighbors' best performances. Each member of the swarm can benefit from other members and a better solution can be found.

The PSO has many similarities with Genetic Algorithms (GAs). The initialization is done by a population of random solutions and then the optimal solution is searched by updating generations. However, different from GA, the PSO has no evolution operators such as crossover and mutation. In PSO, the potential solutions, called particles, fly through the problem space by following the current optimum particles.

In PSO for every iteration, each particle is updated by following two "best" values: the best solution (fitness) achieved so far called pbest and another "best" value tracked by the particle swarm optimizer obtained so far by any particle in the population called gbest. When a particle is part of the population as its topological neighbors, the best value is a local best and is called lbest.

Compared to GA, the PSO is easy to implement and there are few parameters to adjust. PSO has been successfully applied in many areas such as artificial neural network training and fuzzy system control.

2.2 HC Algorithm

The HC is a simple optimization algorithm and can be used to find the best solution from a set of possible solutions. After the initial solution, the algorithm improves the solution by making small changes, which continue until it reaches a local maximum or no further improvement can be made.

The HC can be used for different optimization problems. However, the algorithms can stuck in local maxima and it lacks the diversity in the search space. Therefore, it is often combined with other optimization techniques to improve the search results.

2.3 Genetic Algorithm

Genetic Algorithms (GAs) are a class of search techniques, which uses simplified forms of biological processes. The algorithm first randomly initializes the population and then evaluates the fitness function. Using the fitness value, some members of current population are selected to produce offspring population. After that the current population is replaced with the offspring population. In the next step is evaluated the fitness value for the members of new population. These steps are repeated until a good solution is achieved.

In our simulation system, we consider Distributed GA (DGA), which has a better behaviour than GA and can escape from local optimum. As shown in Fig. 1, each island operates independently, but the individuals are migrated between islands.

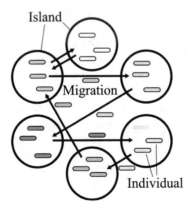

Fig. 1. Model of Migration in DGA.

The main operators of GA are selection, crossover and mutation, which are implemented using different methods. In this paper, we consider BLX-α crossover method, where the area for generation the children is controlled using some variables. For instance, in Fig. 2 there are two parents: $p_1 = (x_1, y_1)$ and $p_2 = (x_2, y_2)$. The value of α parameter is set on both sides of I_x and I_y. Then, the interval for generating the children is $I_x = [x_1, x_2]$ and $I_y = [y_1, y_2]$.

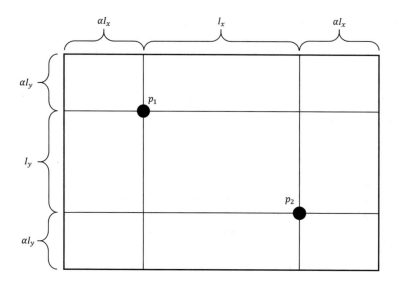

Fig. 2. BLX-α crossover method.

3   Structure of WMN-PSOHCDGA System

We show the flowchart of WMN-PSOHCDGA simulation system in Fig. 3. The initial solution is generated randomly by *ad hoc* methods [14]. Then, it carries out the optimization process for each island. After that the solutions are migrated (swapped) between islands to find a better solution. If the stopping criteria is not achieved the optimization process is continuing, otherwise the best solution is found.

In the evaluation of each islands, we use DGA algorithms. In the case when the solution from DGA part is found, then it goes to PSO part, which updates the velocities and positions of particles by router replacement method. Finally, we use the HC algorithm for improving the convergence of PSO algorithm.

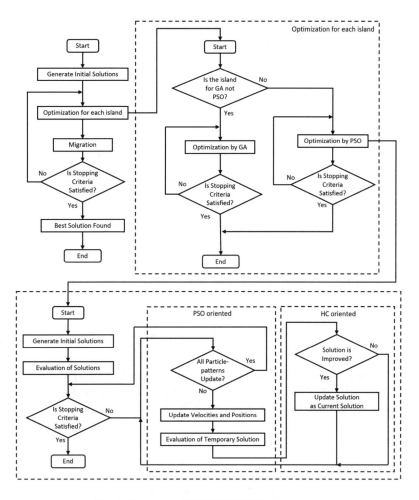

Fig. 3. Flowchart of WMN-PSOHCDGA system.

In WMN-PSOHCDGA system, a particle is considered as a mesh router. The fitness value is calculated by considering the position of mesh routers and mesh clients. The solution of each particle-pattern is shown in Fig. 4. We represent a WMN by a gene and each individual in the population is a combination of mesh routers.

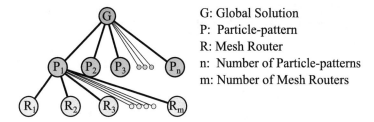

G: Global Solution
P: Particle-pattern
R: Mesh Router
n: Number of Particle-patterns
m: Number of Mesh Routers

Fig. 4. Relationship between particle-patterns and mesh routers.

The fitness function considers SGC (connectivity of mesh routers), NCM (coverage of mesh clients) and NCMCpR (load balancing) parameters as shown in Eq. (1).

$$Fitness = \alpha \times SGC(\mathbf{x}_{ij}, \mathbf{y}_{ij}) + \beta \times NCMC(\mathbf{x}_{ij}, \mathbf{y}_{ij}) + \gamma \times NCMCpR(\mathbf{x}_{ij}, \mathbf{y}_{ij}) \quad (1)$$

Fig. 5. Subway distribution of mesh clients.

In this research work, we consider Subway distribution of mesh clients as shown in Fig. 5. As mesh router replacement method we consider FC-RDVM [11], where V_{max} is the maximum velocity of particles as shown in Eq. (2):

$$V_{max}(k) = \sqrt{W^2 + H^2} \times \frac{T - k}{T + \delta k} \quad (2)$$

where, W and H are width and height of the considered area, respectively. T is the total number of iterations and k shows the current number of iterations, which is varying from 1 to T. Finally, δ is the curvature parameter.

4   Simulation Results

In this section, we investigate the performance of BLX-α crossover method considering FC-RDVM and Subway distribution of mesh clients for small and middle scales WMNs. We carried out many simulations for setting the fitness functions coefficients: $\alpha = 0.1$, $\beta = 0.8$, $\gamma = 0.1$. We show in Table 1 other parameters used for simulations.

Table 1. Simulation parameters.

Parameters	Values	
	Small Scale WMN	Middle Scale WMN
$\alpha : \beta : \gamma$	1 : 8 : 1	
Number of GA Islands	16	
Number of Evolution Steps	9	
Number of Migrations	200	
Number of Mesh Routers	16	32
Number of Mesh Clients	48	96
Mesh Client Distribution	Subway Distribution	
Router Replacement Method	FC-RDVM	
Selection Method	Rulette Selection Method	
Crossover Method	BLX-α	
Mutation Method	Uniform Mutation	

In Fig. 6 are shown the visualization results after optimization for small and middle scale WMNs, respectively. All mesh routers are connected for both scales of WMNs, thus the SGC value is 100%. Also, all mesh clients are covered. However, some mesh routers are concentrated on the right side area in case of small scale WMNs.

In Fig. 7 is shown the NCMC for each router of BLX-α crossover method for small and middle scale WMNs, respectively. As shown also in Fig. 6 all mesh clients are covered.

The relation of standard deviation with the number of updates of BLX-α crossover method for small and middle scale WMNs is shown in Fig. 8, where r indicates the correlation coefficient. The load balancing among routers is better when the standard deviation is a decreasing line. For both small and middle scale WMNs, the standard deviation is a decreasing line. However, for middle scale WMN, we can see that the load balancing is better than small scale WMNs.

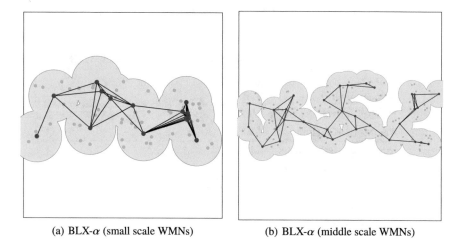

(a) BLX-α (small scale WMNs) (b) BLX-α (middle scale WMNs)

Fig. 6. Visualization results after optimization for small and middle scale WMNs.

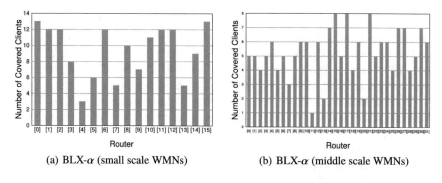

(a) BLX-α (small scale WMNs) (b) BLX-α (middle scale WMNs)

Fig. 7. Number of covered mesh clients for small and middle scale WMNs.

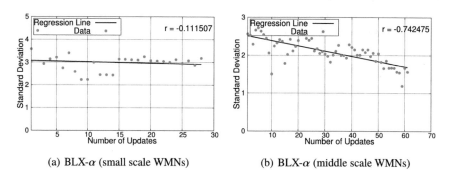

(a) BLX-α (small scale WMNs) (b) BLX-α (middle scale WMNs)

Fig. 8. Relation of standard deviation with number of updates for small and middle scale WMNs.

5   Conclusions

In this paper, we assessed the performance of BLX-α crossover method for different instances of WMNs considering FC-RDVM and subway distribution of mesh clients. We evaluated BLX-α crossover method by computer simulations. The simulation results indicated that for both scales of WMNs, all mesh routers were connected and all clients were covered. However, the load balancing for middle scale WMN was better than small scale WMNs.

In future work, we will consider other crossover, mutation and mesh router replacement methods.

References

1. Barolli, A., Sakamoto, S., Ozera, K., Barolli, L., Kulla, E., Takizawa, M.: Design and implementation of a hybrid intelligent system based on particle swarm optimization and distributed genetic algorithm. In: Barolli, L., Xhafa, F., Javaid, N., Spaho, E., Kolici, V. (eds.) EIDWT 2018. LNDECT, vol. 17, pp. 79–93. Springer, Cham (2018). https://doi.org/10.1007/978-3-319-75928-9_7
2. Franklin, A.A., Murthy, C.S.R.: Node placement algorithm for deployment of two-tier wireless mesh networks. In: Proceedings of Global Telecommunications Conference, pp. 4823–4827 (2007)
3. Girgis, M.R., Mahmoud, T.M., Abdullatif, B.A., Rabie, A.M.: Solving the wireless mesh network design problem using genetic algorithm and simulated annealing optimization methods. Int. J. Comput. Appl. **96**(11), 1–10 (2014)
4. Lim, A., Rodrigues, B., Wang, F., Xu, Z.: k-Center problems with minimum coverage. Theoret. Comput. Sci. **332**(1–3), 1–17 (2005)
5. Maolin, T., et al.: Gateways placement in backbone wireless mesh networks. Int. J. Commun. Netw. Syst. Sci. **2**(1), 44–50 (2009)
6. Muthaiah, S.N., Rosenberg, C.P.: Single gateway placement in wireless mesh networks. In: Proceedings of 8th International IEEE Symposium on Computer Networks, pp. 4754–4759 (2008)
7. Naka, S., Genji, T., Yura, T., Fukuyama, Y.: A hybrid particle swarm optimization for distribution state estimation. IEEE Trans. Power Syst. **18**(1), 60–68 (2003)
8. Sakamoto, S., Kulla, E., Oda, T., Ikeda, M., Barolli, L., Xhafa, F.: A comparison study of simulated annealing and genetic algorithm for node placement problem in wireless mesh networks. J. Mobile Multimedia **9**(1–2), 101–110 (2013)
9. Sakamoto, S., Kulla, E., Oda, T., Ikeda, M., Barolli, L., Xhafa, F.: A comparison study of hill climbing, simulated annealing and genetic algorithm for node placement problem in WMNs. J. High Speed Netw. **20**(1), 55–66 (2014)
10. Sakamoto, S., Oda, T., Ikeda, M., Barolli, L., Xhafa, F.: Implementation and evaluation of a simulation system based on particle swarm optimisation for node placement problem in wireless mesh networks. Int. J. Commun. Netw. Distrib. Syst. **17**(1), 1–13 (2016)
11. Sakamoto, S., Barolli, A., Liu, Y., Kulla, E., Barolli, L., Takizawa, M.: A fast convergence RDVM for router placement in WMNs: performance comparison of FC-RDVM with RDVM by WMN-PSOHC hybrid intelligent system. In: Barolli, L. (eds.) International Conference on Computational Intelligence in Security for Information Systems Conference, vol. 497, pp. 17–25. Springer, Cham (2022). https://doi.org/10.1007/978-3-031-08812-4_3

12. Vanhatupa, T., Hannikainen, M., Hamalainen, T.: Genetic algorithm to optimize node placement and configuration for WLAN planning. In: Proceedings of the 4th IEEE International Symposium on Wireless Communication Systems, pp. 612–616 (2007)
13. Wang, J., Xie, B., Cai, K., Agrawal, D.P.: Efficient mesh router placement in wireless mesh networks. In: Proceedings of IEEE International Conference on Mobile Adhoc and Sensor Systems (MASS-2007), pp. 1–9 (2007)
14. Xhafa, F., Sanchez, C., Barolli, L.: Ad hoc and neighborhood search methods for placement of mesh routers in wireless mesh networks. In: Proceedings of 29th IEEE International Conference on Distributed Computing Systems Workshops (ICDCS-2009), pp. 400–405 (2009)

A Measurement System for Improving Riding Comfortability of Omnidirectional Wheelchair Robot

Keita Matsuo[(⊠)] and Leonard Barolli

Department of Information and Communication Engineering, Fukuoka Institute of Technology
(FIT), 3-30-1 Wajiro-Higashi, Higashi-Ku, Fukuoka 811-0295, Japan
{kt-matsuo,barolli}@fit.ac.jp

Abstract. Recently, there are developed and implemented diverse robots that collaborate with various devices to assist humans. Also, the industrial robots have many applications in factory operations. Additionally, robots like vacuum cleaners, security robots, therapy robots, and wheelchair robots can assist humans in various tasks. Therefore, the utilization of robots and associated technologies have great significance in enhancing the quality of life. In the world, there are over one billion of people with disabilities who depend on wheelchairs for their mobility. In this paper, we consider the riding comfort issue of omnidirectional wheelchair robot. We implemented a measurement system to improve and make more comfortable the riding of omnidirectional wheelchair robot. We carried out some experiments. The experimental results show that 0.7 [s] is a good time for acceleration and deceleration in order to have riding comfort of omnidirectional wheelchair robot.

1 Introduction

According to the WHO (World Health Organization) [1], there are over 1 billion disabled people in the world, which is approximately 15% of the world's population. Recently, many types of facilities and equipments have been developed not only for disabled individuals but also for the elderly population. The wheelchair is one of the most common equipment used in various situations to assist both disabled and elderly individuals.

There are different kinds of wheelchairs which can support humans daily life. For example, wheelchair with navigation system [4, 13, 14], which is able to help users by taking them to a place they want. Head moving interface may drive wheelchair based on inertial sensors [6]. The Brain Computer Interface (BCI) is particularly beneficial for disabled individuals who are unable to move their hands and legs or are essentially experiencing cerebromedullospinal disconnection [2, 18]. The Human Machine Interface (HMI) is using piezoelectric sensors to sense the face and tongue movements [3]. Also, there is an electrooculography based interface for eyes control [5, 7].

In our previous research, we investigated the potential of omnidirectional moving robots for different applications. We utilized the omnidirectional moving robot as a mobile router to provide network support [15–17], and for assisting people who are

L. Barolli (Ed.): EIDWT 2024, LNDECT 193, pp. 353–361, 2024.
https://doi.org/10.1007/978-3-031-53555-0_33

using wheelchair [8,11,12]. In addition, we applied the omnidirectional moving robot as a wheelchair for sports such as wheelchair tennis or badminton [9,10].

Our implemented omnidirectional robot and wheelchair can be used for welfare purposes. If the wheelchair move omnidirectionaly, it is able to give many benefits to users.

In this research work, we propose a measurement system to improve and make more comfortable the riding of omnidirectional wheelchair robot. We carried out some experiments. The experimental results show that when the acceleration and deceleration time is between 0.1 [s] to 0.4 [s], there are significant disparities between the commanded and measured values. However, when the time is between 0.5 [s] and 0.8 [s], the differences are not substantial, particularly after 0.7 [s]. From these results, we decided to use the time 0.7 [s] for the acceleration and deceleration in order to improve and make more comfortable the riding of omnidirectional robot.

The rest of this paper is structured as follows. In Sect. 2, we introduce the proposed omnidirectional wheelchair robot. In Sect. 3, we propose the measurement system for improving riding comfortability of omnidirectional wheelchair robot. In Sect. 4, we explain experimental results. Finally, conclusions and future work are given in Sect. 5.

2   Proposed Omnidirectional Wheelchair Robot

In this section, we introduce the omnidirectional wheelchair robot which we designed and implemented. In Fig. 1(a) is shown the omnidirectional wheelchair robot and in Fig. 1(b) is shown the omniwheel. The omniwheel has 28 small tires, which allows omnidirectional movement in the front, back, right and left directions.

Figure 2 shows an image of moving omnidirectional wheelchair in front of a refrigerator and a sink. In this case, the wheelchair user wants to use the refrigerator, and after that he moves in front of the sink. The omnidirectional wheelchair robot can give a convenient environment for the user.

Figure 3 shows an image of moving omnidirectional wheelchair in a narrow space. The omnidirectional wheelchair robot can move in the narrow space while keeping the same direction. It reduces the time needed to reach a destination since there is not needed to turn and change direction.

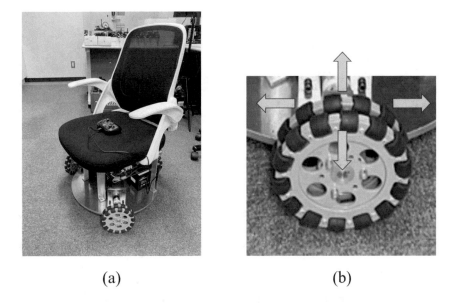

(a)   (b)

Fig. 1. Omnidirectional wheelchair robot and omniwheel.

Fig. 2. Image of moving omnidirectional wheelchair in front of refrigerator and sink.

Fig. 3. Image of moving omnidirectional wheelchair in a narrow space.

Figure 4 shows an image of moving omnidirectional wheelchair in a wide space. When a user needs to utilize a wide area for various tasks, the user will require the ability to move sideways without turning the wheelchair. This omnidirectional wheelchair can provide this functionality.

Fig. 4. Image of moving omnidirectional wheelchair in a wide space.

3 Proposed Measurement System for Riding Comfort of Omnidirectional Wheelchair Robot

In this section, we propose a measurement system to improve and make more comfortable the riding of omnidirectional wheelchair robot. In Fig. 5 is shown the implementation of 3 motors on omnidirectional wheelchair robot, which are placed 120° with each other.

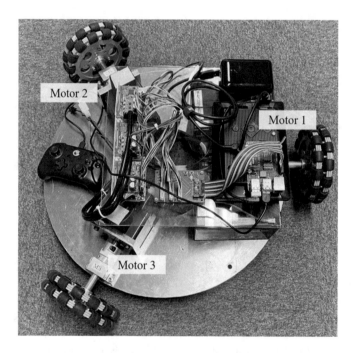

Fig. 5. Implementation of 3 motors on omnidirectional wheelchair robot.

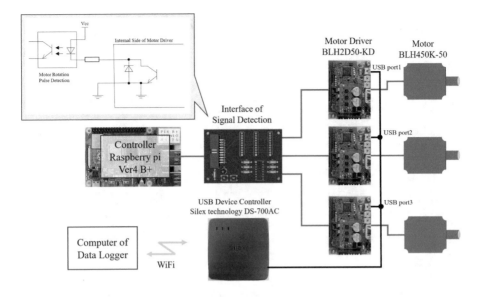

Fig. 6. Measurement system for real motor speed rotation.

In order to improve and improving the riding comfortability of omnidirectional robot, we measure the difference between command rotation value and real motor rotation speed as shown in Fig. 6. We designed and implemented the interface for signal detection from the motors using a photo transistor circuit. Additionally, the USB device controller can transmit information to a computer (Data logger) considering each motor status by using wireless communication, including over or under voltage, current, overload, sensor or driver errors, control mode, and so on.

4   Experimental Results

We investigated the motor accelerating and stopping time. We changed the time range of acceleration and deceleration from 0.0 to 1.0 in order to get experimental data.

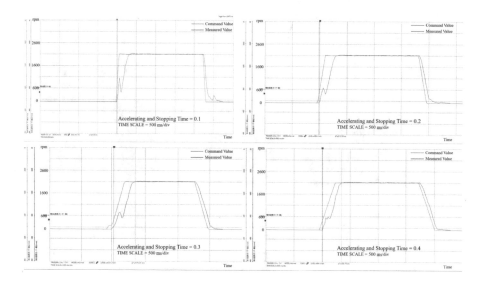

Fig. 7. Experimental results (accelerating and stopping time: 0.1 s–0.4 s).

In Fig. 7 are shown the experimental results when we set the accelerating and stopping time between 0.1 [s] and 0.4 [s]. While, in Fig. 8, the accelerating and stopping time is between 0.5 [s] and 0.8 [s]. Finally, in Fig. 9, the accelerating and stopping time is between 0.9 [s] and 1.0 [s]. The vertical axis shows the rotation speed, while the horizontal axis shows the time. We investigate the difference between command values and measured values.

When the acceleration and deceleration time is between 0.1 [s] to 0.4 [s], we see significant disparities between the commanded and measured values. However, when the time is between 0.5 [s] to 0.8 [s], the differences are not substantial, particularly after 0.7 [s]. The data for time range between 0.9 [s] to 1.0 [s] are almost the same.

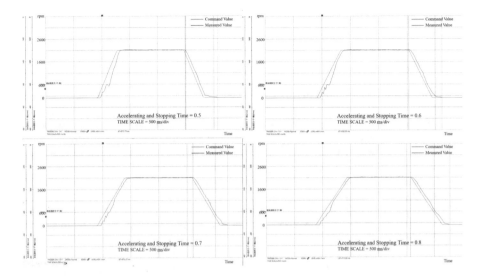

Fig. 8. Experimental results (accelerating and stopping time: 0.5 s–0.8 s).

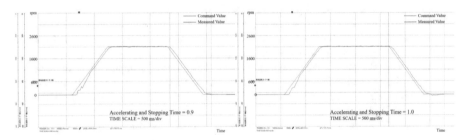

Fig. 9. Experimental results (accelerating and stopping time: 0.9 s–1.0 s).

5   Conclusions and Future Work

In this paper, we explained the current situation of elderly and disabled people in the world and discussed the necessity of wheelchairs. Also, we introduced the applications of omnidirectional wheelchair robot, which can improve the users' quality of life.

We proposed a measurement system to improve and make more comfortable the riding of omnidirectional wheelchair robot. We carried out some experiments. The experimental results show that when the acceleration and deceleration time is between 0.1 [s] to 0.4 [s], there are significant disparities between the commanded and measured values. However, when the time is between 0.5 [s] and 0.8 [s], the differences are not substantial, particularly after 0.7 [s]. From these results, we decided to use 0.7 [s] for acceleration and deceleration time in order to improve the riding comfortability of omnidirectional robot.

In the future work, we would like to investigate the riding comfortability of omnidirectional wheelchair robot for different applications.

Acknowledgements. This work is supported by JSPS KAKENHI Grant Number JP22K11598.

References

1. World health organization. https://www.who.int/
2. Ansari, M.F., Edla, D.R., Dodia, S., Kuppili, V.: Brain-computer interface for wheelchair control operations: an approach based on Fast Fourier Transform and on-line sequential extreme learning machine. Clin. Epidemiol. Global Health **7**(3), 274–278 (2019)
3. Bouyam, C., Punsawad, Y.: Human-machine interface-based wheelchair control using piezo-electric sensors based on face and tongue movements. Heliyon **8**(11), e11679 (2022)
4. Chatterjee, S., Roy, S.: Multiple control assistive wheelchair for lower limb disabilities & elderly people (2021)
5. Choudhari, A.M., Porwal, P., Jonnalagedda, V., Mériaudeau, F.: An electrooculography based human machine interface for wheelchair control. Biocybern. Biomed, Eng. **39**(3), 673–685 (2019)
6. Gomes, D., Fernandes, F., Castro, E., Pires, G.: Head-movement interface for wheelchair driving based on inertial sensors. In: 2019 IEEE 6th Portuguese Meeting on Bioengineering (ENBENG), pp. 1–4 (2019)
7. Kaur, A.: Wheelchair control for disabled patients using EMG/EOG based human machine interface: a review. J. Med. Eng. Technol. **45**(1), 61–74 (2021)
8. Matsuo, K., Barolli, L.: Design and implementation of an omnidirectional wheelchair: control system and its applications. In: 2014 Ninth International Conference on Broadband and Wireless Computing, Communication and Applications, pp. 532–535. IEEE (2014)
9. Matsuo, K., Barolli, L.: Prediction of RSSI by Scikit-learn for improving position detecting system of omnidirectional wheelchair tennis. In: Barolli, L., Hellinckx, P., Enokido, T. (eds.) BWCCA 2019. LNNS, vol. 97, pp. 721–732. Springer, Cham (2020). https://doi.org/10.1007/978-3-030-33506-9_66
10. Matsuo, K., Kulla, E., Barolli, L.: Implementation of a collision avoidance system for machine tennis game. In: Barolli, L. (eds.) Advances in Networked-based Information Systems. NBiS 2023. Lecture Notes on Data Engineering and Communications Technologies, vol. 183, pp. 150–158. Springer, Cham (2023). https://doi.org/10.1007/978-3-031-40978-3_17
11. Mitsugi, K., Matsuo, K., Barolli, L.: A comparison study of control devices for an omnidirectional wheelchair. In: Barolli, L., Amato, F., Moscato, F., Enokido, T., Takizawa, M. (eds.) WAINA 2020. AISC, vol. 1150, pp. 651–661. Springer, Cham (2020). https://doi.org/10.1007/978-3-030-44038-1_60
12. Mitsugi, K., Matsuo, K., Barolli, L.: Evaluation of a user finger movement capturing device for control of self-standing omnidirectional robot. In: Barolli, L., Woungang, I., Enokido, T. (eds.) AINA 2021. LNNS, vol. 227, pp. 30–40. Springer, Cham (2021). https://doi.org/10.1007/978-3-030-75078-7_4
13. Ngo, B.V., Nguyen, T.H., Ngo, V.T., Tran, D.K., Nguyen, T.D.: Wheelchair navigation system using EEG signal and 2D map for disabled and elderly people. In: 2020 5th International Conference on Green Technology and Sustainable Development (GTSD), pp. 219–223. IEEE (2020)
14. Nudra Bajantika Pradivta, I.W., Arifin, A., Arrofiqi, F., Watanabe, T.: Design of myoelectric control command of electric wheelchair as personal mobility for disabled person. In: 2019 International Biomedical Instrumentation and Technology Conference (IBITeC), vol. 1, pp. 112–117 (2019)

15. Toyama, A., Mitsugi, K., Matsuo, K., Barolli, L.: Implementation of a moving omnidirectional access point robot and a position detecting system. In: Barolli, L., Poniszewska-Maranda, A., Park, H. (eds.) IMIS 2020. AISC, vol. 1195, pp. 203–212. Springer, Cham (2021). https://doi.org/10.1007/978-3-030-50399-4_20

16. Toyama, A., Mitsugi, K., Matsuo, K., Barolli, L.: Implementation of control interfaces for moving omnidirectional access point robot. In: Barolli, L., Takizawa, M., Enokido, T., Chen, H.-C., Matsuo, K. (eds.) BWCCA 2020. LNNS, vol. 159, pp. 436–443. Springer, Cham (2021). https://doi.org/10.1007/978-3-030-61108-8_43

17. Toyama, A., Mitsugi, K., Matsuo, K., Kulla, E., Barolli, L.: Implementation of an indoor position detecting system using mean BLE RSSI for moving omnidirectional access point robot. In: Barolli, L., Yim, K., Enokido, T. (eds.) CISIS 2021. LNNS, vol. 278, pp. 225–234. Springer, Cham (2021). https://doi.org/10.1007/978-3-030-79725-6_22

18. Zubair, Z.R.S.: A deep learning based optimization model for based computer interface of wheelchair directional control. Tikrit J. Pure Sci. **26**(1), 108–112 (2021)

A Comparative Study of Four YOLO-Based Models for Distracted Driving Detection

Naoki Tanaka[1], Hibiki Tanaka[1], Makoto Ikeda[2(✉)] ⓘ, and Leonard Barolli[2] ⓘ

[1] Graduate School of Engineering, Fukuoka Institute of Technology,
3-30-1 Wajiro-Higashi, Higashi-Ku, Fukuoka 811-0295, Japan
{mgm22101,mgm23107}@bene.fit.ac.jp
[2] Department of Information and Communication Engineering, Fukuoka Institute
of Technology, 3-30-1 Wajiro-Higashi, Higashi-Ku, Fukuoka 811-0295, Japan
makoto.ikd@acm.org, barolli@fit.ac.jp

Abstract. In general, the accidents are often attributed to the decline of driving skills and reduced attention among the aging population. Moreover, the rise of infotainment features in vehicles has led to loss of concentration while driving. To address this issue, we previously developed an intelligent driver assistance system to identify instances of distracted driving. This paper focuses on enhancing the detection accuracy of distracted driving by utilizing an upgraded version of our system equipped with different detectors. The system is designed to recognize various distracted driving behaviors, particularly focusing on the driver hand movements. From the evaluation, we observed that while models like YOLOv8 show improvements in detecting specific actions like steering wheel and IVI operations, challenges remain in accurately identifying other crucial activities such as cell phone usage and general hand movements.

Keywords: Distracted driving behaviors · Classification · YOLO

1 Introduction

Numerous smartphone applications and in-vehicle infotainment systems, with their services, have increased the risk of driver distraction. Additionally, the progress in multimedia and automated driving technologies has led to an over-reliance and false security in these systems.

Modern vehicles are equipped with various devices and network connections, each serving specific functions [13], which can easily divert the driver attention. Therefore, there are needed sophisticated driver assistance systems and vehicle control technologies to avert accidents [3,6,9,12,18,27]. To meet these needs, both cost-effective devices and reliable distracted driving detection systems are essential.

Artificial Intelligence (AI)-based systems have gained considerable interest across various sectors [5,11,16,24,28]. In edge-focused AI systems, routine training is conducted in the cloud, while real-time detection takes place at the edge [15,19,20,26]. The availability of open datasets, models, and platforms has expedited app development by reducing the time required for application creation [1,2]. In [22], a dataset of

© The Author(s), under exclusive license to Springer Nature Switzerland AG 2024
L. Barolli (Ed.): EIDWT 2024, LNDECT 193, pp. 362–370, 2024.
https://doi.org/10.1007/978-3-031-53555-0_34

distracted driving featuring ten categories is available. However, this dataset is outdated and doesn't consider the impact of vehicles equipped with advanced driver assistance systems or the large displays found in electric vehicles.

In our previous work [14], we proposed an intelligent distracted driving detection system. However, this system only considers YOLOv5m model.

In this paper, we assess the effectiveness of four models implemented in our intelligent driver assistance system, particularly for identifying distracted driving behaviors while moving. We observed that while models like YOLOv8 show improvements in detecting specific actions like steering wheel and IVI operations, challenges remain in accurately identifying other crucial activities such as cell phone usage and general hand movements.

The structure of this paper is organized as follows: Sect. 2 provides an overview of the YOLO series. Section 3 details our method for assisting drivers in detecting distracted behaviors. The evaluation results are discussed in Sect. 4. Finally, the paper concludes with Sect. 5.

2 DNNs and Evolution of YOLO Algorithms

Recently, the applications of Deep Neural Networks (DNNs) have gained significant attention [7, 10]. DNNs are multi-layered networks with intricate hierarchies for feature detection and learning data representations. Representation learning in real-world scenarios involves extracting key information from the observed data [21].

The original YOLO (You Only Look Once) algorithm was developed by Joseph Redmon and his team [17]. But, he stopped the development due to concerns over military applications and privacy issues. The development was then continued by Alexey Bochkovskiy, who introduced YOLOv4 [4] in April 2020. Following this, Glenn Jocher and his team released YOLOv5 [8] in June 2020, which is built on the PyTorch machine learning library. The YOLOv5 release v7.0 includes the instance segmentation feature. The YOLO series further expanded with the announcements of YOLOv6 and YOLOv7 in 2022, and the release of YOLOv8 [25] in January 2023.

The YOLOv8 series offers multiple models, each designed for certain computer vision tasks. These models perform well in addressing a wide range of requirements, covering everything from basic object detection to more advanced tasks including instance segmentation, pose and keypoint recognition, and classification.

In YOLO approach, the entire image is processed as input, divided into a grid, and predictions are made for the entire image in a single step. This approach effectively minimizes background-related errors by leveraging environmental information comprehensively.

3 AI-Enhanced Driver Assistance System

3.1 System Architecture and Design

Figure 1 illustrates the model of our AI-enhanced driver support system. This system consists of an AI-driven application that detects distracted driving behaviors and alerts the driver, as well as an intelligent driving assistant designed specifically for use in vehicles.

For efficient system training, we incorporated an edge computing device, namely the Jetson Xavier NX. This device is internet-connected and employs Web scraping methods to gather training images. We are working on integrating a mechanism that allows periodic in-vehicle training of images when the device is offline, aiming to enhance the training accuracy. The object detection tasks utilize various YOLO models, specifically from v5 and v8 series. These object detection models have good accuracy, accelerated learning rates and minimized overfitting.

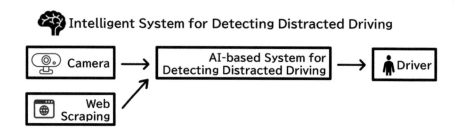

Fig. 1. System model.

3.2 Categorization of Driving Behaviors

In driving scenarios, it is common that both hands are not always on the steering wheel simultaneously and hand positions often vary while driving. Our method categorizes driving behavior into six separate classifications.

1. In-Vehicle Infotainment (IVI) Interaction
2. Steering Wheel Operation with Hands
3. Cell Phone Use with Hands
4. General Hand Movements
5. Operating IVI with Hands
6. False Positives (Predicted) and False Negatives (Actual)

3.3 Dataset Compilation and Model Training for Behavior Detection

In our research, we collect images for both normal and distracted behaviors to assemble a distinct dataset, which includes a total of 1,136 images [23]. We chose YOLO models v5m, v5l, v8m, and v8l for the network model. These models are used over up to 200 epochs. The 'm' and 'l' in each series indicate different parameter sizes, with 'l' being the larger one. The images are processed at a resolution of 960 and the training is conducted in batches of 8.

4   Evaluation Results

4.1   Classification Results and PR Curves

In this section, we analyze the classification results for different models and Precision-Recall (PR) curves of distracted driving detection for batch size 8 and 200 epochs.

Figure 2 displays the classification accuracy of our system for different models. The horizontal axis indicates the actual labels and the vertical axis shows the predicted labels. Accurate detection is achieved when the labels on both axes align. A False Negative (FN) occurs when a positive instance is wrongly identified as negative, whereas a False Positive (FP) occurs when a negative instance is incorrectly labeled as positive. For instance, a background FP happens when an IVI is erroneously identified as if it is present, despite its actual absence. Conversely, a background FN takes place when an IVI that should be detected is overlooked.

For YOLOv5, the v5l model improves the detection accuracy of steering wheel operations by 8% without reducing the accuracy of IVI and cell phone operations. However, the accuracy for general hand detection and operating IVI with hands decreases by 6% and 3%, respectively.

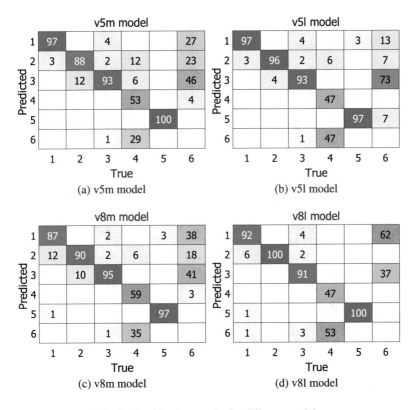

Fig. 2. Classification results for different models.

For YOLOv8, the v8l model shows improved performance in three classes compared to v8m, achieving 100% detection accuracy for steering wheel operations and IVI operations. However, there is a slight decrease in the accuracy for detecting cell phone operations and general hand movements. Overall, the model improves the performance by reducing misclassifications for other classes.

Figure 3 presents PR results, illustrating the relationship between precision and recall rates. In these figures, the precision is plotted on the vertical axis while recall is on the horizontal axis. There is a trade-off between precision and recall, where an increase in recall values typically leads to a decrease in precision.

In PR curve of hand movements, represented by the red line, the PR curve value experiences a sharp decline as recall increases. This pattern is especially prominent with the use of YOLOv5 models. On the other hand, the smoothest behavior of the PR curve is observed when using v8m model. When comparing other classes excluding hand detection, it becomes clear that v8l model exhibits the highest performance.

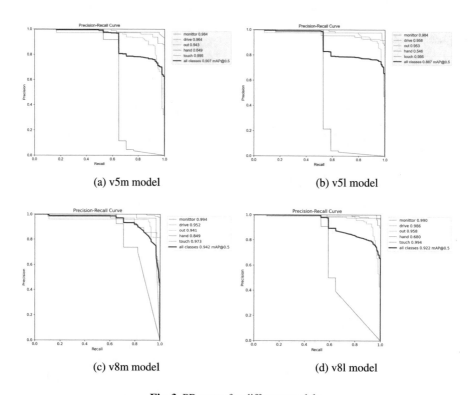

Fig. 3. PR curve for different models.

4.2 Comparative Study of Detected Cases for Four Models

In this section, we compare the performance of four models for distracted driving behaviors using multiple videos or images. The detection results are outputted by model

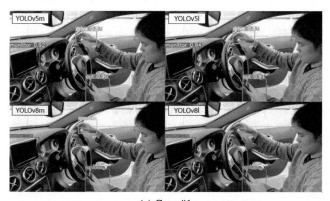

(a) Case #1

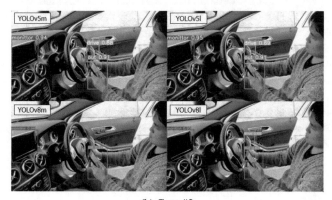

(b) Case #2

(c) Case #3

Fig. 4. Detected cases for different models.

in video format, so we are converting them using ffmpeg to allow frame-by-frame comparison and verification. The text above the rectangular frames in the images indicates the class label and accuracy.

From the results shown in Fig. 4(a), it is observed that YOLOv5 series mistakenly detects the action as driving just before the right hand grips the steering wheel, while YOLOv8m identifies it as a free hand correctly. On the other hand, in Fig. 4(b), YOLOv8m fails to detect the action of the right hand gripping the steering wheel. In Fig. 4(c), which shows an image of the steering wheel being turned, YOLOv8 series correctly identifies the action. These observations indicate that there are still challenges in identification accuracy.

5   Conclusions

In this paper, we presented a comparative study of four YOLO models for distracted driving detection. We have analyzed the classification accuracy, PR curves and practical application of these models in real-world scenarios.

We observed that while models like YOLOv8 show improvements in detecting specific actions like steering wheel and IVI operations, challenges remain in accurately identifying other crucial activities, such as cell phone usage and general hand movements. The comparative analysis indicates different performance of four models, each displaying strengths and weaknesses in different distracted driving scenarios.

In our future work, we would like to address these issues by incorporating tuning and optimization techniques.

References

1. Kaggle: Data science community. https://www.kaggle.com/
2. Roboflow: The world's largest collection of open source computer vision datasets and APIs. https://universe.roboflow.com/
3. Bergasa, L.M., Almeria, D., Almazan, J., Yebes, J.J., Arroyo, R.: DriveSafe: an app for alerting inattentive drivers and scoring driving behaviors. In: Proceedings of the IEEE Intelligent Vehicles Symposium 2014, pp. 240–245 (2014). https://doi.org/10.1109/IVS.2014.6856461
4. Bochkovskiy, A., Wang, C.Y., Liao, H.Y.M.: YOLOv4: optimal speed and accuracy of object detection. Computer Vision and Pattern Recognition (cs.CV), April 2020. https://arxiv.org/abs/2004.10934
5. Chen, G., et al.: NeuroIV: neuromorphic vision meets intelligent vehicle towards safe driving with a new database and baseline evaluations. IEEE Trans. Intell. Transp. Syst. 23(2), 1171–1183 (2022). https://doi.org/10.1109/TITS.2020.3022921
6. Ersal, T., Fuller, H.J.A., Tsimhoni, O., Stein, J.L., Fathy, H.K.: Model-based analysis and classification of driver distraction under secondary tasks. IEEE Trans. Intell. Transp. Syst. 11(3), 692–701 (2010). https://doi.org/10.1109/TITS.2010.2049741
7. Hinton, G.E., Osindero, S., Teh, Y.W.: A fast learning algorithm for deep belief nets. Neural Comput. 18(7), 1527–1554 (2006)
8. Jocher, G.: The project page of Ultralytics YOLOv5 (2020). https://github.com/ultralytics/yolov5/wiki

9. Kandeel, A.A., Elbery, A.A., Abbas, H.M., Hassanein, H.S.: Driver distraction impact on road safety: a data-driven simulation approach. In: Proceedings of the IEEE Global Communications Conference (GLOBECOM-2021), pp. 1–6, December 2021. https://doi.org/10.1109/GLOBECOM46510.2021.9685932

10. Le, Q.V.: Building high-level features using large scale unsupervised learning. In: Proceedings of IEEE International Conference on Acoustics, Speech and Signal Processing 2013 (ICASSP-2013), pp. 8595–8598, May 2013. https://doi.org/10.1109/ICASSP.2013.6639343

11. Li, B., et al.: A new unsupervised deep learning algorithm for fine-grained detection of driver distraction. IEEE Trans. Intell. Transp. Syst. **23**, 1–13 (2022). https://doi.org/10.1109/TITS.2022.3166275

12. Liu, T., Yang, Y., Huang, G.B., Yeo, Y.K., Lin, Z.: Driver distraction detection using semi-supervised machine learning. IEEE Trans. Intell. Transp. Syst. **17**(4), 1108–1120 (2016). https://doi.org/10.1109/TITS.2015.2496157

13. McCall, J.C., Trivedi, M.M.: Driver behavior and situation aware brake assistance for intelligent vehicles. Proc. IEEE **95**(2), 374–387 (2007). https://doi.org/10.1109/JPROC.2006.888388

14. Miwata, M., Tsuneyoshi, M., Ikeda, M., Barolli, L.: Performance evaluation of an AI-based safety driving support system for detecting distracted driving. In: Proceedings of the 16th International Conference on Innovative Mobile and Internet Services in Ubiquitous Computing (IMIS-2022), pp. 10–17, June 2022. https://doi.org/10.1007/978-3-031-08819-3_2

15. Mnih, V., et al.: Human-level control through deep reinforcement learning. Nature **518**, 529–533 (2015). https://doi.org/10.1038/nature14236

16. Poon, Y.S., Lin, C.C., Liu, Y.H., Fan, C.P.: YOLO-based deep learning design for in-cabin monitoring system with fisheye-lens camera. In: Proceedings of the IEEE International Conference on Consumer Electronics (ICCE-2022), pp. 1–4, January 2022. https://doi.org/10.1109/ICCE53296.2022.9730235

17. Redmon, J., Divvala, S., Girshick, R., Farhadi, A.: You only look once: unified, real-time object detection. In: Proceedings of the IEEE Conference on Computer Vision and Pattern Recognition (CVPR-2016), pp. 779–788, June 2016. https://doi.org/10.1109/CVPR.2016.91

18. Shaout, A., Roytburd, B., Sanchez-Perez, L.A.: An embedded deep learning computer vision method for driver distraction detection. In: Proceedings of the 22nd International Arab Conference on Information Technology (ACIT-2021), pp. 1–7, December 2021. https://doi.org/10.1109/ACIT53391.2021.9677045

19. Silver, D., et al.: Mastering the game of Go with deep neural networks and tree search. Nature **529**, 484–489 (2016). https://doi.org/10.1038/nature16961

20. Silver, D., et al.: Mastering the game of Go without human knowledge. Nature **550**, 354–359 (2017). https://doi.org/10.1038/nature24270

21. Simonyan, K., Zisserman, A.: Very deep convolutional networks for large-scale image recognition. In: Proceedings of the 3rd International Conference on Learning Representations (ICLR-2015), May 2015. https://doi.org/10.48550/arXiv.1409.1556

22. State Farm: Dataset of state farm distracted driver detection (2016). https://www.kaggle.com/c/state-farm-distracted-driver-detection/

23. Tanaka, H., Miwata, M., Ikeda, M., Barolli, L.: An enhanced AI-based vehicular driver support system considering hyperparameter optimization. In: Proceedings of the 17th International Conference on Innovative Mobile and Internet Services in Ubiquitous Computing (IMIS-2023), pp. 1–7, July 2023. https://doi.org/10.1007/978-3-031-35836-4_1

24. Ugli, I.K.K., Hussain, A., Kim, B.S., Aich, S., Kim, H.C.: A transfer learning approach for identification of distracted driving. In: Proceedings of the 24th International Conference on Advanced Communication Technology (ICACT-2022), pp. 420–423, February 2022. https://doi.org/10.23919/ICACT53585.2022.9728846

25. Ultralytics: The project page of Ultralytics YOLOv8 (2023). https://github.com/ultralytics/ultralytics

26. Vicente, F., Huang, Z., Xiong, X., la Torre, F.D., Zhang, W., Levi, D.: Driver gaze tracking and eyes off the road detection system. IEEE Trans. Intell. Transp. Syst. **16**(4), 2014–2027 (2015). https://doi.org/10.1109/TITS.2015.2396031

27. Wang, Y.K., Jung, T.P., Lin, C.T.: EEG-based attention tracking during distracted driving. IEEE Trans. Neural Syst. Rehabil. Eng. **23**(6), 1085–1094 (2015). https://doi.org/10.1109/TNSRE.2015.2415520

28. Xing, Y., Lv, C., Wang, H., Cao, D., Velenis, E., Wang, F.Y.: Driver activity recognition for intelligent vehicles: a deep learning approach. IEEE Trans. Veh. Technol. **68**(6), 5379–5390 (2019). https://doi.org/10.1109/TVT.2019.2908425

An Intelligent System Based on CCM, SA and FDTD for Sensor Node Placement Optimization in Wireless Visual Sensor Networks

Yuki Nagai[1], Tetsuya Oda[2(✉)], Chihiro Yukawa[1], Kyohei Wakabayashi[1], and Leonard Barolli[3]

[1] Graduate School of Engineering, Okayama University of Science (OUS), 1-1 Ridaicho, Kita-ku, Okayama 700-0005, Japan
{t22jm23rv,t22jm19st,t22jm24jd}@ous.jp
[2] Department of Information and Computer Engineering, Okayama University of Science (OUS), 1-1 Ridaicho, Kita-ku, Okayama 700-0005, Japan
oda@ous.ac.jp
[3] Department of Information and Communication Engineering, Fukuoka Insitute of Technology, 3-30-1 Wajiro-Higashi-ku, Fukuoka 811-0295, Japan
barolli@fit.ac.jp

Abstract. A Wireless Visual Sensor Network (WVSN) has a mesh mechanism and performs multi-hop communication between multiple sensor nodes equipped with visual sensors. They can be used to monitor infrastructure facilities and rivers. However, the placement of sensor nodes has a significant impact on wireless communication connectivity and transmission loss. In addition, the imaging range of visual sensors is limited, which limits the area that can be imaged by the WVSN. Therefore, to cover all events with the imageable range of the WVSN, it is necessary to decide the optimal placement of sensor nodes in the WVSN and the imaging direction of the visual sensors. Since sensor nodes may be densely placed, it is needed to extend the communication range of sensor nodes and the range over which visual sensors are able to image the entire target area. In this paper, we propose an intelligent system based on CCM, SA and Finite Difference Time Domain Method (FDTD) for sensor node placement optimization in WVSN. We evaluate the proposed system by computer simulations. The simulation results show that the SGC is always maximized and all sensor nodes are connected for all methods. The NCE for CCM at the end of iterations covers 37 [$unit$] events, while for CCM-based SA and AEFSEM all 48 [$unit$] events are covered. For AEFSEM the sensor nodes are spread over a wider area and all events are covered in the imaging range of visual sensors. The ARSE for CCM-based SA at the end of iterations is 57 [%], while for AEFSEM is 72 [%], which shows an improvement of about 15 [%] and the stable communication area is increased.

© The Author(s), under exclusive license to Springer Nature Switzerland AG 2024
L. Barolli (Ed.): EIDWT 2024, LNDECT 193, pp. 371–383, 2024.
https://doi.org/10.1007/978-3-031-53555-0_35

1   Introduction

Wireless Visual Sensor Network (WVSN) is a network that has a mesh mechanism and communicates between multiple sensor nodes equipped with visual sensors through multi-hop communication [1–5]. Therefore, even if some network links fail, the WVSN can continue to communicate via other sensor nodes, enabling stable wireless communication and data collection in a wide area. Since the sensor nodes of the WVSN are equipped with visual sensors and can collect moving images, the WVSN is expected to monitor infrastructure facilities and rivers. However, the placement of sensor nodes has a significant impact on the connectivity and transmission loss of wireless communication [6–11]. Therefore, deciding the placement of sensor nodes is a very important issue [12–16].

We previously proposed the Coverage Construction Method (CCM) [17], CCM-based Hill Climbing (HC) [18] and CCM-based Simulated Annealing (SA) [19, 20] to deal with this problem. In the proposed methods, the imaging range of the visual sensor is limited because the imaging range and viewing angle are decided by the lens and used sensor size. Therefore, it is essential to decide the optimal placement of sensor nodes and the imaging direction of the visual sensor in order that the imaging range of visual sensor cover all events. However, with these methods, sensor nodes may be densely placed and even if they can cover all events, the range in which they can communicate over the entire subject area may be narrow. By maintaining a distance for stable communication and spreading the interval the distance between sensor nodes, it is possible to increase the communication range of sensor nodes and the range in which visual sensors can take images in the entire subject area.

In order to improve the previous proposed methods, we proposed Average Electric Field Strength Expansion Method (AEFSEM) [21], which considers radio wave propagation by electromagnetic wave analysis. In Finite Difference Time Domain (FDTD) method [22–25] the space is divided into a mesh of small rectangular or cubic elements, called cells. The electromagnetic field in each cell is updated sequentially at each time according to equations formulated based on Faraday's law and Ampere's law. In AEFSEM, the FDTD method is used for electromagnetic wave analysis to evaluate the stability of wireless communication.

In this paper, we propose an intelligent system for sensor node placement optimization in WVSN that considers radio propagation by integrating CCM, CCM-based SA and AEFSEM. We evaluate the proposed system by computer simulations. The simulation results show that the SGC is always maximized and all sensor nodes are connected for all methods. The NCE for CCM at the end of iterations covers 37 [unit] events, while for CCM-based SA and AEFSEM all 48 [unit] events are covered. For AEFSEM the sensor nodes are spread over a wider area and all events are covered in the imaging range of visual sensors. The ARSE for CCM-based SA at the end of iterations is 57 [%], while for AEFSEM is 72 [%], which shows an improvement of about 15 [%] and the stable communication area is increased.

The paper is organized as follows. In Sect. 2, we present the proposed intelligent system. In Sect. 3, we show the simulation results. Finally, in Sect. 4, we conclude the paper.

2 Proposed Intelligent System

We show the image of the proposed system in Fig. 1, while Fig. 2 shows its flowchart. The proposed system decides the optimal placement of sensor nodes and the imaging direction of visual sensors based on CCM and CCM-based SA to maximize the Size of Giant Component (SGC) and maximize the Number of Covered Events (NCE) by the imaging range of the WVSN. In addition, the placement of sensor nodes is decided by considering the electric field strength based on radio wave propagation by electromagnetic waves analysis using FDTD method, thus maximizing the stable communication range in the simulation area.

2.1 CCM

In CCM, at the beginning one sensor node is allocated by randomly generating one coordinate and the imaging direction of a visual sensor. This procedure is carried out again to allocate another sensor node. Each sensor node considers the wireless communication range as a circle and performs collision detection with the previously generated sensor node. When a collision with a placed sensor node is detected, a new sensor node is generated. This process continues until the number of sensor nodes reaches a predetermined number or is repeated until an arbitrary number of iterations is achieved. The solution with the highest NCE among the generated solutions is the optimal solution for CCM and the placement of sensor nodes and the imaging direction of the visual sensor are decided.

2.2 CCM-Based SA

CCM-based SA method decides the optimal solution for the sensor node placement by CCM and the imaging direction of visual sensor. Then the SA updates the solution according to the State Transition Probability (STP), which consists of the solution evaluation and temperature.

In CCM-based SA, one sensor node is randomly selected and allocated within the simulation area. Then, the imaging direction of the visual sensor for this node is rotated within an arbitrary range of angles to change the imaging direction and update the solution. After the SGC is maximum, the NCE is calculated. If NCE is increased, then the placement of sensor node and the imaging direction of visual sensor are updated. If the NCE is decreasing or has not been updated, the placement of the sensor nodes and the imaging direction of visual sensor are changed according to STP and the solution is updated. This process is repeated until an arbitrary number of iterations. The solution with the highest NCE at the end of the number of iterations is the optimal solution and the placement of the sensor nodes and the imaging direction of visual sensor are decided.

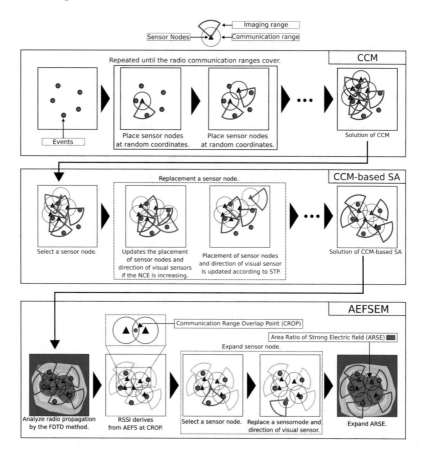

Fig. 1. Structure of proposed system.

2.3   Average Electric Field Strength Expansion Method

Figure 3 shows the flowchart of the FDTD method. The proposed method decides the optimal placement of sensor nodes for a wide area coverage and stable wireless communication. In addition to SGC and NCE, the objective is to have a good RSSI at the Communication Range Overlap Points (CROP) of Area Ratio of Strong Electric field (ARSE).

The ARSE is the rate of the area where Average Electric Field Strength (AEFS) is larger than a given value over the simulation area. The AEFS is the average of the field strength at each time step. While RSSI is measured as AEFS of the CROP of sensor nodes. The proposed method maximizes ARSE and decides the optimal placement of sensor nodes whose RSSI for each CROP is closest to the target value. The target value of RSSI of CROP is set to a value that stabilizes the communication between sensor nodes.

The FDTD method is a numerical analysis technique for solving Maxwell's equations in time domain as shown in Eq. (1), Eq. (2), Eq. (3) and Eq. (4).

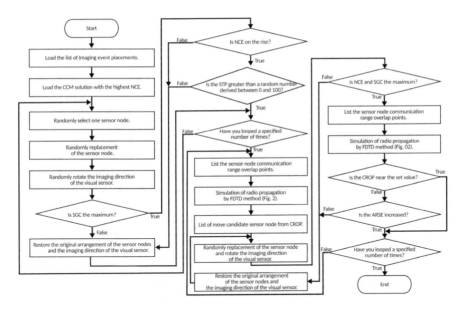

Fig. 2. Flowchart of proposed system.

$$\nabla \times E = -\delta B/\delta t \tag{1}$$

$$\nabla \times H = J + \delta D/\delta t \tag{2}$$

$$\nabla \cdot D = \rho \tag{3}$$

$$\nabla \cdot B = 0 \tag{4}$$

The equations are discretized in time and space with the central difference approximation and the values of the electric and magnetic fields are updated at each time step in two dimensions by Eq. (5) and Eq. (6).

$$
\begin{aligned}
E_z^n(i,j) =& \frac{1 - \frac{\sigma \Delta t}{2\varepsilon}}{1 + \frac{\sigma \Delta t}{2\varepsilon}} E_z^{n-1}(i,j) \\
& + \frac{\delta t/\varepsilon}{1 + \frac{\sigma \Delta t}{2\varepsilon}} \frac{1}{\delta x} \{ H_y^{n-\frac{1}{2}}(i+\frac{1}{2},j) - H_y^{n-\frac{1}{2}}(i,j-\frac{1}{2}) \} \\
& - \frac{\delta t/\varepsilon}{1 + \frac{\sigma \Delta t}{2\varepsilon}} \frac{1}{\delta y} \{ H_x^{n-\frac{1}{2}}(i+\frac{1}{2},j) - H_x^{n-\frac{1}{2}}(i,j-\frac{1}{2}) \}
\end{aligned}
\tag{5}
$$

$$
\begin{aligned}
H_z^{n+\frac{1}{2}}(i+\frac{1}{2},j+\frac{1}{2}) =& H_z^{n+\frac{1}{2}}(i+\frac{1}{2},j+\frac{1}{2}) \\
& - \frac{\delta t}{\mu} \frac{1}{\delta x} \{ E_y^n(i+1,j+\frac{1}{2}) - E_y^n(i,j+\frac{1}{2}) \} \\
& + \frac{\delta t}{\mu} \frac{1}{\delta y} \{ E_x^n(i+\frac{1}{2},j+1) - E_x^n(i+\frac{1}{2},j) \}
\end{aligned}
\tag{6}
$$

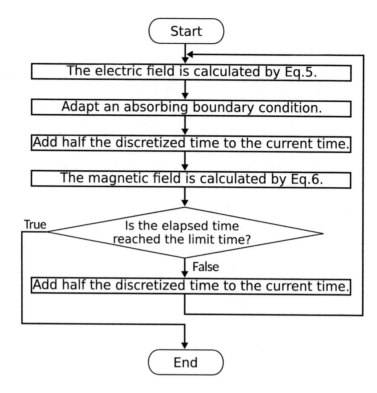

Fig. 3. Flowchart of FDTD method.

In the proposed method, the simulation area is considered as an open area. A second-order Mur Absorption Boundary Condition (ABC) [26] is set, which considers oblique incident waves of Mur ABC. In the FDTD method, the ABC is applied after the electric field and magnetic field calculations are performed. This process is repeated for multiple time steps to simulate the behavior of electromagnetic waves.

The AEFSEM uses as initial solution, the solution of the CCM-based SA. The AEFS is derived by the FDTD method after deciding sensor node placement. After calculating the AEFS, one CROP is randomly selected for the RSSI of each CROP outside the target value and one of the sensor nodes constituting the selected CROP is replaced. In addition, the imaging direction of the visual sensor of the selected sensor node is rotated within an arbitrary range of angles to change the imaging direction. When the SGC after replacing the sensor node is not maximum or the NCE decreases, the sensor node is returned to the state before replacement and the CROP is selected again. If the SGC after relocation is maximum and the NCE is unchanged or increases, the RSSI of the AEFS and CROP with the relocated sensor node are derived based on the FDTD method. If the RSSI of CROP after relocating the sensor node approaches the target value or ARSE increases, the solution is updated and the CROP is selected again.

This process is repeated until an arbitrary number of iterations and the sensor node placement with the highest ARSE is considered as the optimal solution.

3 Simulation Results

We evaluated the performance of the proposed system by computer simulations. In Table 2 are shown the simulation parameters. Event placement is normally distributed within the simulation area. The moving distance of the sensor node at the time of relocation is set in the range of ±5 $[m]$ in the X and Y coordinate direction from the coordinates of the sensor node. The rotation angle to change the imaging direction of the visual sensor is set in the range of ±90 $[deg.]$. The maximum imaging range and angle of view of the visual sensor are set to 20 $[m]$ for the maximum imaging range and 90 $[deg.]$ is set for the angle of view referring to a typical camera and lens.

In Table 1 are shown the parameters of FDTD method. The time domain interval of FDTD method is calculated based on Courant's stability conditions. The ARSE derives the region where AEFS is above −65 $[dBm]$ considered as the strong electric field. The target value of RSSI for CROP to select the sensor node to be replaced is set within ±3 $[dBm]$ with −65 $[dBm]$ as the reference value.

Figure 4, Fig. 5 and Fig. 6 show the simulation results of CCM, CCM-based SA and AEFSEM. While Fig. 4(a), Fig. 5(a) and Fig. 6(a) show the NCMC and SGC of the simulation results of CCM, CCM-based SA and AEFSEM. In Fig. 4(b), Fig. 5(b) and Fig. 6(b) are shown ARSE of the simuration results of CCM, CCM-based SA and AEFSEM.

From Fig. 4(a) and Fig. 5(a), we see that SGC is always maximized 16 $[unit]$ and NCE is 48 $[unit]$, which show that all events are covered with the imaging range of visual sensor. Also, in Fig. 6(a), the SGC and NCE of AEFSEM are always maximized. In Fig. 6(b) ARSE is about 72 [%], which shows an improvement of about 15 [%] from the initial solution (57 [%]). From these simulation results, it can be seen that all events are covered by the imaging range of the visual sensor and the stable communication area is increased.

In Fig. 7, Fig. 8 and Fig. 9 are shown the visualization results of CCM, CCM-based SA and AEFSEM. The ARSE of AEFSEM in Fig. 9 is about 72 [%], which shows the AEFS of the simulated area. From Fig. 7 and Fig. 8, it can be seen the area where the sensor nodes are densely located. On the other hand, in Fig. 9 can be seen that the densely allocated sensor nodes are placed in a wide area and the communication range is expanded to cover all events in the imaging range of the visual sensor.

Table 1. Parameters and values for FDTD method.

Functions	Values
Spatial Discretization Width	1 $[m]$
Speed of Light	2.998×10^8 $[m/s]$
Time Domain Interval	7.586×10^{-5} $[s]$
Frequency	2.4 $[GHz]$
Permittivity	8.854×10^{-12} $[F/m]$
Magnetic Permeability	1.257×10^{-6} $[H/m]$
Electric Power	1 $[mW]$
Number of Loops	100 $[unit]$

Table 2. Parameters and values for simulations.

Functions	Values
Width of Simulation Area	100 $[m]$
Hight of Simulation Area	100 $[m]$
Number of Visual Sensors	16
Radius of Radio Communication Range of Mesh Routers	10 $[m]$
Number of Events	48
Number of Loops for CCM	3000 $[unit]$
Number of Loops for CCM-based SA	10000 $[unit]$
Number of Loops for ARSE Expansion Method	1000 $[unit]$
Imaging Range	20 $[m]$
View Angle	90 $[deg.]$

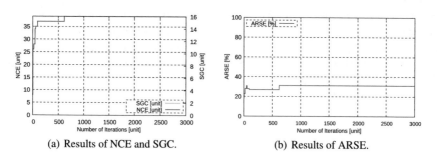

(a) Results of NCE and SGC.   (b) Results of ARSE.

Fig. 4. Simulation results of CCM.

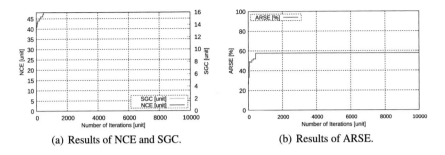

(a) Results of NCE and SGC.   (b) Results of ARSE.

Fig. 5. Simulation results of CCM-based SA.

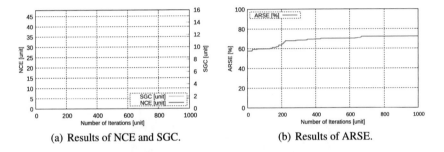

(a) Results of NCE and SGC.   (b) Results of ARSE.

Fig. 6. Simulation results of AEFSEM.

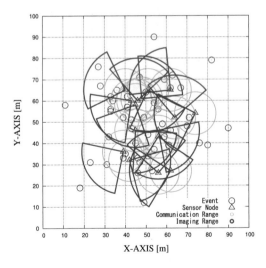

Fig. 7. Visualization results of CCM.

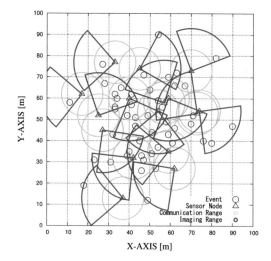

Fig. 8. Visualization results of CCM-based SA.

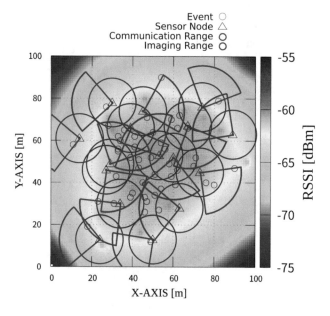

Fig. 9. Visualization results of AEFSEM.

4   Conclusions

In this paper, we propose an intelligent system for sensor node placement optimization in WVSN, which integrates CCM, CCM-based SA and AEFSEM by considering the number of events within the imaging range of visual sensor, placement of sensor nodes and the imaging direction of visual sensors.

From the simulation results, we conclude as follows.

- The SGC is always maximized and all sensor nodes are connected for all methods.
- The NCE for CCM at the end of iterations covers 37 [unit] events, while for CCM-based SA and AEFSEM all 48 [unit] events are covered.
- For AEFSEM the sensor nodes are spread over a wider area and all events are covered in the imaging range of visual sensors.
- The ARSE for CCM-based SA at the end of iterations is 57 [%], while for AEFSEM is 72 [%], which shows an improvement of about 15 [%] and the stable communication area is increased.

In the future, we would like to simulate different scenarios and consider events placed in a 3-dimensional area.

Acknowledgement. This work was supported by JSPS KAKENHI Grant Number JP20K19793.

References

1. Idoudi, M., et al.: Wireless visual sensor network platform for indoor localization and tracking of a patient for rehabilitation task. IEEE Sens. J. **18**(14), 5915–5928 (2018)
2. Peng-Fei, W.: Node scheduling strategies for achieving full-view area coverage in camera sensor networks. Sensors **17**(6), 1303–1321 (2017)
3. Akyildiz, I.F., et al.: Wireless mesh networks: a survey. Comput. Netw. **47**(4), 445–487 (2005)
4. Jun, J., et al.: The nominal capacity of wireless mesh networks. IEEE Wirel. Commun. **10**(5), 8–15 (2003)
5. Oyman, O., et. al.: Multihop relaying for broadband wireless mesh networks: from theory to practice. IEEE Commun. Magaz. **45**(11), 116–122 (2007)
6. Oda, T., et al.: A GA-based simulation system for WMNs: performance analysis of WMN-GA system for different WMN architectures considering DCF and EDCA. In: Proceedings of The 7th International Conference on Intelligent Networking and Collaborative Systems (INCoS-2015), pp. 232–238 (2015)
7. Oda, T., et. al.: Analysis of node placement in wireless mesh networks using Friedman test: a comparison study for Tabu search and Hill Climbing. In: Proceedings of The 9th International Conference on Innovative Mobile and Internet Services in Ubiquitous Computing (IMIS-2015), pp. 133–140 (2015)
8. Oda, T., et al.: Analysis of mesh router placement in wireless mesh networks using Friedman test considering different meta-heuristics. Int. J. Commun. Netw. Distrib. Syst. **15**(1), 84–106 (2015)
9. Oda, T., et al.: Effects of population size for location-aware node placement in WMNs: evaluation by a genetic algorithm-based approach. Pers. Ubiquit. Comput. **18**, 261–269 (2014)

10. T. Oda, et al.: Analysis of mesh router node placement using WMN-GA system considering different architectures of WMNs. In: Proceedings of The 17th International Conference on Network-Based Information Systems (NBiS-2014), pp. 39–44 (2014)

11. Sakamoto, S., et al.: Performance analysis of two wireless mesh network architectures by WMN-SA and WMN-TS simulation systems. J. High Speed Netw. **23**(4), 311–322 (2017)

12. Oda, T., et al.: WMN-GA: a simulation system for WMNs and its evaluation considering selection operators. J. Ambient. Intell. Humaniz. Comput. **4**(3), 323–330 (2013)

13. Ikeda, M., et al.: Analysis of WMN-GA simulation results: WMN performance considering stationary and mobile scenarios. In: Proceedings of The 28th IEEE International Conference on Advanced Information Networking and Applications (IEEE AINA-2014), pp. 337–342 (2014)

14. Oda, T., et al.: Analysis of mesh router placement in wireless mesh networks using Friedman test considering different meta-heuristics. Int. J. Commun. Netw. Distrib. Syst. **15**(1), 84–106 (2015)

15. Oda, T., et al.: A genetic algorithm-based system for wireless mesh networks: analysis of system data considering different routing protocols and architectures. Soft. Comput. **20**(7), 2627–2640 (2016)

16. Sakamoto, S., et al.: Performance evaluation of intelligent hybrid systems for node placement in wireless mesh networks: a comparison study of WMN-PSOHC and WMN-PSOSA. In: Proceedings of The 11th International Conference on Innovative Mobile and Internet Services in Ubiquitous Computing (IMIS-2017), vol. 612, pp. 16–26 (2017)

17. Hirata, A., et al.: Approach of a solution construction method for mesh router placement optimization problem. In: Proceedings of The IEEE 9th Global Conference on Consumer Electronics (IEEE GCCE-2020), pp. 1–2 (2020)

18. Hirata, A., et al.: A coverage construction method based hill climbing approach for mesh router placement optimization. In: Proceedings of the 15th International Conference on Broadband and Wireless Computing, Communication and Applications (BWCCA-2020), vol. 159, pp. 355–364 (2020)

19. Hirata, A., et al.: A voronoi edge and CCM-based SA approach for mesh router placement optimization in WMN: a comparison study for different edges. In: Proceedings of The 36th International Conference on Advanced Information Networking and Applications (AINA-2022), vol. 3, pp. 220–231 (2022)

20. Oda, T.: A delaunay edges and simulated annealing-based integrated approach for mesh router optimization in wireless mesh networks. Sensors **23**(3), 1050 (2023)

21. Nagai, Y.: A CCM, SA and FDTD based mesh router placement optimization in WMN. In: Proceedings of the 17th International Conference on Complex, Intelligent, and Software Intensive Systems, vol. 176, pp. 48–58 (2023)

22. Yee, K.S., Chen, J.S.: The finite-difference time-domain (FDTD) and the finite-volume time-domain (FVTD) methods in solving Maxwell's equations. IEEE Trans. Antennas Propag. **45**(3), 354–363 (1997)

23. Hwang, K.-P., Cangellaris, A.C.: Effective permittivities for second-order accurate FDTD equations at dielectric interfaces. IEEE Microw. Wirel. Compon. Lett. **11**(4), 158–160 (2001)

24. Mahmoud, K.R., Montaser, A.M.: Design of compact mm-wave tunable filtenna using capacitor loaded trapezoid slots in ground plane for 5G router applications. IEEE Access **8**, 27715–27723 (2020)

25. Chen, M., Wei, B., et al.: FDTD complex terrain modeling method based on papery contour map. Int. J. Anten. Propagat. **2023**, 1–10 (2023)
26. Gan, T.H., Tan, E.L.: Mur absorbing boundary condition for 2-D leapfrog ADI-FDTD method. In: Proceedings of the 1st IEEE Asia-Pacific Conference on Antennas and Propagation, pp. 3–4 (2012)

Performance Evaluation of a Cuckoo Search Based System for Node Placement Problem in Wireless Mesh Networks: Evaluation Results for Computation Time and Different Numbers of Nests

Shinji Sakamoto[1]([⊠]), Leonard Barolli[2], and Makoto Takizawa[3]

[1] Department of Information and Computer Science, Kanazawa Institute of Technology,
7-1 Ohgigaoka, Nonoichi, Ishikawa 921-8501, Japan
shinji.sakamoto@ieee.org
[2] Department of Information and Communication Engineering, Fukuoka Institute of Technology, 3-30-1 Wajiro-Higashi, Higashi-Ku, Fukuoka 811-0295, Japan
barolli@fit.ac.jp
[3] Department of Advanced Sciences, Faculty of Science and Engineering, Hosei University, Kajino-Machi, Koganei-Shi, Tokyo 184-8584, Japan
makoto.takizawa@computer.org

Abstract. Wireless Mesh Networks (WMNs) are a technology that is not only cost-effective but also easily deployable and has a high level of robustness. But, WMNs have several issues related with wireless communication. By optimizing the location of mesh routers is possible to deal with these issues. However, the node placement problem is an NP-hard problem. To deal with this problem, we propose and implement an intelligent simulation system based on the Cuckoo Search (CS) algorithm, called WMN-CS. In this paper, we present evaluation results for computation time and different number of nests. The simulation results show that a good performance is archived when the number of nests is more than 60. However, the computation time increases linearly with the increase of the number of nests.

1 Introduction

Wireless Mesh Networks (WMNs) are a technology that is not only cost-effective but also easily deployable and has a high level of robustness. However, when the location of mesh routers is not appropriate, they have several issues such as congestion, interference, decreased throughput, increased packet loss, and increased delay [11].

By optimizing the location of mesh routers is possible to deal with these issues. However, the node placement problem is an NP-hard problem [2], which means that finding optimal solutions is not a realistic strategy [4,7].

L. Barolli (Ed.): EIDWT 2024, LNDECT 193, pp. 384–393, 2024.
https://doi.org/10.1007/978-3-031-53555-0_36

We have proposed an intelligent system based on meta-heuristics for finding better solutions. The proposed system is based on the Cuckoo Search (CS) algorithm and is called WMN-CS. In our previous study, we proposed and implemented WMN-CS [3]. In this study we present evaluation results for computational time and different number of nests.

The rest of this paper is organized as follows. Section 2 provides an overview of related works. The node placement problem in WMNs is defined in Sect. 3. In Sect. 4, we introduce CS algorithm. In Sect. 5, we present WMN-CS simulation system. The simulation findings are reported in Sect. 6. Conclusions and future work are given in Sect. 7.

2   Related Work

There are many research works for resource allocation, node selection, and routing in WMNs [1,14]. Regarding node placement problems, researchers have suggested and considered optimisation models based on Mixed-Integer Linear Programming (MILP), as well as meta-heuristic methods like Hill Climbing (HC), Simulated Annealing (SA), Genetic Algorithm (GA), and Particle Swarm Optimisation (PSO) [5,6,8–10,12].

Heuristic approaches are commonly employed to solve NP-hard problems due to their enhanced time efficiency in comparison to MILP-based techniques [15].

Evaluating solutions for WMNs involves assessing the locations of mesh nodes using simulation results. Heuristic-based solutions show potential for solving the node placement problem in WMNs and other related issues [16].

3   Node Placement Problem in WMNs

For the node placement problem in WMNs, we consider a 2D area which is continuous area and has width and height. Inside of the considered area there are allocated mesh client nodes and mesh router nodes at any location. The main objective is to find the best mesh router node locations that maximise connectivity and user coverage. To evaluate the network connectivity, we analyse a graph of the mesh routers and use the Size of Giant Component (SGC) as metric for measuring network connectivity. The Number of Covered Mesh Clients (NCMC) is used to evaluate the user coverage. The NCMC is calculated based on the number of mesh clients located within the radio coverage area of at least one mesh router node. The network connectivity and user coverage directly influence the network's performance. Therefore, the SGC and NCMC measures are of significant importance.

To establish a formal structure for the problem, we introduce the adjacency matrix of the WMN graph. In this context, the individual nodes represent the routers and client nodes, while the edges symbolise the connections between nodes in the mesh network. Every node is characterised by a triple $v = < x, y, r >$, representing 2D positional coordinates and communication range. A connection is established between two nodes, u and v, if v is inside the communication range of u.

Algorithm 1. Pseudo Code of CS Algorithm.

1: Initialize Parameters:
2:      Computation Time $t = 0$ and T_{max}
3:      Number of Nests $n(n > 0)$
4:      Host Bird Recognition Rate $p_a(0 < p_a < 1)$
5:      Lévy Distribution Scale Parameter $\gamma(\gamma > 0)$
6: Fitness Function to get Fitness Value as f
7: Generate Initial n Solutions S_0
8: **while** $t < T_{max}$ **do**
9:    **while** $i < n$ **do**
10:      $j := i \% \text{len}(S_t)$ // j is the remainder of dividing i by number of solutions.
11:      Generate a new solution S_{t+1}^i from S_t^j by Lévy Flights.
12:      **if** $(f(S_{t+1}^i) < f(S_t^j))$ and $(\text{rand}() < p_a)$ **then**
13:         Discard Solution S_{t+1}^i
14:      **end if**
15:      $i = i + 1$
16:    **end while**
17:    $t = t + 1$
18: **end while**
19: **return** Best solution

4   Cuckoo Search Algorithm

4.1   Outline of Cuckoo Search

A meta-heuristic known as Cuckoo Search (CS) was developed by considering the phenomena of brood parasitism found in certain cuckoo species and served as the conceptual inspiration.

The CS algorithm considers three principles [17].

1. Each cuckoo can lay one egg at a time.
2. The nests with eggs of high quality are considered better and can survive for the next generations.
3. The number of host nests that are accessible remains constant and the host bird can identify the egg laid by a cuckoo with a probability represented by the variable p_a, where p_a is a value between 0 and 1. If the host bird detects the egg presence, it can either get rid of it or abandon the nest and construct a new one in a different location.

We present the pseudo-code of the CS algorithm in Algorithm 1. In CS algorithm, there are three hyperparameters: the number of nests, the host bird recognition rate and the scale parameter of the Lévy distribution.

The CS algorithm begins with initialising the parameters and then calculating each nest's fitness value. Then, it is calculated the most efficient nest from the nest set. After that, the CS try to identify appropriate solutions by repeatedly iterating the process. This iterative process continues until a predetermined stopping criterion is met.

During each iteration, the CS algorithm generates new solutions by employing Lévy flight, a form of stochastic movement. Next, the fitness value of the new solution is computed. When the new solution outperforms the existing solution in terms of quality and effectiveness, the existing solution is replaced by the new solution. But if the current solution is worse than the previous solution, it can be discarded through a stochastic process.

Once that all fitness values are calculated, the optimal solutions are kept while the remaining cuckoo that has not laid its egg exists. Alternative solutions are developed to replace the discarded options. After that, the fitness value of each nest is reassessed. The process continues until the stopping condition is satisfied.

4.2 Lévy Flight

The Lévy flight is utilised in the CS algorithm to generate new solutions using a stochastic process. The Lévy flight is a type of random movement that follows the Lévy distribution and is known for its tendency to exhibit long-tailed behaviour. This characteristic enables CS to explore the search space widely, potentially revealing optimal solutions.

The Lévy flight generally provides short movements. However, it rarely moves in long distances with a low probability. This type of movement is not limited just to theoretical constructions but observed in the natural world. For example, honeybees employ the Lévy flight pattern to locate flower gardens. Honeybees typically engage in short flights to locate a suitable blossom. When unable to identify a satisfactory bloom nearby, they explore long distances to discover a flower garden. This phenomenon is not limited to insects, the humans also display behaviours consistent with the ideas of the Lévy flight.

The Lévy distribution is a continuous probability distribution widely used in probability theory and statistics to model non-negative random variables. It is a specific case of the inverse-gamma distribution.

The probability density function of Lévy distribution is:

$$P(x; \mu, \gamma) = \sqrt{\frac{\gamma}{2\pi}} \frac{e^{-\gamma/2(x-\mu)}}{(x-\mu)^{3/2}}, \tag{1}$$

where μ is a local parameter ($\mu \leq x$) and γ is a scale parameter.

The cumulative distribution function of Lévy distribution is:

$$F(x; \mu, \gamma) = \mathrm{erfc}\left(\sqrt{\frac{\gamma}{2(x-\mu)}}\right), \tag{2}$$

where erfc is the complementary error function.

We can generate values which follows Lévy distribution, by using the inverse transformation method.

$$F^{-1}(x; \mu, \gamma) = \frac{\gamma}{2(\mathrm{erfc}^{-1}(x))^2} + \mu \tag{3}$$

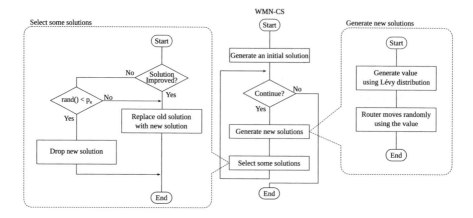

Fig. 1. WMN-CS flowchart.

5 WMN-CS Simulation System

The WMN-CS flowchart is displayed in Fig. 1. In this section, we explain WMN-CS intelligent simulation system especially the startup process, the nesting and egg-laying behaviour and the fitness function.

The WMN-CS algorithm begins by randomly creating the first solution. In this algorithm, solutions are called nests and selected solutions are called eggs. The cuckoo tries to find better nests for laying the eggs. Once the cuckoo identifies a suitable nest for laying the egg, it replaces the existing egg with a value that adheres to the Lévy distribution. This technique mimics the activity of the cuckoo bird, which employs a Lévy flight pattern to select a better nest for egg-laying.

WMN-CS then quantitatively assesses the newly generated solutions using the fitness function, which measures the quality of the nests discovered by cuckoo search. The cuckoo is faced with the decision of whether or not to lay an egg. However, there are instances where the host bird detects that the egg has been replaced, leading to the rejection of the less favourable outcome. This procedure persists until a termination condition is met. After that, the succeeding cohort endeavours to explore novel remedies.

We define the fitness function as:

$$\text{Fitness} = \alpha \times \text{SGC}(\boldsymbol{x}, \boldsymbol{y}, \boldsymbol{r}) + \beta \times \text{NCMC}(\boldsymbol{x}, \boldsymbol{y}, \boldsymbol{r}).$$

The fitness function in WMN-CS is a composite of two metrics: the Size of the Giant Component (SGC) and the Number of Covered Mesh Clients (NCMC). Both of these values are derived from a WMN graph using vectors $\boldsymbol{x}$, $\boldsymbol{y}$, and $\boldsymbol{r}$, which indicate the locations and communication distances of the mesh nodes. The SGC quantifies the connectivity of mesh routers, whereas the NCMC quantifies the levels of user coverage. The fitness function incorporates weight coefficients α and β to equilibrate the influences of SGC and NCMC, respectively.

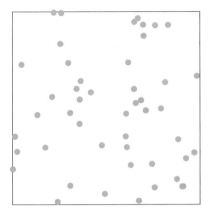

Fig. 2. Uniform distribution.

WMN-CS has the capability to produce various distributions of mesh clients. In this paper, we consider Uniform distribution of mesh clients as shown in Fig. 2. In Uniform distribution, mesh clients are evenly distributed throughout the considered area.

6   Simulation Results

In this section, we present simulation results using WMN-CS. We consider a 2D area with a size of 32 units by 32 units. The simulation parameters for WMN-CS are shown in Table 1. The simulation environment is shown in Table 2.

Table 1. Parameter settings.

Parameters	Values
Clients Distribution	Uniform
Area Size	32×32
Number of Mesh Routers	16
Number of Mesh Clients	48
Total Iterations	2000
Iteration per Phase	10
Number of Nests	From 1 to 100
Radius of a Mesh Router	From 2.0 to 3.0
Fitness Function Weight-coefficients (α, β)	0.7, 0.3
Scale Parameter (γ)	0.09
Host Bird Recognition Rate (p_a)	0.925

Table 2. Computing system environment.

Components	Manufacturers	Models
Motherboard	ASRock	Z790 Pro RS
CPU	Intel	Core i9-13900KF
RAM	Crucial	DDR5-4800 32 GB × 4
OS	Canonical	Ubuntu22.04 LTS

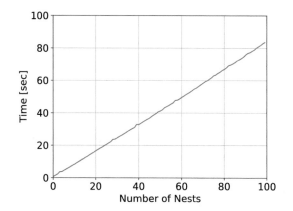

Fig. 3. Simulation results for the computation time.

In our previous research [13], we optimised the Scale Parameter (γ) and the Host Bird Recognition Rate (p_a). To have a comprehensive perspective and mitigate the influence of randomness, we performed 100 simulations for each number of nests.

The computation time for each number of nests is shown in Fig. 3, which is the summation of 100 simulations run. The average time of one simulation can be obtained by dividing 100 by the computation time shown in the figure. We see that the computation time increases linearly with the increase of the number of nests.

For the evaluation, we use the box-plots as shown in Fig. 4. The box shows the Inter Quartile Range (IQR), which is the distance between the third quartile (Q_3) and the first quartile (Q_1). The line in the box shows the median. When data is out of a range 1.5×IQR from the Q_3 and Q_1, we consider the data is outlier. We show whiskers to describe maximum and minimum except outliers.

We show the performance of SGC in Fig. 5. The SGC reaches the maximum value when the number of nests is more than 8. While the NCMC is maximized when the number of nests exceeds 60 as shown in Fig. 6. In this scenario, when the number of nests is more than 60 we get a good performance. However, the computation time increases linearly with the increase of the number of nests. So, we need to consider the trade-off relation between computation time and the number of nests.

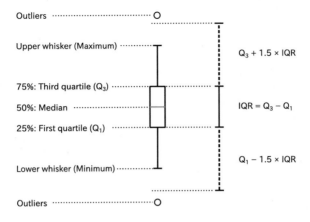

Fig. 4. Example of box-plot.

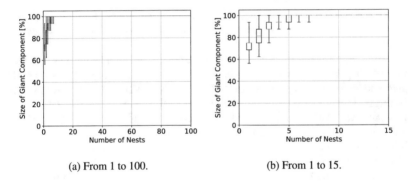

(a) From 1 to 100.   (b) From 1 to 15.

Fig. 5. Simulation results of SGC for different number of nests.

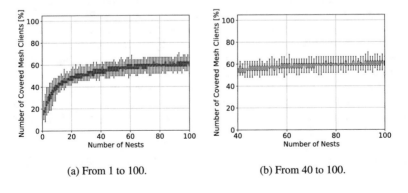

(a) From 1 to 100.   (b) From 40 to 100.

Fig. 6. Simulation results of NCMC for different number of nests.

7 Conclusions

In this work, we proposed and implemented a CS-based system for node placement problem in WMNs called WMN-CS. We conducted simulations to evaluate the computation time and performance of WMN-CS for different number of nests. Simulation results show that the computation time increases linearly with increasing the number of nests and a good performance is archived when the number of nests exceed 60.

In our future work, we will consider the trade-off relation between computation time and the number of nests. Also, we would like to evaluate the proposed system for different distributions of mesh clients.

References

1. Ahmed, A.M., Hashim, A.H.A.: Metaheuristic approaches for gateway placement optimization in wireless mesh networks: a survey. Int. J. Comput. Sci. Netw. Secur. (IJCSNS) **14**(12), 1 (2014)
2. Amaldi, E., Capone, A., Cesana, M., Filippini, I., Malucelli, F.: Optimization models and methods for planning wireless mesh networks. Comput. Netw. **52**(11), 2159–2171 (2008)
3. Asakura, K., Sakamoto, S.: A cuckoo search based simulation system for node placement problem in wireless mesh ntworks. In: Barolli, L. (eds.) Complex, Intelligent and Software Intensive Systems. CISIS 2023. LNDECT, vol. 176, pp. 179–187. Springer, Cham (2023). https://doi.org/10.1007/978-3-031-35734-3_18
4. Basirati, M., Akbari Jokar, M.R., Hassannayebi, E.: Bi-objective optimization approaches to many-to-many hub location routing with distance balancing and hard time window. Neural Comput. Appl. **32**, 13267–13288 (2020)
5. Coelho, P.H.G., do Amaral, J.F., Guimaraes, K., Bentes, M.C.: Layout of routers in mesh networks with evolutionary techniques. In: The 21st International Conference on Enterprise Information System (ICEIS-2019), pp. 438–445 (2019)
6. Elmazi, D., Oda, T., Sakamoto, S., Spaho, E., Barolli, L., Xhafa, F.: Friedman test for analysing WMNs: a comparison study for genetic algorithms and simulated annealing. In: 2015 9th International Conference on Innovative Mobile and Internet Services in Ubiquitous Computing, pp. 171–178. IEEE (2015)
7. Gharehchopogh, F.S., Shayanfar, H., Gholizadeh, H.: A comprehensive survey on symbiotic organisms search algorithms. Artif. Intell. Rev. **53**, 2265–2312 (2020)
8. Lee, S.C., Tan, S.W., Wong, E., Lee, K.L., Lim, C.: Survivability evaluation of optimum network node placement in a hybrid fiber-wireless access network. In: IEEE Photonic Society 24th Annual Meeting, pp. 298–299. IEEE (2011)
9. Lin, C.C.: Dynamic router node placement in wireless mesh networks: a pso approach with constriction coefficient and its convergence analysis. Inf. Sci. **232**, 294–308 (2013)
10. Oda, T., Elmazi, D., Barolli, A., Sakamoto, S., Barolli, L., Xhafa, F.: A genetic algorithm-based system for wireless mesh networks: analysis of system data considering different routing protocols and architectures. Soft. Comput. **20**, 2627–2640 (2016)
11. Qiu, L., Bahl, P., Rao, A., Zhou, L.: Troubleshooting wireless mesh networks. ACM SIGCOMM Comput. Commun. Rev. **36**(5), 17–28 (2006)

12. Sakamoto, S., Oda, T., Ikeda, M., Barolli, L.: Design and implementation of a simulation system based on particle swarm optimization for node placement problem in wireless mesh networks. In: 2015 International Conference on Intelligent Networking and Collaborative Systems, pp. 164–168. IEEE (2015)
13. Sakamoto, S., Asakura, K., Barolli, L., Takizawa, M.: An intelligent system based on cuckoo search for node placement problem in WMNs: tuning of scale and host bird recognition rate hyperparameters. In: Barolli, L. (eds.) Advances on Broad-Band and Wireless Computing, Communication and Applications. BWCCA 2023. LNDECT, vol. 186, pp. 168–177. Springer, Cham (2023). https://doi.org/10.1007/978-3-031-46784-4_15
14. Sanni, M.L., Hashim, A.H.A., Anwar, F., Naji, A.W., Ahmed, G.S.: Gateway placement optimisation problem for mobile multicast design in wireless mesh networks. In: 2012 International Conference on Computer and Communication Engineering (ICCCE), pp. 446–451. IEEE (2012)
15. Seetha, S., Anand John Francis, S., Grace Mary Kanaga, E.: Optimal placement techniques of mesh router nodes in wireless mesh networks. In: Haldorai, A., Ramu, A., Mohanram, S., Chen, M.-Y. (eds.) 2nd EAI International Conference on Big Data Innovation for Sustainable Cognitive Computing. BDCC 2019, pp. 217–226. Springer, Cham (2021). https://doi.org/10.1007/978-3-030-47560-4_17
16. Taleb, S.M., Meraihi, Y., Gabis, A.B., Mirjalili, S., Ramdane-Cherif, A.: Nodes placement in wireless mesh networks using optimization approaches: a survey. Neural Comput. Appl. **34**(7), 5283–5319 (2022)
17. Yang, X.S.: Nature-Inspired Metaheuristic Algorithms. Luniver Press (2010)

Research and Implementation of TFTP Encrypted Traffic Analysis and Attack Technology Based on 4G Man-in-the-Middle

Menglin Ma and Baojiang Cui[✉]

Beijing University of Post and Telecommunications, Haidian District West TuCheng Road 10, Beijing, China
{mamenglin,cuibj}@bupt.edu.cn

Abstract. In the scenario of a 4G man-in-the-middle attack, the analysis of encrypted traffic is a highly complex problem. The attack based on 4G man-in-the-middle only occurs at the protocol level based on DNS protocol, and the implementation conditions are very strict. At the same time, the lack of open source 4G man-in-the-middle attack tools also increases the difficulty of research and limits the attack surface of 4G man-in-the-middle. This paper implements a tool named 4G MITM Attacker, which can carry out 4G man-in-the-middle attacks. Based on the analysis of the TFTP protocol, an analysis method is proposed, which successfully analyzes the TFTP traffic from the encrypted traffic and modifies the traffic with the feature of missing integrity protection of TFTP, resulting in file corruption attacks.

1 Introduction

With the rapid development of mobile communication technology, 5G networks have begun to be deployed worldwide, and 5G has gradually become the choice of mobile phone users. However, 4G is still the dominant communication technology, with several times more base stations and users compared to 5G. 4G has a large user base and a vast number of devices. If there are security problems in 4G, the potential harm is greater, and the impact is wider. Therefore, it is of great significance to do security research based on 4G.

The 4G access network is exposed to the air interface, and anyone with the corresponding radio equipment and related technology can access it, so it is facing a greater security threat. At present, there are many security threats against air interface, including DoS attacks [1], signal masking attacks [2, 3], man-in-the-middle attacks, and so on. Among them, man-in-the-middle attacks can directly sniff user data and can tamper with it, thereby compromising the security of personal information and causing significant harm. This paper primarily focuses on the security of the 4G access network based on the man-in-the-middle mode. It examines the corresponding security analysis technology and explores attack techniques to enhance the security of the 4G access network and protect personal privacy and information security.

In the 4G security problem, the man-in-the-middle attack based on the access network is a very influential attack mode. Man-in-the-middle attack can obtain user data, analyze user behavior, and tamper with user data, making it a topic of high research value.

Implementing a man-in-the-middle attack first requires relevant tools, and there are currently no open-source tools available. Therefore, this paper designs a man-in-the-middle attack system called 4G MITM Attacker.

At present, the research on the man-in-the-middle mode of 4G access network mainly focuses on information tampering, and there is a lack of relevant research on obtaining effective information from encrypted data. And it is very meaningful to obtain the corresponding user behavior information from the encrypted data. This paper studies the encrypted data based on statistical characteristics and TFTP protocol format to obtain relevant and valid information from the encrypted data.

At the same time, the information in TFTP packets can be tampered with, leading to file corruption attacks.

The main contributions of this paper are as follows:

1. Designed a set of man-in-the-middle attack tools, to smoothly implement man-in-the-middle attacks in a 4G environment.
2. Analyzes the encrypted traffic, such as TFTP, to detect the file transfer behavior of users
3. Modify the data packets of the TFTP protocol, which affects the normal use of the service.

This article will be divided into the following sections: background, system overview, system design, experiments, and conclusions. The background section primarily introduces the technologies used in the article. The system overview section provides an overview of the main structure of the 4G MITM Attacker system. The system design focuses on the specific implementation of the 4G MITM Attacker. The experimental part provides information on the environment and experiments required for the experiment steps and other related content. The conclusion section summarizes the conclusions of the experiment and improvements that need to be made in the future.

2 Background

This section mainly introduces the background knowledge related to 4G, including the TFTP protocol.

UE: User Equipment, usually the user's mobile device (such as a cell phone) and SIM card. The UE is used to interact with the base station. UE's protocol stack has protocols such as PHY, MAC, RLC, PDCP, and NAS. UE includes two types of bearers: the control plane and the user plane. The signaling of the control plane has integrity protection, but the signaling of the user plane does not have integrity protection enabled by default. Based on this, the data on the user plane can be tampered with.

eNodeB: A base station in an LTE network that manages wireless resources and encrypts and decrypts data in the air interface. The eNodeB has a layered protocol stack similar to UE. The PDCP layer is the highest-level protocol of the user plane, and the RRC layer is the highest-level protocol of the control plane. The modification of user surface data in this paper focuses on the PDCP layer. Figure 1 shows the LTE protocol stack and the layered protocol situation is shown in the figure.

TFTP: Trivial File Transfer Protocol, a protocol in the TCP/IP protocol family used for simple file transfer between a client and a server, providing a simple and inexpensive file transfer service. The transport layer protocol of TFTP is UDP, which transmits data through the stop-wait protocol. TFTP files are transferred in blocks, but the integrity of the transmitted content is not protected. Therefore, it is easy to modify. TFTP is usually used to update firmware and configurations. If the transferred file is damaged, the firmware upgrade or APP upgrade may fail.

PDCP: Packet Data Convergence Protocol, this protocol is located at the data link layer in the LTE system. The PDCP layer is responsible for header compression and decompression of user plane data, as well as encrypting, decrypting, and ensuring the integrity protection of both user plane and data plane [4]. During the data transmission process, the PDCP layer is responsible for sending and reordering PDUs. It also reorders SDUs to ensure the correct order of transmitted data packets. For timed-out data packets, the PDCP layer will discard them.

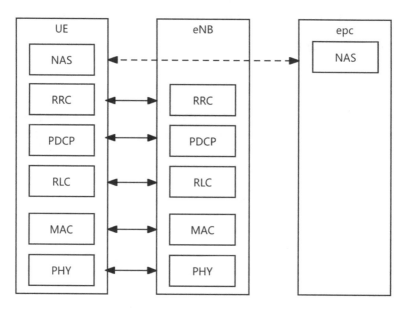

Fig. 1. LTE stack.

3 System View

In the research related to 4G man-in-the-middle attacks, the scope is relatively limited due to the lack of open-source tools. As for the research on the analysis of 4G traffic obtained by middlemen, only the analysis of the DNS protocol is carried out at present, and the identification has certain errors [5]. Analysis of user behavior can determine the volume of VoLTE traffic and obtain information such as the duration of calls and pauses [6]. Other relevant studies are relatively few.

To solve the tool problem, this paper designs a 4G man-in-the-middle attack tool named 4G MITM Attacker. The tool can proxy and transparently forward 4G traffic in the air interface, as well as modify 4G traffic. The tool contains several components, including a fake UE, a fake base station, and a TFTP analyzer. The specific framework is shown in Fig. 2.

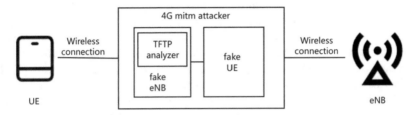

Fig. 2. Structure of 4G MITM Attacker.

Fake Base Station: The fake base station is primarily responsible for attracting nearby mobile phones so that the mobile phone is not connected to the real base station but the fake base station. After receiving the attached signaling and other information from the mobile phone, the fake base station needs to forward the received data packet to the fake UE part, and then the fake UE sends it to the real base station to realize the transparent forwarding of signaling. At the same time, the fake base station will also receive data from the fake UE and transparently forward it to the real phone.

The implementation of the fake base station is mainly based on the srsenb platform in srsran, which implements most of the eNodeB functions in 3GPP [7]. However, srsenb will process the signaling of the mobile phone and interact with the normal protocol stack after the connection with the mobile phone is established, which is not possible in the design of this paper. Therefore, the main modification of srsenb is to control its interaction process and only transparent forwarding without processing the signaling.

Fake UE: The fake UE is mainly responsible for receiving signaling from the real phone of the fake base station, transferring the signaling to the nearby real base station, and receiving the returned data after the real base station has processed the signaling. The returned data is sent by the fake UE to the fake base station.

The implementation of fake UE is mainly based on the srsue platform in srsran, which implements most of the functions of UE in 3GPP. Like srsenb, srsue also needs to control its interaction flow. The connection between the fake base station and the fake UE is established through the socket, and the remote deployment of the fake UE and fake ENB can even be realized under the condition of low network latency.

TFTP Analyzer: This tool is installed on the PDCP layer of the fake base station and can directly analyze the data packets of the PDCP layer. The tool can analyze the user plane traffic at the PDCP layer, analyze the TFTP traffic, and determine the file transfer behavior of the user, as well as the size of the file transferred by the user, the transfer rate, and the transfer time. In addition, the TFTP traffic can be modified so that the receiver does not receive the correct file.

4 System Design

4.1 Fake Base Station Design

The fake base station needs to communicate with the fake UE and the real phone. For the communication on the fake UE side, transparent transmission is mainly done without any processing, and the communication mainly relies on socket connections. In the PDCP layer of the fake base station, socket monitoring is established to obtain the message from the fake UE, which is then forwarded to the real phone. For communication on the real phone side, the wireless connection is mainly used to achieve. The fake base station connects to the phone through the antenna of the USRP B210 device [8]. The USRP needs to be plugged in to boost the antenna's signal, so the phone doesn't connect to the fake base station. After the connection is established, usually the base station will process the signaling from the phone by default. To block signaling processing, the PDCP layer of srsenb needs to perform certain processing tasks, such as blocking the signaling encapsulation and decapsulation process and sending messages to the fake UE via the socket interface.

Since the PDCP layer performs encryption and decryption operations, you need to remove all relevant operations in the PDCP layer to prevent data packets from being modified.

At the same time, in the srsran platform, srsenb will connect to srsepc, the core network, by default, at startup. But man-in-the-middle attacks do not require a core network, so the ability to connect srsenb and srsepc must be removed.

In addition to transparent packet forwarding, the fake base station also analyzes the plaintext signaling of the control surface in the packet. For example, the attachment process contains configuration parameters related to the connection, which affect the stability of the connection. Meanwhile, SR and CQI parameters cannot be obtained directly in plaintext but can be computed using the method described in Cheng [6].

4.2 Fake UE Design

The design of the fake UE is similar to the fake base station, and the fake UE needs to communicate with the fake base station and the real base station. The communication interface is also a socket interface and a wireless interface, not to go into details. To make the fake UE connect to the real base station, the frequency point and PLMN of the fake UE need to be configured. It should be noted that the base station connected to the real phone will also have a frequency and PLMN value. The two values of the fake UE and the real phone must be the same.

4.3 TFTP Analyzer Design

The TFTP analyzer is deployed in the fake base station component of the relay and works at the PDCP layer. TFTP protocol analyzer analyzes user plane data in LTE and mainly processes data in DRB bearer. The deployment in the PDCP layer is primarily because this layer represents the highest level of user data, and the data analysis is relatively simple. For UDP, the PDCP layer only adds a serial number with a fixed number of bytes to the UDP packet. Therefore, the complete UDP packet can be obtained by removing the serial number.

TFTP analyzer mainly contains two functions, 1 is to analyze the TFTP protocol, determine the TFTP traffic within encrypted traffic, and retrieve information such as file transfer time, transfer rate, and file size. 2 is to modify the encrypted traffic. By randomly modifying the file transfer part of the TFTP packet, the file being transmitted by the user can be corrupted.

The first function is mainly implemented according to the characteristics of the TFTP protocol package. The structure of the TFTP frame [9] is shown in Fig. 3.

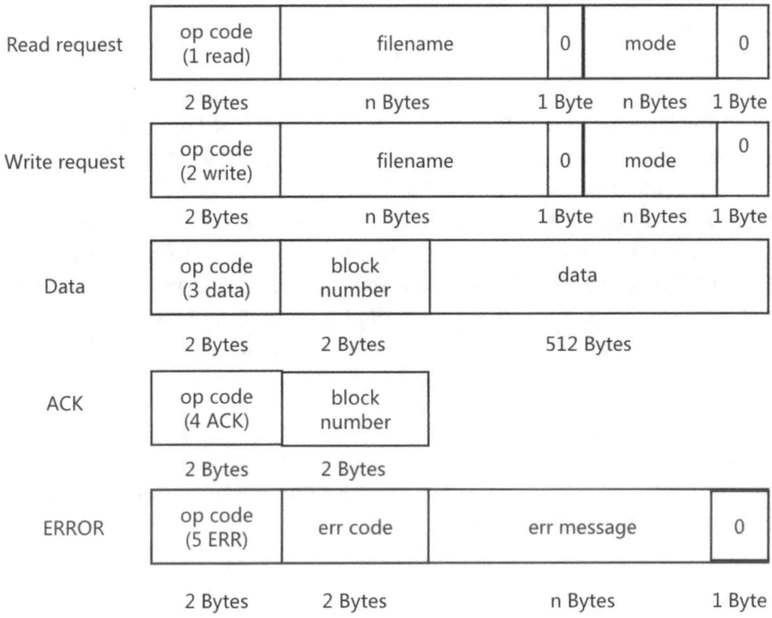

Read request	op code (1 read)	filename	0	mode	0
	2 Bytes	n Bytes	1 Byte	n Bytes	1 Byte

Write request	op code (2 write)	filename	0	mode	0
	2 Bytes	n Bytes	1 Byte	n Bytes	1 Byte

Data	op code (3 data)	block number	data
	2 Bytes	2 Bytes	512 Bytes

ACK	op code (4 ACK)	block number
	2 Bytes	2 Bytes

ERROR	op code (5 ERR)	err code	err message	0
	2 Bytes	2 Bytes	n Bytes	1 Byte

Fig. 3. TFTP frame format.

TFTP has five types of data frames, including read request frame, write request frame, data frame, ACK frame, and error frame. This article focuses on the read data scenario, where the read request involves all frames except the write request.

The main flow of the read request is as follows:

1. When the TFTP client needs to read a file from the TFTP server, it sends a read request. The file name and file schema are included in the request. The size of the read request is affected by the file name and file mode.
2. If the file can be read by the client, the TFTP server returns a data frame numbered "i" starting from 1. If the remaining untransmitted portion of the file is greater than

512 bytes, the data portion of the data frame is 512 bytes. Otherwise, as many bytes as the remaining bytes are transmitted.
3. The TFTP client sends an ACK frame numbered "i".
4. Repeat the steps in 2.

According to the flow of the read request, the corresponding recognition algorithm can be designed, mainly based on the length of the recognition. 4G encryption uses the stream encryption mode, and the plaintext length and ciphertext length are equal. Therefore, the identification based on length is feasible.

According to the read request process, a corresponding identification algorithm can be designed, mainly based on length. 4G encryption uses stream encryption, and the length of the plaintext and the length of the ciphertext are equal. Therefore, identification based on length is feasible.

The identification process is:

1. Read the encrypted traffic of the PDCP layer, remove the sequence number of the PDCP layer, and obtain the complete UDP packet.
2. Take out the data part from the UDP packet according to the offset.
3. Analyze the length of the data segment. If the length is 4 bytes, it is suspected that it may be an ACK frame. If the length of the previous packet of this data packet is 516 bytes, it indicates that the TFTP protocol is downloading the file. At this time, it continues to monitor the subsequent process and extract frames with a length of 516 bytes and 4 bytes, until it finally extracts a confirmation frame of less than 516 bytes and a 4-byte confirmation frame. If the length of the previous packet of the data packet is less than 516 bytes at the beginning, it suggests that the protocol being used may be TFTP protocol and a small file is being downloaded. If the size is greater than 516 bytes, it is not the TFTP protocol.

For the second function, it is relatively easy to implement. After identifying the TFTP data frame, randomly modifying the contents of the data frame can achieve the purpose of destroying the file. TFTP itself does not provide any integrity protection, and any damage that occurs in the channel is undetectable.

5   Experiments

5.1   Experiment Environment

The experiment requires specific hardware to complete, including two ubuntu18.04 hosts, deploying srsenb and srsue respectively, and the two hosts can communicate with each other. Additionally, two USRP B210 devices are needed to send and receive wireless signals. A smartphone with a SIM is also required to access the base station to and the internet. The smartphone should be equipped with the TFTP CS APP. Lastly, a TFTP server must store some test files on the server. The required software and hardware information for the experiment is shown in Table 1.

Table. 1. The required software and hardware information for the experiment

Hardware	Num	Software
Host	2	Ubuntu18.04
USRP B210	2	UHD
SIM Card	1	–
Redmi Phone	1	TFTP CS APP
Server	1	tftpd-hpa

5.2 Experimental Steps

The first step is to enable airplane mode on the smartphone. This will trigger the attachment process once airplane mode is turned off.

The second step is to start the 4G MITM Attacker, turn off the flight mode of the mobile phone, and establish a connection between the mobile phone and the fake base station.

In the third step, when the mobile phone can access the Internet normally, start the TFTP CS APP and download files from the TFTP server. At this time, the TFTP analyzer will automatically analyze the traffic of TFTP transfer files and modify the content in the traffic.

The fourth step is to check the downloaded file and observe whether the downloaded file is damaged.

5.3 Experimental Results

The experimental results mainly include two parts, one is the recognition situation and the other is the modification situation.

This experiment tested 100, 500, and 1000 files respectively. This article stipulates that unidentified files are considered false negatives, incorrectly identified files are false positives, and other files are correctly identified. The results of the experimental false negative rate, false alarm rate, and accuracy are shown in Table 2.

Table. 2. The results of the experimental false negative rate, false alarm rate, and accuracy rate

Num	Accuracy Rate	False Alarm Rate	False Negative Rate
100	81%	14%	5%
500	80.4%	12.6%	7%
1000	80.2%	11.3%	8.5%

Regarding the modification of files, after testing, it was found that the transferred files can be successfully modified. Figure 4 is an example.

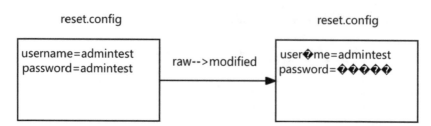

Fig. 4. The transferred file was modified.

6   Conclusion and Discussion

This article examines 4G man-in-the-middle attacks, develops a 4G man-in-the-middle tool named 4G MITM Attacker, and utilizes this tool to analyze, sniff and modify TFTP traffic. Experimental results show that this tool can successfully analyze most of the TFTP traffic and cause damage to the transferred files.

But there are some problems, that is, there is a certain rate of false negatives and false positives. The missed report may be due to the large amount of other traffic in the network and the gap between the data frame and the confirmation frame being too long, causing the TFTP protocol analysis program to fail to successfully analyze this part of the data. The reason for false positives may be that the length of some frames happens to match the length of the data frame and the acknowledgment frame. These are all areas that can be optimized in the future.

This article is an experiment conducted in an ideal situation. Ideal refers to closing all applications that access the network as much as possible. Still, many apps are accessing the network in the background. If the experiment is conducted in a normal environment, the results may be greatly affected.

In future work, we will focus on how to modify the files transferred by TFTP to specific content, such as modifying the values specified in the configuration file, thereby damaging the functions of the APP. At the same time, we will continue to improve the TFTP protocol analysis algorithm to reduce the false negative rate and the false positive rate. In addition, attempts will be made under more complex networks to enhance the capabilities of the TFTP protocol analyzer.

References

1. Yu, C., et al.: Improving 4G/5G air interface security: a survey of existing attacks on different LTE layers. Comput. Netw. **201**, 108532 (2021)
2. Yang, H., et al.: Hiding in plain signal: physical signal overshadowing attack on LTE. In: 28th USENIX Security Symposium (USENIX Security 19), pp. 55–72 (2019)
3. Erni, S., et al.: Adaptover: Adaptive overshadowing of LTE signals. arXiv preprint (2021)
4. Packet Data Convergence Protocol (PDCP) specification (3GPP TS 38.323 version 16.2.0 Release16). https://www.etsi.org/deliver/etsi_ts/138300_138399/138323/16.02.00_60/ts_138 323v160200p.pdf
5. Rupprecht, D., et al.: Breaking LTE on layer two. In: 2019 IEEE Symposium on Security and Privacy (SP), pp. 1121–1136. IEEE (2019)

6. Cheng, Z., et al.: Watching your call: breaking VoLTE privacy in LTE/5G networks. arXiv preprint arXiv:2301.02487 (2023)
7. Open Source SDR LTE Software Suite. https://github.com/srsran/srsRAN_4G
8. Ettus Research USRP B210. https://www.ettus.com/product/details/
9. RFC 1350 - The TFTP Protocol (Revision 2). https://datatracker.ietf.org/doc/html/rfc1350

4G Access Network Protection and Compliance Detection Based on Man-in-the-Middle Model

Ziyang Chen, Baojiang Cui[✉], and Zishuai Cheng

School of Cyberspace Security, Beijing University of Posts and Telecommunications, Haidian District Xitucheng Road No. 10, Beijing, China
{chenzy,cuibj}@bupt.edu.cn

Abstract. The access network is the most important part of the 4G mobile communication network, connecting user equipment with E-UTRAN Node B. Several mechanisms are designed for the protection of this connection, but the access network still suffers from man-in-the-middle (MITM) problem, which is one of traditional network security threats, causing privacy leak and data tampering. This paper proposes a new protection method against the MITM problem of 4G access network. By analyzing the feature of this problem, we find out the difference between regular connection and MITM connection and use this as a signal of suffering MITM attack. Besides, we discover the potentiality for the MITM model in the 4G access network which can be used to build a compliance detection system in a more privacy-friendly way. Finally, several experiments have been accomplished to verification of the protection method and compliance detection system.

1 Introduction

In recent years, the global wireless communications industry has been developing in the direction of mobility, broadband, and IP, and competition among major 4G mobile communication operators in the mobile communications industry has become increasingly fierce. With the vigorous development of new mobile application formats, traditional communication networks that support voice, text messaging and other functions are no longer enough to meet people's daily needs, making 4G mobile communication systems with higher data transmission rates and lower transmission delays possible. To meet the increasing business needs of users, the 4G systems of major operators have completed large-scale deployment and application.

The Security protection of 4G mobile communication networks is urgently needed, and the access network, one of the most important parts of 4G mobile communication networks, must also be completely protected. Compared with 2G and 3G networks, the 4G access network has added more strategies to improve security in all aspects. However, due to the openness of wireless channels and the development of open-source radios, access network security threats still exist.

MITM model is a traditional security technology. The MITM attack deriving from it can theoretically target almost all channels that transmit information. MITM attackers can eavesdrop on all information transmitted through the system channel, so it is difficult to detect them during operation.

L. Barolli (Ed.): EIDWT 2024, LNDECT 193, pp. 404–414, 2024.
https://doi.org/10.1007/978-3-031-53555-0_38

It is very necessary to analyze the security of MITM attacks on 4G mobile communication networks and provide protective measures. On the other hand, the concealment and data sniffing capabilities of MITM technology also enable it to have security analysis capabilities. It can be designed to provide a non-inductive compliance detection tool to the system under controllable conditions.

The purpose of this paper is to analyze the security threats of the 4G access network, implement a MITM technology prototype system for the 4G access network, propose a protection plan against the MITM model, and implement a 4G access network compliance detection program based on the MITM model.

The structure of this paper is formed as below: the second section discusses the related work of the MITM study. The third section introduces the structure of access network protection based on the MITM model, which includes the features of the MITM and the solution based on this feature. The fourth section introduces the basic information to build an access network compliance detection based on the MITM model system and the structure of this system. The fifth section introduces the experiment to verify the access network protection and the access network compliance detection, which consists of the lab requirements, the plan for the experiment, and the result. The sixth section is the conclusion of this research.

2   Related Work

There is current research on the access network MITM model, which focuses on the attack method.

Using the MITM model to design a network attack system has made a huge process. By designing and implementing an access network MITM attacking system, hackers are able to execute network attacks such as wireless tapping and data tampering. Regarding the MITM threats of the 4G access network, in [1] the authors proposed a 4G MITM attack scheme, but this scheme is difficult to implement. Based on the lack of integrity protection of the 4G data plane, in [2] the authors used 4G MITM to carry out chosen ciphertext attacks to achieve DNS request redirection attacks. Based on [2], in [3] the authors extended the 4G second-layer protocol vulnerability to the third layer, completely breaking the two-way authentication. In [4], the authors proposed identification attacks, downgrade attacks, and battery drain attacks on cellular devices based on the unprotected device registration truncation of device function information, achieving imperceptible attacks on all LTE devices. Literature [5] focuses on the VoLTE system and shows two types of privacy attacks, including a VoLTE/NR activity monitoring attack and an identity recovery attack. The researchers also advanced the MITM attack system to improve its reliability.

In summary, the research on 4G access network MITM model analysis has little content on protection and compliance testing. Research on 4G MITM attacks is flawed and no prototype system is usable. The researchers focus on using it as a means of attack. Although some literature proposes to provide integrity protection for the user plane, this can only avoid data tampering but cannot fundamentally solve the MITM problem. As a highly concealed attack method, the potential of MITM for compliance detection has not yet been discovered. Combining the MITM model with compliance detection can check for security threats by verifying that systems comply with standards with little impact on operations. Therefore, conducting a security analysis on the access network MITM and proposing targeted protection measures is necessary.

3 Access Network Protection Based on the MITM Model

This chapter discusses the protection scheme for 4G access network MITM equipment at the signaling level. According to the 3GPP standard, 4G protocol signaling can be divided into multiple processes, including attachment, location update, release, etc. In order to prevent MITM devices from threatening user privacy, this research mainly focuses on the attachment process, because the security context is implemented during the attachment process. The attachment process includes necessary parameters for establishing a security context, wireless connection parameters, etc. This research analyzes these specific parameters and proposes a detection method for the 4G access network MITM model.

3.1 Analysis of the Access Network MITM Model

The MITM connection in the access network can be very powerful, mainly because of two reasons: I. The openness of the wireless channel, by which any attacker can directly access the network, establishes a wireless connection to deceive the victim; II. Unmodified forwarding of data means that the data flow can transparently forward the upper-layer protocols of the mobile communication network, so it is difficult to detect most protocol fields. This research mainly analyzes the characteristics of the 4G access network MITM system. Based on the comparative analysis of channel characteristic parameters, it detects the parameter fields that must be modified to establish the MITM and completes the design and experimental environment for the 4G mobile communication network MITM protection scheme.

3.1.1 Structure of MITM Model in Access Network

The structure of the MITM model in an access network always contains three parts, which are introduced as the fake UE module in this article, the fake eNB module, and the data exchange module. MITM model running between real user's equipment (real UE) and E-UTRAN Node B (real eNB). The fake UE module aims to communicate with real E-UTRAN and the fake eNB module aims to communicate with real UE. The data exchange module has several functions used by both of the fake modules to exchange data got by them. The structure of the MITM model in the access network is shown in Fig. 1.

We use a computer and a Universal Software Radio Peripheral (USRP) to simulate UE and eNB. The MITM system is built with a computer and two USRPs. The left one in the picture represents a fake eNB module and the right one represents a fake UE module. The data exchange module is running on the computer.

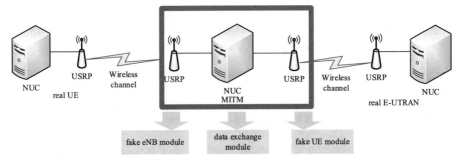

Fig. 1. Structure of MITM model in access network.

3.1.2 Feature of MITM Model in Access Network

After comparing several parameters, we find out that the access network MITM attack system has a problem called "fake UE being interfered with and accessing the fake eNB". This issue is discussed below.

The 4G access network MITM attack system needs to attract real UE to access, so the fake eNB needs to provide higher signal strength. However, the fake UE may also access the fake eNB due to higher signal strength, causing the MITM attack to fail. In order to avoid this situation, the fake eNB and fake UE in MITM need to use different operating frequencies. Usually, mobile communication network operators support multiple different operating frequencies, so this solution can attract real UE access without affecting the operation of fake UE.

Because actual operating frequencies are a float type, which is hard to transfer, E-UTRA Absolute Radio Frequency Channel Number (EARFCN) is used to indicate the frequency of the channel, by which the system can calculate the actual frequency in LTE. Here are some EARFCN and the frequency they represent in Table 1.

Table 1. EARFCN and the frequency.

EARFCN	operating frequency
1300	1815
1506	1835.6
3350	3350

To solve the "fake UE being interfered with and accessing the fake eNB" problem, the EARFCN of the fake UE side must be different from the EARFCN of the fake eNB side.

3.2 Design of Access Network Protection Based on MITM Model

The EARFCNs between fake UE and fake eNB are different, which means the EARFCN fake UE sharing with real E-UTRAN does not equal the EARFCN fake eNB sharing with real UE. However, the real UE should share the same EARFCN with real E-UTRAN, so this can be used to detect whether the channel suffers from the MITM model attack.

Here is the scheme planning for protection based on the MITM model.

1) Real UE initializes the UE protocol stack, and waits for the cell selection to be completed;
2) After completing the cell selection, Real UE checks the selected physical layer parameters, records the downlink EARFCN as *e1*, and then waits for the security context established;
3) After the security context is established, real UE obtains the downlink EARFCN of the eNB from the signaling, recording as *e2*;
4) The real UE compares *e1* with *e2*;
5) If the result *e1* does not equal *e2*, the real UE will report to the user or disconnect.

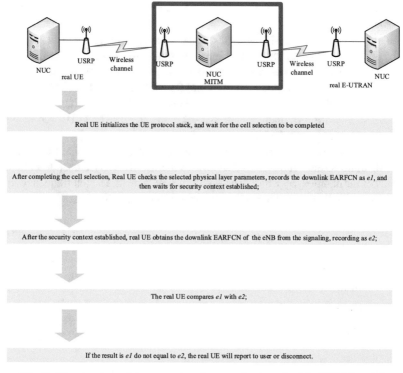

Fig. 2. The structure of Access network protection based on the MITM model.

Because the attacker deploying MITM can change the data, the downlink EARFCN of the eNB must be sent after the security context have already been set to avoid against hacker changing it to escape from this protection.

Considering the real eNB always has several UE attaching to it, it increases the load of eNB to be too heavier. The protection program is mainly implemented on UE. Here in Fig. 2 showing the protection process. The main part of the protection program is implemented on real UE.

4   Access Network Compliance Detection Based on MITM Model

The 4G access network MITM model has the potential to conduct compliance detection of 4G access networks. The design of this compliance detection technology is mainly based on system security parameters. System security parameters can be divided into two aspects: UE and eNB security parameters.

For the UE, the UE will provide the UE network capability field in the Attach request signaling, which contains all security algorithms supported by the UE. The 4G mobile communication network mainly contains three security algorithms: snow 3G, AES, and ZUC. However, when specifically configured, these algorithms are divided into encryption algorithms (represented by EEAX, where X is an integer 0–7) and integrity Protection algorithms (represented by EIAX, where X is an integer 0–7). The device supports these two algorithms respectively. The content of the UE network capability field packet is as Fig. 3. Each bit corresponds to an algorithm 1 means supporting and 0 means not supporting.

```
v UE network capability
    Length: 2
    1... .... = EEA0: Supported
    .0.. .... = 128-EEA1: Not supported
    ..0. .... = 128-EEA2: Not supported
    ...0 .... = 128-EEA3: Not supported
    .... 0... = EEA4: Not supported
    .... .0.. = EEA5: Not supported
    .... ..0. = EEA6: Not supported
    .... ...0 = EEA7: Not supported
    0... .... = EIA0: Not supported
    .1.. .... = 128-EIA1: Supported
    ..1. .... = 128-EIA2: Supported
    ...1 .... = 128-EIA3: Supported
    .... 0... = EIA4: Not supported
    .... .0.. = EIA5: Not supported
    .... ..0. = EIA6: Not supported
    .... ...0 = EIA7: Not supported
```

Fig. 3. Algorithms that UE may support.

For the eNB, the eNB assigns the security algorithm used for this connection to the UE in the Security Mode Command, which is also divided into an encryption algorithm and an integrity protection algorithm. There are two integers used here. The first one is used to represent the ciphering algorithm (EEA) and the second one is used to represent the integrity protection algorithm (EIA). This picture represents that the eNB assigns EEA0 and EIA2 to UE for connection. This configuration is located in the securityAlgorithmConfig field, shown in Fig. 4.

```
securityAlgorithmConfig
  cipheringAlgorithm: eea0 (0)
  integrityProtAlgorithm: eia2 (2)
```

Fig. 4. The algorithms that eNB may assign.

Both fields interact before the security context is established, so compliance detection is possible in any case. This compliance detection mainly finds whether the UE can support all three security algorithms; and whether the eNB has allocated the suitable encryption algorithm and integrity protection algorithm.

It should be added that the 4G mobile communication system has two security parameters, EEA0 and EIA0. These two security parameters respectively indicate that no encryption algorithm and no integrity protection algorithm are used. The 4G standard only prohibits the use of EIA0 during signaling transmission, but EEA0 can be selected for both signaling and data. For the data plane, the 4G standard allows manufacturers to use EIA0 parameters to reduce system overhead, but this has security risks and is unacceptable for users with high-security requirements. This compliance detection design will use both EEA0 and EIA0 as non-compliance cases. The UE is allowed to support these two algorithms, but the eNB is not allowed to select these two algorithms.

Message types field in the signaling package can be used to distinguish different signaling. The message types fields correspond to the Attach request and Security Mode Command in Table 2.

Table 2. Message types field in signaling package.

EARFCN	Message types field
Attach request	01000001
Security Mode Command	01011101

Figure 5 exhibits the construct of access network compliance detection based on the MITM model, which is completely implemented on the MITM system. First, the researchers must compose the rule of compliance detection. Meanwhile, the MITM system should extract every field relating to this compliance detection. Then, the system detects whether the system fits the rules. Finally, a report of this detection will be made to discuss the conclusion of this detection.

5 Experimental Results and Evaluation

5.1 Experiment Environment

For hardware environment of the experiment as shown in Fig. 1, a computer and a USRP are used to simulate the real UE, and another computer and USRP compose the real eNB. On the same computer as the real eNB, a core network of LTE is running. The MITM model system is deployed on a computer with two USRPs, one of which is used by fake eNB and the other is used by fake UE.

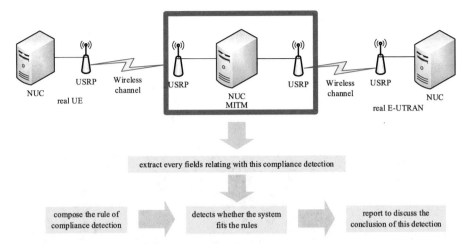

Fig. 5. The structure of access network compliance detection based on the MITM model.

This paragraph discusses the software environment of the experiment. The MITM system is programmed based on discussion in literature [5]. The other part of the software environment in this experiment is shown in Table 3.

Table 3. Software environment for experiment.

Experiment Entity	Open Source
real UE	srsRAN
real eNB	srsRAN
core network	open5gs

In Table 2, srsRAN is a famous open-source radio project that consists of UE, eNB, and EPC. The UE and eNB in srsRAN are configured in the computers of the real UE and the real eNB to provide an LTE protocol stack. The core network is based on the open-source mobile telecommunication network project named open5gs. The same identity of the user used in this experiment is configured in the core network and real UE.

5.2 Experiment of Access Network Protection Based on MITM Model

As we discussed in Sect. 3, the code of the real UE is changed to fit the protection structure.

In the function called cell_select, the srsRAN program provides a function that can check whether the cell selection process has been done, so this can be a semaphore of complete cell configuration. After ensuring that the configuration of the cell has been initialized, the system will read the current EARFCN and write it into a file. Then, the system will wait for the signaling named RRC reconfiguration. To complete the protection, the EARFCN of eNB used in this connection must be inserted into this signaling, the reason of which is that RRC reconfiguration is after the signaling named Security Mode Complete and gets integrity protection. As receiving RRC reconfiguration, the real UE will read the EARFCN sent by the real eNB and compare it with the EARFCN reading from the file recording the EARFCN of UE. If the result is that these two EAR-FCNs are different, the MITM model system can be found and, in this experiment, we show the result on the Command Prompt.

Here is what the Command Prompt is like after deploying the protection in Fig. 6.

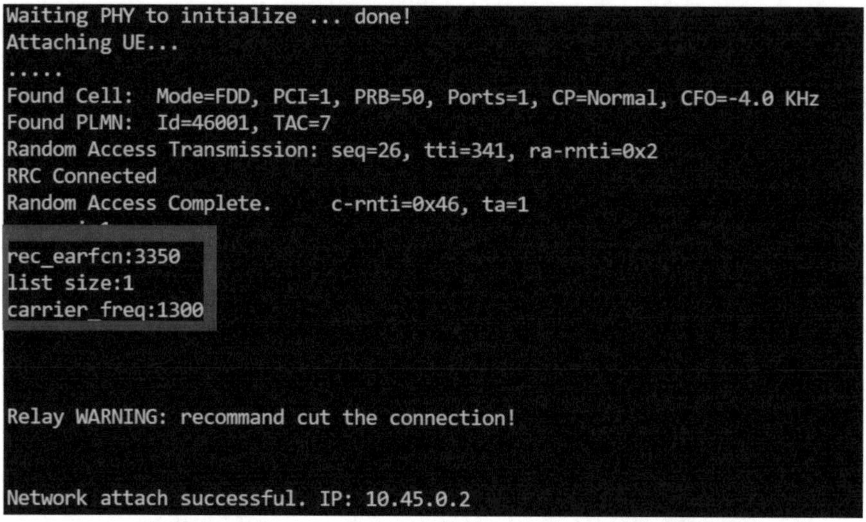

```
Waiting PHY to initialize ... done!
Attaching UE...
.....
Found Cell:  Mode=FDD, PCI=1, PRB=50, Ports=1, CP=Normal, CFO=-4.0 KHz
Found PLMN:  Id=46001, TAC=7
Random Access Transmission: seq=26, tti=341, ra-rnti=0x2
RRC Connected
Random Access Complete.      c-rnti=0x46, ta=1

rec_earfcn:3350
list size:1
carrier_freq:1300


Relay WARNING: recommand cut the connection!

Network attach successful. IP: 10.45.0.2
```

Fig. 6. The experiment result of the protection against the MITM model system.

In Fig. 6, there is a red square highlighting the important information about this protection. The field named "rec_earfcn" means the EARFCN of the real UE, and the other field named "carrier_freq" is the EARFCN of the real eNB. The list size field means how many pairs of EARFCNs have been recorded, which in this experiment is one. After that, the Command Prompt showed a warning because the EARFCNs did not match. This experiment has already shown that the protection structure can detect and defend the MITM model system.

5.3 Experiment of MITM Model Access Network Compliance Detection

As we discussed in Sect. 4, the code of the MITM model system is changed to fit the protection structure.

The MITM model system can transfer most of the signaling on the LTE air interface, so during this process, the system is able to obtain all the signal messages. When the system gets a new message, it will compare the Message Types field of it and filter the Attach Request and Security Mode Command.

As for the Attach Request message, UE network capability is going to be checked. In this experiment, we check all three cipher algorithms and all three integrity protection algorithms. UE which does not support any one of them is regarded as Error.

As for the Security Mode Command message, the securityAlgorithmConfig field is going to be checked. In this experiment, we check the cipher algorithm and the integrity protection algorithm which are configured to the real UE by the real eNB. If the real eNB configures EEA0 or EIA0 to UE, it will be regarded as a WARNING.

In Fig. 7, we show the command prompt after compliance detection of UE.

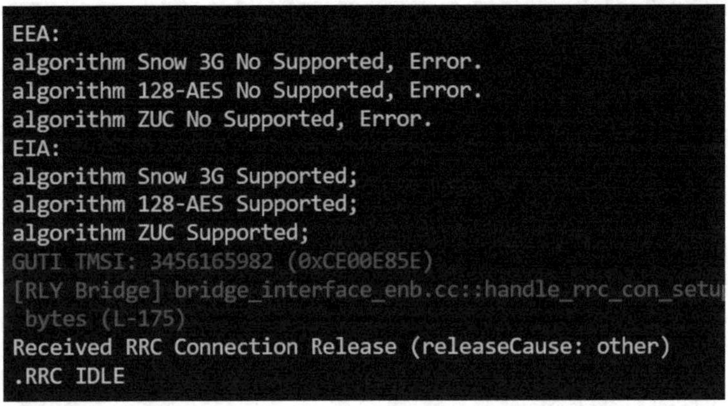

```
EEA:
algorithm Snow 3G No Supported, Error.
algorithm 128-AES No Supported, Error.
algorithm ZUC No Supported, Error.
EIA:
algorithm Snow 3G Supported;
algorithm 128-AES Supported;
algorithm ZUC Supported;
GUTI TMSI: 3456165982 (0xCE00E85E)
[RLY Bridge] bridge_interface_enb.cc::handle_rrc_con_setu
 bytes (L-175)
Received RRC Connection Release (releaseCause: other)
.RRC IDLE
```

Fig. 7. Result of access network MITM model compliance detection system.

In order to test the compliance detection system, the algorithms of the real UE are configured so that all the cipher algorithms are not supported and all the integrity protection algorithms are supported. In Fig. 7, the MITM model compliance detection system can disclose the configuration of the real UE algorithms.

In Fig. 8, we show the command prompt after compliance detection of eNB has been deployed.

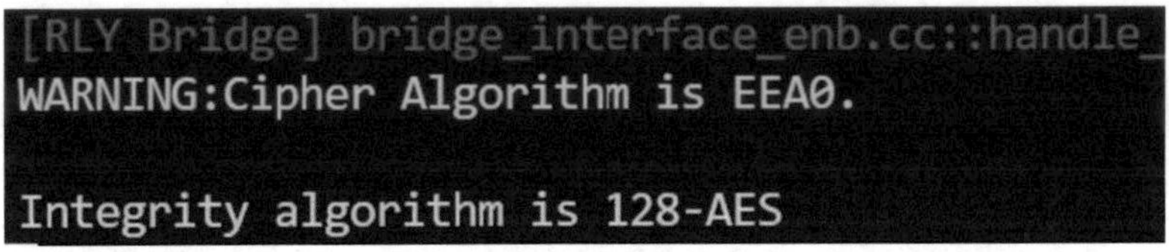

```
[RLY Bridge] bridge_interface_enb.cc::handle_
WARNING:Cipher Algorithm is EEA0.

Integrity algorithm is 128-AES
```

Fig. 8. Result of access network MITM model compliance detection system.

In order to test the compliance detection system, the configuration of the algorithm from the real eNB is configured that the cipher algorithm is EEA0 and the integrity protection algorithm is EIA1. In Fig. 8, the MITM model compliance detection system can disclose the configuration from the real eNB algorithms. The integrity algorithm, which is 128-AES in Fig. 8, is what the real algorithm of the EEA1 represented.

6 Conclusions

Today, the MITM model in the 4G access network is not only a threat requiring protection but also has the capacity to achieve new security goals, such as compliance detection. In this research, we investigated several articles related to the MITM model in 4G access network, finding out that there was still a lack of method to protect against the MITM model system in 4G security construct and the capability of the access network MITM model to achieve new security goal did not fully use. According to the investigation, a protection method aiming at the MITM model system and a compliance detection system based on the MITM model was proposed. After an explanation of these two systems, we designed and achieved the experiments to verify the protection method and the compliance detection system. This research can also be the foundation of further research aiming at the MITM model in the 4G access network.

References

1. Wu, T., Gong, G.: The weakness of integrity protection for LTE. In: Proceedings of the Sixth ACM Conference on Security and Privacy in Wireless and Mobile Networks (2013)
2. Rupprecht, D., et al.: Breaking LTE on layer two. In: 2019 IEEE Symposium on Security and Privacy (SP). IEEE (2019)
3. Rupprecht, D., et al.: IMP4GT: IMPersonation Attacks in 4G NeTworks. NDSS (2020)
4. Shaik, A., et al.: New vulnerabilities in 4G and 5G cellular access network protocols: exposing device capabilities. Proceedings of the 12th Conference on Security and Privacy in Wireless and Mobile Networks (2019)
5. Cheng, Z., et al.: Watching your call: breaking VoLTE privacy in LTE/5G networks. arXiv preprint arXiv:2301.02487 (2023)

Energy Sharing System Among Vehicles on a Vehicular Network

Walter Balzano[(✉)], Antonio Lanuto , Erasmo Prosciutto,
and Biagio Scotto di Covella

University of Naples, Federico II, Naples, Italy
walter.balzano@gmail.com

Abstract. In today's mobility landscape, characterised by an accelerated transition towards electric vehicles, significant challenges emerge related to the management of battery recharging. This paper proposes a solution for energy sharing between electric vehicles, inspired by the effectiveness of wireless charging technologies for smartphones and integrating the use of VANET networks. Through communication techniques such as V2V and V2G, we aim to develop a system that not only alleviates charging issues, but also promotes more sustainable mobility and optimised use of energy resources.

Keywords: Energy Sharing · VANETs · Client-Server

1 Introduction

The rise of electric vehicles (EVs) represents a major step towards more sustainable mobility. An MIT study, 'Insights Into Future Mobility', compares vehicles in different configurations, showing that battery-powered all-electric vehicles emit about 200 g of CO2 per kilometre, significantly less than the more than 350 g per kilometre of petrol-powered vehicles [5]. EVs help improve air quality by reducing or eliminating emissions of pollutants such as nitrogen oxides and particulate matter, which are common in combustion vehicles. But the advantages of electric vehicles do not stop at environmental aspects alone, but also issues such as lower operating costs. A study from the U.S. Department of Energy Office of Scientific and Technical Information highlights that the estimated scheduled maintenance cost for a light-duty battery-electric vehicle totals 6.1 cents per mile, whereas for a conventional ICE vehicle, it's 10.1 cents per mile [8].

Energy sharing among vehicles is an innovative concept that aims to reduce the carbon footprint of transportation. In this model, vehicles share their excess energy with other vehicles or the electricity grid, which can help reduce greenhouse gas emissions and promote the use of renewable energy sources.

Electric vehicles are the primary vehicles that can share energy among each other. This is made possible through the use of advanced technology, such as vehicle-to-grid (V2G) [4], vehicle-to-vehicle (V2V) [6] and wireless charging [1]. V2G technology allows electric vehicles to share their excess energy with the electricity grid, which can help balance the supply and demand of electricity. When an electric vehicle is parked, it can be connected to the grid and supply electricity during times of peak demand. This can help prevent blackouts and other energy-related problems.

V2V technology enables vehicles to share energy with one another through a wireless communication system. This allows vehicles to exchange energy while driving, which can extend the range of electric vehicles and reduce the need for frequent charging. The technology behind V2V energy sharing is still in development, but it has the potential to revolutionize the transportation industry.

Wireless charging allows electric vehicles to charge wirelessly, eliminating the need for cords or charging ports. It can be done in a parking lot, a garage, or even on the road. This technology is still in the early stages of development, but it has the potential to make charging electric vehicles more convenient and accessible.

Energy sharing via electric vehicles can optimize the use of existing infrastructure, reducing the need for new construction. This system allows vehicles to feed energy into the grid during peak demand, alleviating the need for new power plants and transmission lines. This approach not only generates significant economic savings but also helps to decrease the environmental impact of building new infrastructure, promoting more efficient and sustainable use of energy resources.

Energy sharing can also help reduce the cost of electric vehicles ownership by allowing owners to earn money by sharing their excess energy. This can create a new revenue stream for owners and reduce the cost of owning an electric vehicle. As the technology behind energy sharing becomes more advanced and widely adopted, the potential benefits for owners are expected to increase.

There are currently about one billion cars on the road in the world (source: report.rai), of which only 0.2% are electric. Recent research [3] has shown that sales of electric cars are increasing exponentially year by year. Currently, the market for electric cars is booming, but many users and potential buyers are still hesitant about buying an electric car, mainly because of the timing of charging the batteries of electric vehicles. In fact, electric vehicles currently require at least 30 min to recharge up to 80% using the fastest charging stations. And it is in this scenario that we introduce a possible solution to make recharging faster and less dependent on fixed charging stations.

2   Related Work

In recent years, the scientific community's interest in the concept of energy sharing between vehicles has grown significantly. This has led to various research papers exploring various perspectives and approaches to optimise the energy efficiency and reduce the environmental impact of electric vehicles. Some of these facets are addressed, for example, in a recent paper [7], in which the authors examine the main issues related to energy exchange between vehicles, such as energy exchange prices, vehicle parameters, and user privacy and security issues.

A key issue concerns the development and regulation of energy exchange costs and markets. In this regard, the article [9] aims to optimally implement vehicle-to-vehicle (V2V) energy exchange in the Internet of Things for Vehicles (IoEV) environment, making use of a blockchain-based system.

Vehicle-to-vehicle energy sharing is also based on the idea of a world in which each vehicle can recharge its energy in a fast and distributed manner. An interesting contribution to this topic is presented in the article [10], which explores the study of an interactive network between buildings and vehicles, using energy flexibility evaluation criteria that can be quantified.

In our paper, a crucial aspect is to understand the relevant energy sources in the system. In particular, we take inspiration from the approach outlined in the landmark article [2], in which the most relevant web sources are carefully chosen through a detailed analysis of a hypergraph in a social multimedia network. This approach could provide a meaningful starting point for the selection of vehicles, enabling the identification of those that offer an energy situation consistent with the user's needs.

3   The Idea Behind the Energy Sharing

To understand the idea behind the energy sharing system to be developed, let us first introduce an example and then go into the specifics later. The basic idea is developed from what already exists: the exchange of energy between two smartphones using the physical principle of electromagnetic induction. Just as two smartphones exchange energy, replacing mobile phones with two cars, we hypothesise an exchange of energy between neighbouring cars. Specifically, we imagine that we are within a vanet network, and we imagine that we have access to information such as vehicle location, vehicle energy level, and also the destination and relative route being taken. With this information, a vehicle that is in our system and needs a significant amount of energy, either to complete its journey or to arrive safely at the next charging station, can request and receive energy from a nearby vehicle willing (according to certain parameters later explained) to give up energy. Two types of energy exchange systems are hypothesised:

- Normal Sharing System
- Emergency Sharing System

These two systems work simultaneously, but have two different tasks. In particular:

1. The first assumes a system condition where the user would not need energy to complete the journey or to arrive at the first charging station, but may request extra energy from another car so as not to have to stop at the charging station to recharge the car or for his or her own safety in completing the journey.

2. The second, on the other hand, assumes an emergency situation in which the car to be assisted does not have enough energy to reach the nearest charging station and may, as a result, stop suddenly. In this case, the system sends a car (or a user) to'rescue' in order to help the car in trouble and provide it with the energy it needs to get to the next charging station.

In both cases, the system takes into account a number of values that allow it to determine which car is suitable for the energy exchange.

4   Development Environment

The implementation of the energy sharing system in a Vehicle Ad-hoc Network (VANET) requires each car to have the capability to set its travel route autonomously. This enables the system to calculate the energy threshold of each vehicle, a concept we define as **T.E.S.** (Threshold Energy Sharing). The T.E.S. is a critical value, representing the ratio of a car's remaining energy in terms of kilometers to the preset energy level required to reach the next charging station. This equation is pivotal:

$$T.e.s = \frac{E_{Remaining-Km-Possibly}}{E_{Remaining-Km-C.Station}} \tag{1}$$

This system's scalability and adaptability are significant, as it can dynamically map the number of cars in need of energy and those capable of assisting nearby. A unique feature is the dynamic calculation interval, Δ_t, allowing T.E.S. values to be updated based on vehicle density and network traffic patterns.

4.1   Enhanced Vehicle Status Dynamics

Each vehicle's status within the network is more than just its T.E.S. value. We introduce a comprehensive variable named"**Status of Car**," encompassing various operational states. The enhanced states are:

- **Ready** - Vehicle is fully prepared for energy exchange.
- **Pairing** - Vehicle is matched with a partner and is preparing for energy transfer.
- **Charging** - Vehicle is actively engaged in energy exchange.
- **Not Ready** - Vehicle is currently unable to participate in energy sharing.
- **Navigation** - Vehicle is in transit, potentially adjusting its route based on T.E.S. and network conditions.
- **Standby** - Vehicle is in a low-power state, conserving energy but still connected to the network.

5   Refined System Structure

Our energy sharing system's structure leverages a sophisticated multigraph approach, as depicted in Fig. 1. This structure not only manages the distribution of
vehicles and charging stations but also incorporates real-time traffic data and
environmental factors, making the system more responsive and efficient. The
graph G=(V, E) now includes dynamic nodes and edges, reflecting the ever-
changing nature of the network.

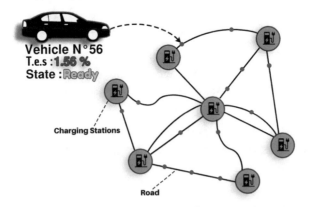

Fig. 1. Advanced Multi-graph Representation of the Energy Sharing Algorithm

6   Enhanced Algorithmic Approach

The algorithm, at its core, is designed to be highly adaptable, catering to varying
scenarios within the network. We distinguish between two primary modes of
operation:

6.1   Enhanced Normal-Sharing

This mode represents the standard operational state. Here, the categorization of
vehicles is refined:

- **Energy Provide Car** - These are high-capacity vehicles, potentially with
  renewable energy generation capabilities (like solar panels), serving as mobile
  charging stations.
- **Client Normal Car** - Regular users of the network, now with enhanced
  decision-making algorithms for selecting optimal energy providers based on
  efficiency, cost, and proximity.

6.2   Advanced Emergency-Sharing

This mode is activated under critical conditions. The system becomes more proactive in this state, with the following refined categories:

- **Client Emergency Car** - Cars in critical energy states can now activate an advanced AI-assisted emergency protocol, optimizing their chances of receiving timely assistance.
- **Neutral Car** - These vehicles act as intermediaries, capable of relaying information or even small amounts of energy to support the network.
- **Energy Special Car** - In emergencies, these vehicles can act as mobile energy hubs, providing substantial support to those in dire need.

This enhanced approach ensures that the energy sharing system is not only efficient and scalable but also robust and capable of adapting to various scenarios, ranging from everyday commutes to emergency situations.

7   Case Study: Implementing an Advanced VANET-Based Energy Sharing System in an Urban Environment

7.1   Scenario Overview

In this case study, we examine the deployment and operation of an advanced Vehicle Ad-hoc Network (VANET) energy sharing system in a bustling urban setting. The scenario chosen is a typical weekday in a densely populated city center, characterized by heavy traffic, a diverse mix of electric vehicles (EVs), and a network of charging stations. The focus is on demonstrating the system's ability to facilitate efficient energy distribution among EVs, adapt to varying energy demands throughout the day, and respond effectively to emergency energy needs.

7.2   Dynamic Energy Management During Peak Hours

During the morning rush hour, a high volume of EVs enters the city. The system actively monitors and calculates the Threshold Energy Sharing (T.E.S.) values for each vehicle. Vehicles with higher T.E.S. values are identified as potential energy providers, while those with lower values are marked as potential energy receivers. As the day progresses, the system continuously updates these values and vehicle statuses, optimizing energy distribution and ensuring that vehicles with lower energy reserves can receive support from those with surplus energy.

7.3   Emergency Energy Sharing in Critical Situations

A critical test of the system occurs when an EV, heavily utilized throughout the day, finds its energy reserves depleted, insufficient to reach the nearest charging station. The driver activates the Emergency-Request-Energy-Sharing (E.R.E.S.)

protocol. The system quickly classifies the vehicle as a Client Emergency Car and broadcasts an emergency signal to nearby vehicles. An Energy Special Car with a high T.E.S. value responds and navigates to the client's location. A secure energy transfer is conducted, enabling the client car to safely reach a charging station or its destination.

This case study shows a potential application of the advanced VANET-based energy sharing system in an urban context. The system not only improves the efficiency of energy distribution among electric vehicles, but also demonstrates its robustness in responding to emergency situations. The knowledge gained in this scenario provides valuable data for further refinement of the system. Future research could explore integration with smart city infrastructure, improvement of artificial intelligence algorithms for predictive energy management, and incorporation of renewable energy sources to further increase the sustainability and efficiency of the system.

8   Economic Structuring of the Energy Exchange System

The energy sharing system just described also lends itself well to developments in the economic field and is well suited to a profitable energy exchange not only for those who receive energy, but also for those who give it away. It is therefore intended to present a possible basis for the regulation of remunerated energy exchange. In particular, in the pay system we can banally distinguish two categories of users: the buyer and the seller of energy. Specifically, the pay system is managed by an algorithm which, on the basis of certain parameters later explained, provides the cost that a buyer user must give up in order to provide himself with the amount of energy required to complete his journey. With this in mind, a higher reward must be imagined for those who perform an emergency sharing system, compared to those who perform a normal sharing system.

On the other hand, we have some rules for even cars that can actually perform energy exchange. As shown in Fig. 2, a client car in emergency mode can request energy from both cars in emergency mode and normal mode. Unlike a client car in normal mode can only request energy with provider cars in normal mode.

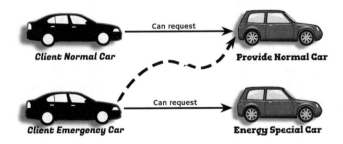

Fig. 2. System exchange rules.

9   Future Uses

Let us conclude the discussion so far by introducing several possible scenarios in which energy sharing could be applied. In particular, we want to consider how energy sharing can have a very strong source of economic development, as it is possible to imagine energy exchange both as an effective help for those who receive energy, but also as a source of economic gain for those who give energy. With this in mind, it is not far-fetched to imagine possible investments by energy companies in this new market. Indeed, assuming that the surrendered energy is somehow bought by the receiving users, it is also possible to imagine the introduction of **_M.E.R._** cars (mobile energy rechargers) into the energy sharing system. **_M.E.R._**'s are nothing but new types of cars, equipped with any type of battery, and entering our system in the form of mobile recharging stations. These cars could be operated by external companies, which, for profit, would introduce new charging methods for users into the system. We can therefore say that energy sharing winks a lot at being a revenue-generating system, not only for companies that want to invest in it, but also for individual users who decide to be 'special' users, i.e. users with **_M.E.R._** cars. Obviously, energy sharing being a new system for recharging cars, it can be declined in various fields and various uses.

References

1. Ahmad, A., Alam, M.S., Chabaan, R.: A comprehensive review of wireless charging technologies for electric vehicles. IEEE Trans. Transport. Electrific. **4**(1), 38–63 (2017)
2. Amato, F., Castiglione, A., Moscato, V., Picariello, A., Sperl, G.: Multimedia summarization using social media content. Multim. Tools Appl. **77**, 07 (2018)
3. International Energy Agency. Global EV Outlook 2022. (2022). Accessed: [insert date here]
4. Kempton, W., Tomić, J.: Vehicle-to-grid power fundamentals: Calculating capacity and net revenue. J. Power Sources **144**(1), 268–279 (2005)
5. MIT Climate Portal. Are electric vehicles definitely better for the climate than gas-powered cars? (2022)
6. Molisch, A.F., Tufvesson, F., Karedal, J., Mecklenbrauker, C.F.: A survey on vehicle-to-vehicle propagation channels. IEEE Wirel. Commun. **16**(6), 12–22 (2009)
7. Shurrab, M., Singh, S., Otrok, H., Mizouni, R., Khadkikar, V., Zeineldin, H.: An efficient vehicle-to-vehicle (v2v) energy sharing framework. IEEE Internet Things J. **9**(7), 5315–5328 (2022)
8. U.S. Department of Energy, Office of Scientific and Technical Information. Maintenance costs comparison: electric vehicles and internal combustion engine vehicles (2021). Accessed 11 Dec 2023
9. Wang, Y., et al.: A fast and secured vehicle-to-vehicle energy trading based on blockchain consensus in the internet of electric vehicles. IEEE Trans. Veh. Technol. **72**(6), 7827–7843 (2023)
10. Zhou, Y., Cao, S., Kosonen, R., Hamdy, M.: Multi-objective optimisation of an interactive buildings-vehicles energy sharing network with high energy flexibility using the pareto archive NSGA-II algorithm. Energy Convers. Manage. **218**, 113017 (2020)

Proposal of Indoor AR Navigation System Using SLAM for Location Search

Yusuke Gotoh[1](✉), Ryusei Tsunetomo[2], and Kiki Adhinugraha[3]

[1] Faculty of Environmental, Life, Natural Science and Technology, Institute of Academic and Research, Okayama University, Okayama, Japan
y-gotoh@okayama-u.ac.jp

[2] Graduate School of Natural Science and Technology, Okayama University, Okayama, Japan

[3] Department of Computer Science and Information Technology, La Trobe University, Melbourne, Australia

Abstract. In recent years, the increased computing power of mobile terminals has enabled users to recognize images accurately and in detail. In particular, augmented reality (AR) spaces created from camera shots by mobile terminals are being applied to navigation technology that guides users to their destinations indoors. Unlike outdoor systems, most indoor AR navigation systems cannot use the Global Positioning System (GPS). Therefore, AR navigation systems based on simultaneous localization and mapping (SLAM) have been proposed. SLAM uses physical information such as AR markers and distinctive shapes to simultaneously estimate the self-position of a terminal used indoors and create an environment map, which gives map information of the area around a movement path. However, this system requires the proper placement of a large number of AR markers. In addition, as the travel distance increases, errors occur in the self-position estimation. In this paper, we propose an indoor AR navigation system that uses multiple environment maps based on both SLAM and AR markers. The proposed system recognizes areas where navigation is possible by applying the environment maps created using SLAM, and then it presents a route in AR that excludes object areas that would present obstacles when moving indoors. Accordingly, users can move to their destinations while avoiding such obstacles. We designed and implemented a prototype of the proposed system and evaluated its performance. From the evaluation results, we confirmed that the proposed system can identify multiple rooms using AR markers by performing self-position estimation based on the feature points given in the environment map.

1 Introduction

In recent years, the increased computing power of mobile terminals has enabled users to recognize images accurately and in detail. In particular, augmented reality (AR) spaces created from camera shots by mobile terminals are being applied to navigation technology that guides users to their destinations indoors.

In the field of indoor AR navigation, image recognition technology based on deep learning has developed along with the increasing computational power of mobile terminals. Recently, an indoor AR navigation system using simultaneous localization and

L. Barolli (Ed.): EIDWT 2024, LNDECT 193, pp. 423–434, 2024.
https://doi.org/10.1007/978-3-031-53555-0_40

mapping (SLAM) [1] has been proposed, which simultaneously estimates the self-position of the terminal and creates map information (i.e., an environment map) consisting of feature points and sensor data around the moving path. For example, Pokémon GO [2], released by Niantic in 2016, uses SLAM-based AR functionality to enhance the integration of virtual reality and reality. The SLAM-based AR function uses the device's camera and built-in sensors, such as its gyroscope, to intuitively present the location of objects to the user. This makes it easy for users to understand the actual environment they are perceiving and supports their actions in a wide range of fields such as advertising, navigation, and medical care.

In many cases, the Global Positioning System (GPS) cannot be used for indoor AR navigation systems because they are used in environments where there are many obstacles and the range of radio waves is limited. Therefore, it is necessary to recognize the user's current location using AR markers, distinctive graphics, or SLAM.

In this paper, we propose an indoor AR navigation system that can use multiple environment maps by leveraging both SLAM and AR markers. The proposed system recognizes navigable areas by applying an environment map created using SLAM and presents a route on AR that excludes object areas that may present obstacles when moving indoors. Consequently, the proposed system enables users to move to their destinations while avoiding obstacles.

The remainder of the paper is organized as follows. In Sects. 2 and 3, we explain AR technology and SLAM. Related works are introduced in Sect. 4. We design and implement the proposed system in Sects. 5 and 6. Finally, we offer our conclusions in Sect. 7.

2   AR Technology

In recent years, many augmented reality (AR) applications have become popular. For example, Pokémon Go [2] uses AR technology in the monster capture screen displayed on the user's mobile device. In general, virtual reality (VR) is a visual technology that presents images of a virtual world to the user. On the other hand, AR is a visual technology that superimposes additional information onto images of the real world, enabling users to better understand the real world in detail and easily.

With the spread of smartphones, AR functions are used not only for delivering videos and games but also for business tasks and daily-life activities. The 5th generation mobile communication system (5G) can provide users with large-volume content such as video and 3D CG images at high transmission speeds.

In AR navigation, the user acquires his or her location using GPS and AR markers and then searches for a route to the destination as well as sightseeing content for the area. AR navigation can be used in two types of locations: indoors and outdoors. The following sections describe each of them in turn.

2.1   Outdoor AR Navigation

Location-based AR applications use sensors or GPS built into the terminal to acquire the user's current location. Data and images corresponding to the user's location acquired with the location-based type are displayed on the user's device. In DRAGON QUEST WALK [3], a location- and GPS-based AR application released in 2019, monsters appear on the display in response to the user's location. In a similar manner, an outdoor AR navigation system uses the location obtained from the device's built-in sensors and GPS to superimpose a route to the destination onto a real-world image captured by the camera.

PinnAR [4] is a typical AR navigation system. PinnAR acquires the device's position in real time between the starting point and the destination and then guides the user to the destination.

2.2 Indoor AR Navigation

Indoor AR applications use a vision-based system that acquires the current position by analyzing images of the real world acquired by a camera using image recognition and spatial recognition technologies. The vision-based approach consists of two types: a marker type that uses AR markers and a markerless type that recognizes real objects and spaces instead of markers. IKEA Place [5], released in 2017, is a markerless type that uses a camera and sensors mounted on the terminal to determine the size of a room and identify objects in order to place virtual furniture within the AR space.

2.3 Challenges of Indoor AR Navigation

Indoor AR navigation systems require higher accuracy than outdoor AR navigation systems because they are used in environments having narrower roads and more complex travel routes. However, conventional GPS technology is difficult to use in indoor AR navigation systems due to the poor reception of GPS signals indoors and large errors in locating the current position. Therefore, AR markers and pedestrian dead reckoning (PDR) have been proposed as techniques to obtain the user's current position.

AR Marker

An AR marker is a specific image or photo that is captured by a camera when using an AR application. AR markers are used to place 3D models, play audio and video, and present route guidance to the destination. AR navigation uses AR markers to perform ID recognition, calculate the camera's position and orientation, and present the user with route guidance corresponding to the AR markers.

In the case of indoor AR navigation, many AR markers are installed when trying to increase the accuracy of a guided route, and this can degrade the surrounding physical landscape. On the other hand, AR markers placed in consideration of the surrounding landscape may degrade recognition accuracy due to the addition of data other than those data needed for route guidance. Furthermore, the installation of a large number of AR markers imposes a heavy burden on system administrators.

PDR

Pedestrian dead reckoning (PDR) uses acceleration, magnetic, and angular velocity sensors on a mobile device to detect a pedestrian's posture, movement, direction, and range

Fig. 1. Environment for feature point detection

Fig. 2. Feature points detected by Visual-SLAM

of motion and then accumulate the amount of movement per step to estimate the current position. The use of a smartphone equipped with a PDR-capable sensor incurs a cumulative error in the amount of movement. Therefore, it is necessary to compensate for this accumulated error by using sparse position data over long distances [6].

3   SLAM

Simultaneous localization and mapping (SLAM) [1] technology is widely used in the development of autonomous robots and self-driving technology. SLAM is also used in drones and cleaning robots. Users can adopt SLAM to perform functions such as automatic cleaning and charging.

SLAM creates an environment map and estimates a self-position simultaneously by measuring the distance and direction of an object with a laser scanner or camera. To improve the accuracy of self-position estimation, image processing and point cloud processing must be performed in short cycles. This increases the computational load and imposes a heavy load on the device. When SLAM is used on a single mobile device, a large error in self-position estimation may cause the device to judge an unobstructed area as an obstacle. To solve this problem, Nihonyanagi et al. proposed a method [7] to more accurately detect obstacles by applying SLAM on two mobile devices. In addition, when SLAM is used in a large area, there is a cumulative error due to self-position estimation. Therefore, for long-distance travel, it is necessary to compensate for the accumulated error by incorporating a process called loop closure.

Figures 1 and 2 show examples of SLAM-based measurements. Figure 1 shows the environment for feature point detection, and Fig. 2 shows the feature points detected by Visual-SLAM for Fig. 1. From these figures, it is observed that feature point detection can detect not only differences in visible color but also differences in shape due to surface irregularities. SLAM uses a laser scanner, ultrasonic sensor, camera, and Wi-Fi wireless environment to measure the distance to surrounding obstacles and create a map of the surrounding environment while also estimating the user's position.

SLAM is widely used as a technology to achieve automatic driving. In particular, the function of avoiding obstacles is used in the automatic cleaning and charging of cleaning robots used in ordinary households.

3.1 Examples of SLAM Applications

Visual-SLAM [1] detects feature points and transforms their coordinates via a computer vision library such as OpenCV for each frame of video captured by the camera, mainly for low-speed autonomous robot behavior and AR recognition. In addition, Visual-SLAM simultaneously creates an environment map and estimates the self-position. However, sensor data contain many errors, and these errors follow a normal distribution. Therefore, by fusing multiple sensor data with a Kalman filter [8] that uses a normal distribution, the errors in self-position estimation are reduced, the data are mapped, and a landmark object represented by a point is created on the environment map. By leveraging this environment map, Visual-SLAM can estimate the camera's posture and distance traveled.

4 Related Works

In many cases, GPS cannot be used for indoor positioning because the signals from satellites, which are the signals' sources, do not reach the device. Therefore, it is difficult to estimate a user's current location. To solve this problem, many research approaches have been pursued for indoor AR navigation using indoor location technology.

4.1 Indoor Location Estimation Method

Manabe et al. proposed a method [9] using Wi-Fi round trip time (Wi-Fi RTT) based on Wi-Fi access points (APs) for indoor location discovery. The Wi-Fi RTT value varies depending on the presence of obstacles, the location of the APs, and the frequency band. Therefore, when the APs are installed at a sufficient height and free of obstructions, the user can estimate his/her location with high accuracy. However, as the distance from the AP increases, the estimation error increases. In this case, the cost increases as more APs are installed at regular intervals. In addition, the user needs to know the locations of all APs when this method is used to obtain indoor locations.

Yamasaki et al. proposed a self-positioning estimation method [10] that divides previously captured video at regular intervals and creates a database based on feature points detected at each frame. On one computer, feature points in the video acquired from the camera are detected and compared with the feature point data in the database. The self-position is estimated with high accuracy by simultaneously estimating the self-position using SLAM on a separate computer. However, this method requires the use of two computers because the two processes cannot be performed on the same computer.

Amano et al. [11] used two types of location maps in their AP-based location search: a virtual AP on a building wall, based on GPS-available outdoor observation data, and a

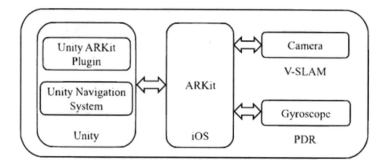

Fig. 3. Structure of conventional system

relative AP based on indoor observation data. The user's current position indoors is estimated by converting the location map of the relative APs to a global coordinate system
based on the position of the virtual APs. By using Wi-Fi installed indoors, the indoor
AP locations can be estimated without using indoor observation locations or anchors.
Since the user does not need to set up markers or create an environment map, this approach reduces the processing burden. However, the error in self-position estimation is
larger because it does not use indoor observation locations.

4.2   Indoor AR Navigation

Rustagi et al. [12] created a room's floor plan and updated MapBox APIs with room lay-
outs, wall locations, guidance routes, landmarks, and destinations. By creating a tileset,
which is a data resource for tiles, the user can be guided to the destination accurately.
However, since the map is created on Unity, this imposes a significant burden.

Verma et al. proposed a method [13] for converting an uneven 3D model and creating a map by applying the height map technique to a planar drawing of a building. The
user's self-position is estimated by comparing the feature points of the building between
the plan view and the 3D model. However, this method requires pre-processing to apply
the height map technique to the building's plan view.

In displaying routes using AR navigation, many AR navigation applications show
the shortest route from the user's current position to the destination, and thus the route
may be displayed in an area that is a blind spot to the user. Gerstweiler et al. proposed
a method [14] to display the route within the user's field of view. However, the map
required for AR navigation is based on a vectorized planar drawing. Therefore, this
method cannot be used when such a drawing is not available.

The indoor AR navigation method [15] proposed by Romli et al. uses AR markers
that are pre-imaged images of objects that exist in the real world, rather than QR codes
or distinctive shapes that are commonly used as AR markers. Therefore, this method
does not degrade the surrounding scenery of tourist attractions and buildings.

For indoor locations, the method [16] proposed by Ng et al. uses magnetic fields,
Wi-Fi signals, and inertial sensors. This method improves the accuracy of pathfinding
by using the ant colony optimization (ACO) algorithm.

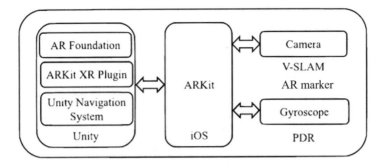

Fig. 4. Structure of proposed system

5   Design

5.1   Assumed Environment

In this paper, we propose an indoor AR navigation system that uses Visual-SLAM and AR markers to search for the self-position in indoor AR navigation. The proposed system uses an iPad terminal and the Unity game engine to create an AR environment that combines images from the real world and the virtual world using Visual-SLAM. By creating an environment map on the terminal, the proposed system can be used in places such as shopping malls and underground shopping arcades. By using AR markers to switch the environment map, the proposed system can be used in buildings consisting of multiple rooms and multiple levels.

5.2   System Configuration

The proposed system uses AR Foundation [17] and the Unity Navigation System [18] to configure AR navigation. First, the system allocates an area that is determined to be flat by recognizing the space in AR Foundation. The user acquires planar data and creates an environment map by pressing a part of the planar area displayed on the screen of the mobile device. For AR marker detection, the proposed system uses the Image Tracking feature [19] of AR Foundation. For route guidance, the system uses the environment map to estimate the current position by SLAM. Next, based on the obtained position, the system searches for a guided route to the destination (LandMark) on the Unity Navigation System and presents it to the user.

Figures 3 and 4 show the structures of the conventional system and the proposed system, respectively. In Fig. 3, the conventional system uses ARKit, which permits the use of Visual-SLAM and PDR. In addition, the conventional system uses the Unity Navigation System to implement navigation functions. On the other hand, in Fig. 4, the proposed system uses AR Foundation instead of ARKit. Since AR Foundation includes all of the functions of ARKit, the proposed system can be used in the same way as the conventional system. The navigation function is performed using the Unity Navigation System, as in the conventional system. AR markers are used to switch the environment map.

Table 1. Measurement environment

Machine	OS	Windows 10 Pro
		Mac OS Catalina Version 10.15.7
	IDE	Unity 2019.4.11f1 Personal
		Visual Studio Community 2019 Version 16.6.1
		Xcode Version 12.1
	Plugin	AR Foundation Version 2.1.10
		ARKit XR Plugin Version 2.1.10
		NavMesh Extension
Device	iPad Air	iPad 13.1.1

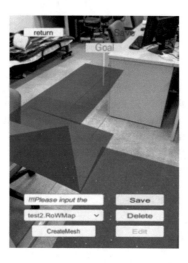

Fig. 5. Function for saving environment map

AR Foundation

To achieve the basic functionality of AR navigation, we created an AR space using the SLAM-enabled AR Foundation. Compared to ARToolKit, ARCore, ARKit, and Vuforia, AR Foundation is an easier way to build a development environment and develop a system to create AR spaces for multiple devices. Unity is cross-platform, so it supports not only Vuforia and iOS-based ARKit but also Android-based ARCore plug-ins. Furthermore, since AR Foundation includes all ARKit and ARCore features, it can be used on both iOS and Android platforms, except for a few features. The proposed system uses AR Foundation and the ARKit XR Plugin in Unity to perform planar recognition, feature point detection, and image recognition using AR Foundation.

Unity Navigation System

The Unity Navigation System is mainly used in games developed in Unity. For navigation, the system stores the surface of the user's walkable plane (NavMesh) as polygons.

Fig. 6. Function for loading environment map

When navigating the user, the system uses the polygons closest to the starting point and the LandMark, and it moves through the contiguous neighboring polygons in order from the starting polygon to the LandMark polygon. The A algorithm [20] is used for nearest neighbor search. In this paper, a static environment map created in advance is used to find the path.

6   Implementation

We implemented the proposed system based on the design described in Sect. 5. The computer environment of the proposed system is shown in Table 1. The functions carried out by the proposed system are described in turn in the following sections.

6.1   Function for Saving Environment Map

Figure 5 shows an example of using the function for saving the environment map. By selecting an area on the screen of the mobile device, the user adds NavMesh vertices and LandMarks to create an environment map. The environment map is divided into two files: one to store the NavMesh and LandMark coordinates and the other to store the detected feature point data.

6.2   Function for Loading Environment Map

Figure 6 shows an example of using the function for loading the environment map. To load an environment map specified by the AR marker recognition function, NavMesh and LandMark are created to build a search environment. Next, the proposed system estimates its location by detecting feature points and comparing them with the feature point data contained in the environment map.

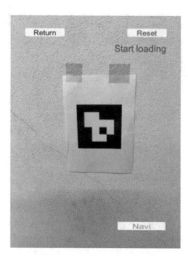

Fig. 7. Function for recognizing AR markers

Fig. 8. Function for presenting guided routes

6.3   Function for Recognizing AR Markers

Figure 7 shows an example of using the function for recognizing AR Markers. When a user recognizes a specific AR marker through the camera mounted on the mobile device, the environment map corresponding to the AR marker is specified by using the Image Tracking function in AR Foundation. By using multiple types of AR markers, it is possible to switch between multiple environment maps, which can be used in multiple rooms and multi-story buildings.

6.4   Function of Presenting Guided Routes

Figure 8 shows an example of using the function for presenting guided routes. After recognizing the AR marker, the proposed system presents a route from the user's current position to LandMark when the user presses the AR route presentation button in the lower-right corner of the screen.

7   Conclusion

In this paper, we proposed an indoor AR navigation system that can use multiple environment maps by adopting both SLAM and AR markers. The proposed system recognizes navigable areas by applying an environment map created using SLAM and then presents a route on AR that excludes object areas that would be obstacles when moving indoors. In the design and implementation of the proposed system, we used an iPad device and the Unity game engine to create an AR environment that fuses images of the real world with those of a virtual world using Visual-SLAM. By creating an environment map on the device, the proposal system can be used in places such as shopping malls and underground shopping arcades. In addition, by using AR markers to switch the environment map, this system can be used in buildings consisting of multiple rooms and multiple levels.

In the future, we will evaluate the loading time of the environment map, the processing time of self-position estimation, and the processing time of switching between multiple environment maps using AR markers to confirm the effectiveness of the proposed system.

Acknowledgement. This work was supported by JSPS KAKENHI Grant Numbers JP21H03429 and JP22H03587, the JGC-S Scholarship Foundation, and the JSPS Bilateral Joint Research Project (JPJSBP120229932).

References

1. Dissanayake, G., Newman, P., Cleark, S., Durrant-Whyte, H., Csorba, M.: A solution to the simultaneous localization and map building (SLAM) problem. IEEE Trans. Robot. Autom. **17**(3), 229–241 (2001)
2. Pokémon GO. https://www.pokemongo.jp. Accessed 19 Nov 2023
3. DRAGON QUEST WALK. https://play.google.com/store/apps/details?id=com.square_enix. android_googleplay.dqwalkj&hl=en_US&pli=1. Accessed 19 Nov 2023
4. PinnAR. https://www.pinnar.net. Accessed 19 Nov 2023
5. IKEA application, Google Play. https://play.google.com/store/apps/details?id=com.ingka. ikea.app&hl=ja&gl=US. Accessed 19 Nov 2023
6. Nozaki, J., Hiroi, K., Kaji, K., Kawaguchi, N.: Compensation scheme for PDR using sparse location and error model. In: Proceedings of 2017 ACM International Symposium on Wearable Computers (UbiComp 2017), pp. 587–596 (2017). https://doi.org/10.1145/3123024. 3124406
7. Nihonyanagi, K., Katsuma, R., Yasumoto, K.: Reducing falsely-detected feature points of SLAM by estimating obstacle-free area for RCMSs. In: Adjunct Proceedings of 2021 International Conference on Distributed Computing and Networking (ICDCN 2021), pp. 80–85 (2021). https://doi.org/10.1145/3427477.3428187

8. Irie, K., Sugiyama, M., Tomon, M.: A dependence maximization approach towards street map-based localization. In: Proceedings of 2015 IEEE/RSJ International Conference on Intelligent Robots and Systems (IROS), pp. 1–8 (2015). https://doi.org/10.1109/IROS.2015.7353898

9. Manabe, T., Aihara, K., Kojima, N., Hirayama, Y., Suzuki, T.: A design methodology of Wi-Fi RTT ranging for lateration IEICE Trans. Fundamentals **E104-A**(12), 1704–1713 (2021). https://doi.org/10.1587/transfun.2020EAP1129

10. Yamasaki, K., Shishido, H., Kitahara, I., Kameda, Y.: Evaluation for harmonic location estimation system of image retrieval and SLAM. In: International Workshop on Advanced Image Technology (IWAIT), pp. 1–6 (2020). https://doi.org/10.1117/12.2566223

11. Amano, T., Kajita, S., Yamaguchi, H., Higashino, T., Takai, M.: A crowdsourcing and simulation based approach for fast and accurate Wi-Fi radio map construction in urban environment. In: Proceedings of IFIP Networking 2017 (Networking 2017), pp. 1–9 (2017). https://doi.org/10.23919/IFIPNetworking.2017.8264841

12. Rustagi, T., Yoo, K.: Indoor AR navigation using tilesets. In: Proceedings of 24th ACM Symposium on Virtual Reality Software and Technology (VRST 2018), vol. 54, pp. 1–2 (2018). https://doi.org/10.1145/3281505

13. Verma, P., Agrawal, K., Sarasvathi, V.: Indoor navigation using augmented reality. In: Proceedings of 2020 4th International Conference on Virtual and Augmented Reality Simulations (ICVARS 2020), pp. 58–63 (2020). https://doi.org/10.1145/3385378.3385387 (2020)

14. Gerstweiler, K.P., Kaufmann, H.: DARGS: dynamic AR guiding system for indoor environments. Computers **7**(5), 1–19 (2018). https://doi.org/10.3390/computers7010005

15. Romli, R., Razali, A.F., Ghazali, N.H., Hanin, N.A., Ibrahim, S.Z.: Mobile Augmented Reality (AR) marker-based for indoor library navigation. IOP Conf. Ser. Mater. Sci. Eng. **767**(012062), 1–9 (2020). https://doi.org/10.1088/1757-899X/767/1/012062

16. Ng, X.H., Lim, W.N.: Design of a mobile augmented reality-based indoor navigation system. In: Proceedings of 2020 4th International Symposium on Multidisciplinary Studies and Innovative Technologies (ISMSIT), pp. 1–6 (2020). https://doi.org/10.1109/ISMSIT50672.2020.9255121

17. About AR Foundation, AR Foundation. https://docs.unity3d.com/Packages/com.unity.xr.arfoundation@3.1/manual/. Accessed 19 Nov 2023

18. Navigation System in Unity, Unity. https://docs.unity3d.com/2019.4/Documentation/Manual/nav-NavigationSystem.html. Accessed 19 Nov 2023

19. AR tracked image manager, AR Foundation. https://docs.unity3d.com/Packages/com.unity.xr.arfoundation@4.0/manual/tracked-image-manager.html. Accessed 19 Nov 2023

20. Inner Workings of the Navigation System, Unity. https://docs.unity3d.com/2019.4/Documentation/Manual/nav-InnerWorkings.html. Accessed 19 Nov 2023

A Fuzzy-Based System for Real-Time Volume Control of Electric Bass Guitars

Genki Moriya[1], Kyohei Wakabayashi[1], Yuki Nagai[1], Chihiro Yukawa[1],

Tetsuya Oda[2(✉)], and Leonard Barolli[3]

[1] Graduate School of Science and Engineering, Okayama University of Science (OUS),
1-1 Ridaicho, Kita-ku, Okayama 700-0005, Japan
{r23smq2au,t22jm24jd,t22jm23rv,t22jm19st}@ous.jp
[2] Department of Information and Computer Engineering,
Okayama University of Science (OUS), 1-1 Ridaicho, Kita-ku, Okayama 700-0005, Japan
oda@ous.ac.jp
[3] Department of Information and Communication Engineering,
Fukuoka Institute of Technology, 3-30-1 Wajiro-higashi, Higashi-ku, Fukuoka 811-0295, Japan
barolli@fit.ac.jp

Abstract. The Direct Injection (DI) is used by electric bass guitar players during live house performances and recordings. A DI is an impedance converter for connecting an instrument such as an electric guitar, electric bass guitar, or electronic keyboard directly to a mixer or interface. When an electric bass guitars output is connected to a DI, the mixer or loudspeaker may be damaged when the volume is high and is generated loud noise. Therefore, Public Address (PA) engineers shut off the input signal to the mixer. There is a rule that the electric bass guitar should only be connected after signaling to the engineer during live performances and recordings. However, beginners or people who do not know the rules sometimes connect an electric bass guitar without signaling the engineer and the mixer and speakers are sometimes destroyed by the loud noise input. In this paper, we propose a real-time volume control based on fuzzy system for electric bass guitars to prevent damage to equipment such as mixers and loudspeakers. The experimental results show that the proposed system is able to automatically control the volume of electric bass guitar when playing sound.

1 Introduction

In music production, automatic composition, automatic mix engineering [1–6] and other automation [7–10] based on artificial intelligence are actively being studied and are expected to be utilized in live performance and recording of electric bass guitar. The live performance or recording of an electric bass guitar is done using Direct Injection (DI) [11], sending the output from the electric bass guitar to a mixer. The DI is an impedance converter for connecting electronic instruments such as electric guitars, electric bass guitars and keyboards directly to a mixer or interface. The output signals from most electronic instruments are unbalanced signals, which are prone to noise and using a shorter cable improves signal quality. On the other hand, balanced signals are less susceptible to noise, even used in long cables. From a noiseless point of view, balanced signals are essential for Public Address (PA) engineers and recording engineers. The DI can convert an unbalanced signal to a balanced signal.

There are active and passive types of electric bass guitars. The active type incorporates a preamp, requiring a power battery to produce sound. In contrast, the passive type does not require a power source. The output of the active type has low impedance, while the output of the passive type has high impedance. In the case when the high impedance signal output from a passive type electric bass guitar is directly connected to a mixer or interface, problems such as noise generation and attenuation of high frequencies will occur. Therefore, it is necessary to convert the high impedance signal to a low impedance signal by using DI.

When receiving the output of an electric bass guitar as an input to a DI, the mixer and loudspeaker may be damaged if the volume is high. One of the factors that can cause high volume noise is the connection of the electric bass guitar with DI. Therefore, the PA engineers shut off the input signal to the mixer. In live performances and recordings, as a rule connection of the electric bass guitar with DI should be done after signaling the engineer. However, beginners or people who do not know the rules sometimes connect an electric bass guitar without signaling the engineer and the mixer and speakers are sometimes destroyed by the loud noise input.

Depending on the electric bass guitar playing style, high volume is instantaneously input to the DI and mixer. For example, there are variations of volume in the slap style, fingerstyle and chord style of playing. In particular, techniques such as thumbing where the strings are struck with the thumb and popping where the strings are pulled with the index or middle finger, in slap playing style, result in momentary high volumes.

In this paper, we propose a fuzzy system for real-time volume control of electric bass guitars using a compressor and automatic volume control to prevent the damage of equipments such as mixers and loudspeakers. The proposed system automatically controls the volume of the electric bass guitar sound using compressor and motor fader fuzzy-based control [8, 12–14]. Also, visualization of the volume control by the motor fader is expected to provide feedback to the performer on the amount of adjustment. Furthermore, the system can control the volume regardless of the electric bass guitar playing style.

The structure of the paper is as follows. In Sect. 2, we present the proposed system. In Sect. 3, we describe the experimental results. Finally, conclusions and future work are given in Sect. 4.

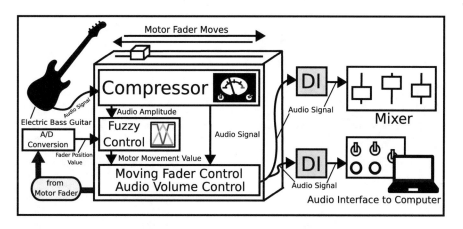

Fig. 1. Overview of proposed system.

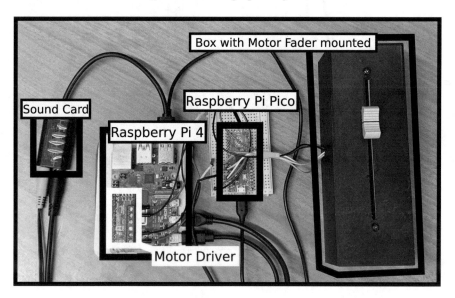

Fig. 2. Hardware configuration of proposed system.

2   Proposed System

In this section, we present the proposed system. Figure 1 shows the overview of proposed system. The proposed system controls the volume of electric bass guitar output by using a compressor and a motor fader. The motor fader is controlled by fuzzy inference by changing the position of fader knob.

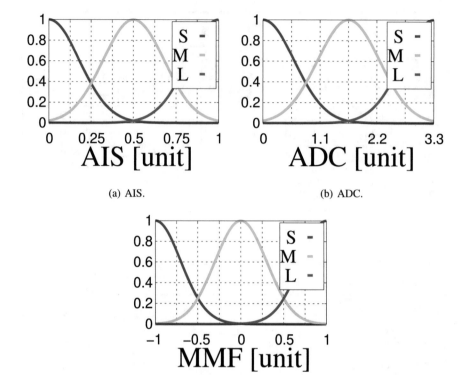

(a) AIS.   (b) ADC.

(c) MMF.

Fig. 3. Membership functions of proposed system.

Table 1. FRB.

AIS	ADC	MMF
S	S	L
S	M	L
S	L	M
M	S	M
M	M	M
M	L	M
L	S	M
L	M	S
L	L	S

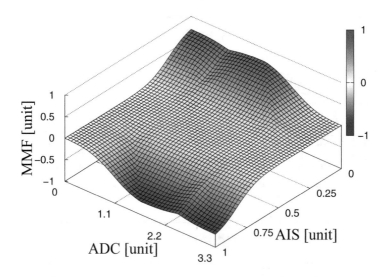

Fig. 4. Output results of fuzzy control.

Figure 2 shows the hardware configuration of the proposed system. The proposed system consists of a Raspberry Pi 4, Raspberry Pi pico, MAX14870 dual motor driver, sound card and a box with motor fader.

The operation of the proposed system is explained in following. The electric bass guitar sound is initially input into a sound card connected to the Raspberry Pi 4. Then, it is processed by a compressor. The Raspberry Pi 4 provides a 3.3 V voltage to the motor fader. The Raspberry Pi Pico captures A/D conversion value at the center pin of the motor fader and transmits it to the Raspberry Pi 4 via serial communication. This value serves as the fader knob position, which can be estimated from the A/D conversion value. The Raspberry Pi 4 adjusts the volume of the processed sound based on the motor fader knob position controlled by fuzzy inference [15–18].

Figure 3 shows the membership functions and Table 1 show the Fuzzy Rule Base (FRB). The input parameters are the Amplitude of Input Sound (*AIS*) and the A/D Conversion value (*ADC*), while the output parameter is the Movement of Motor Fader (*MMF*) to control the position of the motor fader knob (volume control). In Table 1 and Fig. 3, *S* indicates small, *M* indicates middle and *L* indicates large. When the amplitude of input sound is small, *AIS* is small, the fader knob is at the lowest position, which means *ADC* is small. For example, if *AIS* is *S* and *ADC* is *S*, it is estimated that the volume is small. So, the fader will be moved to the top in order to increase the volume. Thus, the *MMF* outputs will be *L*. The compressor reduces the amplitude of audio signal when it exceeds a certain threshold value. In this way, the audio level is normalized and signal clipping and distortion are prevented.

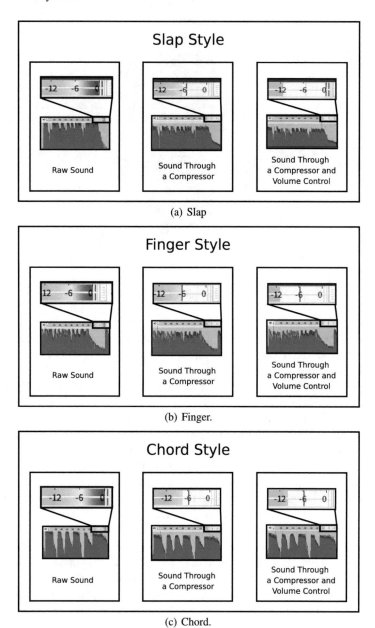

Fig. 5. Experimental result when moving position of motor fader knob.

3   Experimental Results

In this section, we discuss the experimental results. Figure 4 shows the fuzzy control output results for *MMF* decision. It can be seen that the output value is changing adaptively.

The electric bass guitar converts the vibration of the strings into an electrical signal using a microphone called a pickup. The electric signal from the electric bass guitar is very weak, so it is amplified by a bass amplifier to a level that can be the output through loudspeakers. Then, the DI converts the signal from a high impedance signal to a low impedance signal.

The sound of the performer by electric bass was recorded and used as the sound source for the experiment. We consider three performance styles: slap, finger and chord. We compare and evaluate the waveforms of the sound source with volume control by the proposed system with compressor only and without any control.

Figure 5 shows the visualization results of three waveforms for each of three playing styles. The waveforms were visualized with Audacity. In Fig. 5, the original sound of three styles has a waveform peak reaching 0 [*dB*], which causes sound cracking. In the case when using only the compressor, the peak of the waveform is suppressed to about −6 [*dB*]. When the volume is controlled by the proposed system, the peak of the waveform is suppressed to about −12 [*dB*] [19].

Considering the same recorded sound source, the volume of the electric bass guitar sound was controlled by the motor fader control. By changing the position of the motor fader knob, the volume of the electric bass guitar performance can be adjusted.

From experimental results, we found that the proposed system can suppress the sound peaks when the amplitude of the input signal increases rapidly. In addition, the proposed system can automatically adjust the volume of the electric bass guitar when playing sound by controlling the motor fader using fuzzy control.

4   Conclusions

In this paper, we proposed a real-time volume control of electric bass guitars based on fuzzy control. The sound of a performer by electric bass was recorded and used as the sound source for the experiment. We considered three performance styles: slap, finger and chord. We compared and evaluated the waveforms of the sound source with volume control by the proposed system with compressor only and without any control. From the experimental results, we conclude that the proposed system can control the volume of electric bass guitar performance sound by fuzzy control. Also, the proposed system can suppress the peaks of the performance sound by using a compressor and automatic volume control.

In the future, we would like to improve the proposed system and develop an automatic volume control device to replace the DI when sending the performance sounds to a mixer or interface.

Acknowledgement. This work was supported by JSPS KAKENHI Grant Number JP20K19793.

References

1. Steinmetz, C.J., et al.: Automatic multitrack mixing with a differentiable mixing console of neural audio effects. In: Proceedings of IEEE International Conference on Acoustics, Speech and Signal Processing, pp. 71–75 (2021)
2. Koo, J., et al.: Music mixing style transfer: a contrastive learning approach to disentangle audio effects. In: Proceedings of IEEE International Conference on Acoustics, Speech and Signal Processing, pp. 1–5 (2023)
3. Gonzalez, E.P., Reiss, J.D.: A real-time semiautonomous audio panning system for music mixing. EURASIP J. Adv. Signal Process. 1–10 (2010)
4. Scott, J., et al.: Automatic multi-track mixing using linear dynamical systems. In: Proceedings of the 8th Sound and Music Computing Conference, Padova, pp. 12–17 (2011)
5. Gonzalez, E.P., Reiss, J.: Automatic gain and fader control for live mixing. In: Proceedings of IEEE Workshop on Applications of Signal Processing to Audio and Acoustics, pp. 1–4 (2009)
6. Reiss, J.D.: Intelligent systems for mixing multichannel audio. In: Proceedings of IEEE 17th International Conference on Digital Signal Processing, pp. 1–6 (2011)
7. Malgaonkar, S., et al.: An AI based intelligent music composing algorithm: Concord. In: Proceedings of IEEE International Conference on Advances in Technology and Engineering, pp. 1–6 (2013)
8. Jarrah, A., et al.: Automotive volume control using fuzzy logic. J. Intell. Fuzzy Syst. **18**(4), 329–343 (2007)
9. Moffat, D., Sandler, M.: Automatic mixing level balancing enhanced through source interference identification. In: 146th Audio Engineering Society Convention, pp. 1–5 (2019)
10. De Man, B.: Towards a better understanding of mix engineering. Ph.D. thesis, Queen Mary University of London (2017)
11. Garrett, B.: Computer technology in modern music: a study of current tools and how musicians use them (2013)
12. Sanghoon, J., et al.: A fuzzy inference-based music emotion recognition system. In: 2008 5th International Conference on Visual Information Engineering, pp. 673–677 (2008)
13. Varun, O., et al.: Heuristic design of fuzzy inference systems: a review of three decades of research. Eng. Appl. Artif. Intell. **85**, 845–864 (2019)
14. Le Carrou, J.L., et al.: Influence of the player on the dynamics of the electric guitar. J. Acoust. Soc. Am. **146**, 3123–3130 (2019)
15. Yukawa, C., et al.: Evaluation of a fuzzy-based robotic vision system for recognizing microroughness on arbitrary surfaces: a comparison study for vibration reduction of robot arm. In: Proceedings of NBiS-2022, pp. 230–237 (2022)
16. Saito, N., et al.: Approach of fuzzy theory and hill climbing based recommender for schedule of life. In: Proceedings of IEEE LifeTech-2020, pp. 368–369 (2020)
17. Matsui, T., et al.: FPGA implementation of a fuzzy inference based quadrotor attitude control system. In: Proceedings of IEEE GCCE-2021, pp. 691–692 (2021)
18. Yukawa, C., et al.: Design of a fuzzy inference based robot vision for CNN training image acquisition. In: Proceedings of IEEE GCCE-2021, pp. 806–807 (2021)
19. Aisha, A., et al.: Audacity software analysis in analyzing the frequency and character of the sound spectrum. Jurnal Penelitian Pendidikan IPA **8**(1), 177–182 (2022)

Implementation of a Low-Cost and Intelligent Driver Monitoring System for Engine Locking in the Presence of Alcohol Detection or Drowsiness in Drivers

Evjola Spaho[✉], Orjola Jaupi, and Anxhela Cala

Faculty of Information Technology, Polytechnic University of Tirana, Mother Teresa Square, No. 4, 1001 Tirana, Albania
{espaho,orjola.jaupi,anxhela.cala}@fti.edu.al

Abstract. In this paper we propose a low-cost Driver Monitoring System (DMS) designed to reduce accidents caused by drunkenness and sleepiness and to enhance road safety. The main components of the system include an alcohol detection sensor (MQ-3), an IR sensor, an Arduino Uno, a motor driver, a DC motor, and an LCD. The system assesses the driver's condition to identify states that may lead to potential accidents and notifies the driver of potential dangers. Additionally, it has the capability to power off the engine if the driver is identified as drunk and/or sleepy. The driver's condition is analyzed using both the alcohol sensor and the IR sensor. As per the proposed system, the engine will automatically shut down, preventing the driver from operating the car upon detection of drunkenness and/or sleepiness.

1 Introduction

Numerous fatal road accidents result from either drowsy driving or driving under the influence of alcohol. The number of cars has increased too much, causing too much traffic and inducing distraction and fatigue in the drivers, making road safety a worldwide concern. Another factor related directly to road safety is alcoholic drivers. Alcohol remains a reason for brain function reduction and brings problems with thinking and reasoning which all play a crucial role in vehicle safe operation. Death rates because of drunk driving is different for every country and changes every year [1].

To improve road safety, it is very important to monitor drivers during their trip and to help avoid accidents during their drunk driving. It is advised that all new cars should have a system to monitor alcoholic drivers to alert them when need to, and lately not allow them to drive at all.

Driver Assistance Systems (DAS) and Driver Monitoring Systems (DMS) are intelligent systems that can support drivers in diverse ways.

DAS are used to enhance safety and to improve the driving experience. These systems use innovative technologies, artificial intelligence, and sensors to assist drivers.

DMS are designed to monitor and assess the drivers' conditions, actions, and behaviors. These systems use sensors to assess drowsiness, vital signs, and other driver conditions. DMS can detect drivers' distraction and create alerts to refocus in driving when

L. Barolli (Ed.): EIDWT 2024, LNDECT 193, pp. 443–451, 2024.
https://doi.org/10.1007/978-3-031-53555-0_42

smartphones or other entertainment systems are used. DMS is used for monitoring and ensuring the driver's attentiveness and safety. Its focus is to monitor the driver's state and behavior, issuing alerts and interventions when necessary [2, 3].

DMS can play a key role in detecting drivers' drowsiness. DMS with sensors and actuators can be used to detect closed eyelids and alert the driver to prevent accidents from exhaustion. DMS can also be used in identifying in drivers' potential signs of impairment due to alcohol consumption.

Many new car models are equipped with a DMS. However, there are many old cars that are not equipped with these systems. Advanced DMS are expensive and there is a need for low-cost systems that can be used in older vehicles.

The collaboration between DMS and intelligent traffic control systems [4] is a significant step toward creating safer, more efficient road networks. The microcontrollers [5] function as the intermediary between the DMS sensors and other systems within the vehicle, processing the incoming data, making decisions, and initiating appropriate responses to ensure the driver's safety.

The development of cost-effective DMSs using sensors and actuators represents an innovative approach to address this issue. These systems aim to reduce accidents and promote responsible driving practices by actively monitoring the driver's behavior.

In this paper, we propose a cost-effective DMS that monitors the driver's condition and controls the engine in cases where sleepiness and/or drunkenness is detected. Four different scenarios are considered:

1. 1.When the driver's alcohol level is above the norm and the driver is not sleepy
2. When the driver's alcohol level is above the norm and the driver is sleepy
3. When the driver's alcohol level is normal, and the driver is sleepy
4. When everything is normal, the driver's alcohol level is normal, and the driver is not sleepy.

The remainder of this paper is as follows. Section 2 presents related work for DAS and DMS, and explains innovative techniques used to detect a driver's distraction and fatigue. The proposed system is described in Sect. 3. Section 4 presents experimental evaluation and discussion. Finally, conclusions and a brief description of future work are given in Sect. 5.

2   Related Work

In recent years, there has been a growing emphasis on developing advanced Driver Monitoring Systems (DMS) aimed at enhancing road safety by detecting and mitigating issues such as driver drowsiness and drunkenness. This section reviews the related work for the development of DMS for identification of drowsiness and drunk driving.

In [6] authors proposed a Drowsy Driver Detection system using facial movement throughout machine learning techniques. The proposed system is evaluated in a driving computer game. Diverse environmental conditions, including variations in lighting, weather, and angles, can pose challenges for machine learning models and impact the system's accuracy. Different people have unique facial movements and expressions even under similar levels of drowsiness, and this adds complexity to the system's accuracy.

The efficiency of real-time processing is vital for driver safety systems, but the computational demands of running machine learning algorithms may affect system speed and responsiveness.

In [7] authors proposed a vision-based driver drowsiness detection system. The system estimates the drowsiness level of a driver based on facial expressions. The lighting compensation process is applied, and the system takes into consideration various illumination techniques. However, video stability techniques are not applied.

In [8] the authors introduce a driver drowsiness and monitoring system designed to assist drivers, with a primary goal of mitigating fatal crashes attributed to drowsiness and distraction. The system employs an analysis of eye blinks and yawn frequency to detect drowsiness, while distraction is addressed through head pose estimation and eye tracking. An alarm is activated if any of these conditions are detected. In this work, the proposed monitoring system does not detect drunk driving.

In [9] authors proposed a DDS. This system monitors the driver's eyes and sounds an alarm if the driver becomes drowsy. However, this system does not consider alcohol detection.

In [10] authors developed and implemented a system for detecting and alerting, in real-time, the driver's level of fatigue. The system uses video monitoring of the driver's eyes position, when open or closed, using a machine learning object detection algorithm. If the driver keeps the eyes closed for more than a certain amount of time the system triggers a set of warning lights and sounds. However, the proposed system does not take into consideration other parameters of the body condition.

In [11] authors propose a DMS based on fuzzy logic and a testbed. The system considers different parameters like Vehicle's Environment Temperature, Noise Level, Heart Rate, Respiratory Rate to decide Driver's Situational Awareness. Based on the system output, a smart box informs the driver and provides assistance. However, external factors that may affect the system are not considered.

In [12] authors proposed a low-cost remote driver sleep monitoring system. It is a control system that measures and analyzes drivers' movements. Authors designed a headset case that monitors the driver and sends an alarm if it predicts that the driver is sleeping. The system is based on angle values created by the driver's head. If the angle varies from 10° to 20°, it is identified as sleeping state for roll, pitch, and yaw angles. Bluetooth standard was used to transfer angle values and status information collected from the designed wearable headset to the mobile application. Mobile application is used to sends a notification to the driver when sleep status is detected and sends an informative SMS to the headquarters if the driver is unresponsive to this notification.

In [13] authors proposed a design monitoring system for logistics truck drivers. The focus is to reduce logistical truck driver's accidents and improve driver safety. Design thinking is used as a method for creating an IoT prototype. It offers real time analysis and results during driving to prevent accidents.

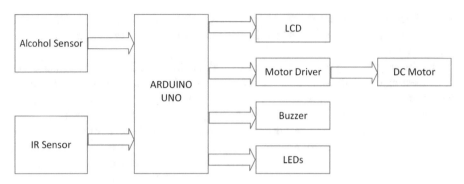

Fig. 1. The main components of the system.

A blinking eye sensor is used to check the eye status. A limit value is set and when it is exceeded, a sound will notify the sleepy driver. The proposed DMS can provide real-time reports on vehicles.

In [14] authors implemented an IoT based device using Raspberry Pi3 that monitors the drivers face while driving the vehicle. The device is built with Raspberry Pi3 attaching a camera and a buzzer. It collects data while the driver is driving and sends them to a trained model to identify if there is any sign of drowsiness. It produces warnings and instructions for the driver to stop the vehicle if there is a sign of drowsiness.

In this work, we propose a simple DMS based on sensors and actuators for detection of drowsiness and drunkenness that notifies the driver of potential danger and shuts down the engine.

3   Proposed System

In this paper we propose and create a prototype of an alcohol and drowsiness detection system. Main components of the system are Arduino Uno, Alcohol Sensor (MQ-3), Motor driver, IR sensor, buzzer, DC motor, LEDs, potentiometer, LCD, resistors, and power supply, and are presented in Fig. 1.

IR sensor is used to monitor the driver's blink rate. An alerting mechanism gives warnings to the driver throughout buzzer when abnormal situation is detected.

Arduino Uno is connected to the alcohol sensor, which gives the value of the alcohol level detected and IR sensor which detects drowsiness. The MQ-3 alcohol sensor operates on 5V DC and heating consumption approximately 750 mW. It can detect alcohol concentrations ranging from 25 to 500 ppm. The threshold value of BAC considered is 0.8 mg/L. Above this level, drivers may manifest euphoria, sedation, impaired coordination, decreased sensory responses to stimuli or decreased judgement.

The implemented prototype is shown in Fig. 2.

The flowchart of the system is shown in Fig. 3. In our system, if the alcohol level is above the norm, the vehicles engine will shut down, a red LED light will light up and the buzzer will sound an alarm.

Fig. 2. Implemented system.

Fig. 3. The flowchart of the system.

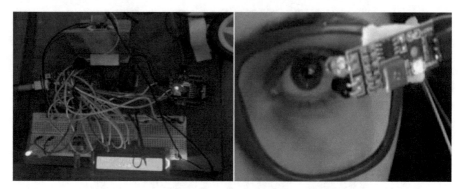

Fig. 4. Scenario 1 where the driver is drunk.

In another case, when the alcohol level is at the determined normal rate, the green LED light stays on, the buzzer will be off, and the vehicle's engine will turn on to function normally.

The same logic applies to the detection of drowsiness. If the driver keeps eyes closed for more than 5 s, the alarm sounds and engine shutdown cycle will start.

The detection of drowsiness means the LED light that is located behind the sensor turns green when it detects sleep and does not light up when everything is normal.

4 Experimental Evaluation and Discussion

The driver's eye movements and blink rate are monitored by IR sensor to detect signs of drowsiness. The alcohol sensor will detect the presence of alcohol. An alerting system will notify in real time the driver when an abnormal situation is detected.

The goal of the proposed system is to detect alcohol, display the value of this detection on the LCD or notify driver drowsiness. For alcohol detection, we set a threshold value. Several scenarios have been considered by the system.

Scenario 1: When the alcohol sensor has detected a higher alcohol level than the defined threshold, and the IR sensor has not detected drowsiness, our system will display on the LCD the word DRUNK! as well as the alcohol level. The red LED will light up and the buzzer will make a sound, and the DC motor will stop. In Fig. 4 is presented this scenario where the driver is drunk and not sleepy.

Scenario 2: When alcohol and sleepiness are detected, our system will display on the LCD "Drunk & Sleepy", show the alcohol level, light up the red LED, the buzzer will make sound and DC motor will stop. Figure 5 presents this scenario where the driver is drunk and sleepy.

Scenario 3: When no alcohol is detected but only drowsiness, the writing "SLEEPY!" appears on the LCD, the red LED and the buzzer are on, meanwhile the DC motor is off. Figure 6 shows this scenario where the driver is not drunk but is sleepy.

Scenario 4: When neither alcohol nor drowsiness is detected, the text "NORMAL!" appears on the LCD, only the green LED light stays on and the DC motor works normally. Figure 7 shows this scenario where the situation is normal.

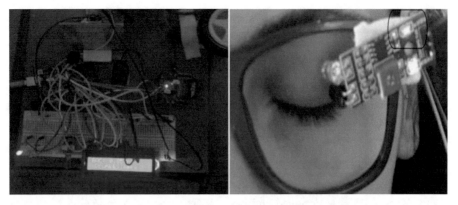

Fig. 5. Scenario 2 where the driver is drunk and sleepy.

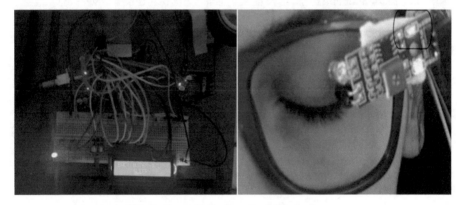

Fig. 6. Scenario 3 where the driver is sleepy.

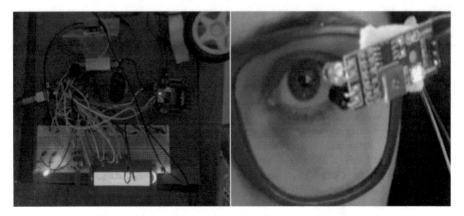

Fig. 7. Scenario 4 where the driver is in normal condition.

Advantages of the implemented system:

– Alcohol detection with engine lock can help avoid accidents because of alcohol use.
– The proposed system is low-cost and can be used in old vehicles.

Limitations of the implemented system:

– The driver can hide his breath of alcohol by not allowing the sensor to measure the real alcohol level.
– Drivers should wear glasses with the IR sensor all the time to detect drowsiness.

5 Conclusions

In this paper a DMS is designed and implemented using a combination of sensors, actuator, and an Arduino platform. The developed system is an effective and affordable DMS, able to monitor and respond to driver behavior in real-time and with the potential to significantly enhance road safety.

The proposed system evaluates the driver's condition using alcohol sensor and IR sensor and controls the engine. Its performance is evaluated for different scenarios.

The proposed system is cost-effective and able to detect, alert drivers and control the engine in cases of drunken and drowsy state while driving.

In the future we would like to further improve the proposed DMS with new sensors, cameras, and new features. We would like to consider the application of artificial intelligence techniques for abnormal situation detection and implementing machine learning algorithms for pattern recognition to reduce false positives and false negatives.

Acknowledgments. Authors would like to thank Polytechnic University of Tirana for providing financial support for this research work.

References

1. Isong, B., Khutsoane, O., Dladlu, N.: Real-time monitoring and detection of drinkdriving and vehicle over-speeding. Int. J. Image Graph. Signal Process. **9**(11), 1–9 (2017). https://doi.org/10.5815/ijigsp.2017.11.01
2. Dong, Y., Hu, Z., Uchimura, K., Murayama, N.: Driver inattention monitoring system for intelligent vehicles: a review. IEEE Trans. Intell. Transp. Syst. **12**(2), 596–614 (2011). https://doi.org/10.1109/TITS.2010.2092770
3. Begum, S.: Intelligent driver monitoring systems based on physiological sensor signals: a review. In: International Conference on Intelligent Transportation Systems (ITSC), IEEE, pp. 282–289. IEEE (2013). https://doi.org/10.1109/ITSC.2013.6728246
4. Opoku, D., Kommey, B.: FPGA-based intelligent traffic controller with remote operation mode. Int. J. Innov. Technol. Interdiscipl. Sci. **3**(3), 490–500 (2020). https://doi.org/10.15157/IJITIS.2020.3.3.490-500
5. Mohapatra, B.N., Mohapatra, R.K., Mirpagar, M., Quershi, A.: Ultrasonic based easy parking system based on Microcontroller. Int. J. Innov. Technol. Interdiscipl. Sci. **3**(2), 429–434 (2020). https://doi.org/10.15157/IJITIS.2020.3.2.429-434

6. Vural, E., Cetin, M., Ercil, A., Littlewort, G., Bartlett, M., Movellan, J.: Drowsy driver detection through facial movement analysis. In: Lew, M., Sebe, N., Huang, T.S., Bakker, E.M. (eds) Human–Computer Interaction. HCI 2007. LNCS, vol 4796, pp. 6–18. Springer, Heidelberg (2007). https://doi.org/10.1007/978-3-540-75773-_2
7. Yao, K.P., Lin, W.H., Fang, C.Y., Wang, J.M., Chang, S.L., Chen, S.W.: Real-time vision-based driver drowsiness/fatigue detection system. In: 71st Vehicular Technology Conference pp. 1–5. IEEE (2010). https://doi.org/10.1109/VETECS.2010.5493972
8. Rozali, R.A.F., et al.: Driver drowsiness detection and monitoring system (DDDMS). Int. J. Adv. Comput. Sci. Appl. **13**(6), 1–7 (2022). https://doi.org/10.14569/IJACSA.2022.0130691
9. Kannan, R., Jahnavi, P., Megha, M.: Driver drowsiness detection and alert system. In: 2023 IEEE International Conference on Integrated Circuits and Communication Systems (ICI-CACS), Raichur, India, pp. 1–5 (2023). https://doi.org/10.1109/ICICACS57338.2023.101 00316
10. Adochiei, R., et al.: Drivers' drowsiness detection and warning systems for critical infrastructures. In: 2020 International Conference on e-Health and Bioengineering (EHB), Iasi, Romania, pp. 1–4 (2020). https://doi.org/10.1109/EHB50910.2020.9280165
11. Bylykbashi, K., Qafzezi, E., Ikeda, M., Matsuo, K., Barolli, L.: Fuzzy-based driver monitoring system (FDMS): implementation of two intelligent FDMSs and a testbed for safe driving in VANETs. Future Gener. Comput. Syst. **105**, 665–674 (2020). https://doi.org/10.1016/j.future.2019.12.030
12. Eksi, Z., Camgöz, A.N., Özarslan, M., Sözer, E., Çelebi, M.F., Feyzioglu, A.: A low-cost remote driver sleep monitoring system. J. Mechatron. Artif. Intell. Eng. **3**(2), 101–106 (2022). https://doi.org/10.21595/jmai.2022.22975
13. Dewi, R., Suzianti, A., Puspasari, M.: Design of driver monitoring system for logistics truck with design thinking approach. In: Proceedings of the 4th Asia Pacific Conference on Research in Industrial and Systems Engineering, May 2021, pp. 481–487 (2021). https://doi.org/10.1145/3468013.3468645
14. Patil, S., Mujawar, A., Kharade, K., Kharade, S.K., Katkar, S.V., Kamat, R.K.: Drowsy driver detection using Opencv and raspberry Pi3. Webology. **19**(2), 6003–6010 (2022)

An Accident Detection System for Private Lavatories Using Fuzzy Control and Thermal Camera

Tomoaki Matsui[1]([✉]), Tetsuya Oda[2], Kyohei Wakabayashi[3], Yuki Nagai[3], Chihiro Yukawa[3], and Leonard Barolli[4]

[1] Graduate School of Science and Engineering, Okayama University of Science (OUS), 1-1 Ridaicho, Kita-ku, Okayama 700–0005, Japan
r23smf3vv@ous.jp

[2] Department of Information and Computer Engineering, Okayama University of Science (OUS), 1-1 Ridaicho, Kita-ku, Okayama 700–0005, Japan
oda@ous.ac.jp

[3] Graduate School of Engineering, Okayama University of Science (OUS), 1-1 Ridaicho, Kita-ku, Okayama 700–0005, Japan
{t22jm24jd,t22jm23rv,t22jm19st}@ous.jp

[4] Department of Information and Communication Engineering,
Fukuoka Institute of Technology, 3-30-1 Wajiro-higashi, Higashi-ku, Fukuoka 811-0295, Japan
barolli@fit.ac.jp

Abstract. In this paper, we propose an accident detection system for private lavatories (toilets) using fuzzy control and thermal camera. In the proposed system, fuzzy control input values are obtained from a low-resolution thermal camera and the output value is used for detecting the accidents or falls of persons. The proposed Fuzzy-based system uses simplified fuzzy control to reduce the computational cost. Therefore, the proposed system can be implemented in small and power-saving devices. From the evaluation result, we found that the proposed system can detect falls or accidents from low-resolution thermal camera.

1 Introduction

Accidents such as falls of elderly people, hospitalized patients and people with disabilities are one of the social problems that need to be solved. Such accidents need to be detected as soon as possible in order to prevent further problems [1,2]. However, because private lavatories are closed environments, the detection of accidents may be delayed [3]. The detection systems for such accidents include camera surveillance [4–6], wearable terminal surveillance [7–9] and thermal camera surveillance [10–12].

Camera surveillance provides reliable information, but privacy issues makes it impractical. Accident detection using wearable terminals places the burden to wearing the device by the lavatory user. Thermal imaging camera surveillance can detect accidents while maintaining privacy if low-resolution sensors are used. Therefore, there are several accident detection systems that use low-resolution sensors.

L. Barolli (Ed.): EIDWT 2024, LNDECT 193, pp. 452–459, 2024.
https://doi.org/10.1007/978-3-031-53555-0_43

In recent years, these detection systems use computationally demanding algorithms such as Convolutional Neural Networks (CNN) [13] [10–12]. We consider the application of fuzzy control [14–18] to detect private lavatory accidents from low-resolution images. The Fuzzy Logic (FL) and Fuzzy Control (FC) can incorporate experts knowledge and rules in the system [14]. It also has the advantage of low computational load [19], so it can be operate under various computer resource constraints. Therefore, the system can be implemented with more compactly and in low-power-consumption devices.

In our system, we determine fuzzy membership functions and Fuzzy Rule Base (FRB) based on experiments in a virtual lavatory environment. The proposed system normalizes the values obtained from the thermal camera to values between 0 and 255 in order to operate without problems in environments of various temperatures. The sum of a specific range of normalized values is then obtained and used as input values for fuzzy control.

In this paper, we propose an accident detection system for private restrooms or lavatories using FC and thermal cameras. The fuzzy control parameters are determined based on experiments in a virtual toilet environment.

The paper is organized as follows. In Sect. 2, we present the proposed system. In Sect. 3, we describe the environment in which the experiments are performed and the simulation results of the fuzzy control. Finally, conclusions and future work are given in Section 4.

2 Proposed System

In the proposed system, we use a mlx90640 thermal camera. It takes pictures with 24×32 pixels and runs at 3.3 [V] We adjusted the FC parameters in an experimental environment that replicates an actual private lavatory (toilet). The proposed system has a thermal camera installed at the same height as the toilet seat. The position of the thermal camera was determined by considering the patient posture. In this position, we got the thermal image features for both tipping over and not tipping over.

In the proposed system, the maximum value of each pixel is 255 and the minimum value is 0, which are relatively normalized values, in order to obtain values that are independent of the temperature environment of the lavatory. The maximum value was set to 255 because an unsigned 8-bit integer type was used to reduce the amount of data. Let x and y be the locations of the pixels in the thermal camera and C be the temperature in degrees Celsius obtained at each pixel, then the value for each pixel is obtained as follows.

$$a_{(x,y)} = (C_{(x,y)} - C_{max}) \times 255/(C_{max} - C_{min})$$

Table 1. FRB.

No.	S_1	S_2	S_3	Fall risk	No.	S_1	S_2	S_3	Fall risk	No.	S_1	S_2	S_3	Fall risk
1	H	H	H	L	10	M	H	H	L	19	L	H	H	M
2	H	H	M	VL	11	M	H	M	L	20	L	H	M	L
3	H	H	L	VL	12	M	H	L	VL	21	L	H	L	L
4	H	M	H	L	13	M	M	H	M	22	L	M	H	H
5	H	M	M	L	14	M	M	M	L	23	L	M	M	H
6	H	M	L	VL	15	M	M	L	L	24	L	M	L	L
7	H	L	H	M	16	M	L	H	H	25	L	L	H	VH
8	H	L	M	M	17	M	L	M	H	26	L	L	M	VH
9	H	L	L	M	18	M	L	L	M	27	L	L	L	M

where C_{max} is the maximum value of all pixels and C_{min} is the minimum value of all pixels. The higher is the relative temperature and closer to 255, the lower is the temperature and closer to 0.

In general, the temperature of human body is higher than the temperature inside the private toilet, so the value is higher at certain positions of the human body. In addition, the overall value of $a_{(x,y)}$ is found to be slightly larger when there is no person in the thermal camera field of view.

We are focusing on falls and accidents in a private lavatory (toilet). The proposed system uses simplified fuzzy control system [20] to detect accidents involving falls in private lavatories. The input values S_1, S_2, and S_3 of the fuzzy control system are the sum of $a_{(x,y)}$ in the range shown in Fig. 1. Based on the data obtained from the experiments, we defined the fuzzy membership function shown in Fig. 2 and the FRB shown in Table 1. The FRB is designed such that the smaller the values of S_1 and S_2 and the larger the value of S_3, the closer the posture is to a tipping posture and is more dangerous situation.

The proposed FL-based system acquires thermal camera images every second and decides the output value *Fall risk*, which is used detect a fall or accident. The output values *Fall risk* are considered as risk values, which are used to prevent false detection of falls and are changed (increased or decreased) by FC. When the risk value reaches 1.0, the system notify the manager or other people that the person has fallen. If the values of temperatures momentarily leads to a situation where it is possible to be in a tipping posture, it will not cause false detection. When the risk value decreases and approaches zero, the values of temperature are considered to be a non-tipping situation.

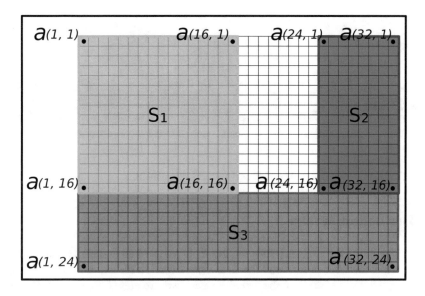

Fig. 1. Ranges of thermal images: S_1, S_2 and S_3.

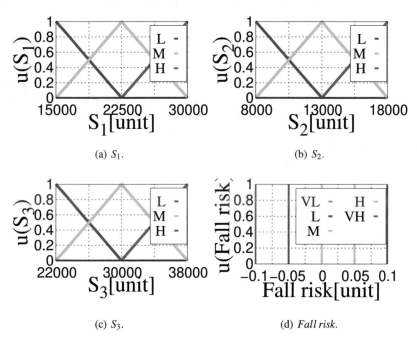

(a) S_1.

(b) S_2.

(c) S_3.

(d) *Fall risk*.

Fig. 2. Membership functions of FC-based system.

 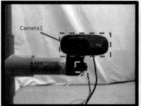 

(a) Entire experimental envi- (b) Camera1 and Thermal (c) Position of Camera1 and
ronment.   Camera.   Camera2.

Fig. 3. Experimental Environment.

3   Evaluation Results

Figure 3(a) shows the experimental environment. We use a chair instead of a toilet seat. Figure 3(b) shows the camera and thermal camera used in the experiment. To compare the thermal image results with the actual images, two cameras (Camera1 and Camera2) are set to take images at the same time

Figure 3(c) shows the position of Camera1 and Camera2. Camera1 shows the image taken by camera in Fig. 3(b) and Camera2 shows the actual image so that the entire image is visible. Camera1 is used for comparison with the thermal image and Camera2 is used to capture the entire posture.

Figure 4 shows the output result of FC-based system for each input value. From the results, we see that the lower are the values of S_1 and S_2, and the higher is the value of S_3, the higher is the output value. This indicates that the output is designed as expected. Figure 5, 6, 7, 8 and 9 show the images of Camera1 and Camera2 and the corresponding thermal images. While, Table 2 shows examples of input values and output results for each case. Figure 5 shows the case of sitting on the toilet seat and Fig. 6 the case of standing in front of the toilet seat. In both cases the output value is negative.

Figures 7 and 8 are two different falling postures. Also, the case of Fig. 8, which is only slightly visible on thermal cameras, the output values is positive. The case where no one is within range of the thermal camera is shown in Fig. 9, which has a negative output value.

From the above results, we can see that the output value is positive in the case of a fall and negative in all other cases. Also, the increase or decrease in risk value is under control.

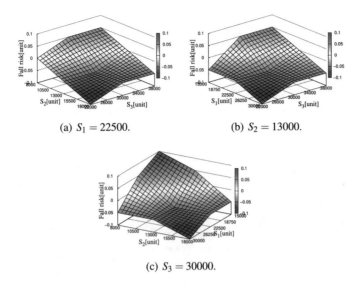

(a) $S_1 = 22500$.   (b) $S_2 = 13000$.

(c) $S_3 = 30000$.

Fig. 4. Output results of FC-based system.

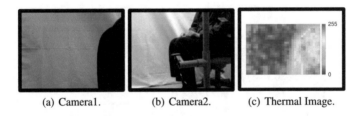

(a) Camera1.   (b) Camera2.   (c) Thermal Image.

Fig. 5. Case 1: Sitting on toilet seat.

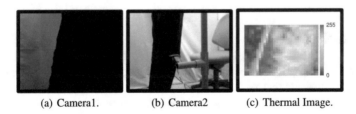

(a) Camera1.   (b) Camera2   (c) Thermal Image.

Fig. 6. Case 2: Standing in front of toilet seat.

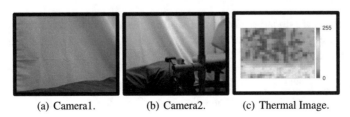

(a) Camera1.   (b) Camera2.   (c) Thermal Image.

Fig. 7. Case 3: Type 1 of falling down.

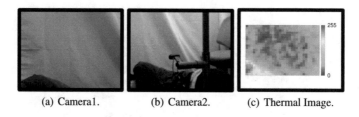

(a) Camera1.   (b) Camera2.   (c) Thermal Image.

Fig. 8. Case 4: Type 2 of falling down.

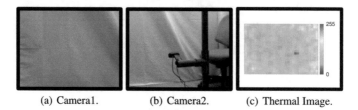

(a) Camera1.   (b) Camera2.   (c) Thermal Image.

Fig. 9. Case 5: There is no person in view of thermal camera.

Table 2. Input and output values of fuzzy control in Fig. 5, 6, 7, 8 and 9

	S_1	S_2	S_3	Fall risk
Case 1	9,855	20,986	25324	−0.041399296211964406
Case 2	30,070	17,062	35479	−0.06105804438205279
Case 3	11,978	7,907	40,471	0.1
Case 4	12,256	8,152	22,922	0.01482558562817634
Case 5	26,654	17,073	29,238	−0.0657727563351981

4   Conclusions

In this paper, we proposed an accident detection system for private lavatories (toilets) using fuzzy control and thermal camera. The proposed FC-based system uses simplified fuzzy control to reduce the computational cost. Therefore, the proposed system can be implemented in small and power-saving devices. From the evaluation result, we found that the proposed system can detect falls or accidents from low-resolution thermal camera.

In the future, we intend to develop other intelligent systems for different accidents. In addition, we will implement an alert system to notify the system administrator or manager when a fall is detected.

Acknowledgement. This work was supported by JSPS KAKENHI Grant Number JP20K19793 and Grant for Promotion of OUS Research Project (OUS-RP-23-4).

References

1. Bagala, F., et. al.: Evaluation of accelerometer-based fall detection algorithms on real-world falls. PLoS One. **7**, e37062 (2012)
2. Igual, R., et al.: Challenges, issues and trends in fall detection systems. Biomed. Eng. Online **12**(1), 1–24 (2013)
3. Inamasu, J., Miyatake, S.: Cardiac arrest in the toilet: clinical characteristics and resuscitation profiles. Environ. Health Prev. Med. **18**, 130–135 (2013)
4. Rougier, C., et al.: Robust video surveillance for fall detection based on human shape deformation. IEEE Trans. Circuits Syst. Video Technol. **21**(5), 611–622 (2011)
5. Auvinet, E., et al.: Fall detection with multiple cameras: an occlusion-resistant method based on 3-D Silhouette vertical distribution. IEEE Trans. Inf. Technol. Biomed. **15**(2), 290–300 (2010)
6. Ramirez, H., et al.: Fall detection and activity recognition using human skeleton features. IEEE Access **9**, 33532–33542 (2021)
7. chen, X., et. al.: Subject-independent slow fall detection with wearable sensors via deep learning. In: Proceedings of The 2020 IEEE SENSORS, pp. 1–4 (2020)
8. Beak, W., et. al.: Real life applicable fall detection system based on wireless body area network. In: Proceedings of the IEEE 10-th Consumer Communications and Networking Conference, pp. 62–67 (2013)
9. Hu, K., et al.: Application for detecting falls for elderly persons through internet of things combined with pulse sensor. Sens. Mater. **35**(11), 3655–3669 (2023)
10. Tateno, S., Miyatake, S.: Privacy-preserved fall detection method with three-dimensional convolutional neural network using low-resolution infrared array sensor. Sensors **20**(20), 5957 (2020)
11. Sixsmith, A., et. al.: An unobtrusive fall detection system using low resolution thermal sensors and convolutional neural networks. In: 2021 43rd Annual International Conference of the IEEE Engineering in Medicine & Biology Society, pp. 6949–6952 (2021)
12. Kawashima, T., et. al.: Action recognition from extremely low-resolution thermal image sequence, In: Proceedings of the 14-th IEEE International Conference on Advanced Video and Signal Based Surveillance, pp. 1-6 (2017)
13. Li, Z., et al.: A survey of convolutional neural networks: analysis, applications, and prospects. IEEE Trans. Neural Netw. Learn. Syst. **33**(12), 6999–7019 (2021)
14. Passino, K.M., et al.: Fuzzy Control, vol. 42. Addison-Wesley, Boston (1998)
15. Mamdani, E.H.: Application of fuzzy algorithms for control of simple dynamic plant. In: Proceedings of the Institution of Electrical Engineers, vol. 121, No.12, pp. 1585–1588 (1974)
16. Matsui, T., et. al.: FPGA implementation of a fuzzy inference based quadrotor attitude control system. In: Proceedings of the IEEE 10-th Global Conference on Consumer Electronics, pp. 770–771 (2021)
17. Matsui, T., et. al.: FPGA implementation of a interval type-2 fuzzy inference for quadrotor attitude control. In: Proceedings of the 10-th International Conference on Emerging Internetworking, Data & Web Technologies, pp. 357–365 (2022)
18. Matsui, T., et. al.: FPGA implementation of a interval type-2 fuzzy inference based wildfire monitoring system. In: Proceedings of the IEEE 11-th Global Conference on Consumer Electronics, pp. 876–877 (2022)
19. Munakata, T., Jani, Y.: Fuzzy systems: an overview. Commun. ACM **37**(3), 69–76 (1994)
20. Mizumoto, M.: Realization of PID controls by fuzzy control methods. Fuzzy Sets Syst. **70**(2–3), 171–182 (1995)

A Fuzzy-Based System for Decision of Compressed Image Degree Using Cluster Load Level

Kenya Okage[1], Tetuya Oda[1(✉)], Yuki Nagai[2], Chihiro Yukawa[2],
Kyohei Wakabayashi[2], and Leonard Barolli[3]

[1] Department of Information and Computer Engineering,
Okayama University of Science (OUS), 1-1 Ridaicho, Kita-ku, Okayama 700-0005, Japan
t20j020ok@ous.jp, oda@ous.ac.jp

[2] Graduate School of Engineering, Okayama University of Science (OUS),
1-1 Ridaicho, Kita-ku, Okayama 700–0005, Japan
{t22jm24jd,t22jm23rv,t22jm19st}@ous.jp

[3] Department of Information and Communication Engineering,
Fukuoka Institute of Technology, 3-30-1 Wajiro-higashi, Higashi-ku, Fukuoka 811-0295, Japan
barolli@fit.ac.jp

Abstract. In recent years, floods caused by overflowing rivers due to heavy rains have caused significant damages. To prevent human casualties in the event of a disaster, it is necessary to provide appropriate information for prompt evacuation. However, in general the municipalities that provide evacuation information use low computational resource computers. So, there is a risk that the servers will overload by concentration of evacuees when a disaster occurs, causing delays or stoppages of evacuation information. Therefore, clustering of computers with low computational resources as a single computational resource is very important. Also, the continuous provision of map image of evacuated areas can be possible by adaptively compressing map image based on the server load condition. In this paper, we propose a Fuzzy-based system for decision of the compressed map image degree of evacuated areas during disasters using cluster load level. The experimental results show that the proposed system can change the compression of map image of evacuated areas according to the load level of the cluster system. Also, the evacuation route information displayed in the map image of evacuated areas is sufficiently readable.

1 Introduction

In recent years, heavy rainfall has become more frequent around the world with climate change [1]. In addition, with the intensification of disasters, floods caused by overflowing rivers by torrential rains have caused significant damages. To prevent human casualties in the event of a disaster, it is necessary to provide appropriate information before, during and after a disaster to evacuate the people promptly [2]. However, municipalities in general provide evacuation information by using low computational resource computers, which may degrade the performance with the aging of components such as CPU

L. Barolli (Ed.): EIDWT 2024, LNDECT 193, pp. 460–469, 2024.
https://doi.org/10.1007/978-3-031-53555-0_44

and memory. Also, there is a risk that the servers will be overloaded by the concentration of evacuees when a disaster occurs, causing delays or stoppages of the information provided by the local governments. This may cause difficulties to evacuees to make quick and appropriate actions by themselves. Therefore, it is necessary to provide continuous information to evacuees even with systems using low computational resource computers.

The Fuzzy Logic (FL) [3–7] can be used for adaptively controlling different systems. Also, there are many studies for clustering of low computing resource computers [8–12]. A High Availability (*HA*) cluster system is a collection of computers that operate as a single system by connecting multiple nodes, increasing system availability by minimizing system downtime in the event of a failure.

It is possible to increase the availability of systems that provides evacuation information for municipalities by making effective use of low computing resource computers by using the *HA* cluster. We consider to decrease the cluster load by controlling compression level of the map image of evacuated areas based on the cluster load level [13–17]. Thus, even when the cluster load increases, it is possible to continuosly provide the map image of evacuated areas to evacuees.

In this paper, we propose a clustering system using low computational resource computers that continuously can provide information by adaptively changing the compression level of the map image of evacuated areas based on the cluster load level calculated by fuzzy control. In the proposed system, a single computing resource is constructed by clustering low computing resource computers in a *HA* cluster. By increasing the availability of the system, the appropriate information can be continuously provided to evacuees.

The structure of the paper is as follows. In Sect. 2, we present the proposed system. In Sect. 3, we describe the experimental results. Finally, conclusions and future work are given in Sect. 4.

2   Proposed System

Figure 1 shows the image of the proposed system. The proposed system is a *HA* cluster system composed of low computational resource computers and decides the Cluster Load Level (*CLL*) based on fuzzy control. The *CLL* is expressed as a value between 0 to 1.0 [*unit*]. After selecting an active node of a *HA* cluster, the degree of image compression is decided by *CLL* value calculated by fuzzy control. Then, the compressed map image of evacuated areas is sent to evacuees.

Figure 2 shows the membership functions of the fuzzy-based system. While in Fig. 2(a), Fig. 2(b) and Fig. 2(c) are shown the membership functions of the input parameters and Fig. 2(d) shows the membership function of the output parameter. The Fuzzy Rule Base (FRB) used for control is shown in Table 1.

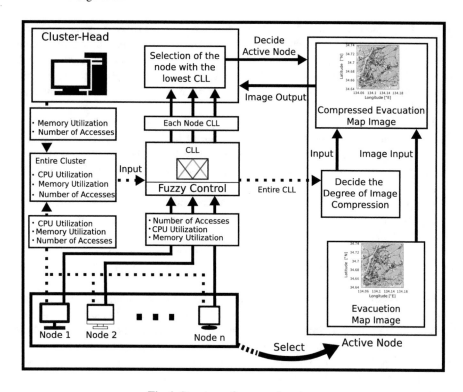

Fig. 1. Structure of proposed system.

The proposed system considers three parameters for deciding the cluster load level. The CPU utilization and memory utilization indicate the number of executed processes. While, when a disaster occurs the number of accesses is an indicator to evaluate the degree of cluster load. Therefore, the CPU Utilization (*CU*), Memory Utilization (*MU*) and Number of Accesses (*NA*) are the input parameters and *CLL* is the output parameter of fuzzy-based control system.

In the proposed system, *CU* valuses are considered from 0 to 100 [%], *MU* from 0 to 100 [%] and *NA* from 0 to 3000 [*unit*]. The input parameters have three levels: High (*H*), Middle (*M*) and Low (*L*), while the output parameter has five levels: *H*, Middle High (*MH*), *M*, Middle Low (*ML*) and *L*. When the *CLL* value is low, the system provides images with good quality. While when the *CLL* value is high, the system reduces the quality of the map image of evacuated areas by compressing the image, thus reducing the image size.

Table 2 shows the image quality after compression according to *CLL*, which is calculated by Fuzzy-based control system at each *HA* cluster node. Then each node sends *CLL*, total memory, memory usage and *CU* to the cluster head. The cluster head decides the node with the lowest *CLL* from the *CLL* received from each node as the active node and the other nodes as standby nodes. The cluster head calculates the *MU* and *CU* of

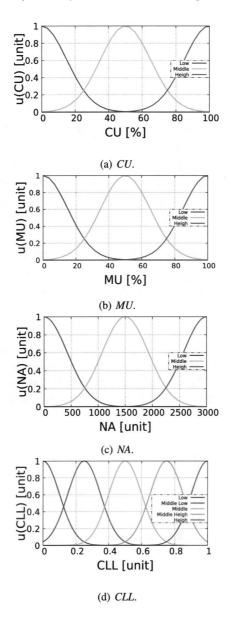

(a) *CU*.

(b) *MU*.

(c) *NA*.

(d) *CLL*.

Fig. 2. Membership functions.

the entire cluster system based on the total amount of memory, memory usage and *CU* received from each node. The *CLL* of the entire cluster system is calculated using *MU*, *CU* and *NA* of the entire cluster.

Table 1. FRB.

CU	MU	NA	CLL	CU	MU	NA	CLL	CU	MU	NA	CLL
H	H	H	H	M	H	H	H	L	H	H	MH
H	H	M	H	M	H	M	MH	L	H	M	M
H	H	L	MH	M	H	L	MH	L	H	L	M
H	M	H	H	M	M	H	MH	L	M	H	ML
H	M	M	MH	M	M	M	M	L	M	M	L
H	M	L	MH	M	M	L	M	L	M	L	L
H	L	H	MH	M	L	H	ML	L	L	H	L
H	L	M	M	M	L	M	L	L	L	M	L
H	L	L	M	M	L	L	L	L	L	L	L

Table 2. Compression results of map image of evacuated areas.

	$0 \leq CLL < 0.2$	$0.2 \leq CLL < 0.4$	$0.4 \leq CLL < 0.6$	$0.6 \leq CLL < 0.8$	$0.8 \leq CLL \leq 1.0$
Image quality [%]	100	80	60	40	20

The cluster head sends the map image of evacuated areas and the derived *CLL* of the entire cluster system to the active node. The receiving node compresses the image based on the map image of evacuated areas and the *CLL* of the entire cluster system. After compression, the active node sends the map image of evacuated areas to the cluster head.

3　Experimental Results

Figure 3 shows the output results of Fuzzy-based control system. While Fig. 3(a) shows the results of *CLL* for *NA* of 1 [*unit*], Fig. 3(b) for *NA* of 1000 [*units*], Fig. 3(c) for *NA* of 2000 [*units*] and Fig. 3(d) for *NA* of 3000 [*units*]. From the experimental results, it can be seen that when *NA* is large furthermore both *CU* and *MU* are large, *CLL* becomes large.

Table 3. Specifications of computers configuring cluster system.

Computer	OS	CPU	Memory
Cluster head	Ubuntu 18.04	Intel Core i7-870 2.93GHz	DDR3-1333 $2GB \times 2$
Node 1	Ubuntu 18.04	Intel Core i7-870 2.93GHz	DDR3-1333 $2GB \times 2$
Node 2	Ubuntu 16.04	Intel Core i7-870 2.93GHz	DDR3-1333 $2GB \times 2$

Table 4. Compression results of map image of evacuated areas.

	$0 \leq CLL < 0.2$	$0.2 \leq CLL < 0.4$	$0.4 \leq CLL < 0.6$	$0.6 \leq CLL < 0.8$	$0.8 \leq CLL \leq 1.0$
Image Quality [%]	100	80	60	40	20
Image Data Size [KB]	233	121	86	69	48

Table 3 shows the specifications of computers configuring the cluster system. The *HA* cluster system is created by three low computational resource computers, while the network structure is a tree structure. The target area of the map image of evacuated areas is about 9 [km] square centered at 34.690 [$\circ N$] latitude and 134.120 [$\circ E$] longitude at Yoshii River in Setouchi City, Okayama Prefecture in Japan. The map image is obtained from OpenStreetMap (OSM).

Table 4 shows the compression results of the map image of evacuated areas. While, in Fig. 4 are shown the visualization results of the compressed map image of evacuated areas, where Fig. 4(a) shows the original image, Fig. 4(b) shows the image with the image quality changed to 80 [%], Fig. 4(c) shows the image with the image quality changed to 60 [%], Fig. 4(d) shows the image with the image quality changed to 40 [%], and Fig. 4(e) shows the image with the image quality changed to 20 [%].

From the experimental results, we see that the amount of data on the image can be reduced by lowering the quality of the image. Also, it can be seen that the evacuation route information displayed in the map image of evacuated areas is sufficiently readable.

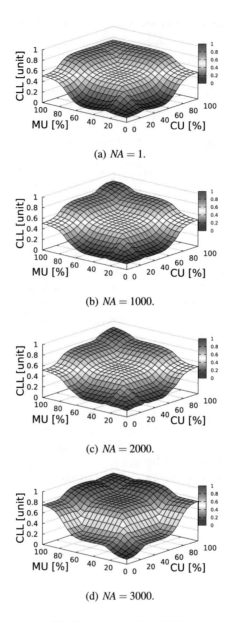

(a) *NA* = 1.

(b) *NA* = 1000.

(c) *NA* = 2000.

(d) *NA* = 3000.

Fig. 3. Output results of *CLL*.

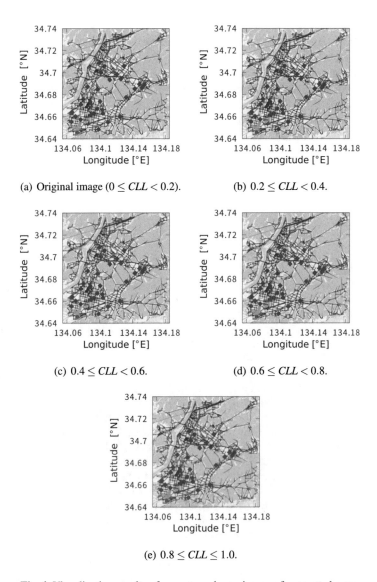

(a) Original image ($0 \leq CLL < 0.2$).

(b) $0.2 \leq CLL < 0.4$.

(c) $0.4 \leq CLL < 0.6$.

(d) $0.6 \leq CLL < 0.8$.

(e) $0.8 \leq CLL \leq 1.0$.

Fig. 4. Visualization results of compressed map image of evacuated areas.

4   Conclusions

In this paper, we proposed a clustering system using low computational resource computers that continuously provides information by adaptively changing the compression level of the map image of evacuated areas based on the cluster load level using fuzzy control. The experimental results have shown that the amount of data on the image can be reduced by lowering the quality of the image. Also, the evacuation route information displayed in the map image of evacuated areas is sufficiently readable.

In the future, we would like to develop a new system by forecasting the cluster load level.

Acknowledgement. This work was supported by JSPS KAKENHI Grant Number JP20K19793.

References

1. International federation of red cross and red crescent societies. World Disasters Report (2022) 2022
2. Okage, K., et al.: A route conditions based adaptive route planning system considering Yoshii river floods. In: Proceedings of the IEEE 12-th Global Conference on Consumer Electronics (IEEE GCCE-2023), pp. 921-922 (2023)
3. Saito, N., et al.: Design of a fuzzy inference based robot vision for CNN training image acquisition. In: Proceedings of the IEEE 10-th Global Conference on Consumer Electronics (IEEE GCCE-2021), pp. 806-807 (2021)
4. Matsui, T., et al.: FPGA implementation of a fuzzy inference based quadrotor attitude control system. In: Proceedings of the IEEE 10-th Global Conference on Consumer Electronics (IEEE GCCE-2021), pp. 691-692 (2021)
5. Yukawa, C., et al.: Design of a robot vision system for microconvex recognition. In: Barolli, L., Kulla, E., Ikeda, M. (eds.) Advances in Internet, Data & Web Technologies. Lecture Notes on Data Engineering and Communications Technologies, vol. 118, pp. 366–374. Springer, Cham (2022). https://doi.org/10.1007/978-3-030-95903-6_39
6. Lata, S., et al.: Fuzzy clustering algorithm for enhancing reliability and network lifetime of wireless sensor networks. IEEE Access **8**, 66013–66024 (2020)
7. Moriya, G., et al.: A web streaming system for electric guitar and its improving QoS based on fuzzy control. In: Proceedings of the IEEE 11-th Global Conference on Consumer Electronics (IEEE GCCE-2022), pp. 893-894 (2022)
8. Hayashi, K., et al.: A Fuzzy control based cluster-head selection and CNN distributed processing system for improving performance of computers with limited resources. In: Barolli, L. (ed.) Advances on P2P, Parallel, Grid, Cloud and Internet Computing. Lecture Notes in Networks and Systems, vol. 571, pp. 232–239. Springer, Cham (2022). https://doi.org/10.1007/978-3-031-19945-5_23
9. Jader, O.H., et al.: A state of art survey for web server performance measurement and load balancing mechanisms. Int. J. Sci. Technol. Res. **8**(12), 535–543 (2019)
10. Chandak, A., et al.: Dynamic load balancing of virtual machines using QEMU-KVM. Int. J. Comput. Appl. **46**(6), 10–14 (2012)
11. Polepally, V., et al.: Dragonfly optimization and constraint measure-based load balancing in cloud computing. Clust. Comput. **22**, 1099–1111 (2019)
12. Lapegna, M., et al.: Clustering algorithms on low-power and high-performance devices for edge computing environments. Sensors **21**(16), 5395 (2021)

13. Liu, D., et al.: Content-based light field image compression method with gaussian process regression. IEEE Trans. Multimedia **22**(4), 846–859 (2020)
14. Ma, S., et al.: Image and video compression with neural networks: a review. IEEE Trans. Circuits Syst. Video Technol. **30**(6), 1683–1698 (2020)
15. Mentzer, F., et al.: High-fidelity generative image compression. In: Advances in Neural Information Processing Systems, pp.11913-11924 (2020)
16. Song, M., et al.: Variable-rate deep image compression through spatially-adaptive feature transform. In: Proceedings of the IEEE/CVF International Conference on Computer, pp. 2380-2389 (2021)
17. Zhuo, C., et al.: Toward intelligent sensing: intermediate deep feature compression. IEEE Trans. Image Process. **29**, 2230–2243 (2020)

A Fuzzy Control Based Method for Imaging Position Decision and Its Performance Evaluation

Chihiro Yukawa[1], Tetsuya Oda[2(✉)], Yuki Nagai[1], Kyohei Wakabayashi[1], and Leonard Barolli[3]

[1] Graduate School of Engineering, Okayama University of Science (OUS), 1-1 Ridaicho, Kita-ku, Okayama 700-0005, Japan
{t22jm19st,t22jm23rv,t22jm24jd}@ous.jp
[2] Department of Information and Computer Engineering, Okayama University of Science (OUS), 1-1 Ridaicho, Kita-ku, Okayama-shi 700-0005, Japan
oda@ous.ac.jp
[3] Department of Informention and Communication Engineering, Fukuoka Institute of Technology, 3-30-1 Wajiro-Higashi-ku, Fukuoka 811-0295, Japan
barolli@fit.ac.jp

Abstract. Unmanned Aerial Vehicles (UAVs) are utilized in various fields such as aerial shots, transportation, spraying chemicals in agriculture and surveys of plant growth in forestry. In particular, aerial photography by drones is used for 3D surveying. Compared to surveying from the ground, 3D surveying with a drone provides more accurate surveying and a wider area can be surveyed in a shorter period of time. Also, drones can be used to survey inaccessible areas while flying. The point cloud data obtained from 3D surveying are used in various fields such as topographical surveying, disaster assessment, and archaeological site surveying. Currently, the Structure from Motion (SfM) is the most widely used 3D surveying method in terms of technology and price. The SfM is a 3D surveying technique that uses multiple images of an object. Therefore, it is important that the captured images contain many feature points and an appropriate focal distance. In this paper, we propose a fuzzy control based imaging position decision method and 3D measurement method using 2DLiDAR. The experimental results show that the proposed system can provide 3D measurement using 2DLiDAR and can decide imaging position based on fuzzy control.

1 Introduction

Unmanned Aerial Vehicles (UAVs) are utilized in various fields such as aerial shots [1,2], transportation [3,4], spraying chemicals in agriculture [5,6] and surveys of plant growth in forestry [7,8]. Also, it can be applied to infrastructure facilities [9,10], building surveillance [11,12], and progress management on the site for construction [13,14]. Various types of UAVs have been proposed and developed in recent years to address a wide variety of problems and applications [15,16]. In addition, Autonomous Aerial Vehicles (AAVs) are adaptive vehicles using deep reinforcement learning [17,18] and

L. Barolli (Ed.): EIDWT 2024, LNDECT 193, pp. 470–479, 2024.
https://doi.org/10.1007/978-3-031-53555-0_45

implemented with intelligent algorithms such as obstacle avoidance based on computer vision technology [19–21] for image recognition.

The aerial photography by drones is used for 3D surveying. Compared to surveying from the ground, 3D surveying has more accurate surveying and a wider area can be surveyed in a shorter period of time. Also, drones can be used to survey inaccessible areas while flying. The point cloud data obtained from 3D surveying are used in various fields such as topographical surveying, disaster assessment, and archaeological site surveying. It can be also applied to Simultaneous Localization and Mapping, which enables automated driving, Automated Guided Vehicles, robotics, and AR/VR/MR.

Laser scanners, laser measurement, Mobile Mapping System (MMS), and other methods are used in 3D surveying. However, they use specialized equipment and require specialized skills. Also, these equipments are expensive and requires a large amount of money for measurement. Therefore, Structure from Motion (SfM) technique is the most widely used 3D surveying method in terms of technology and price.

The SfM is a 3D surveying technique that uses multiple images of an object [22–24]. It creates a 3D point cloud from the feature points that match the captured images. Therefore, it is important that the captured images contain many feature points and an appropriate focal distance. Thus, it is necessary to determine the appropriate imaging point.

In this paper, we propose a fuzzy control based imaging position decision method and 3D measurement method using 2DLiDAR. The experimental results show that the proposed system can provide 3D measurement using 2DLiDAR and can decide imaging position based on fuzzy control.

The structure of the paper is as follows. In Sect. 2, we present the proposed system. In Sect. 3, we discuss the experimental results. Finally, conclusions and future work are given in Sect. 4.

2 Proposed System

The proposed system uses 2DLiDAR to perform 3D measurements and determines the imaging position based on 3D measurement results.

2.1 3D Measurement Method Using 2DLiDAR

The LiDAR emits a laser beam and measures the distance based on the reflected light. By using 2DLiDAR can be survey two-dimensional plane by emitting a rotating laser beam. Thus, 3D measurement can be achieved using 2DLiDAR by increasing the number of rotation axes using a servo motor.

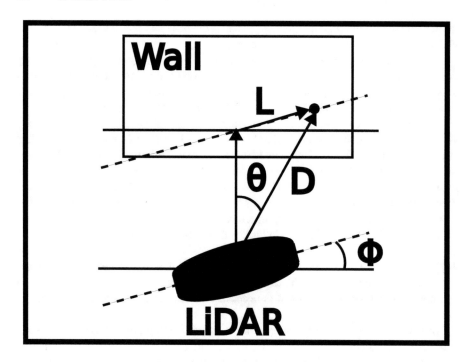

Fig. 1. 3D measurement method using 2DLiDAR.

Figure 1 shows 3D measurement method using 2DLiDAR. The angle of 2DLiDAR is $\theta \in \{0 < \mathbb{R} < 2\pi\}$, the angle of the servo motor is $\varphi \in \{0 < \mathbb{R} < \pi\}$, and the distance measured by 2DLiDAR is $D \in \mathbb{R}_+$. The target point is determined by Eq. (1), Eq. (2), Eq. (3) and Eq. (4).

$$L = D \times \sin \theta \qquad (1)$$

$$x = L \times \cos \varphi \qquad (2)$$

$$y = D \times \cos \theta \qquad (3)$$

$$z = L \times \sin \varphi \qquad (4)$$

Based on the frequency of x, y, and z axis, the coordinate with the highest frequency is abstracted from the point cloud as a face.

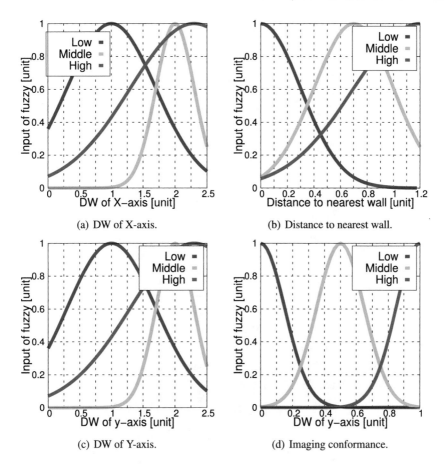

(a) DW of X-axis.

(b) Distance to nearest wall.

(c) DW of Y-axis.

(d) Imaging conformance.

Fig. 2. Membership function.

2.2 Fuzzy Control Based Method for Imaging Position Decision

The proposed system determines the imaging position based on the results of 3D measurement method. The imaging position is determined by applying fuzzy control to all positions in the 3D space. The membership functions fare shown in Fig. 2 and the Fuzzy Rule Base is shown in Table 1. The input parameters are Distance from Wall (DW) of X-axis and Distance to Nearest Wall (DNW) and DW of Y-axis. The output parameter is Imaging Conformance (IC) used for imaging position. Based on the output value, imaging is performed at the point of highest imaging conformance.

Table 1. FRB.

DW of X-axis	DNW	DW of Y-axis	IC
High	High	High	High
High	High	Middle	High
High	High	Low	Low
High	Middle	High	High
High	Middle	Middle	High
High	Middle	Low	Low
High	Low	High	Low
High	Low	Middle	Low
High	Low	Low	Low
Middle	High	High	Middle
Middle	High	Middle	Middle
Middle	High	Low	Low
Middle	Middle	High	Middle
Middle	Middle	Middle	Middle
Middle	Middle	Low	Low
Middle	Low	High	Low
Middle	Low	Middle	Low
Middle	Low	Low	Low
Low	High	High	Middle
Low	High	Middle	Low
Low	High	Low	Low
Low	Middle	High	Low
Low	Middle	Middle	Low
Low	Middle	Low	Low
Low	Low	High	Low
Low	Low	Middle	Low
Low	Low	Low	Low

Table 2. Components used for experiment.

Components	Models
2DLiDAR	RPLiDAR A1
Motor	S03N 2BBMG

3   Experimental Results

Figure 3 shows the experimental environment while Table 2 shows the components used for experiment.

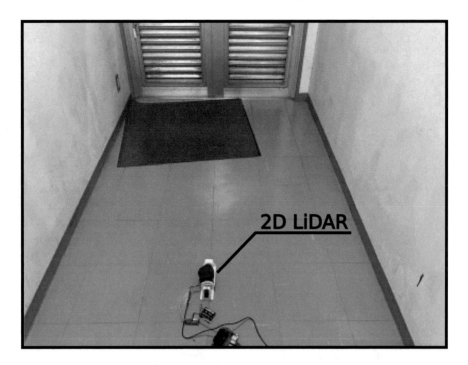

Fig. 3. Experimental environment.

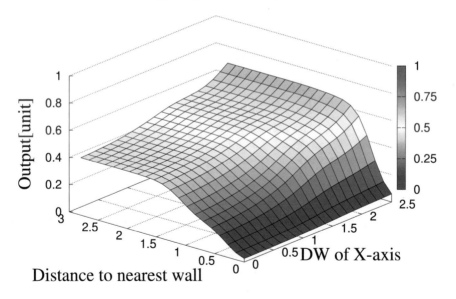

Fig. 4. Results when DW of Y-axis is 2.5.

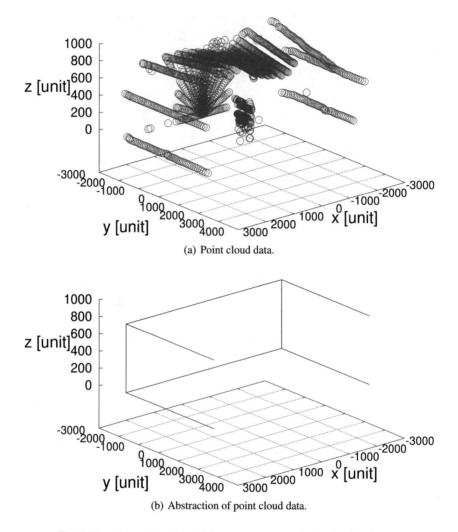

(a) Point cloud data.

(b) Abstraction of point cloud data.

Fig. 5. Experimental results of 3D measurement method using 2DLiDAR.

Figure 4 shows the results for imaging position decision method when DW of Y-axis is 2.5. It can be seen that if the DW of X-axis increases, the imaging conformance increases.

Figure 5 shows the results of 3D measurement method using 2DLiDAR. While Fig. 6 shows the experimental results of fuzzy control based imaging position decision method considered the results of Fig. 5(b). From Fig. 6 can be seen that imaging conformance increases with distance from the floor and ceiling. Figure 7 shows a snapshot of high imaging suitability.

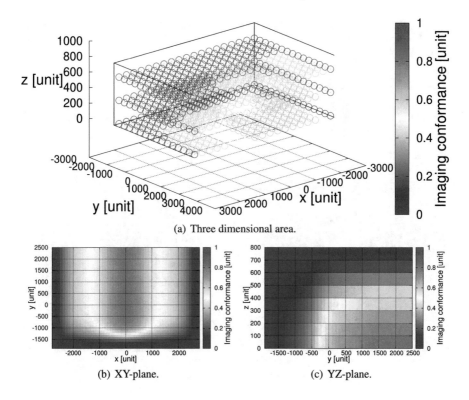

(a) Three dimensional area.

(b) XY-plane.   (c) YZ-plane.

Fig. 6. Experimental results of proposed method considering Fig. 5(b).

Fig. 7. A snapshot of high imaging suitability.

4   Conclusions

In this paper, we proposed a fuzzy control based imaging position decision method. The experimental results show that the proposed system can provide 3D measurement using 2DLiDAR and can decide imaging position based on fuzzy control.

In the future, we would like to improve the SfM method.

Acknowledgement. This work was supported by JSPS KAKENHI Grant Number 20K19793.

References

1. Mademli, I., et. al.: Challenges in autonomous UAV cinematography: an overview. In: 2018 IEEE International Conference on Multimedia And Expo (ICME), pp. 1-6 (2018)
2. Mademli, I., et al.: Autonomous unmanned aerial vehicles filming in dynamic unstructured outdoor environments [applications corner]. IEEE Signal Process. Mag. **36**(1), 147–153 (2018)
3. Thiels, A., et al.: Use of unmanned aerial vehicles for medical product transport. Air Med. J. **34**(2), 104–108 (2015)
4. Villa, K.D., et al.: A survey on load transportation using multirotor UAVs. J. Intell. Robot. Syst. **98**, 267–296 (2020)
5. Huang, Y., et al.: Development of a spray system for an unmanned aerial vehicle platform. Environ. Pract. **25**(6), 803–809 (2009)

6. Faiçal, S., et al.: The use of unmanned aerial vehicles and wireless sensor networks for spraying pesticides. J. Syst. Architect. **40**, 393–404 (2014)
7. Mohan, M., et al.: UAV-supported forest regeneration: current trends, challenges and implications. Remote Sens. **13**(13), 2596 (2021)
8. de Castro, I., et al.: UAVs for vegetation monitoring: overview and recent scientific contributions. Remote Sens. **13**(11), 2139 (2021)
9. Shvetsova, S., et al.: Safety when flying unmanned aerial vehicles at transport infrastructure facilities. Transp. Res. Procedia., 141-145 (2015)
10. Jofré-Briceño, C., et al.: Implementation of facility management for port infrastructure through the use of UAVS, photogrammetry and BIM. Sensors **21**(19), 6686 (2021)
11. Qazi, S., et al.: UAV based real time video surveillance over 4G LTE. In: 2015 International Conference on Open Source Systems & Technologies (ICOSST), vol. 13, no. 11, pp. 141-145 (2015)
12. Anwar, N., et al.: Construction monitoring and reporting using drones and unmanned aerial vehicles (UAVs). In: The Tenth International Conference on Construction in the 21st Century (CITC-10), vol. 8, no. 3, pp. 2-4 (2018)
13. Li, Y., et al.: Applications of multirotor drone technologies in construction management. Int. J. Constr. Manag. **19**(5), 401–412 (2019)
14. Irizarry, J., et al.: Exploratory study of potential applications of unmanned aerial systems for construction management tasks. J. Manag. Eng. **32**(3), 05016001 (2016)
15. Bustamante, M., et al.: Design and construction of a UAV VTOL in ducted-fan and tilt-rotor configuration. In: 2019 16th International Conference on Electrical Engineering, Computing Science and Automatic Control (CCE), pp. 1-6 (2019)
16. Zong, J., et al.: Evaluation and comparison of hybrid wing VTOL UAV with four different electric propulsion systems. Aerospace **8**(9), 256 (2021)
17. Akter, R., et al.: CNN-SSDI: convolution neural network inspired surveillance system for UAVs detection and identification. Comput. Netw. **201**, 108519 (2021)
18. Benjdira, B., et al.: Car detection using unmanned aerial vehicles: comparison between faster R-CNN and YOLOV3. In: 2019 1st International Conference on Unmanned Vehicle Systems-Oman (UVS), pp. 1-6 (2019)
19. Kanellakis, C., et al.: Survey on computer vision for UAVs: current developments and trends. J. Intell. Robot. Syst. **87**, 141–168 (2017)
20. Bouguettaya, A., et al.: A review on early wildfire detection from unmanned aerial vehicles using deep learning-based computer vision algorithms. Signal Process. **190**, 108309 (2022)
21. Cazzato, D., et al.: Survey of computer vision methods for 2D object detection from unmanned aerial vehicles. J. Imaging **6**(8), 78 (2020)
22. Saito, N., et. al.: Approach of fuzzy theory and hill climbing based recommender for schedule of life. In: Proceedings of LifeTech-2020, pp. 368-369 (2020)
23. Ozera, K., et al.: A fuzzy approach for secure clustering in MANETs: effects of distance parameter on system performance. In: Proceedings of IEEE WAINA-2017, pp. 251-258 (2017)
24. Elmazi, D., et al.: Selection of secure actors in wireless sensor and actor networks using fuzzy logic. In: Proceedings of BWCCA-2015, pp. 125-131 (2015)

Comparative Analysis of 1D-CNN and 2D-CNN for Network Intrusion Detection in Software Defined Networks

Sami Alsaadi[✉], Tertsegha J. Anande, and Mark S. Leeson

School of Engineering, University of Warwick, Coventry, UK
{sami.alsaadi,tertsegha-joseph.anande,mark.leeson}@warwick.ac.uk

Abstract. A network intrusion detection system (NIDS) is crucial for computer networks by identifying and protecting against malicious activities and abnormal attacks. Recently, there have been several attempts to utilise convolutional neural networks (CNNs) for NIDS in Software Defined Networks (SDNs). However, these have not investigated the performance of 1D-CNN and 2D-CNN in terms of parameter number and complexity time for the same structure of layers. As a result, any comparisons may well lead to incompatible and inaccurate results. In this paper, we investigate the use of CNNs for NIDS and compare various performance parameters and time complexity for intrusion detection events. The results show that a 1D-CNN achieves 99.32% classification accuracy, surpassing the 2D-CNN and other evaluated methods. On the other hand, 2D-CNN offers less computation time when compared with 1D-CNN indicating better performance.

1 Introduction

Traditional computer networks are typically decentralized, consisting of various devices such as routers, switches, and firewalls. They are complex, with multiple protocols and management tasks to ensure smooth operation. However, Software-Defined Networking (SDN) represents a new paradigm. SDN architectures decouple the control layer from the data layer, allowing direct network programming and abstracting the underlying infrastructure using protocols such as OpenFlow. SDN is a promising networking platform that disrupts current network architectures [1,2] that has received attention from the networking community but little from the security community [2].

SDN's structural design consists of three layers: the data plane (DP), control plane (CP), and application plane (AP); Fig. 1 shows an overview of the SDN architecture. The CP serves as the control point, managing flow control and enabling developers to control the network using mechanisms and deploy intelligent networks [3,4]. One critical network security application in SDN is the Network Intrusion Detection System (NIDS), a type of Intrusion Detection System (IDS) that monitors network traffic to detect and alert administrators about threats [5]. Several network threat detection methods have been developed, which are classified into (i) statistical techniques; (ii) data-mining techniques; (iii) machine learning (ML) and Deep learning (DL) based

© The Author(s), under exclusive license to Springer Nature Switzerland AG 2024
L. Barolli (Ed.): EIDWT 2024, LNDECT 193, pp. 480–491, 2024.
https://doi.org/10.1007/978-3-031-53555-0_46

techniques [6]. A NIDS can be applied using two detection methods: signature-based and anomaly-based. The former relies on a database of known attack signatures and matches them against incoming traffic [7]. While effective, this method requires large storage and cannot detect novel threats. In contrast, the latter focuses on identifying abnormal behaviour by learning patterns from network data [8], often using ML/DL techniques [9]. Several ML models, including Decision Tree Split [10], and Random Forest Modeling [11], have shown promise in securing SDNs. However, such methods depend heavily on feature engineering, which becomes inadequate for real-time attack detection due to the rapid evolution of attacker techniques [12].

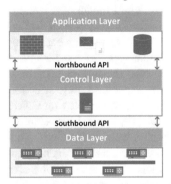

Fig. 1. The SDN architecture

DL, particularly CNNs, offers a promising solution to IDS and enhances the intrusion detection rate by automatic correlation discovery in raw data. In this research, we make the following contributions: I) investigation of the specific parameters for 1D-CNN and 2D-CNN structures; II) evaluation of classification performance parameters between 1D-CNNs and 2D-CNNs grayscale image input in CNNs and comparison of the results with other ML methods; III) assessment of computational complexity between 1D-CNN and 2D-CNN for determining optimal computational time.

2   Related Work

DL is a branch of ML based on learning data representations [13]. It has been successfully applied in many fields, such as speech recognition [14], natural language processing (NLP) [15] and image classification [16]. Elements such as increased data size, enhanced regularization and optimization methods, and significant new technologies, such as Graphics Processing Units (GPUs), have contributed to the success of DL algorithms that should now improve intrusion detection, increase detection attack rates and decrease detection times.

It is highly likely that fully connected DL techniques will offer a more effective solution for implementing NIDS, since they contain a large number of training parameters. The problem can be resolved by CNNs through weight sharing in the convolution kernels or the filter layer and by introducing sparse connectivity [17]. CNNs are a type of feed-forward artificial neural network (NN) responsible for mapping feature vectors from input data and improving the accuracy of data classification. They are valuable in identifying previously unseen attacks such as Zero-day attacks during detection.

In [18], a new learning classifier for network anomaly detection using a CNN based on a modified *Le-Net-5* architecture incorporating the *Softmax* function. This used a behaviour-based three-phase (feature extraction; model training; model verification) classification approach for network intrusion detection. In the initial phase of feature reduction, the information gain (IG) method was applied to reduce the number of features from 41 to 32. The classification object was dissimilar types of attacks and it resulted in an accuracy rate of 99.65%. Nonetheless, CNNs offer the advantage of automatically extracting essential data features without requiring a specific feature selection strategy. However, the initial use of IG for feature reduction before employing CNNs can potentially increase the model's complexity. Hu *et al.* [19] applied the adaptive synthetic sampling (ADASYN) method to improve the performance of the CNN. This approach is specifically engineered to address imbalanced datasets by introducing new instances for minority classes, thereby ensuring that the model does not exhibit a bias toward the more prevalent samples. Moreover, [19] utilized the split convolution module SPC-CNN technique to solve the problem of inter-channel information redundancy and enhance feature variety. This SPC-CNN model had an accuracy of 83.83%, which was 4.35% greater than the standard CNN on the *NSL-KDD* dataset[20]. Furthermore, [19] also applied the SPC-CNN method to address issues related to redundant inter-channel information and improve the diversity of features. An accuracy of 83.83% was achieved, which was improved on the standard CNN's accuracy on the *NSL-KDD* dataset by 4.35%. A novel CNN DL model, based on the *Lenet-5* typical model and known as the CNN-IDS model, has also been presented [21]. To meet the requirements of the CNN model, initial reduction techniques were implemented, including the use of principal component analysis (PCA) and autoencoder algorithms. PCA played a crucial role in reducing the dimensionality of the input vector, transforming it from 1×122 to 1×100. Subsequently, the traffic features were transformed into a matrix format to match the image shape of either 11×11 or 10×10 before being fed into the CNN's input layer. The *Lenet-5* model provided the basis for the preprocessing technique to remove irrelevant features and data from network traffic [21]. While achieving 94% accuracy with the *KDDCUP99* dataset [20], the accuracy for the User-to-Root attack class (focused on gaining unauthorized access to local root privileges [22]) and the Remote-to-Local attack class (intending to secure unauthorized remote access by unregistered users to the computer system [22]) remained notably low, at 20.61% and 18.96%, respectively.

When contrasted with alternative classification methods, CNN exhibits a relatively short computational process. However, most previous studies rely on outdated datasets, such as *KDDCUP99* and *NSL-KDD*, which are not compatible with modern NIDS [20]. *KDDCUP99* and *NSL-KDD* are widely categorized as unsuitable for modern network traffic due to the continuous evolution and increasing complexity of new attack types, making them more challenging to detect. Furthermore, the datasets *KDDCUP99* and *NSL-KDD* were formulated using conventional networks, not in SDN environments. The Openflow protocol in SDN presents unique challenges that traditional network protocols do not encounter, primarily due to the separation of the CP from the DP.

The previous work points to significant potential benefits from the further investigation and development of a NIDS by using ML and DL, regarding its advantages in learning new information from raw data. Therefore, this work intends to fill the gap by comparing the performance of 1D-CNN and 2D-CNN in terms of parameter numbers and complexity time in the same structure of layers for NIDS-based SDN.

3 Methodology

The architectures of both 1D- and 2D-CNNs have identical numbers of layers except for the dimensions of the input and the filter. A simple CNN design comprises convolutional and pooling layers that are alternated, followed by one or more fully connected layers. The CNN architecture has four types of layers in this model: convolution, max pooling for feature extraction, Flatten and fully connected. This section also briefly discusses the most widely used evaluation metrics.

3.1 1D-CNN Architecture

The 1D-CNN classifier created is shown in Fig. 2 (a) which include:

- Two 1D convolutional layers with 32 and 64 filters, respectively; filter size 3 and stride 1 are utilized for each layer. The main aim of the convolutional layers is to detect the existence of a set of features from the input. Such layers take their inputs as 1D arrays. To calculate the number of parameters, it is necessary to perform the weight and bias calculation, namely *filter number* x *filter size* + *filter size* x *stride* giving $3 \times 32 + 3 \times 1 = 128$ parameters for the first convolutional layer. In the second convolutional layer, there are similarly $3 \times 64 \times 32 + 3 \times 1 = 6208$ parameters.
- Two 1D max-pooling layers are used after each convolutional layer to increase the richness of features with size 2. There is no parameter calculation in this layer because it is only extracting low-level features such as the edges, points and so on.
- The output from the max pooling layer is flattened for a fully connected layer with 100 neurons.
- In the fully connected layer, the size is 100. Therefore, there are $100 \times 2 + 2 \times 1 = 202$ parameters. The nonlinear activation function ReLU is used for all layers before the output layer.
- To address over-fitting, the contribution of some neurons towards the next layer is dropped (ignored) in a dropout layer, whilst leaving other neurons untouched [23]. Here, a dropout value of 0.5 is used after the second max pool and also after the fully-connected layer.

- Softmax classification is used in the output layer with 2 output classes to classify the network traffic into normal (1) or anomalous traffic (0); it also uses binary cross-entropy as a loss function.

The input (1D array) size is 77×1 for a 1D-CNN. As a result, the total number of 1D-NN parameters is 115438, as explained in Fig. 2 (a) and the steps above.

3.2   2D-CNN Architecture

- Two 2D convolutional layers with 32 and 64 filters, respectively; filter size 3×3 and stride 1 are utilized for each layer, taking a 2D input array. Each convolutional filter is convolved across the width and height of the 2D array. The main difference from 1D is the 2D filter size which decreases the dominance of the input data. To calculate the number of parameters in the first layer, we use the same mathematical calculation in a 1D-CNN but we multiply the filter size twice as it is 2D size; this produces $(3 \times 3 \times 32) + (1 \times 1 \times 32) = 230$ parameters. The second convolutional layer has $(3 \times 3 \times 32 \times 64) + (1 \times 1 \times 64) = 18496$ parameters.
- Two size two 2D max-pooling layers are used after each convolutional layer to shrink the vectors while maintaining their relevance. Also, there are no new parameters in this layer as it just pools the maximum features to different positions.
- The output from the second max pooling layer is flattened, into one vector dimension with $2 \times 1 \times 64 = 128$ parameters for a 100-neuron fully connected layer, in which there are $100 \times 128 + 100 \times 1 = 12900$ parameters. The nonlinear activation function ReLU is used for all layers before the output layer
- The dropout layer is used after the second max pool and also after the fully connected layer to prevent the over-fitting problem.
- A Softmax function with 2 output classes normal and Anomaly is used on the output layer. Thus, there will be $2 \times 100 + 2 \times 1 = 202$ parameters for this layer.

The input (2D array) size is 11×7. As a result, the total number of 2D-CNN parameters 31918, as explained in Fig. 2 (b). Comparing the parameter numbers for both a 1D-CNN and a 2D-CNN, we see that the latter needs fewer parameters than the former. Thus, a 2D-CNN will consume less computational space than a 1D-CNN during training.

3.3   Evaluation Criteria

The evaluation metrics most commonly employed to evaluate the effectiveness of ML and DL techniques in NIDS are utilized., such as accuracy, precision, recall, and F-score. These metrics are represented by equations (1-5).

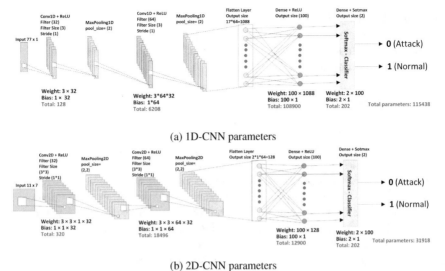

(a) 1D-CNN parameters

(b) 2D-CNN parameters

Fig. 2. Model structure and parameters

$$Precision = \frac{TP}{TP+FP} \quad (1) \qquad Recall = \frac{TP}{TP+FN} \quad (2)$$

$$False\ alarm\ rate = \frac{FP}{FP+TN} \quad (3) \quad True\ negative\ rate = \frac{TN}{TN+FP} \quad (4)$$

$$Accuracy = \frac{TP+TN}{TP+TN+FP+FN} \quad (5)$$

4   Dataset

We utilised a new updated SDN attack-specific dataset, namely InSDN [24], to evaluate the performance of our proposed model. This dataset is considered to be one of the first solutions that produces a comprehensive set of traffic parameters. Utilizing non-specialized SDN datasets may lead to compatibility issues since the attack vector deployment should consider the new architecture of the network [25]. The InSDN dataset contains several types of normal traffic applications (i.e., HTTP, HTTPS, DNS, Email, FTP and SSH). More details concerning these features can be found in [26]. The training data comprised 103073 records (57956 for BENIGN and 45117 for attack), while the testing data had 27697 samples (10468 for BENIGN and 17229 for attack (new subset attack)). 20% of the training data was selected for validation. Moreover, the attack samples utilised during the testing phase came from a different distribution to the training samples. There were 50230 records in the testing dataset and 135870 records in the training dataset.

4.1 Dataset Pre-processing Steps

1. The dataset contained the socket features ('Flow ID', 'Src IP', 'Src Port', 'Dst IP', 'Dst Port', 'Timestamp'). We removed these to prevent model bias towards those features. This left a remainder of 77 features.
2. We used max-min normalization to map the value of features between [0,1]. The normal label was designated as 1 and all attack labels were assigned a value of 0.
3. To balance the dataset, we used the undersampling and oversampling methods for attack classes (i.e. 136848 records), equalizing the normal class.
4. The label traffic features were converted into numerical data [1,0].
5. The one-dimensional traffic was converted into a 2D array of 7×11.

5   Simulation Results and Analysis

All our simulations employed the Keras Library in Python. They were executed using a machine that includes 11th Gen Intel(R) Core(TM) i7-1185G7 @ 3.00GHz 1.80 GHz (8 cores), Windows 10 operating system.

A comprehensive evaluation of the performance of 1D-CNN, 2D-CNN and other classical ML classifiers including support vector machines (SVM), random forest (RF), Gaussian naive Bayes (GaussianNB), decision tree (DT) and k-nearest neighbor classifier (KNN) is shown in Table 1. The models used in the 1D- and 2D-CNNs are depicted in Fig. 2 whereas all other classifiers use the default parameter settings from the scikit-learn Python toolbox.

In all models, the normal class was classified somewhat lower than the attack class. This is because throughout the training and assessment phases, the average class size was small. The classifier could not learn the behaviour of raw data from the small number. Thus, further evaluation was undertaken to compare the CNNs models, namely the receiver operating characteristics curve and the confusion matrix.

5.1 Classification Performance

Returning to Table 1, it may be observed that CNNs performed better than all traditional ML algorithms. Such classical techniques try to extract information from traffic data through complex feature engineering (i.e. manually extracted features). This can cause loss of information in the original data, resulting in low judgment accuracy, which makes them unsuitable for deployment in actual situations.

Table 1. Evaluation of CNN models compared to other classical ML classifiers

Approach	Precision		Recall		F-score	
	Attack	Normal	Attack	Normal	Attack	Normal
SVM	99.00	58.00	69.00	98.00	81.00	73.00
Random Forest	99.00	68.00	79.00	99.00	88.00	80.00
GaussianNB	98.00	70.00	82.00	97.00	89.00	81.00
Decision Tree	99.00	77.00	87.00	99.00	93.00	87.00
KNeighbors	99.00	86.00	93.00	99.00	96.00	92.00
2D-CNNs	99.31	91.04	95.78	98.47	97.51	94.61
1D-CNNs	99.33	99.93	99.97	98.45	99.65	99.18

Moreover, the 1D-CNN delivered greater accuracy than the 2D-CNN at 99.5% and 96.6%, respectively as depicted in Fig. 3. This additional power in traffic classification in the 1D-CNN over the 2D-CNN most likely arises because the input data for traffic are one-dimensional.

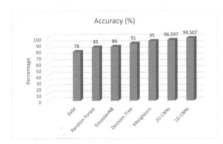

Fig. 3. Accuracy for Comparing different Classification Algorithms

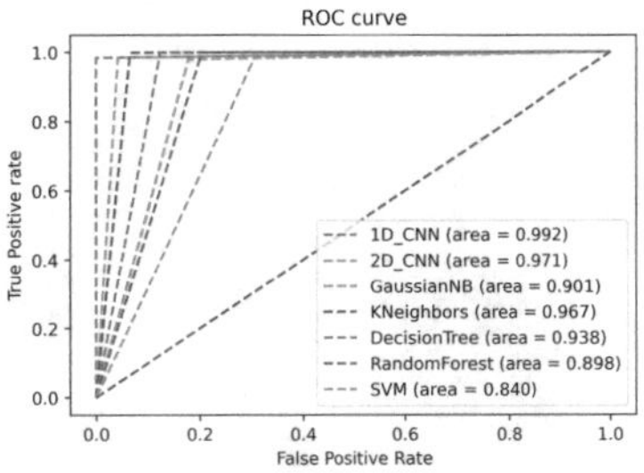

Fig. 4. ROC curve

5.2   Computational Time

The computational time for CNNs is influenced by factors such as network structure, input and filter size, and layer count. Python's *'Time'* module was employed to measure this time. Training time was measured from the algorithm initiation to completion of the compile and train phase, while testing time involved the prediction function. Figure 5 and 6, display the training and testing times for all the algorithms, providing insight into their execution efficiency.

Despite the shorter execution times of classic ML algorithms compared to CNN models, their performance is not satisfactory to use them in real-world NIDS case scenarios. In contrast, the 2D CNN model exhibits significantly reduced CPU execution times in both the training and testing phases, specifically 136.38 s and 2.249 s, respectively. This may be attributed to its lower parameter count in comparison to the 1D CNN.

Fig. 5. Training Time for Comparing Different Algorithms

Fig. 6. Testing Time for Comparing different Algorithms

5.3 Receiver Operating Characteristic (ROC)

A ROC curve, as shown in Fig. 4, was used to assess the models' performance. The area in the ROC curve (abbreviated to *AUC*) measures the model's overall ability to distinguish between attack and normal traffic. A larger AUC signifies better discrimination, while a value close to zero means that it was failing to correctly identify the traffic. The 1D_ and 2D-CNN gave values of 99.2% and 97.1%, respectively, followed by KNN at 96.7%, indicating that percentages of positive and negative rates were successfully separated. In contrast, all other classifiers provided low AUC values and thus inferior performance.

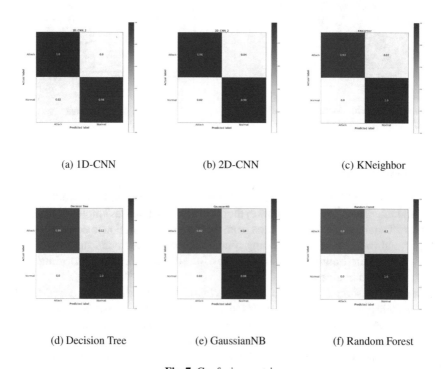

(a) 1D-CNN   (b) 2D-CNN   (c) KNeighbor

(d) Decision Tree   (e) GaussianNB   (f) Random Forest

Fig. 7. Confusion matrix

5.4   Confusion Matrix

A confusion matrix summarizes the actual and predicted target from the machine learn-ing models. For the best performance, the on-diagonal values should be as high as pos-sible, leaving few in the off-diagonal positions. The binary confusion matrices for all models are shown in Fig. 7 (a)-(f). These confirm that the 1D-CNN offers the best per-formance among all the other algorithms used in the confusion matrices.

6   Discussion

The 1D-CNN has demonstrated superior performance on network traffic compared to 2D-CNN and other, shallow, classifiers. The original network flow contained one-dimensional vector data, so converting the network traffic into two-dimensional data had a high computational complexity. Thus, a 1D-CNN is advantageous and thus prefer-able to 2D counterparts in dealing with 1D signals. This is because first, the compu-tational complexity of 1D-CNN is significantly lower than 2D-CNN the former just requires simple array operations; second, there is a smaller number of hidden layers and neurons in a 1D-CNN compared to a 2D-CNN. This relatively shallow architecture is able to learn challenging tasks involving 1D signal traffic for which a 2D-CNN will usually require a deeper architecture. However, the computational time of a 2D-CNN is relativity smaller than that of a 1D-CNN. This is because the total number of parameters in the former is some 96,320 lower than in the latter.

The issue of a new dataset development or the availability of network data may face difficulties related to privacy or legal matters. Furthermore, one of the challenges and future research directions is the utilization of the same dataset for the testing distribu-tion as the training dataset [27]. Evaluating proposed models using such methods and datasets is not a suitable technique for the anomaly detection environment, since many simple algorithms can achieve high accuracy (99% or higher) as a result of over-fitting.

Thus, in common with many ML scenarios, the best method for assessing the effec-tiveness of intrusion detection models is to check how they perform when used with an unseen dataset not used during training. In addition, for a fair performance compar-ison of two algorithms, they should have the same structure as investigated here and achieved in our comparison of 1D- and 2D-CNN models.

Although the accuracies obtained using our shallow 1D- and 2D-CNNs were 99.5% and 96.6%, respectively, this is not convincing enough for application in real scenarios. We are working on further investigation of our CNN model structure by extending our simulation to apply multi-class classification of the attacks. This will add degrees of freedom to our system and go beyond purely binary classification.

7   Conclusions

We have compared the parameter numbers and the performance of 1D- and 2D-CNNs for binary traffic classification to address SDN security issues. For a fair comparison, both CNN models had the same layer structure except for details of the inputs and the filters. The performance of both was conducted with a focus on accuracy and time complexity. The simulation results indicated that the 1D-CNN achieved an accuracy of 99.50% in classifying new attacks compared to the 96.59% value for the 2D-CNN. Both CNNs performed significantly better than a range of shallow classifiers, confirmed by ROCs and confusion matrices. However, the 2D-CNN required less time than the 1D-CNN because of its lower number of parameters.

References

1. Shan-Shan, J., Ya-Bin, X.: The APT detection method based on attack tree for SDN. In: Proceedings of the 2nd International Conference on Cryptography, Security and Privacy (2018). https://doi.org/10.1145/3199478.3199481
2. Yoon, C., Park, T., Lee, S., et al.: Enabling security functions with SDN: a feasibility study. Comput. Netw. **85**, 19–35 (2015). https://doi.org/10.1016/j.comnet.2015.05.005
3. Tang, T.A., Mhamdi, L., McLernon, D., et al.: Deep learning approach for network intrusion detection in software defined networking. In: 2016 International Conference on Wireless Networks and Mobile Communications (WINCOM) (2016). https://doi.org/10.1109/wincom.2016.7777224
4. Park, Y., Kengalahalli, N.V., Chang, S.-Y.: Distributed security network functions against botnet attacks in software-defined networks. In: 2018 IEEE Conference on Network Function Virtualization and Software Defined Networks (NFV-SDN) (2018). https://doi.org/10.1109/nfv-sdn.2018.8725657
5. Le, D.-H., Tran, H.-A.: A novel machine learning-based network intrusion detection system for software-defined network. In: 2020 7th NAFOSTED Conference on Information and Computer Science (NICS) (2020). https://doi.org/10.1109/nics51282.2020.9335863
6. Kumar, S.: Survey of current network intrusion detection techniques, pp. 1-18. Washington University in St. Louis (2007)
7. Novaes, M.P., Carvalho, L.F., Lloret, J., Proença, M.L.: Adversarial deep learning approach detection and defense against DDoS attacks in SDN environments. Futur. Gener. Comput. Syst. **125**, 156–167 (2021). https://doi.org/10.1016/j.future.2021.06.047
8. Yang, Y., Xu, W., Hou, P., et al.: Improving maize grain yield by matching maize growth and solar radiation. Sci. Rep. (2019). https://doi.org/10.1038/s41598-019-40081-z
9. Said Elsayed, M., Le-Khac, N.A., Dev, S., Jurcut, A.D.: Network anomaly detection using LSTM based autoencoder. In: Proceedings of the 16th ACM Symposium on QoS and Security for Wireless and Mobile Networks (2020). https://doi.org/10.1145/3416013.3426457
10. Rai, K., Devi, M.S., Guleria, A.: Decision tree based algorithm for intrusion detection. Int. J. Adv. Netw. Appl. **7**, 2828–2834 (2016)
11. Farnaaz, N., Jabbar, M.A.: Random forest modeling for network intrusion detection system. Procedia Comput. Sci. **89**, 213–217 (2016). https://doi.org/10.1016/j.procs.2016.06.047
12. ElSayed, M.S., Le-Khac, N.-A., Albahar, M.A., Jurcut, A.: A novel hybrid model for intrusion detection systems in SDNS based on CNN and a new regularization technique. J. Netw. Comput. Appl. **191**, 103160 (2021). https://doi.org/10.1016/j.jnca.2021.103160
13. Lateef, A., Al-Janabi, S., Al-Khateeb, B.: Survey on intrusion detection systems based on deep learning. Periodic. Eng. Nat. Sci. **7**, 1074–1095 (2019)
14. Hinton, G., Deng, L., Yu, D., et al.: Deep neural networks for acoustic modeling in speech recognition: the shared views of four research groups. IEEE Signal Process. Mag. **29**, 82–97 (2012). https://doi.org/10.1109/msp.2012.2205597
15. Collobert, R., Weston, J., Bottou, L., et al.: Natural language processing (almost) from scratch. J. Mach. Learn. Res. **12**, 2493–2537 (2011)

16. Krizhevsky, A., Sutskever, I., Hinton, G.E.: ImageNet classification with deep convolutional neural networks. Commun. ACM **60**, 84–90 (2017). https://doi.org/10.1145/3065386
17. Gu, J., Wang, Z., Kuen, J., et al.: Recent advances in convolutional neural networks. Pattern Recogn. **77**, 354–377 (2018). https://doi.org/10.1016/j.patcog.2017.10.013
18. Lin, W.-H., et al.: Using convolutional neural networks to network intrusion detection for cyber threats. In: 2018 IEEE International Conference on Applied System Invention (ICASI) (2018). https://doi.org/10.1109/icasi.2018.8394474
19. Hu, Z., Wang, L., Qi, L., et al.: A novel wireless network intrusion detection method based on adaptive synthetic sampling and an improved convolutional neural network. IEEE Access **8**, 195741–195751 (2020). https://doi.org/10.1109/access.2020.3034015
20. Divekar, A., et al.: Benchmarking datasets for anomaly-based network intrusion detection: KDD Cup 99 alternatives. In: 2018 IEEE 3rd International Conference on Computing, Communication and Security (ICCCS) (2018). https://doi.org/10.1109/cccs.2018.8586840
21. Xiao, Y., Xing, C., Zhang, T., Zhao, Z.: An intrusion detection model based on feature reduction and convolutional neural networks. IEEE Access **7**, 42210–42219 (2019). https://doi.org/10.1109/access.2019.2904620
22. Komar, M., et al.: High performance adaptive system for cyber attacks detection. In: 2017 9th IEEE International Conference on Intelligent Data Acquisition and Advanced Computing Systems: Technology and Applications (IDAACS) (2017). https://doi.org/10.1109/idaacs.2017.8095208
23. Koivu, A., Kakko, J.-P., Mäntyniemi, S., Sairanen, M.: Quality of randomness and node dropout regularization for fitting neural networks. Expert Syst. Appl. **207**, 117938 (2022). https://doi.org/10.1016/j.eswa.2022.117938
24. Elsayed, M.S., Le-Khac, N.-A., Jurcut, A.D.: INSDN: a novel SDN intrusion dataset. IEEE Access **8**, 165263–165284 (2020). https://doi.org/10.1109/access.2020.3022633
25. Ring, M., et al.: A survey of network-based intrusion detection data sets. Comput. Secur. **86**, 147–167 (2019). https://doi.org/10.1016/j.cose.2019.06.005
26. Zoppi, T., Ceccarelli, A., Bondavalli, A.: Unsupervised algorithms to detect zero-day attacks: strategy and application. IEEE Access **9**, 90603–90615 (2021). https://doi.org/10.1109/access.2021.3090957
27. Thakkar, A., Lohiya, R.: A survey on intrusion detection system: feature selection, model, performance measures, application perspective, challenges, and future research directions. Artif. Intell. Rev. **55**, 453–563 (2021). https://doi.org/10.1007/s10462-021-10037-9

Analysis and Development of a New Method for Defining Path Reliability in WebGIS Based on Fuzzy Logic and Dispersion Indices

Walter Balzano[✉], Antonio Lanuto, Erasmo Prosciutto, and Biagio Scotto di Covella

University of Naples, Federico II, Naples, Italy
walter.balzano@gmail.com

Abstract. When planning a trip, we wish to receive a precise itinerary, taking into account various factors such as traffic, distance, types of roads. However, we often have to deal with unreliable and inaccurate information. Evaluating the reliability of a route is a crucial aspect to improve the quality of navigation services and guarantee the user an experience that is not only effective but also efficient in terms of cost and travel time. The problem was approached with an innovative solution, which uses fuzzy logic and dispersion indices to measure variations in the traffic situation at different times of the day and on different days of the week, with tests carried out on real routes collected by the Google Maps platform API.

Keywords: Path Reliability in WebGIS · Fuzzy Logic · Dispersion Indices

1 Introduction

The advent of mobile devices and real-time navigation applications such as Google Maps has made navigation an essential part of daily life. However, one of the main limitations of these applications is the absence of a *reliability* attribute for a given route. Currently, navigation applications only provide an estimate of travel time, which can cause problems for users who need to plan their time, especially in situations where time is limited or when there are delays.

To ensure the safety and punctuality of users, it is critical to define the *reliability* of a route. A more reliable route is less prone to unexpected traffic delays or traffic jams, which means users will arrive at their destination more accurately than a less reliable route. In addition, the definition of route *reliability* can be used to determine the safety of the route itself, such as avoiding roads with a high accident rate.

© The Author(s), under exclusive license to Springer Nature Switzerland AG 2024
L. Barolli (Ed.): EIDWT 2024, LNDECT 193, pp. 492–501, 2024.
https://doi.org/10.1007/978-3-031-53555-0_47

The travel time estimation provided by navigation applications does not take into account route variability. Two alternative routes might have the same travel time estimate, but one might have greater variability than the other. This means that a route with a higher *reliability* might be preferred over one with a lower value because it offers greater safety and punctuality.

The following sections are divided as follows: in Sect. 2 the background and the related works are described; in Sect. 3 the proposed solution is presented with an in depth look at the structure of the algorithm; Sect. 4 deals with the evaluation with the results of the simulation of the algorithm execution in the city of Naples; in Sect. 5 some future developments are introduced; finally in Sect. 6 the conclusions are explained.

2   Background

Current research is focusing on analyzing the *reliability* of routes in transportation networks, using measures such as the probability of arriving at the destination on time. To do this, scholars have developed algorithms based on stochastic networks [3], which represent a type of mathematical modeling that uses probability to describe dynamic systems. In essence, these networks use a graphical representation to describe the relationships between system variables and their transition probabilities over time.

However, these algorithms have some limitations. First, they require a large number of calculations to determine the transition probabilities, especially for systems with a large number of possible states and transitions. This can make them difficult to use for large or real-time problems. Second, they may have difficulty handling imprecise or non-quantitative uncertainties, and they may provide inaccurate results if the data change.

Despite this, these algorithms mainly focus on the *reliability* of route time estimation, using algorithms based on historical data and statistical analysis. However, route *reliability* studies are still under development and have not yet been widely adopted in GPS navigation because they cannot predict extraordinary events or traffic conditions that may affect the actual travel time.

2.1   Related Work

The problem of road traffic congestion is still one of the most important challenges that many researchers are working on (see [1] for a survey), and there are many studies on how to optimise the distribution of vehicles on city streets, an example of which is the [5], where the problem is addressed by posing an isomorphism with resource management in operating systems, or in [4] were an innovative hierarchical model, based on two abstraction levels, prevents the traffic congestion. Moreover, as the population increases over the years, as seen in [16], so does the number of vehicles.

The problem of road congestion is closely linked to the problem of route *reliability*, that has been addressed several times from different aspects [15].

Many of these systems use stochastic networks to create mathematical modelling that uses probability to describe dynamic systems [6,14].

Another interesting aspect to note is that analyzing the *reliability* of routes in transportation networks also shares methodological parallels with advancements in other computational fields. For instance, similar to the use of stochastic networks in route reliability, Quantum Natural Language Processing (QNLP) also represents an innovative blend of complex theoretical models with practical applications. In QNLP, the principles of quantum mechanics are applied to language processing tasks, demonstrating the integration of advanced theoretical frameworks with real-world problems, akin to our approach in route analysis using stochastic networks [8].

Moreover, the challenge in route *reliability* of handling imprecise or non-quantitative uncertainties is reminiscent of the challenges addressed in the multilingual BERT (mBERT) language model study. The mBERT research investigates cross-lingual syntactic transfer, revealing how complex syntactic knowledge in one language can be applicable in other languages. This kind of cross-domain application of theoretical models is analogous to our approach in route *reliability*, where we deal with the variability and unpredictability of real-world traffic conditions [9].

In conclusion, the widespread use of fuzzy logic in these fields such as, routing algorithm [7], shortest path [11], has led us to the use of such technology to better address the *reliability* problem.

2.2 The Key Role of Variance for Reliability

Variance is a widely used statistical measure in computer science. It allows us to represent the dispersion of data with respect to the mean, giving us an estimate of the variability of the data. For task management of a processor, scheduling algorithms use *variance* to evaluate adaptivity. The goal is to minimize the waiting time for task execution.

Choosing a scheduling algorithm with a lower *variance*, even if it results in a slight increase in average wait time, means opting for a more adaptive solution. In this way, tasks will run more smoothly and predictably, which is essential for optimal utilization of processor resources.

Variance plays a key role in the design of processor scheduling algorithms. This has been demonstrated by several studies, including [10] and [12]. *Variance* minimization can improve the quality of service (QoS) of systems involved in task scheduling, and it can significantly affect system performance and user satisfaction. Therefore, consideration of *variance* in the design of scheduling algorithms is a crucial aspect of ensuring stable and predictable performance.

3   Proposed Solution

In the previous section, we explored how *variance* plays a key role for process scheduling algorithms in terms of *reliability*. Similarly, the proposed system uses fuzzy logic to define the *reliability* of a path based on the *standard deviation* (i.e. the mean square deviation) of the travel time estimates. It represents precisely the dispersion of values around the mean and can be used as a measure of the *variance* of the data. The mean square deviation is equal to the square root of the *variance*, which allows the increasing and decreasing characteristics of the original function to be maintained, thus preserving information on the properties of the data distribution. Therefore, using the *standard deviation* to assess the *reliability* of transport routes is a consistent and reasoned choice, as it allows us to describe the dispersion of travel times with respect to the mean, and to assess the regularity of the route. This means that the system will be able to assess its variability, providing a more complete view of the route. In this way, it will be possible to more consciously choose the most suitable route to take, rather than opting for the one with the lowest estimated travel time.

We aim to highlight the significance of route *reliability* by employing Google APIs to build a comprehensive data-set stored in a database. The data-set will track traffic patterns on different routes during various times of the day. Although the starting and finishing points of each route will be the same, each route will differ in terms of its trajectory and traffic. We will then perform a process of converting the data into fuzzy logic for each route, analysing the *standard deviation* from the average travel time. These results will be represented using graphs to facilitate visualisation and better understanding of the information. This approach will help us to develop a more accurate understanding of route *reliability* and enable us to identify key factors that impact travel time.

3.1   Path Reliability Estimation Algorithm

In this section, we will take an in-depth look at how the *Path Reliability Estimation Algorithm* works. The algorithm consists of two main parts, each of which plays an important role in data processing:

1. **The route reliability storage part** - Here we will define the structure of the database that handles the *reliability* estimation data. It will also be important to analyse the manipulation of this data, together with the management of the *WebGIS* calls needed to derive the estimates.
2. **The data processing part** - It plays a crucial role in processing the data collected in the first part. Using fuzzy logic, we will process the data to obtain a more accurate estimate of path *reliability*. Finally, we will transform the processed data into graphs for a more intuitive visualisation.

3.1.1   Route Reliability Storage

In this section, we will delve into the process of data collection for determining route *reliability* estimates. To do so, we will examine two fundamental structures that govern this process:

- **The first structure** is a simple custom-built database that stores the routes to be analyzed along with some crucial values. Each route has a base time in the absence of traffic, and is associated with a data set that takes into account the estimated values of time with traffic on different days. These values will allow us to subsequently analyze the *reliability* of each route.
- **The second structure** is a Python file that utilizes the Google Maps *Distance Matrix* API to store the values of duration and expected duration in traffic for a given route. This part of the algorithm retrieves the data received from Google and processes it in the format required by our database. This operation is repeated every 5 min in the background.

3.1.2   Data Processing

The algorithm implements a fuzzy system to assess the *reliability* of a travel route, considering two main factors; *distance* and *standard deviation* percentage from the average estimate of travel time as a function of time. Using these values, we can construct three models as a function of *distance*, *standard deviation* and route *reliability*.

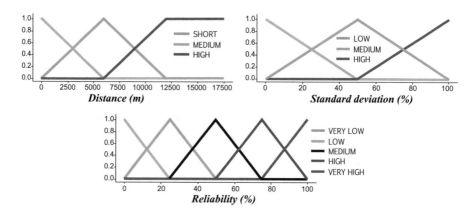

Fig. 1. Linguistic variables of the fuzzy system.

As can be seen in the Fig. 1 each of the first two characterise the main *reliability* model by going to base our inference engine on the variability of the data obtained. The results of the algorithm are then subsequently aggregated, using the centre of area (COA) method. Finally, graphs are generated to facilitate data analysis [2,8,9,13].

4   Evaluation

In this section we want to provide some simulation results of the algorithm execution, as well as to analyse the different results obtained on different routes in the city of Naples. In order to obtain a simulation faithful to a real situation of path *reliability*, we chose four paths with different trajectories all starting and arriving from the same point. We analysed these four routes for thirty days, for a total of over **300,000 records**, which allowed us to analyse certain points in detail.

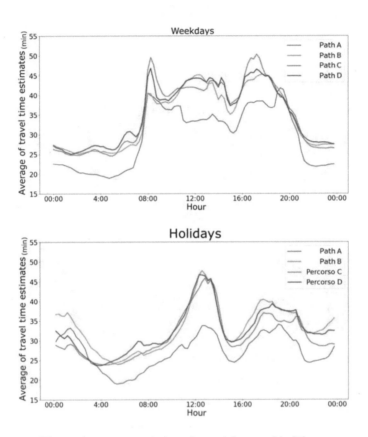

Fig. 2. Average travel times in weekdays and holidays.

As can be seen from the graphs of the *average travel time* estimates, Fig. 2, **Route A**, despite being the route with the maximum traffic-free duration, is definitely *the most advantageous*, since it maintains an average that is almost always at least *five minutes less than the other routes in all time slots*, both on weekdays and holidays, with the gap reaching as much as fifteen minutes at 1:00 PM local time on holidays.

Given the unpredictability of variables such as traffic, bad weather, traffic accidents, and so on, it is not possible to be certain of arriving in the times indicated by the Google Maps estimate. In this sense, the *standard deviation* graphs in Fig. 3 showed that although in some places one route may be better than another in terms of *average travel time* estimates, it recorded a higher percentage of *standard deviation* from the average. This means that the variability of this route is higher than that of the others.

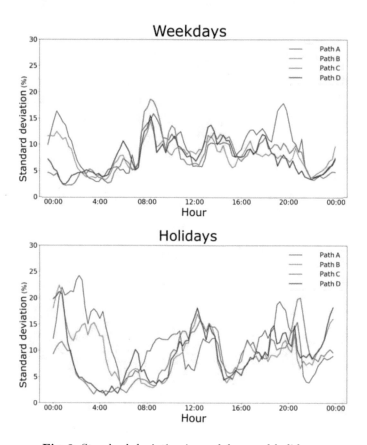

Fig. 3. Standard deviation in weekdays and holidays.

In the images, it can be seen that the *standard deviation* trend is much more erratic than that of the *average travel time* estimates, synonymous with the fact that for a more conscientious choice of which route to take, we cannot consider only its estimate. In fact, **Route A** records a *standard deviation* percentage of **25%**, far higher than the others, on holidays at about 02:30 AM. It is most likely that this is due to the large amount of traffic on "Via Caracciolo" at that time; for that matter, the *standard deviation* percentages of **Route C** and **Route D** turn out to be **5%** as they pass through "Via Manzoni."

It is also possible to observe that **Route A**, just before 8:00 PM, has a *standard deviation* percentage that is *almost twice that of the other routes.*

In the *reliability* graphs of the Fig. 4, what has been said so far about the variability of the routes studied is evident: *greater variability resulted in lower*

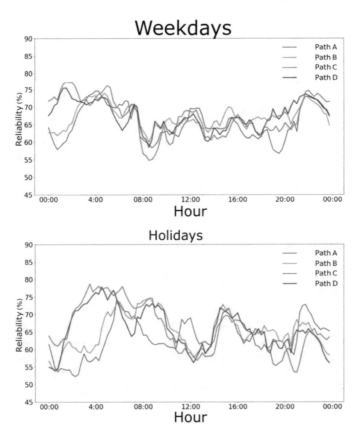

Fig. 4. Routes reliability in weekdays and holidays.

reliability and vice versa. From this point of view, if one were at 4:00 AM on a public holiday and had the four routes analyzed, it would be preferable to choose **Route C**, despite it having a higher *travel time estimate* than **Route A**, as the *reliability* values are about **80%** and **57%**, respectively.

It is evident from the results that there is no unambiguous answer to whether it makes sense to choose a route with lower travel time estimate and higher *standard deviation*, however, *more tools can be provided to the user so that he or she can choose the route to take more conscientiously.*

5   Future Works

Regarding the future evolution of the application, there are several possibilities for improving the efficiency and effectiveness of its functionality:

- **Implementation of a more universal and versatile fuzzy logic** that can be applied in contexts other than urban settings. This would broaden the utility of the application and make it better suited to the needs of a wide range of users.

- **Optimization of the source code** in order to improve the performance and the scalability of the application. This could involve the introduction of more efficient programming techniques or the use of external libraries to simplify the development process and improve software efficiency.
- **Implementation of the application on real *WebGIS*** such as Google Maps and Waze. In particular, it would be important for these services to add a new attribute that defines the adaptability of a route, similar to how they implemented the attribute that defines which route is less fuel-intensive. This would allow this functionality to be integrated into applications already used by millions of users, improving the user experience.

6   Conclusions

The objective of this paper is to demonstrate that *the choice of a route to follow should not be based solely on the estimated travel time proposed by Google Maps.* In fact, in addition to the duration of the route, it is also important to consider its variability, i.e., its ability to be predictable and not undergo unpredictable changes over time. In fact, due to factors such as traffic, bad weather and traffic accidents, the estimate provided by the ***WebGIS*** may be inaccurate and the proposed route may be less reliable than it initially seems. The goal is to provide the user with tools to make a more informed choice, taking into account the variability of travel time and its *reliability*. In this way, the user will be able to choose the route best suited to his or her needs, even taking into account of any delays or unforeseen events that could affect the duration of the route.

References

1. Afrin, T., Yodo, N.: A survey of road traffic congestion measures towards a sustainable and resilient transportation system. Sustainability **12**(11), 4660 (2020)
2. Amato, F., Castiglione, A., Moscato, V., Picariello, A., Sperlì, G.: Multimedia summarization using social media content. Multimedia Tools Appl. **77**, 07 (2018)
3. Arun Prakash, A.: Algorithms for most reliable routes on stochastic and time-dependent networks. Transp. Res. Part B Methodol. **138**, 202–220 (2020)
4. Balzano, W., Lanuto, A., Mascia, C., Stranieri, S.: Hierarchical VANET: a traffic congestion management approach based on critical points. In: Barolli, L. (ed.) 3PGCIC 2022. LNCS, vol. 571, pp. 192–202. Springer, Cham (2023). https://doi.org/10.1007/978-3-031-19945-5_19
5. Balzano, W., Prosciutto, E., di Covella, B.S., Stranieri, S.: A resource allocation technique for VANETs inspired to the banker's algorithm. In: Barolli, L. (ed.) 3PGCIC 2022. LNCS, vol. 571, pp. 222–231. Springer, Cham (2023). https://doi.org/10.1007/978-3-031-19945-5_22
6. Gaonkar, R.S.P., Nigalye, A.V., Pai, S.P.: Possibilistic approach for travel time reliability evaluation. Int. J. Math. Eng. Manage. Sci. **6**(1), 223 (2021)

7. Ghalavand, G., Ghalavand, A., Dana, A., Rezahosieni, M.: Reliable routing algorithm based on fuzzy logic for mobile ad hoc network. In: 2010 3rd International Conference on Advanced Computer Theory and Engineering (ICACTE), vol. 5, pp. V5–606. IEEE (2020)
8. Guarasci, R., De Pietro, G., Esposito, M.: Quantum natural language processing: challenges and opportunities. Appl. Sci. **12**(11), 5651 (2022)
9. Guarasci, R., Silvestri, S., De Pietro, G., Fujita, H., Esposito, M.: Bert syntactic transfer: a computational experiment on Italian, French and English languages. Comput. Speech Lang. (71/101261) (2022)
10. Jebreen, A.: On minimum variance CPU-scheduling algorithm for interactive systems using goal programming. Int. J. Comput. Appl. **135**, 51–59 (2016)
11. Keshavarz, E., Khorram, E.: A fuzzy shortest path with the highest reliability. J. Comput. Appl. Math. **230**(1), 204–212 (2009)
12. Mahapatra, S., Dash, R.R., Pradhan, S.K.: A heuristic job scheduling algorithm for minimizing the waiting time variance. In: 2015 International Conference on Futuristic Trends on Computational Analysis and Knowledge Management (ABLAZE), pp. 446–451 (2015)
13. Minutolo, A., Guarasci, R., Damiano, E., De Pietro, G., Fujita, H., Esposito, M.: A multi-level methodology for the automated translation of a coreference resolution dataset: an application to the Italian language. Neural Comput. Appl. **34**(24), 22493–22518 (2022)
14. Pan, F., Morton, D.P.: Minimizing a stochastic maximum-reliability path. Networks Int. J. **52**(3), 111–119 (2008)
15. Rakha, H.A., El-Shawarby, I., Arafeh, M., Dion, F.: Estimating path travel-time reliability. In: 2006 IEEE Intelligent Transportation Systems Conference, pp. 236–241. IEEE (2006)
16. Statista. Proportion of population in cities worldwide from 1985 to 2050 (2022)

Object Detection and Speech Recognition Based Motion Analysis System for Pointing and Calling

Kyohei Wakabayashi[1], Chihiro Yukawa[1], Yuki Nagai[1], Tetsuya Oda[2(✉)], and Leonard Barolli[3]

[1] Graduate School of Engineering, Okayama University of Science (OUS),
1-1 Ridaicho, Kita-ku, Okayama 700–0005, Japan
{t22jm24jd,t22jm19st,t22jm23rv}@ous.jp
[2] Department of Information and Computer Engineering,
Okayama University of Science (OUS), 1-1 Ridaicho, Kita-ku, Okayama 700–0005, Japan
oda@ous.ac.jp
[3] Department of Informention and Communication Engineering,
Fukuoka Insitute of Technology, 3-30-1 Wajiro-Higashi-ku, Fukuoka 811-0295, Japan
barolli@fit.ac.jp

Abstract. Pointing and calling is a method for ensuring safety during working by pointing at work objects or tools and verbally calling and telling the situation in order to predict accidents and raise safety awareness. Therefore, pointing and calling are implemented at production sites to reduce human error and prevent accidents. This method can be used for detecting processing deficiencies and confirming the number of products to be shipped. However, due to a lack of work experience and insufficient recognition of the work importance, there are many cases where the correct motion and procedures are not followed. To ensure the safety of workers in a factory, it is necessary to continuously monitor their actions. Also, the instructors spend a lot of time for monitoring. In this paper, we propose a motion analysis system for pointing and calling based on object detection and speech recognition. The proposed system takes into account the safety measures in order to prevent accidents and injuries at production sites where people are involved. The experimental results show that the proposed system correctly detects the motion sequence of pointing, looking and speaking.

1 Introduction

The highest priority on the production site is to ensure worker safety, which is directly related with the improving production efficiency and product quality. Workplace accidents caused by safety issues can result in human, material and production losses, making safety assurance an urgent issue in production sites in all industries.

In recent years, many disabled people are working in factories, where soldering and milling are performed as part of manual labor. However, the manual processes require learning and repeating the same tasks, which is time-consuming. At the same time, there are concerns about human error resulting from accidents, lack of safety checks and inexperience.

L. Barolli (Ed.): EIDWT 2024, LNDECT 193, pp. 502–511, 2024.
https://doi.org/10.1007/978-3-031-53555-0_48

One method to maintain worker concentration in their work is pointing and calling, which is a method for ensuring safety by pointing at work objects or tools, and verbally calling and telling the situation in order to predict accidents and raise safety awareness. Therefore, pointing and calling are implemented at production sites to reduce human error and prevent accidents. This method can be used for detecting processing deficiencies and confirming the number of products to be shipped. However, due to a lack of work experience and insufficient recognition of the work importance, there are many cases where the correct motion and procedures are not followed.

To ensure the safety of workers in factories, it is necessary to continuously monitor their actions. Also, the instructors spend a lot of time monitoring. Sometime the instructors may be unable to work on other duties, which may affect the overall work, placing a heavy burden on the instructors and supervisors. Therefore, it is necessary to both ensure worker safety and reduce the burden on the instructors.

In our previous work, we proposed a soldering danger detection system [1, 2], a soldering motion monitoring system [3], a haptics based soldering training system [4–6], a whole body motion analysis system for people with disabilities [7] and some systems considering ambient intelligence [8–14].

In this paper, we propose a motion analysis system for pointing and calling based on object detection and speech recognition. The proposed system uses a depth camera to capture the color and depth of image of a worker during pointing and calling. The system also recognizes the object of pointing and calling and also recognizes the worker voice. The proposed system takes into account the safety measures in order to prevent accidents and injuries at production sites. The proposed system can also recognize correct pointing and calling based on object detection, speech recognition and posture estimation [15–19]. Furthermore, the system detects the direction of the index finger during pointing and calling. To evaluate the proposed system, we carried out experiments for the case of a worker sitting in a chair. From the experimental results, we conclude that the proposed system can detect the pointing and calling based on the depth camera and microphone. Also, the proposed system can support people with disabilities for safely work.

The structure of the paper is as follows. In Sect. 2, we describe the proposed system. In Sect. 3, we present the experimental results. Finally, conclusions and future work are given in Sect. 4.

Table 1. Motion sequence for pointing and calling.

State	Contents of Motion
Procedure 1	Looking at the object and pointing
Procedure 2	Without changing gaze, pull your fingers up to your ears
Procedure 3	Point again and call the state of the object

2   Proposed System

In this section, we present the proposed system. Figure 1 shows the structure of proposed system. The safety of workers is the first priority in order to ensure continuous operations in production plants. The proposed system detects whether the pointing and calling are performed correctly by analyzing the actions of workers and supports the safety of instructors and workers. The proposed system evaluates the pointing and calling motions performed by the worker based on the sequence of motions as shown in Table 1. A camera captures moving images of the worker performing the pointing and calling, estimates the worker posture and recognizes the object of the pointing and calling. It also uses a microphone to recognize the worker voice associated with the motion.

The proposed system detects whether the fingertips and gaze point to the object correctly or not and whether the worker speaks correctly or not in order to evaluate the pointing and calling motion. The motion analysis system capures a full-color image expressed in red, green and blue and depth that represents the distance from the camera to the image target.

The proposed system estimates the worker posture in three dimensions based on two dimensional keypoint coordinates obtained from MediaPipe [20, 21], which is capable to make the skeletal estimation, and the distance from the depth camera to each keypoint. The proposed system detects the correctness of the pointing based on the relative positions of the fingertips and the object as shown in Fig. 2. The pointing object is detected in three dimensions by using two dimensional coordinates and depth [22] of the bounding box using the object detection method YOLOv5 [23]. The center coordinates of pointing and calling objects are decided from the bounding box using YOLOv5. Then, a spherical zone is generated from the center coordinates and judged based on the Euclidean distance from the fingertip and the angle to the target. The radius of the sphere is 1.1 times the size of the detection target.

The proposed system is also able to detect whether the gaze direction is correct or incorrect for the pointing and calling target. By gaze estimation detects whether the worker is looking at the target by deciding the direction of the face and the vector to the pointing and calling target from the two-dimensional key point coordinates of the left and right eyes and nose. In addition, it detects the correctness or incorrectness of speech that needs to be made at the same time as pointing. The Sound is captured using a microphone to measure the sound pressure. Then, the Julius [24], which is a speech recognition engine, analyzes the content of speech. Finally, the proposed system judges that pointing and calling are correctly performed and executed.

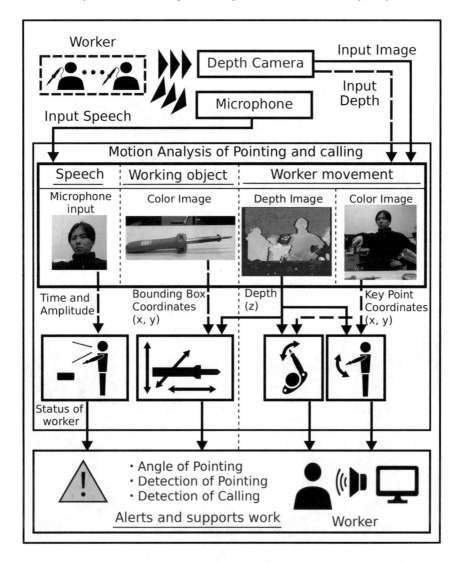

Fig. 1. Structure of proposed system.

3 Experimental Results

This section presents the experimental results. In order to evaluate the performance of the proposed system, an experiment to detect the correct pointing and calling motions is conducted when the proposed system is placed in front of the target and worker carring out the pointing and calling.

Figure 3 shows the experimental environment. The experimental scenario consists of a correct sequence of pointing and calling motions, incorrect motions and correct actions. Also, the correct motion is performed in the range of approximately 10 [s]

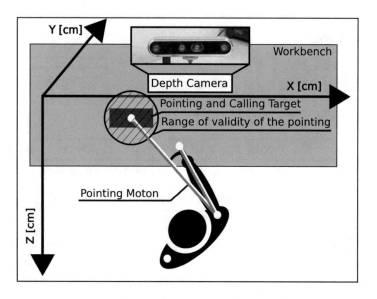

Fig. 2. Measuring of pointing by depth camera.

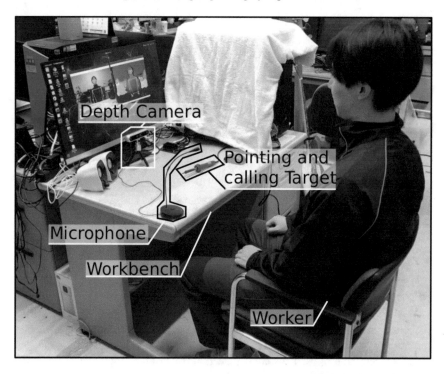

Fig. 3. Experimental environment.

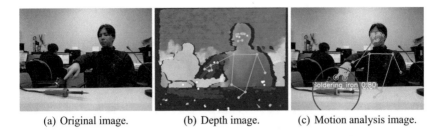

(a) Original image.   (b) Depth image.   (c) Motion analysis image.

Fig. 4. Correct motion procedure 1.

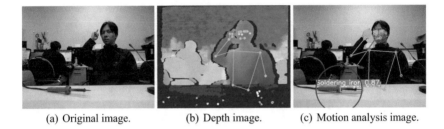

(a) Original image.   (b) Depth image.   (c) Motion analysis image.

Fig. 5. Correct motion procedure 2.

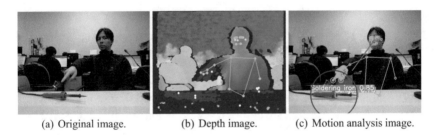

(a) Original image.   (b) Depth image.   (c) Motion analysis image.

Fig. 6. Correct motion procedure 3.

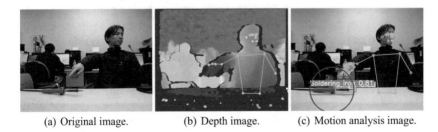

(a) Original image.   (b) Depth image.   (c) Motion analysis image.

Fig. 7. Incorrect motion procedure 1.

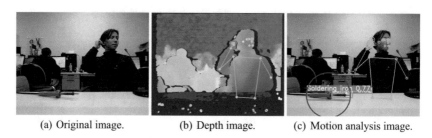

(a) Original image.   (b) Depth image.   (c) Motion analysis image.

Fig. 8. Incorrect motion procedure 2.

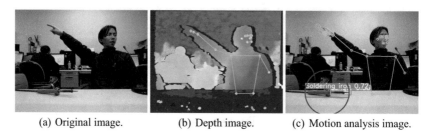

(a) Original image.   (b) Depth image.   (c) Motion analysis image.

Fig. 9. Incorrect motion procedure 3.

to 20 [s], 25 [s] to 35 [s] and 40 [s] to 50 [s]. The correct speech is "Soldering Iron YOSHI" and the incorrect is "Soldering OK". The experiment time is 60 [s] and the worker condition is captured at 1.0 [s] intervals.

Figure 4, Fig. 5 and Fig. 6 show examples of correct actions. While, examples of incorrect actions are shown in Fig. 7, Fig. 8 and Fig. 9. Figure 4(a), Fig. 5(a), Fig. 6(a), Fig. 7(a), Fig. 8(a) and Fig. 9(a) are the original images and Fig. 4(b), Fig. 5(b), Fig. 6(b), Fig. 7(b), Fig. 8(b) and Fig. 9(b) are the depth images. While, Fig. 4(c), Fig. 5(c), Fig. 6(c), Fig. 7(c), Fig. 8(c) and Fig. 9(c) are images with applied motion analysis.

Experimental results of the detection of pointing and calling motions using the proposed system are shown in Fig. 10. The correct motion is designated as True and the incorrect motion is designated as False. Figure 10(a) shows the visualization results of sound pressure and correct speech content detection. It can be seen that the sound pressure increases when the worker speaks. The speech content is judged when it exceeds 50 [dB]. The correct speech is detected at about 15 [s] and 45 [s] and speech is detected at about 30 [s]. Figure 10(b) shows the visualization results of gaze detection at the pointing and calling target. It can be seen that the subject looks at the target correctly from about 10 [s] to 20 [s] and from about 40 [s] to 50 [s], while the subject looks away from the target from about 25 [s] to 35 [s]. Figure 10(c) shows the visualization results of the detection of pointing at the target. It can be seen that the finger is pointing at the target in the same way as speech and gaze based on the angle and distance of the fingertip to the target. Figure 10(d) shows the visualization results of the detection of motion sequence of pointing and calling. It can be seen that the workers state changes

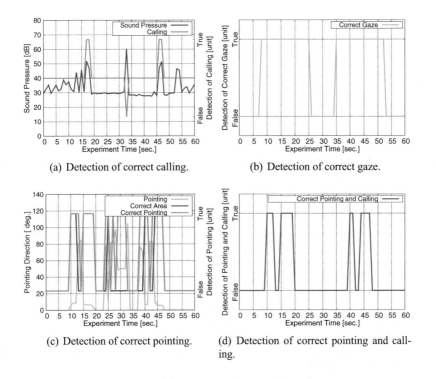

(a) Detection of correct calling.

(b) Detection of correct gaze.

(c) Detection of correct pointing.

(d) Detection of correct pointing and calling.

Fig. 10. Experimental results.

with time and that the system can detect whether the correct pointing and calling is being performed or not.

The experimental results show that the proposed system correctly detects the motion sequence of pointing, looking and speaking.

4   Conclusions

In this paper, we proposed a motion analysis system for pointing and calling based on object detection and speech recognition. The proposed system can detect whether the pointing and calling are performed correctly by analyzing the actions of workers and supports the safety of instructors and workers. The proposed system evaluates the pointing and calling motions performed by the worker based on the sequence of motions. We carried out experiments for the case of a worker sitting in a chair. From the experimental results, we conclude as following.

- The proposed system can detect the pointing and calling based on the depth camera and microphone.
- The proposed system is an effective system for providing a good safety during pointing and calling.
- The proposed system can support people with disabilities for safely work.

In the future, we will consider different scenarios to evaluate the proposed system. We also would like to improve the detection accuracy of the proposed system.

Acknowledgement. This work was supported by JSPS KAKENHI Grant Number JP20K19793.

References

1. Yasunaga, T., et al.: Object detection and pose estimation approaches for soldering danger detection. In: Proceedings of The IEEE 10-th Global Conference on Consumer Electronics, pp. 776–777 (2021)
2. Yasunaga, T., et al.: A soldering motion analysis system for danger detection considering object detection and attitude estimation. In: Proceedings of The 10-th International Conference on Emerging Internet, Data & Web Technologies, pp. 301–307 (2022)
3. Toyoshima, K., et al.: Analysis of a soldering motion for dozing state and attention posture detection. In: Proceedings of 3PGCIC-2022, pp. 146–153 (2022)
4. Toyoshima, K., et al.: Proposal of a haptics and LSTM based soldering motion analysis system. In: Proceedings of The IEEE 10-th Global Conference on Consumer Electronics, pp. 1–2 (2021)
5. Toyoshima, K., et al.: Design and implementation of a haptics based soldering education system. In: Proceedings of IMIS-2022, pp. 54–64 (2022)
6. Toyoshima, K., et al.: Experimental results of a haptics based soldering education system: a comparison study of RNN and LSTM for detection of dangerous movements. In: Proceedings of INCoS-2022, pp. 212–223 (2022)
7. Toyoshima, K., et al.: A soldering motion analysis system for monitoring whole body of people with developmental disabilities. In: Proceedings of AINA-2023, pp. 38–46 (2023)
8. Obukata, R., et al.: Design and evaluation of an ambient intelligence testbed for improving quality of life. Int. J. Space-Based Situated Comput. **7**(1), 8–15 (2017)
9. Oda, T., et al.: Design of a deep Q-network based simulation system for actuation decision in ambient intelligence. In: Proceedings of AINA-2019, pp. 362–370 (2019)
10. Obukata, R., et al.: Performance evaluation of an AmI testbed for improving QoL: evaluation using clustering approach considering distributed concurrent processing. In: Proceedings of IEEE AINA-2017, pp. 271–275 (2017)
11. Yamada, M., et al.: Evaluation of an IoT-based e-learning testbed: performance of OLSR protocol in a NLoS environment and mean-shift clustering approach considering electroencephalogram data. Int. J. Web Inf. Syst. **13**(1), 2–13 (2017)
12. Hirota, Y., et al.: Proposal and experimental results of an ambient intelligence for training on soldering iron holding. In: Proceedings of BWCCA-2020, pp. 444–453 (2020)
13. Hirota, Y., et al.: Proposal and experimental results of a DNN based real-time recognition method for Ohsone style fingerspelling in static characters environment. In: Proceedings of The IEEE 9-th Global Conference on Consumer Electronics, pp. 476–477 (2020)
14. Oda, T., et al.: Design and implementation of an IoT-based e-learning testbed. Int. J. Web Grid Serv. **13**(2), 228–241 (2017)
15. Toshev, A., Szegedy, C.: DeepPose: human pose estimation via deep neural networks. In: Proceedings of The 27-th IEEE/CVF Conference on Computer Vision and Pattern Recognition (IEEE/CVF CVPR-2014), pp. 1653–1660 (2014)
16. Haralick, R., et al.: Pose estimation from corresponding point data. IEEE Trans. Syst. **19**(6), 1426–1446 (1989)
17. Fang, H., et al.: RMPE: regional multi-person pose estimation. In: Proceedings of the IEEE International Conference on Computer Vision, pp. 2334–2343 (2017)

18. Xiao, B., et al.: Simple baselines for human pose estimation and tracking. In: Proceedings of the European Conference on Computer Vision (ECCV), pp. 466–481 (2018)
19. Martinez, J., et al.: A simple yet effective baseline for 3d human pose estimation. In: Proceedings of the IEEE International Conference on Computer Vision, pp. 2640–2649 (2017)
20. Lugaresi, C., et al.: MediaPipe: a framework for building perception pipelines. arXiv preprint arXiv:1906.08172 (2019)
21. Antonio, M., et al.: Real-time upper body detection and 3D pose estimation in monoscopic images. In: European Conference on Computer Vision, pp. 139–150 (2006)
22. Andriyanov, N., et al.: Intelligent system for estimation of the spatial position of apples based on YOLOv3 and real sense depth camera D415. Symmetry **14**(1) (2022)
23. Yang, D., et al.: Research of target detection and distance measurement technology based on YOLOv5 and depth camera. In: 2022 4th International Conference on Communications, Information System and Computer Engineering (CISCE), pp. 346–349 (2022)
24. Paplu, S., et al.: Utilizing semantic and contextual information during human-robot interaction. 2021 IEEE International Conference on Development and Learning (ICDL), pp. 1–2 (2021)

Trust Zone Model with the Mandatory Access Control Model

Shigenari Nakamura[1]([⊠]) and Makoto Takizawa[2]

[1] Tokyo Denki University, Tokyo, Japan
`s.nakamura@mail.dendai.ac.jp`
[2] Hosei University, Tokyo, Japan
`makoto.takizawa@computer.org`

Abstract. In order to protect internal networks of enterprises from any threats, a perimeter model has been widely used. Here, every access from subjects inside the perimeter is assumed to be trustworthy and the access is allowed without any conditions. Hence, a threat enters inside the perimeter once, the effect of the threat spreads inside the perimeter easily. In order to protect internal networks from any threats, a ZT (Zero Trust) model is proposed. Here, it is assumed that every access to an object is not trustworthy. Hence, every access is checked whether or not the access can be allowed to be performed. Although the ZT model makes the system secure, it is not easy to change the present system model to the ZT model for many enterprises. In order to make the shift to the ZT model smooth, a TZ (Trust Zone) model is proposed. In the TZ model, every object belongs to a trust zone. The authorization process in the TZ model is more simple than the ZT model. Hence, the TZ model is known as an intermediate stage of the shift to the ZT model. However, it is not discussed how to prevent illegal information flow in the TZ model. In this paper, we newly propose a TZMAC (Trust Zone with Mandatory Access Control) model to prevent illegal information flow in the TZ model. Here, the trust zones in the TZ model are used as the security classes of the MAC (Mandatory Access Control) model. Each subject also belongs to a security class. If a subject accesses an object, the relation between the security classes of the subject and object is also checked. Since the accesses implying illegal information flow are not allowed to be performed, the system is made more secure.

Keywords: Trust zone model · Zero trust model · Mandatory access control model · Information flow

1 Introduction

In order to protect internal networks of enterprises from any threats, the perimeter model has been widely used. In the perimeter model, it is assumed that no threats can enter inside a perimeter. Here, every access from subjects inside the perimeter is assumed to be trustworthy and the access is allowed to be performed without any conditions. Hence, a threat enters inside the perimeter once, the effect of the threat spreads inside

L. Barolli (Ed.): EIDWT 2024, LNDECT 193, pp. 512–521, 2024.
https://doi.org/10.1007/978-3-031-53555-0_49

the perimeter easily. Nowadays it is impossible to protect internal networks from any threats by only the perimeter model.

In order to protect internal networks from any threats, a ZT (Zero Trust) model [5, 15] is proposed. Here, it is assumed that every access to an object is not trustworthy differently from the perimeter model. Hence, it is checked whether or not an access can be allowed to be performed on receipt of the access in the ZT model. Access control models [3, 17, 18] which are widely used in information systems are also considered in the ZT model. Even if a subject is issued an access right to access an object in an access control model, the access may be rejected. That is because not only the access right issued to the subject but also other information on the access, e.g. information on the subject, a device used by the subject, and the network, are checked to make an authorization decision.

However, it is not easy to change the present system model to the ZT model for many enterprises. Hence, a TZ (Trust Zone) model [2] is proposed to make the shift to the ZT model smooth. In the TZ model, every object belongs to a trust zone. The trust levels of objects in the same trust zone are same. An authorization decision is made for each trust zone. Here, the authorization process in the TZ model is more simple than the ZT model. Hence, the TZ model is known as an intermediate stage of the shift to the ZT model. In this paper, we consider the TZ model.

In access control models, subjects may get data via other subjects and objects even if the subjects are issued no access right to get the data, i.e. illegal information flow may occur [1]. It is critical to prevent illegal information flow to make information systems secure. In our previous studies [6, 8], types of protocols to prevent illegal information flow in database and P2PPSO (Peer-to-Peer Publish/Subscribe with Object concept) systems are proposed. Furthermore, protocols for the IoT (Internet of Things) [14] are proposed in papers [7, 9, 10]. The protocols are improved [11, 12] and evaluated in terms of the electric energy consumption [13].

In the TZ model, it is not discussed how to prevent illegal information flow. In order to prevent illegal information flow, we consider a MAC (Mandatory Access Control) model [4, 16]. Here, every entity such as subject or object belongs to a security class. Access rules are given based on the relations among the security classes of subjects and objects. If subjects manipulate objects in the MAC model, illegal information flow does not occur.

In this paper, we newly propose a TZMAC (Trust Zone with Mandatory Access Control) model to prevent illegal information flow in the TZ model. Here, the trust zones in the TZ model are used as the security classes of the MAC model. Each subject also belongs to a security class. If a subject accesses an object, the relation between the security classes of the subject and the object is also checked.

In Sect. 2, we discuss a system model. In Sect. 3, we newly propose the TZMAC model to prevent illegal information flow in the TZ model. In Sect. 4, we show an example of the TZMAC model.

2 System Model

2.1 TZ (Trust Zone) Model

Nowadays infrastructures of enterprises get complex. The infrastructures are composed of several internal networks. Here, objects are located in not only local networks but also outside the local networks such as mobile devices and cloud services. Although a perimeter model has been widely used to protect internal networks from any threats, the threats can enter inside the perimeter more easily than before due to the change of infrastructures. In the perimeter model, it is assumed that no threats can enter inside a perimeter. Here, every access from subjects inside the perimeter is assumed to be trustworthy and the access is allowed to be performed without any conditions. If a threat enters inside the perimeter, the effect of the threat spreads inside the perimeter easily. Hence, it is impossible to protect internal networks from any threats by only the perimeter model.

In order to protect internal networks from any threats, a ZT (Zero Trust) model [5, 15] is proposed. Here, it is assumed that every access to an object is not trustworthy differently from the perimeter model. Hence, it is checked whether or not an access can be allowed to be performed on receipt of the access in the ZT model. In the access control model which is used in the ZT model, subjects are issued access rights. Here, an access right is a pair $\langle o, op \rangle$ of an object o and an operation op. In this paper, each subject sb_i is assumed to support operations *read* and *write* for objects. However, even if a subject is issued an access right to access an object in the access control model, the access may be rejected. That is because not only the access right issued to the subject but also other information on the access, e.g. information on the subject, a device used by the subject, and the network, are checked to make an authorization decision.

In the ZT model, there are the PDP (Policy Decision Point) and PEP (Policy Enforcement Point). If a subject sb_i accesses an object o, the access arrives at the PEP. The PEP forwards the access to the PDP. Next, the PDP checks the access and makes an authorization decision in a trust algorithm. Then, the PDP informs the PEP of the decision. If the access is allowed to be performed, the PEP enables the connection between the subject sb_i and the object o. Otherwise, the access is rejected and the PEP terminates the connection.

In order to make authorization decisions, a trust algorithm is performed in the PDP. The trust algorithm is the process to decide whether or not each access from a subject sb_i is allowed to be performed. Here, not only an access right issued to a subject sb_i but also the following information on the access are checked:

- Subject: The information shows who the subject sb_i is, e.g. account ID, department, and job category.
- Access: The information shows the situation of the access, e.g. time and location.
- Device: The information shows the status of the device used by the subject sb_i, e.g. owner, OS version, patch level, and software installed.
- Object: The information shows the requirements of the object o accessed by the subject sb_i, e.g. data sensitivity and device configuration.
- Threat: The information shows the feeds on general threats, e.g. signature of active malware.

There are some types of methods to implement the trust algorithm. In this paper, we consider the criteria based trust algorithm [15]. Here, the above information is compared with the criteria configured by the enterprise. If all the criteria are met, the access is allowed to be performed. Otherwise, the access is rejected.

In the ZT model, it is ideal that a PEP is placed for each object as shown in Fig. 1. Here, the more number of objects there are, the more number of PEPs are placed. Hence, it is not easy to change the present system model to the ZT model for many enterprises. In order to make the shift to the ZT model smooth, a TZ (Trust Zone) model [2] is proposed. In the TZ model, every object belongs to a trust zone. The trust levels of objects in the same trust zone are same. Since a PEP is placed for each trust zone, the PDP makes an authorization decision for each trust zone as shown in Fig. 2. Here, the authorization process in the TZ model is more simple than the ZT model. Hence, the TZ model is known as an intermediate stage of the shift to the ZT model. In this paper, we consider the TZ model.

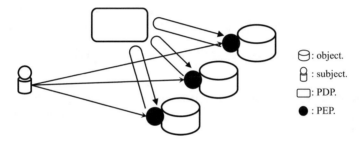

$\bigcirc$: object.

$\bigcap$: subject.

$\square$: PDP.

$\bullet$: PEP.

Fig. 1. ZT model.

2.2   MAC (Mandatory Access Control) Model

Suppose there are a pair of objects o_1 and o_2. The objects o_1 and o_2 are manipulated by a pair of subjects sb_i and sb_j. The subject sb_i is issued a pair of access rights $\langle o_1, read \rangle$ and $\langle o_2, write \rangle$ and another subject sb_j is issued an access right $\langle o_2, read \rangle$. First, the subject sb_i reads data x from the object o_1. Next, the subject sb_i writes the data x to the object o_2. Finally, the subject sb_j reads data y from the object o_2. Here, the data y may include the data x and the subject sb_j can get the data x. This means, the subject sb_j can get data of the object o_1 via the subject sb_i and the object o_2 although the subject sb_j is not issued the access right $\langle o_1, read \rangle$. Here, illegal information flow [1] occurs. It is critical to prevent illegal information flow to make information systems secure.

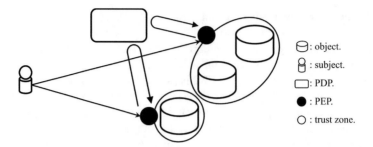

Fig. 2. TZ model.

In order to prevent illegal information flow, a MAC (Mandatory Access Control) model [4, 16] is proposed. Here, every entity such as subject or object belongs to a security class sc. Let sc_k be a security class where an entity e_k belongs. There is a precedent relation "$\rightarrow$" among the security classes. An information flow relation between a pair of entities e_k and e_l is defined based on a pair of security classes sc_k and sc_l as follows:

Definition 1. *Information can flow from an entity e_k to an entity e_l iff (if and only if) $sc_k \rightarrow sc_l$.*

The security classes partially ordered in the precedent relation "$\rightarrow$" are specified in a lattice. In the MAC model, access rules are given so that the precedent relation "$\rightarrow$" among security classes is satisfied.

Definition 2. *A subject sb of a security class sc_k can write data to an object o of a security class sc_l iff $sc_k \rightarrow sc_l$.*

Definition 3. *A subject sb of a security class sc_k can read data from an object o of a security class sc_l iff $sc_l \rightarrow sc_k$.*

Definition 4. *A subject sb of a security class sc_k can read and write data from and to an object o, i.e. modify data of an object o, of a security class sc_l iff $sc_k \rightarrow sc_l$ and $sc_l \rightarrow sc_k$.*

Suppose there are a pair of objects o_1 and o_2. The object o_1 belongs to a security class sc_k and another object o_2 belongs to a security class sc_l. The objects o_1 and o_2 are manipulated by a pair of subjects sb_i and sb_j. The subject sb_i belongs to the security class sc_k and another subject sb_j belongs to the security class sc_l. Let $sc_l \rightarrow sc_k$ hold. Here, the subject sb_i is allowed to read data from the object o_1 but not allowed to write data to the object o_2. Although the subject sb_i can read data x from the object o_1, the subject sb_i cannot write the data x to the object o_2. Hence, the subject sb_j cannot read the data x of the object o_1 from the object o_2. Furthermore, the subject sb_j is not allowed to read data from the object o_1 because $sc_k \rightarrow sc_l$ does not hold. Therefore, illegal information flow does not occur in the MAC model.

3   TZMAC (Trust Zone with Mandatory Access Control) Model

In the ZT model, the PDP makes an authorization decision for an access in a trust algorithm and informs the PEP of the decision. If the access is allowed to be performed, the PEP enables the connection between a subject and an object. Otherwise, the access is rejected and the PEP terminates the connection. A PEP is placed for each trust zone in the TZ model. Hence, the PDP makes an authorization decision for each trust zone. Every object belongs to a trust zone. If a pair of objects belong to the same trust zone, the trust levels of the objects are same.

In the trust algorithm, many types of information on an access is checked. Hence, many types of threats are prevented in the TZ model. However, it is not discussed how to prevent illegal information flow. In order to prevent illegal information flow, we consider a MAC (Mandatory Access Control) model. Here, every entity such as subject or object belongs to a security class. Access rules are given so that the precedent relation "$\rightarrow$" among security classes is satisfied. If subjects manipulate objects in the MAC model, illegal information flow does not occur.

In this paper, we newly propose a TZMAC (Trust Zone with Mandatory Access Control) model to prevent illegal information flow in the TZ model. Here, the trust zones in the TZ model are used as the security classes of the MAC model. This means, security classes where objects belong are decided based on the trust zones where the objects are. Each subject also belongs to a security class. If a subject accesses an object, the relation between the security classes of the subject and object is also checked.

Figure 3 shows the trust algorithm in the TZMAC model. Suppose a subject sb_i accesses an object o. The subject sb_i belongs to a security class sc_k and the object o belongs to a security class sc_l. In the trust algorithm, the pair of security classes sc_k and sc_l of the subject sb_i and the object o are used. It is decided how the subject sb_i is allowed to manipulate the object o based on the relation between the security classes sc_k and sc_l. If the relation between the security classes sc_k and sc_l is not satisfied by the access from the subject sb_i, the access is rejected even if the other criteria are met. Therefore, illegal information flow can be prevented in the TZMAC model.

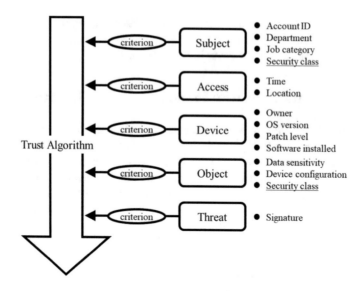

Fig. 3. Trust algorithm in the TZMAC model.

4   Example

We consider an example as shown in Fig. 4. Suppose there are three security classes sc_H, sc_M, and sc_L. The trust levels of the security classes sc_H, sc_M, and sc_L are "High", "Medium", and "Low", respectively. It is assumed that $sc_L \rightarrow sc_M$ and $sc_M \rightarrow sc_H$ hold. The objects o_1, o_2, and o_3 belong to the security classes sc_H, sc_M, and sc_L, respectively. The pair of subjects sb_i and sb_j belong to the security classes sc_H and sc_L, respectively.

Suppose the subject sb_i of the security class sc_H tries to read data form the object o_2 of the security class sc_M. The relation between the security classes sc_H and sc_M is checked in the trust algorithm. Since $sc_M \rightarrow sc_H$, the subject sb_i is allowed to read data from the object o_2. In this paper, we consider the criteria based trust algorithm. Hence, if the other criteria are also met in the trust algorithm, the access form the subject sb_i is allowed to be performed. Similarly, the following accesses are allowed to be performed if the other criteria are also met in the trust algorithm:

- The subject sb_i is allowed to read data from the object o_3 because $sc_L \rightarrow sc_H$.
- The subject sb_j is allowed to write data to the object o_1 because $sc_L \rightarrow sc_H$.
- The subject sb_j is allowed to write data to the object o_2 because $sc_L \rightarrow sc_M$.
- The subject sb_i is allowed to read and write data from and to the object o_1, i.e. modify data of the object o_1, because $sc_H \rightarrow sc_H$.
- The subject sb_j is allowed to modify data of the object o_3 because $sc_L \rightarrow sc_L$.

Suppose the subject sb_j reads data from the object o_3 and then writes the data to the object o_2. Here, although the subject sb_i can read the data of the object o_3 reading the data from the object o_2, illegal information flow does not occur because $sc_L \rightarrow sc_H$ holds. The subject sb_i is not allowed to write data to the objects o_2 and o_3 because sc_H

$\rightarrow sc_M$ and $sc_H \rightarrow sc_L$ do not hold. Although the subject sb_i is allowed to write data to the object o_1, the subject sb_j is not allowed to read data from the object o_1 because $sc_H \rightarrow sc_L$ does not hold. Hence, the subject sb_j cannot get data of the object o_1. Similarly, the subject sb_j cannot get data of the object o_2. Therefore, illegal information flow does not occur in the TZMAC model.

Even if the relation between the security classes of a subject and an object is satisfied by an access from the subject, i.e. illegal information flow does not occur, the access is rejected if another criterion is not met in the trust algorithm in the TZMAC model.

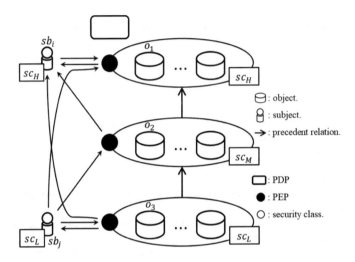

Fig. 4. TZMAC model.

5   Concluding Remarks

In order to protect internal networks of enterprises from any threats, a ZT (Zero Trust) model is proposed. Here, it is assumed that every access to an object is not trustworthy. A PDP (Policy Decision Point) makes an authorization decision for every access in a trust algorithm and informs a PEP (Policy Enforcement Point) of the decision. The PEP enables or terminates the connection between a subject and an object according to the decision. Although it is ideal that a PEP is placed for each object, it is not easy for many enterprises. Hence, a TZ (Trust Zone) model is proposed to make the shift to the ZT model smooth. Here, every object belongs to a trust zone and a PEP is placed for each trust zone. The PDP makes an authorization decision for each trust zone.

In the trust algorithm, it is not discussed how to prevent illegal information flow. In order to prevent illegal information flow, we consider a MAC (Mandatory Access Control) model. Here, every entity such as subject or object belongs to a security class. Access rules are given so that the relation among security classes of a subject and an object is satisfied. If subjects manipulate objects in the MAC model, illegal information flow does not occur.

In this paper, we newly propose a TZMAC (Trust Zone with Mandatory Access Control) model to prevent illegal information flow in the TZ model. Here, the trust zones in the TZ model are used as the security classes of the MAC model. This means, security classes where objects belong are decided based on the trust zones where the objects are. The relation between the security classes of the subject and object is also checked in the trust algorithm. Since the accesses implying illegal information flow are not allowed to be performed, the system is made more secure.

Acknowledgements. This work is supported by Japan Society for the Promotion of Science (JSPS) KAKENHI Grant Number JP23K16887.

References

1. Denning, D.E.R.: Cryptography and Data Security. Addison Wesley, Boston (1982)
2. Department of Homeland Security, Cybersecurity & Infrastructure Security Agency: Trusted internet connections (tic) 3.0 core guidance documents (2021). https://www.cisa.gov/resources-tools/resources/trusted-internet-connections-tic-30-core-guidance-documents
3. Fernandez, E.B., Summers, R.C., Wood, C.: Database Security and Integrity. Adison Wesley, Boston (1980)
4. Ferraiolo, D.F., Kuhn, D.R., Chandramouli, R.: Role-Based Access Control, 2nd edn. Artech, Norwood (2007)
5. Kindervag, J., Balaouras, S., Mak, K., Blackborow, J.: No more chewy centers: the zero trust model of information security. Technical report, Forrester Research (2016)
6. Nakamura, S., Duolikun, D., Enokido, T., Takizawa, M.: A read-write abortion protocol to prevent illegal information flow in role-based access control systems. Int. J. Space-Based Situat. Comput. **6**(1), 43–53 (2016)
7. Nakamura, S., Enokido, T., Takizawa, M.: Information flow control based on the CAPBAC (capability-based access control) model in the IoT. Int. J. Mob. Comput. Multimedia Commun. **10**(4), 13–25 (2019)
8. Nakamura, S., Enokido, T., Takizawa, M.: Information flow control in object-based peer-to-peer publish/subscribe systems. Concurr. Comput. Pract. Exp. **32**(8) (2020)
9. Nakamura, S., Enokido, T., Takizawa, M.: Implementation and evaluation of the information flow control for the internet of things. Concurr. Comput. Pract. Exp. **33**(19) (2021)
10. Nakamura, S., Enokido, T., Takizawa, M.: Information flow control based on capability token validity for secure IoT: implementation and evaluation. Internet Things **15**, 100,423 (2021)
11. Nakamura, S., Enokido, T., Takizawa, M.: Traffic reduction for information flow control in the IoT. In: Proceedings of the 16th International Conference on Broad-Band Wireless Computing, Communication and Applications, pp. 67–77 (2021)
12. Nakamura, S., Enokido, T., Takizawa, M.: Capability token selection algorithms to implement lightweight protocols. Internet Things **19**, 100,542 (2022)
13. Nakamura, S., Enokido, T., Takizawa, M.: Assessment of energy consumption for information flow control protocols in IoT devices. Internet Things **24**, 100,992 (2023)
14. Oma, R., Nakamura, S., Duolikun, D., Enokido, T., Takizawa, M.: An energy-efficient model for fog computing in the internet of things (IoT). Internet Things **1–2**, 14–26 (2018)
15. Rose, S., Borchert, O., Mitchell, S., Connelly, S.: Zero trust architecture. Technical report, National Institute of Standards and Technology (2020). https://csrc.nist.gov/publications/detail/sp/800/207/final

16. Sandhu, R.S.: Lattice-based access control models. IEEE Comput. **26**(11), 9–19 (1993)
17. Sandhu, R.S., Coyne, E.J., Feinstein, H.L., Youman, C.E.: Role-based access control models. IEEE Comput. **29**(2), 38–47 (1996)
18. Yuan, E., Tong, J.: Attributed based access control (ABAC) for web services. In: Proceedings of the IEEE International Conference on Web Services (ICWS'05), p. 569 (2005)

A DPI-Based Network Traffic Feature Vector Optimization Model

Yuqing Zhao, Baojiang Cui[(✉)], Jun Yang, and Meiyi Jiang

Beijing University of Posts and Telecommunications, Beijing, China
{galadriel,cuibj,junyang,myjiang}@bupt.edu.cn

Abstract. Network traffic features play a key role in network anomaly detection, which is of great significance to ensure normal network service. Conventional network traffic feature extraction focuses on the network layer to obtain features such as IP source, destination address, timestamp and so on, which describe the status involved in the transmission process of messages, but it lacks the connection with specific application behavior. At the same time, huge feature data will also burden the detection work. In view of the above situation, this paper proposes a DPI-based network traffic feature vector optimization model——DRFV optimization model.

This model combines the DPI technology to expand the feature extraction of the original traffic to the application layer, realize the traffic analysis of the application layer data, and expand and increase the feature vector dimension. After obtaining abundant features, the random forest model based on Bagging thought is used to classify the features, obtain feature effect ranking, and select the optimized feature vector according to the conditions required for network anomaly detection, so as to achieve the goal of dimension reduction and optimization, and obtain the feature vector with more research significance. The network anomaly detection model uses the model proposed in this paper to optimize the traffic feature extraction, which can obtain better results in detection results and operating performance. It has a significant improvement in accuracy, F1 value and running time.

1 Introduction

With the development of information technology and computers, the Internet has been widely used in various fields of social life, affecting people's work and life. Network technology is a double-edged sword, which not only facilitates people's life, but also brings dangers and challenges. Many criminals spread computer viruses, attack network systems and affect the stable operation of the network through the Internet. Therefore, NAD(Network Anomaly Detection) has become an indispensable step to ensure the normal operation of network services. As the network environment becomes more complex and the means of network attacks become more diverse, the traditional rule-based detection methods can no longer meet the detection needs of new network anomalies such as zero-day attacks. In order to face the rapidly updated network state more flexibly, machine learning has been studied and applied in the field of NAD in recent years.

L. Barolli (Ed.): EIDWT 2024, LNDECT 193, pp. 522–531, 2024.
https://doi.org/10.1007/978-3-031-53555-0_50

Through the analysis of network data and model training, the effect of anomaly detection is achieved. In the process of network transmission and exchange, traffic, as the carrier of information, contains a large number of valuable data. Traffic features are the key entry data for network anomaly detection. However, due to the complexity of network environment and the uncertainty of data traffic, NAD is currently facing the problem with difficult detection of high-dimensional features and incomplete information of low-dimensional features. At the same time, the uneven distribution of types and abnormal features also brings difficulties to network anomaly detection.

Network traffic feature extraction is an important step in NAD, which is of great significance to realize network anomaly detection and ensure the normal service of network. Conventional network traffic feature extraction focuses on the Network layer, and obtains such features as IP source address, destination address, timestamp and protocol information about this layer, etc. that describe the status of messages during transmission. The attacks and exceptions that can be detected by this method are mainly reflected in the message format, while the connection between context and specific application behavior is lacking. The data that can better reflect the position of traffic message in the whole flow or behavior distribution is mainly distributed in the Application layer.

In view of the above situation, this paper proposes a network traffic feature vector optimization model based on DPI(Deep Packet Inspection) technology——DRFV(DPI-RandomForest Feature Vector) optimization model, which is optimized in two stages of feature extraction and feature selection. The DRFV model, based on the existing traffic features and combined with DPI technology, extends the feature extraction of the original traffic to the Application layer, realizes the traffic analysis of the application layer data, expands and increases the dimension of the feature vector, so as to solve the problem of incomplete coverage of the feature vector. After obtaining rich features, DRFV uses the random forest model based on Bagging thought to train features, obtain feature importance ranking, then filters according to the conditions required for network anomaly detection to obtain the optimized feature vectors, which achieves the purpose of reducing the dimension and optimizing the content of feature vectors. The work has a promoting effect on the development of network traffic feature study, and also has significance for the development of network anomaly detection.

In the first chapter, this paper expounds the background and significance of the research, and briefly explains the main content. In the second chapter, the main progress of related research in recent years is introduced and the main problems solved by the model proposed in this paper are explained. In Sect. 3, methodology, the model is summarized, and the theories and methods used are introduced. Section 4 describes the relevant content of the experiment, including the detailed description of the dataset and model, and carries out the analysis and comparison of the experimental results. Finally, in Sect. 5, the research content and results of this paper are summarized.

2   Related Work

Network traffic feature extraction and selection is an important means for network operators to achieve network traffic monitoring and then effectively manage the network. In the past two decades, scholars have continuously improved the ability and performance of network anomaly detection by studying traffic feature vectors. After rule-based traffic classification gradually fails to adapt to the development of the network, machine learning is applied to network anomaly detection, and feature extraction plays a key role.

The early work of network traffic feature extraction is mainly aimed at a certain kind of malicious attack traffic, such as the statistical historical traffic feature extraction proposed by Guang Cheng et al. in 2003 in the high-speed network model [6]. In 2006, Zhixin Sun et al. proposed the feature extraction algorithm AFCCA based on field and traffic clustering [14], which can obtain the effective features of DOS/DDOS attacks and dynamically make filtering strategies and Yun et al. proposed a feature extraction method based on network protocol and field semantics [17], which promoted the identification of malicious traffic, but the traffic feature information it obtained still has limitations. As the network environment becomes complex and abnormal behaviors become diversified, the feature extraction for a specific abnormal activity can no longer meet the demand. In order to enable the traffic feature extraction to contain more network information, in 2015, G Alotibi et al. proposed a behavior based traffic feature extraction method to judge the characteristics from the perspective of user activities [2], which has achieved ideal results in this field, but its application scope needs to be widened. With the application of deep learning in network anomaly detection, the range of network traffic feature extraction gradually expands to the combination of intra packet and inter packet features to improve the dimension of feature vector and make it more suitable for feature training under the model.

However, with the rapid increase of traffic data volume and the increase of feature vector dimensions, the huge redundant data also brings difficulties to the detection performance. Therefore, the selection and optimization of features has also been put into research. In 2022, N Yoshimura et al. proposed a network traffic feature extraction and anomaly detection method based on DOC-IDS deep learning [16], which simplifies the classification task of multi class labeled traffic and reduces the load brought by eigenvalue design and extraction, but it can still be improved in feature selection; In 2023, T Yang et al. proposed AutoEncoder algorithm for feature selection to improve the identification accuracy of traffic [15], but there may be large differences for data with large distribution deviation. In general, we hope to get more features and information in the aspect of traffic feature extraction, but we don't want the huge data load to bring burden and offset error to the detection.

Therefore, based on these two considerations, this paper proposes an optimization model of DRFV. Compared with the existing feature extraction, DRFV uses DPI technology to increase the dimensional features of the interaction between the application layer and the process, and obtains more information in the feature extraction process; At the same time, the random forest based on Bagging thought is used as the feature set to optimize the dimensionality reduction, which ensures the detection performance to a certain extent. The proposed DRFV optimization model promotes the research and development of network traffic feature vector, and also has some significance for the research of network anomaly detection.

3   Methodology

3.1   Overview

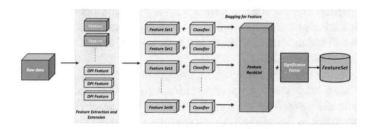

Fig. 1. Overview of DRFV optimization model.

The DRFV (DPI-RF Feature Vector) optimization model proposed in this article is shown in Fig. 1. For the original traffic data, the model performs two levels of feature extraction on it. Firstly, the existing methods mainly focus on feature extraction of network packet fields and information, obtaining a relatively common feature vector set of *Features*. Secondly, based on this, DPI technology is used to conduct in-depth analysis and statistics of traffic, obtaining a deeper feature set of *DPI Features* that focuses more on application services and flow information. The model obtains as much information as possible during the feature extraction stage to expand the dimensionality of the feature set. Next, we will use Bagging based Random Forest to perform dimensionality reduction optimization on the feature set, which is mainly achieved through feature selection. Randomly partition the newly obtained original feature set, and each partition generates a decision tree. The classifier further calculates it to achieve random forest feature sorting, ultimately leaving more meaningful features as optimization results. Overall, the DRFV model can be divided into two parts: feature extraction and feature selection. It increases content as much as possible during the feature extraction stage, and selects effective features as much as possible during the feature selection stage. From a dimensional perspective, it involves first increasing dimensionality and then reducing dimensionality for optimization. While enhancing the importance of the information contained in the feature set, it also avoids the complex redundancy of feature vectors through selection. Next, we will provide a detailed introduction to the technologies and implementation processes involved in these two parts.

3.2 Feature Extraction Based on DPI

DPI is a traffic detection and control technology based on the application layer [5]. Enterprises and Internet service providers use it to identify and prevent network attacks, track user behavior, prevent malware, and monitor network traffic. DPI uses the OSI model application layer to extract statistical information and analyze, find, and organize data packets as needed. Unlike traditional data filtering, DPI can extract data other than the packet header, and expand the traditional monitoring scope of the network layer and transport layer to the application layer data [8]. In addition to the actual data, it also includes metadata that identifies the traffic source, content, destination address, and other important information.

This model uses the DPI analysis idea. Compared with the data results of flow statistics, it pays more attention to the process characteristics obtained in the in-depth analysis [1]. From the perspective of feature extraction, it expands the extraction range of flow characteristics to the application layer, enriches the traditional feature extraction dimensions, and makes the feature coverage include the behavior information of the application layer. Based on the feature extraction of traditional network traffic, DRFV uses DPI to deeply analyze the packet, obtain more feature information, and realize the dimensionality increase compared with the traditional feature set on the feature dimension.

3.3 Bagging for Features

Bagging is the representative of parallel integrated learning [3]. Its basic process is to sample a sample set of several training samples from the sample set, train a base learner based on this sample set, repeat it for many times, and then combine the trained multiple learners [4]. When outputting the forecast, it will vote and output the classified tasks. The dimension reduction method used in this paper is realized through Random Forest. Random Forest is an extended variant of Bagging [11]. The Bagging structure is constructed by decision tree based learners, and random factors are introduced into the training process [13]. First, it selects some subsets of features from all feature sets, and then selects the best feature from this subset as the partition feature of the current node of the decision tree [10].

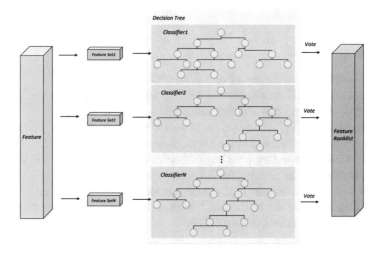

Fig. 2. Rank the features.

Figure 2 describes the process of obtaining feature importance ranking through random forests. After features are extracted from the traffic data, different feature sets are obtained by resampling the original feature set *Feature*, as is shown in Eq. (1).

$$Feature = \{FeatureSet_1, FeatureSet_2, ..., FeatureSet_N\} \tag{1}$$

Classifiers are trained separately on these new training sample sets. Each classifier corresponds to a decision tree. Multiple decision trees can prevent over fitting. Noise interference calculation is conducted for each sub feature set supplement through OOB(Out of Bag) [9].

$$\{FeatureSet - FeatureSet_i\} \tag{2}$$

If the OOB accuracy rate drops significantly after adding noise, as is shown in Eq. (2), it indicates that the feature is of high importance. Different decision trees have different prediction effects. Finally, the final feature importance ranking is obtained by voting the prediction results of all trees, which provides a selection basis for subsequent optimization and dimension reduction. The calculation formula for judging the importance of a feature is shown in Eq. (3):

$$X = \frac{\sum(e_{i2} - e_1)}{T} \tag{3}$$

4   Experiment

4.1   Datasets

The traffic data set used in the experiment is the CIC-IDS2017 intrusion detection evaluation data set [12]. CIC(Canadian Institute for Cybersecurity) has evaluated 11 existing data sets since 1998 and believes that most data sets are outdated and unreliable. In contrast, the CIC-IDS2017 dataset performs well in terms of traffic diversity and capacity. It is generated based on real natural background traffic, and constructs the abstract behavior of 25 users, covering a variety of known attacks, similar to real world data. Its data collection was as of July 7th, 2017, including five days of data traffic. Except for Monday, which is a normal day, attacks implemented during the other four days include violent FTP, violent SSH, Dos, Heartbleed, web attacks, infiltration, botnets, and DDoS, with data sizes exceeding 60GB.

4.2   Feature Extraction

The feature extraction section first uses CICFlowMeter as a tool to extract basic traffic features, mainly focusing on data packets and flow data, such as packet length, flow rate, and packet header features, totaling 78 dimensional vectors. The Table 1 shows the main kinds of the features.

Table 1. Features Extracted by CICFlowMeter

Packet	Flow	Port
Fwd/Back Packets	Flow Duration	Dst Port
Packet Length	Flow Bytes/s	
Header Length	Flow Packets/s	
Packet Size	Flow IAT	
Active Packets	Subflow Packets	
Idle Packets	Init Win bytes	

Then, through the nDPI tool, the traffic data was analyzed, and more features related to the data flow and application layer were obtained [7]. The feature vectors, such as protocol, port, and the interaction behavior between the server and the client, were supplemented, with a total of 52 dimensions. The relevant information related to these characteristics can be roughly divided into four parts, as shown in Table 2.

Table 2. Feature Extracted by nDPI

Protocol	Flow	Port	Time
Protocol	Flow id	Dst Port	First seen
nDPI protocol	c-s Bytes/s	Src Port	Last seen
Proto by IP	s-c Bytes/s	Server name	Duration
TLS Info	c-s Packets/s		
HTTP User Agent	s-c Packets/s		
Idle Packets	Goodput Ratio		
QUIC Info	ACK/PSH/RST/SYN		

From Table 2, it can be seen that the feature information extracted through DPI has increased significantly in terms of protocol services, such as protocol types, HTTP/TLS/QUIC and other related information. At the same time, there is further recognition and parsing of related network packets. In addition, there have been new features in the field of flow information, which have been refined into information statistics between the client and server, and even extracted the flow position of flow data packets in the transmission protocol, such as ACK/PSH/RST/SYN and other related information of data packets.

4.3   Feature Selection

The importance ranking of random forest features based on Bagging is performed on the above 130 dimensional vector, and the correlation calculation is performed using OOB to obtain the data and ranking. In order to prevent the selection of features from depending on the data characteristics and thus affecting the model preference, and at the same time, it is more explanatory in theory, the model adds some necessary features on the basis of the features with the highest ranking results, and finally obtain the optimized feature list. The top 10 of the list is as shown in Table 3.

As shown in Table 3, in addition to the feature values extracted from network packet information, many features related to application services and flow information also hold high importance, even ranking in the top ten. This indicates that these vectors extracted based on DPI technology are also quite important. Due to limited space, Table 3 only displays the top ten ranked features. The optimization model in this paper retains features with the importance calculation value $X \geq 0.001$ as the optimized vector, with a total of 99 dimensions. The effectiveness will be tested in the following experiments.

Table 3. Top10 Features

Rank	Feature	X
1	Fwd IAT Max	0.0883
2	Init Win bytes forward	0.0598
3	Flow IAT Max	0.0439
4	c-s good bytes	0.0385
5	Fwd IAT Mean	0.0382
6	Bwd URG Flags	0.0281
7	Flow IAT Std	0.0270
8	Flow IAT Mean	0.0252
9	s-c pkts	0.0251
10	Bwd Header Length	0.0237

4.4 Results

In order to test whether the DRFV optimization model performs well, we will use the previously optimized feature set in multiple detection models to compare the performance of data and running time. The results are shown in Table 4.

Table 4. Optimization Results

Algorithm	Precision	F1	Time(s)
RF(Before)	0.9700	0.9530	12.0
RF (After)	**0.9756**	**0.9533**	**11.7**
XGBoost(Before)	0.9640	0.9770	33.7
XGBoost (After)	**0.9644**	**0.9772**	**27.8**
Whisper(Before)	0.9533	0.9421	18.1
Whisper (After)	**0.9634**	**0.9435**	**18.0**

It can be seen that the feature set processed by the DRFV optimization model can guarantee the original precision, and even perform better in some models. At the same time, the time performance is improved. In terms of reducing the performance of time, the model with more complex feature processing process is significantly improved, while the model more dependent on flow aggregation and other methods is slightly improved.

5   Conclusion

This paper proposes a DPI-based optimization model for network traffic——DRFV optimization model, which is suitable for feature processing of network anomaly detection. In the network traffic feature extraction phase, the use of DPI technology increases the dimension of acquiring features. Compared with the existing feature extraction, it increases the data analysis and application process information statistics of the application layer, making the feature set contain more information; In the feature selection stage, the random forest based on Bagging thought is used to achieve feature importance ranking, and the results are used as the basis for feature selection to achieve dimension reduction optimization. While obtaining more effective feature information, the detection load is reduced, which promotes the research of feature vector optimization and network anomaly detection.

References

1. Alonso, G., et al.: DPI: the data processing interface for modern networks. In: CIDR 2019 Online Proceedings, p.11 (2019)
2. Alotibi, G., Li, F., Clarke, N., Furnell, S.: Behavioral-based feature abstraction from network traffic. In: ICCWS 2015-The Proceedings of the 10th International Conference on Cyber Warfare and Security, pp. 1–9 (2015)
3. Breiman, L.: Bagging predictors. Mach. Learn. **24**, 123–140 (1996)
4. Bühlmann, P., Yu, B.: Analyzing bagging. Ann. Stat. **30**(4), 927–961 (2002)
5. Bujlow, T., Carela-Español, V., Barlet-Ros, P.: Independent comparison of popular DPI tools for traffic classification. Comput. Netw. **76**, 75–89 (2015)
6. Cheng, G., Gong, J., Ding, W.: A real-time anomaly detection model based on sampling measurement in a high-speed network. J. Software **14**(3), 594–599 (2003)
7. Deri, L., Martinelli, M., Bujlow, T., Cardigliano, A.: NDPI: open-source high-speed deep packet inspection. In: 2014 International Wireless Communications and Mobile Computing Conference (IWCMC), pp. 617–622. IEEE (2014)
8. Ghosh, A., Senthilrajan, A.: Classifying network traffic using DPI and DFI. Int. J. Sci. Technol. Res. **8**(11), 1019 (2019)
9. Matthew, W., et al.: Bias of the random forest out-of-bag (OOB) error for certain input parameters. Open J. Stat. **2011** (2011)
10. Paul, A., Mukherjee, D.P., Das, P., Gangopadhyay, A., Chintha, A.R., Kundu, S.: Improved random forest for classification. IEEE Trans. Image Process. **27**(8), 4012–4024 (2018)
11. Rigatti, S.J.: Random forest. J. Insur. Med. **47**(1), 31–39 (2017)
12. Rosay, A., Cheval, E., Carlier, F., Leroux, P.: Network intrusion detection: a comprehensive analysis of CIC-IDS2017. In: 8th International Conference on Information Systems Security and Privacy, pp. 25–36. SCITEPRESS-Science and Technology Publications (2022)
13. Speiser, J.L., Miller, M.E., Tooze, J., Ip, E.: A comparison of random forest variable selection methods for classification prediction modeling. Expert Syst. Appl. **134**, 93–101 (2019)
14. Sun, Z., Tang, Y., Zhang, W., Gong, J., Wang, R.: A router anomaly traffic filter algorithm based on character aggregation. J. Software **17**(2), 295–304 (2006)
15. Yang, T., Jiang, R., Deng, H., Tang, X.: A network traffic identification method based on autoencoder-a feature selection algorithm. J. Phys. Conf. Ser. **2593**, 012007 (2023)
16. Yoshimura, N., Kuzuno, H., Shiraishi, Y., Morii, M.: DOC-IDS: a deep learning-based method for feature extraction and anomaly detection in network traffic. Sensors **22**(12), 4405 (2022)
17. Yun, X., Wang, Y., Zhang, Y., Zhou, Y.: A semantics-aware approach to the automated network protocol identification. IEEE/ACM Trans. Networking **24**(1), 583–595 (2015)

A Hill Climbing System for Optimizing Component Selection of Multirotor UAVs

Nobuki Saito[1], Tetsuya Oda[2(✉)], Yuki Nagai[3], Kyohei Wakabayashi[3], Chihiro Yukawa[3], and Leonard Barolli[4]

[1] Graduate School of Science and Engineering, Okayama University of Science (OUS), 1-1 Ridaicho, Kita-ku, Okayama 700–0005, Japan
r23sde7zn@ous.jp
[2] Department of Information and Computer Engineering, Okayama University of Science (OUS), 1-1 Ridaicho, Kita-ku, Okayama 700–0005, Japan
oda@ous.ac.jp
[3] Graduate School of Engineering, Okayama University of Science (OUS), 1-1 Ridaicho, Kita-ku, Okayama 700–0005, Japan
{t22jm23rv,t22jm24jd,t22jm19st}@ous.jp
[4] Department of Information and Communication Engineering, Fukuoka Institute of Technology, 3-30-1 Wajiro-higashi, Higashi-ku, Fukuoka 811-0295, Japan
barolli@fit.ac.jp

Abstract. The propulsion system in multirotor Unmanned Aerial Vehicle (UAV) affects the flight speed, operating time and payload of UAVs. The propulsion system consists of a motor, propeller, electronic speed controller, power distribution board and battery. When implementing a multirotor UAV, the selection of components is very important in order to achieve the operational objectives. Among the components, the propulsion system significantly affects the operation time of UAVs. In the component selection process, the components should be combined to meet the required performance. Also, it should be considered the compatibility between components. This requires much time and effort to select the appropriate combination of components. Therefore, it is needed an efficient method for selection of most suitable components for achieving the operation objectives. In this paper, we propose an optimisation method based on Hill Climbing (HC) algorithm for selecting the components of a propulsion system in a multirotor UAVs. For simulation of component selection we considered three types of multirotor UAVs: quadrotor, hexacopter and octocopter. The evaluation results show that each component is compatible and the proposed system selected a good combination of components.

1 Introduction

The multirotor Unmanned Aerial Vehicles (UAVs) will be utilised in a variety of fields such as transportation [1], pesticide spraying [2], surveying [3], security [4] and so on. The multirotor UAVs use rotary wings to control flight and is equipped with multiple rotors. They are classified into tricopters [5], quadrotors [6] and hexacopters [7] depending on the number of rotors. The multirotor UAVs are capable of vertical takeoff and horizontal movement, as well as hovering capabilities that allow them to operate from a fixed point. There are many research works that consider to expand the functions of UAVs for different applicatons [8–10]. In order to spread the range of applications for multi-rotor UAVs, it is necessary to implement multirotor UAVs that can fly for a long time, are expandable and have large payloads.

When implementing a multirotor UAV, the selection of components is very important in order to achieve the operational objectives. Among the components, the propulsion system significantly affects the operation time of UAVs [11] and payload [12]. Many multirotor UAVs [13–16] operate using electricity and the propulsion system consists of a motor, propeller, battery, Electronic Speed Controller (ESC) and Power Distribution Board (PDB).

In the component selection process, the components should be combined to meet the required performance, based on a vast amount of data collected through market research. Also, it should be considered the compatibility between components. This requires much time and effort to select the appropriate combination of components. When increasing the number of motors, it is possible to increase the payload and stability, which increases the power consumption. So, the operating time will decrease. When the arrangement of motors differ, the appropriate components also have different combinations. Therefore, it is needed an efficient method for selection of most suitable components for achieving the operation objectives.

There are different optimization methods for selecting the components such as Simulated Annealing [17], Greedy Algorithms [18] and Genetic Algorithms (GA) [19], Particle Swarm Optimization (PSO) [20], Heuristic Algorithms [21], Analytic Hierarchy Process (AHP) [22] and Fuzzy Theory (FT) [23].

In this paper, we propose an optimisation method based on Hill Climbing (HC) algorithm for selecting the components of a propulsion system in a multirotor UAVs. For simulation of component selection, we considered three types of multirotor UAVs: quadrotor, hexacopter and octocopter. The evaluation results show that each component is compatible and the proposed system selected a good combination of components.

The structure of the paper is as follows. In Sect. 2, we show the component configuration of multirotor UAV. In Sect. 3, we present the proposed system. In Sect. 4, we discuss the simulation results. Finally, conclusions and future work are given in Sect. 5.

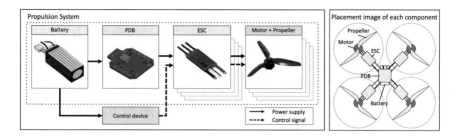

Fig. 1. Component of multirotor UAV.

2   Component Configuration of Multirotor UAV

In this section, we explain the component configuration of multirotor UAV. We consider three types of multirotor UAVs: a quadrotor with four rotors, a hexacopter with six rotors and an octocopter with eight rotors [24]. The components of a multirotor are: the propulsion system, control device and frame. Therefore, in the process of implementing a UAV, it is essential to select appropriate components of the propulsion system according to the operational purpose of the UAV.

The selections of components requires a lot of time and effort because are needed repeated calculations to decide whether the required performance can be achieved by combined the components based on the huge amount of data collected through market research. In addition, the evaluation and selection processes are performed on multiple items such as operating time, payload and size.

Figure 1 shows the configuration of the propulsion system and placement image of each component. The propulsion system consists of a motor, propeller, Power Distribution Board (PDB), Electronic Speed Controller (ESC), and battery. The control device consists of sensors, communication equipment and a single-board computer such as Jetson or Raspberry Pi. The frame fixes the components and control devices. It uses carbon plates and carbon pipes that are light and rigid. When the frame size increases, the weight also increases. In addition, the motor fixing point is moved away from the center of gravity in the body, increasing the moment of inertia during flight, which may decrease the manoeuvrability. When the distances between the motors are short, the thrust force is attenuated due to loss, so it is necessary to reduce the size of the frame and the thrust loss.

Figure 2 shows the placement of the motor and frames length in quadrotor, hexacopter and octocopter. In this paper, the motor and propeller of each aircraft is installed on the same plane. In Fig. 2, the green line shows the distance between the motors and the blue line shows the frame which is fixed the motor. The distance between motors is 2.4 times of the radius of propeller when the loss of thrust is minimal [25]. When the distance between motors divided by 2 is r and the radius of the propeller is P_r, then $r = 2.2P_r$. When the number of motors is N_m, the total length of frame $F_{total\ length}$, $F_{total\ length}$ are $r\sqrt{2}N_m$ for

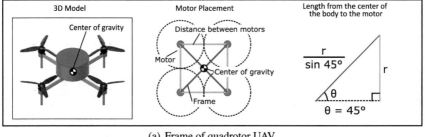

(a) Frame of quadrotor UAV.

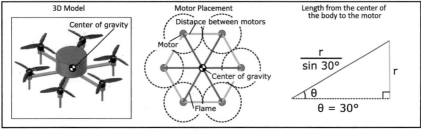

(b) Frame of hexacopter UAV.

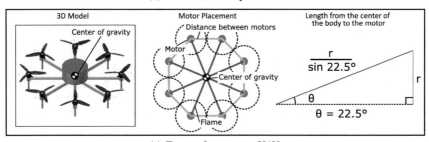

(c) Frame of octocopter UAV.

Fig. 2. Frame of multirotor UAVs.

quadrotor, $2rN_m$ for hexacopter and $\frac{\sqrt{2-\sqrt{2}}}{2}rN_m$ for octocopter. In addition, in the case of a frame using a pipe, the frame weight F_w is calculated by Eq. (1).

$$F_w = \rho \pi F_{total\ length} \left(r_{outer}^2 - r_{inner}^2 \right) \tag{1}$$

In Eq. (1), r_{outer} and r_{inner} are the outer radius and inner radius of the pipe, while ρ is the mass of material per unit volume. It is possible to optimize the components considering the frame weight by calculating the frame size according to the propeller size and number of motors.

Table 1. Components and parameters.

Parameters	Values
Multirotor	Model, Propulsion system, Total weight, Max. thrust of single motor, Compatibility
Propulsion system	Motor, Propeller, Battery, PDB, ESC
Model of multirotor	Number of Motors, Number of Propellers, Number of Batteries, Number of PDBs, Number of ESCs
Motor	Weight, Shaft diameter, Max. current, Voltage, KV value
Propeller	Weight, Shaft diameter, Diameter, Pitch
Battery	Weight, Voltage, Capacity, Discharge Rate
PDB	Weight, Number of max. battery connections, Number of max. ESC connections, Min. voltage, Max. voltage, Max. current
ESC	Weight, Number of max. motor connections, Min. voltage, Max. voltage, Max. current
Compatibility	Propeller-Motor, Motor-Battery, Motor-ESC, Motor-PDB

3   Proposed System

In this section, we present the proposed system. For UAV, the propulsion system has a significant impact on the performance including flight speed, operating time, payload and mobility. The propulsion system consists of the motor, propeller, ESC, PDB and battery. For component selections, it is necessary to consider the compatibility between components such as drive voltage and the number of connectable components. The components to be selected and their parameters are shown in Table 1.

The proposed system considers quadrotors, hexacopters, and octocopters. Therefore, the number of motors and propellers can be 4, 6 or 8. The number of batteries is 1, or 2 if the PDB allows batteries to be connected in parallel. The number of PDBs is 1 or 0 if the ESC is 4in1 type and the battery is connected directly to the ESC. In addition, the number of ESCs is 1, 4, 6 or 8. If the number of motors is 4, it is possible to select the 4in1 type ESC and the number of ESCs is 1.

The proposed system is based on HC algorithm [26] for selecting components of the propulsion system in the multirotor UAVs. Alg. 2 shows the flow of the proposed system. The proposed system aims to maximize the Operating Time of Multirotor (OTM) and the Payload of Multirotor (PM). When all parameters included in compatibility are *True*, which means that the selected components are compatible and executable. The components compatibility in the proposed system is shown in Fig. 3.

The OTM is calculated by Eq. (2). When operating a multirotor, it is necessary to make a flight plan and consider whether the OTM is sufficient to carry out the flight plan. On the other hand, the current consumption varies depend-

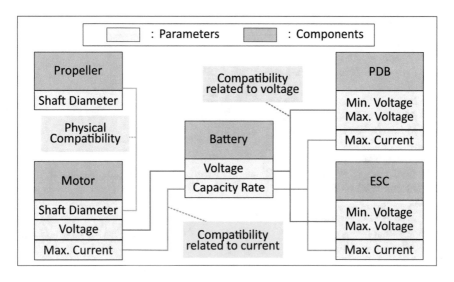

Fig. 3. Components compatibility.

Algorithm 1. HC-based component selection.

Input: Data list of Motor, Propeller, Battery, PDB and ESC.
1: Types of multirotor ← Quadrotor, Hexacopter or Octocopter.
2: C_w ← *any weight*
3: Set the Solution list
4: **for** $i = 0 \ldots$ Number of iterations **do**
5:     Generate M_i, P_i, B_i, PDB_i, E_i as random index.
6:     multirotor ← $Motor[M_i], Propeller[P_i], Battery[B_i], PDB[PDB_i], ESC[E_i]$.
7:     **if** Alg: 2 input multirotor = True. **then**
8:         Calculate the OTM and PM.
9:         $Eval$ ← $OTM \times C_{OTM} + PM \times C_{PM}$.
10:        **if** Length of *Solution list* = 0. **then**
11:            *Solution list* add to $\{Eval, M_i, P_i, B_i, PDB_i, E_i\}$ as initial solution.
12:        **else if** $Eval$ in *Solution list*[*Solution list*.length] < $Eval$. **then**
13:            *Solution list* add to $\{Eval, M_i, P_i, B_i, PDB_i, E_i\}$ as current solution.
14:     $i \leftarrow i + 1$

---

ing on flight conditions such as ascending, descending, horizontal movement and hovering. Therefore, by using C_c as the current consumption coefficient and considering the ratio of the average current consumption to the maximum current consumption, it is possible to be closer tp the actual operating time. In Eq. (2), 0.001 is multiplied to convert the unit of battery capacity from mAh to Ah and 60 is multiplied to convert the unit of operating time from hour to minute. The PM is calculated by Eq. (3). The total weight of mobility used to derive PM is calculated by Eq. (4).

Algorithm 2. Check of component compatibility.

Input: Multirotor.

1: **if** $Motor_s = Prop._s$ **then**
2:     Propeller-Motor $\leftarrow True$.
3: **else**
4:     Propeller-Motor $\leftarrow False$.
5: **if** $Motor_v = Batt._v \cap \frac{Motor_{mc} \times N_m}{Batt._c} \leq Batt._d$ **then**
6:     Motor-Battery $\leftarrow True$.
7: **else**
8:     Motor-Battery $\leftarrow False$.
9: **if** $ESC_{min.v} \leq Motor_v \leq ESC_{max.v} \cap Motor_{max.c} \leq ESC_{max.c}$ **then**
10:     Motor-ESC $\leftarrow True$
11: **else**
12:     Motor-ESC $\leftarrow False$
13: **if** $PDB_{min.v} \leq Motor_v \leq PDB_{max.v} \cap Motor_{max.c} \leq PDB_{max.c} \cap N_b \leq PDB_{num.batt.}$ **then**
14:     Motor-PDB $\leftarrow True$
15: **else**
16:     Motor-PDB $\leftarrow False$
17: **if** $Compatibility$ of all components $= True$ **then**
18:     **return** True.
19: **else**
20:     **return** False.

$$OTM = \frac{B_v B_c N_b \times 0.001}{M_v M_{max.\ c} C_c N_m} \times 60 \tag{2}$$

$$PM = \frac{N_m \times C_t \rho n^2 D^4}{2} - T_w \tag{3}$$

$$T_w = M_w N_m + P_w N_p + E_w N_e + B_w N_b + P_{dbw} N_{pdb} + C_w + F_w \tag{4}$$

In Eqs. (2), (3) and (4), B_v is voltage of battery, B_c is capacity of battery, N_b is number of batteries, M_v is voltage of motor, $M_{max.c}$ is max current of motor, C_c is current coefficient, N_m is number of motors. While, C_t is thrust coefficient, ρ is air density, n is rotations per minute of the motor (derived from KV value multiplied by motor voltage), and D is diameter of propeller. In addition, T_w is total weight of mobility, M_w is a motor weight, P_w is propeller weight, N_p is number of propellers, E_w is ESC weight, N_e is number of ESCs, B_w is battery weight, P_{dbw} is PDB weight, N_{pdb} is number of PDBs, C_w is control unit weight and F_w is frame weight. C_w is derived from the total weight of the sensors, single board computer, communication devices, camera and other devices to be installed on the multirotor. Also, F_w is derived according to the frame weight calculation.

4   Simulation Results

In this section, we discuss the simulation results of the proposed system.

Table 2. Simulation parameters.

Parameters	Values
Number of motor data [Times]	1000
Number of propeller data [Times]	1000
Number of battery data [Times]	1000
Number of PDB data [Times]	1000
Number of ESC data [Times]	1000
Number of iteration [Unit]	500000
Weight of Controller [g]	1000

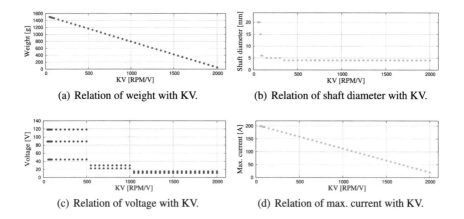

(a) Relation of weight with KV.   (b) Relation of shaft diameter with KV.

(c) Relation of voltage with KV.   (d) Relation of max. current with KV.

Fig. 4. Distribution of motor parameters.

The simulation parameters are shown in Table 2, while Figs. 4, 5, 6 and 7 show the distribution of each generated component parameter for the proposed system. In addition, the capacity rate of battery, max. current, min. and max. voltage in PDB and ESC are randomly generated with uniform distribution.

Figure 8 shows the results of the OTM and PM transitions of each UAV for each iteration. It can be seen that both OTM and PM are in increasing trend for each UAV. The OTM is the highest for quadrotor type and PM is the highest for octocopter. This is because the quadrotor UAV have fewer motors and consume less total current during flight than hexacopters and octocopters, so the operating time is longer. In addition, octocopters have a large number of motors and can obtain more thrust, so they have a large payload.

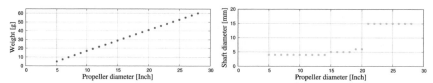

(a) Relation of weight with propeller diameter. (b) Relation of shaft diameter with propeller diameter.

Fig. 5. Distribution of propeller parameters.

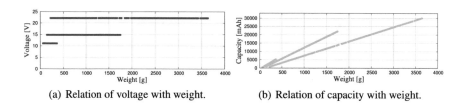

(a) Relation of voltage with weight.   (b) Relation of capacity with weight.

Fig. 6. Distribution of battery parameters.

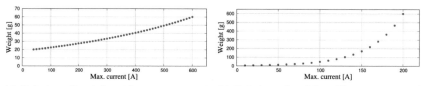

(a) Relation of weight with max. current in (b) Relation of weight with max. current in PDB.   ESC.

Fig. 7. Distribution of PDB and ESC parameters.

Table 3 shows the parameters of the components at the last iteration. It can be seen that each component is compatible and a good combination has been selected. The proposed system selects the combination of components that generate the higher thrust with low power consumption to maximize OTM and PM. Also, almost the same parameters of components were selected for motors and propellers.

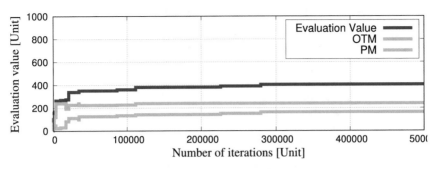

(a) Simulation results of quadrotor.

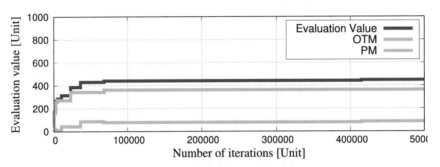

(b) Simulation results of hexacopter.

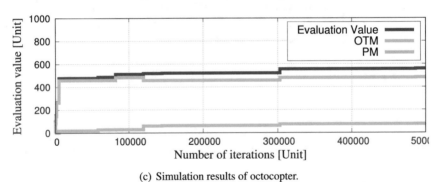

(c) Simulation results of octocopter.

Fig. 8. Results of OTM and PM for each iteration.

Table 3. Final solution of each component for quadrotor.

Parameters		Quadcopter	Hexacopter	Octocopter
Model of multirotor	Number of Motors [Unit]	4	6	8
	Number of Propellers [Unit]	4	6	8
	Number of Batteries [Unit]	1	1	1
	Number of PDBs [Unit]	1	1	1
	Number of ESCs [Unit]	4	6	8
Motor	Weight [g]	50.0	50.0	50.0
	Shaft Diameter [mm]	4.0	4.0	4.0
	Max Current [A]	20.0	20.0	20.0
	Voltage [V]	14.8	14.8	14.8
	KV Value [RPM / V]	2000	2000	2000
Propeller	Weight [g]	26.5	26.5	26.5
	Shaft Diameter [mm]	4.0	4.0	4.0
	Diameter [Inch]	14.0	14.0	14.0
	Pitch [Inch]	4.6	7.1	6.7
Battery	Weight [g]	1744.0	1363.0	1607.0
	Voltage [V]	14.8	14.8	14.8
	Capacity [mAh]	22000	17000	20200
	Capacity Rate [Unit]	120	75	75
PDB	Weight [g]	22.1	50.3	58.8
	Max. Connections of Batteries [Unit]	2	1	1
	Max. Connections of ESCs [Unit]	4	4	8
	Min. and Max. voltage [V]	7.4, 88.8	7.4, 44.4	7.4, 22.2
	Max. current [A]	90.0	510.0	590.0
ESC	Weight [g]	132.3	6.4	8.3
	Max. connections of motors [Unit]	1	1	1
	Min. and Max. voltage [V]	14.8, 29.6	7.4, 14.8	14.8, 44.4
	Max current [A]	140.0	20.0	30.0

5   Conclusions

In this paper, we proposed an optimisation method based on HC algorithm for selecting the components of a propulsion system in a multirotor UAV. We performed the simulation of component selection for three types of multirotor UAVs using proposed system. From simulation results, we conclude as follows.

- Considering OTM, the quadrotor had the highest value and the octocopter had the smallest value.
- Considering PM, the octocopter had the highest value and the quadrotor had the smallest value.
- The proposed system selected the combination of components that generate the higher thrust with low power consumption to maximize OTM and PM.

In the future work, we would like to improve the proposed system and considers different scenarios.

Acknowledgement. This work was supported by JSPS KAKENHI Grant Number JP20K19793 and JSPS KAKENHI Grant Number JP23KJ2123.

References

1. Bacelar, T., et al.: On-board implementation and experimental validation of collaborative transportation of loads with multiple UAVs. Aerosp. Sci. Technol. **107**, 106284 (2020)
2. Hafeez, A., et al.: Implementation of drone technology for farm monitoring & pesticide spraying: a review. Inf. Process. Agric. **10**(2), 192–203 (2023)
3. Casella, E., et al.: Accuracy of sand beach topography surveying by drones and photogrammetry. Geo-Mar. Lett. **40**(2), 255–268 (2020)
4. Pablo, G.A., et al.: Monitoring traffic in future cities with aerial swarms: developing and optimizing a behavior-based surveillance algorithm. Cogn. Syst. Res. **54**, 273–286 (2019)
5. Nam, K.J., et al.: Tri-copter UAV with individually tilted main wings for flight maneuvers. IEEE Access **8**, 46753–46772 (2020)
6. Jiang, F., et al.: Design, implementation, and evaluation of a neural-network-based quadcopter UAV system. IEEE Trans. Industr. Electron. **67**(3), 2076–2085 (2020)
7. Senthilkumar, S., et al.: Design, dynamics, development and deployment of hexacopter for agricultural applications. In: 2021 International Conference on Advancements in Electrical, Electronics, Communication, Computing and Automation (ICAECA-2021), pp. 1–6 (2021)
8. Gray, A.G., et al.: Design and flight testing of a UAV with a robotic arm. In: Proceedings of 2023 IEEE Aerospace Conference, pp. 1–13 (2023)
9. Ali, Z.A., Zhangang, H.: Maneuvering control of hexrotor UAV equipped with a cable-driven gripper. IEEE Access **9**, 65308–65318 (2021)
10. Ouyang, Z., et al.: Control of an aerial manipulator using a quadrotor with a replaceable robotic arm. In: Proceedings of 2021 IEEE International Conference on Robotics and Automation (ICRA-2021), pp. 153–159 (2021)
11. Amici, C., et al.: Review of propulsion system design strategies for unmanned aerial vehicles. Appl. Sci. **11**(11), 5209 (2021)
12. Piljek, P., et al.: Method for characterization of a multirotor UAV electric propulsion system. Appl. Sci. **10**(22), 8229 (2020)
13. Servais, E., et al.: Ground control of a hybrid tricopter. In: Proceedings of 2015 International Conference on Unmanned Aircraft Systems (ICUAS), pp. 945–950 (2015)
14. Badr, S., et al.: A design modification for a quadrotor UAV: modeling, control and implementation. Adv. Robot. **33**(1), 13–32 (2019)
15. Wang, R., et al.: An actuator fault detection and reconstruction scheme for hexrotor unmanned aerial vehicle. IEEE Access **7**, 93937–93951 (2019)
16. Zhu, H., et al.: Design and assessment of octocopter drones with improved aerodynamic efficiency and performance. Aerosp. Sci. Technol. **106**, 106206 (2020)
17. Franzin, A., Stützle, T.: Revisiting simulated annealing: a component-based analysis. Comput. Oper. Res. **104**, 191–206 (2019)
18. Dai, H., et al.: Learning combinatorial optimization algorithms over graphs. Adv. Neural. Inf. Process. Syst. **30**, 1–11 (2017)
19. Benabbou, N., et al.: An interactive regret-based genetic algorithm for solving multi-objective combinatorial optimization probleme. In: Proceedings of the AAAI Conference on Artificial Intelligence, vol. 34, no. 03, pp. 2335–2342 (2020)

20. Xu, X., et al.: CS-PSO: chaotic particle swarm optimization algorithm for solving combinatorial optimization problems. Soft. Comput. **22**, 783–795 (2018)
21. Halim, A.H., Ismail, I.: Combinatorial optimization: comparison of heuristic algorithms in travelling salesman problem. Arch. Comput. Methods Eng. **26**, 367–380 (2019)
22. Balaji, M., et al.: An application of analytic hierarchy process in vehicle routing problem. Period. Polytech. Transp. Eng. **47**(3), 196–205 (2019)
23. Kong, D., et al.: A decision variable-based combinatorial optimization approach for interval-valued intuitionistic fuzzy MAGDM. Inf. Sci. **484**, 197–218 (2019)
24. Zhu, H., et al.: Design and assessment of octocopter drones with improved aerodynamic efficiency and performance. Aerosp. Sci. Technol. **106**, 106206 (2020)
25. Shukla, D., Komerath, N.: Multirotor drone aerodynamic interaction investigation. Drones **2**(4), 43 (2018)
26. Goswami, S., Chakraborty, S., Guha, P., Tarafdar, A., Kedia, A.: Filter-based feature selection methods using hill climbing approach. In: Li, X., Wong, K.-C. (eds.) Natural Computing for Unsupervised Learning. USL, pp. 213–234. Springer, Cham (2019). https://doi.org/10.1007/978-3-319-98566-4_10

Optimization of Deer Repellent Devices Placement Based on Hill Climbing and Acoustic Ray Tracing

Sora Asada[1], Yuki Nagai[2], Kyohei Wakabayashi[2], Chihiro Yukawa[2], Tetsuya Oda[3(✉)], and Leonard Barolli[4]

[1] Graduate School of Science and Engineering, Okayama University of Science (OUS), Okayama, 1-1 Ridaicho, Kita-ku, Okayama 700–0005, Japan
r23smk5vb@ous.jp

[2] Graduate School of Engineering, Okayama University of Science (OUS), Okayama, 1-1 Ridaicho, Kita-ku, Okayama 700–0005, Japan
{t22jm23rv,t22jm24jd,t22jm19st}@ous.jp

[3] Department of Information and Computer Engineering, Okayama University of Science (OUS), 1-1 Ridaicho, Kita-ku, Okayama 700–0005, Japan
oda@ous.ac.jp

[4] Department of Informention and Communication Engineering, Fukuoka Insitute of Technology, 3-30-1 Wajiro-Higashi-ku, Fukuoka 811-0295, Japan
barolli@fit.ac.jp

Abstract. Deer repelling sound is affected by the attenuation of reflected sound due to terrain undulations and air absorption. Therefore, sound wave propagation analysis can be used to decide the location of devices. This paper deals with the optimization of deer repellent device placement using the mountaineering and acoustic ray tracking methods. A Digital Elevation Model (DEM) for mid-mountainous areas is used for the placement of deer repellent devices takinginto account the undulations of the terrain. We use OpenStreetMap (OSM) to get the location of deers in their living environment. Simulation results show that the proposed system is able to optimize the placement of deer repellent devices by maximizing network connectivity and deer repellent sound range.

1 Introduction

In Japan, deers cause extensive damage to agriculture and forestry. They are commonly found in hilly and mountainous areas with agricultural fields and wooded areas. Generally, the damages from deers are prevented by using fences. However, the fences must be laid continuously over a wide area. Also, in many mountainous areas where deer damage needs to be prevented the fences are maintained and managed by elderly people. Therefore, the maintenance and management of fences is a burden for residents in these areas. There are some other measures for preventing the damages by hunting and repelling with pest animal lights, but the damage are enormous.

© The Author(s), under exclusive license to Springer Nature Switzerland AG 2024
L. Barolli (Ed.): EIDWT 2024, LNDECT 193, pp. 545–554, 2024.
https://doi.org/10.1007/978-3-031-53555-0_52

One of the countermeasures for deer damage is deer repellent with sound. It has been reported that the deers can be repelled at high frequencies by ultrasonic sounds that humans cannot hear [1–3]. This approach can repel the deers to a distance that the deer-repellent sound can reach. Therefore, sound based deer repellent devices can be used instead of fences and reduce the maintenance and management burden. The ultrasonic deer-repellent sound is not a problem because humans cannot hear the sound, but the high-frequency sound may cause noise damage to residents. Therefore, deer repellent and the noise damage to residents caused by the repellent sound causes can be reduced by considering the hearing ranges of humans and deers.

Few previous studies on the hearing range of deers have been conducted. In [4], a hearing range study by operand conditioning has been conducted for white-tailed deers, which are similar in body size to deers. In the study, white-tailed deers were found to have a hearing range of 115 [Hz] to 54 [kHz] at a sound pressure level of 60 [dB]. This sound pressure level is commonly used when comparing the hearing ranges of various species. By increasing the sound pressure level, the hearing range is extended from 32 [Hz] (96.5 [dB]) to 64[kHz] (93[dB]).

The comparison of hearing ranges of white-tailed deer and humans is shown in Table 1 [5]. It can be seen that the hearing sensitivity for each frequency is better than that of humans for high frequencies above 8 [kHz]. In addition, the sika deer and white-tailed deer are considered to have similar hearing ranges.

Deer repellent devices can be wirelessly connected between them to build a sensor actuator network based on a mesh mechanism, which enables the detection of device failures and can coordinate the operation among devices [6–9]. However, the placement of each deer repellent device in the sensor actuator network affects the connectivity and transmission loss of wireless communication [10]. In addition, the optimization of the deer repellent device placement requires consideration of the effect of attenuation by reflections of deer repellent sound caused by terrain [11]. Also, the deer-repellent sound in outdoor environments is affected by air absorption [12]. Therefore, for the placement of deer repellent devices can be done by conducting sound wave propagation analysis considering the effects of medium reflection and air absorption [13].

In this paper, we propose a system for the optimization of deer repellent devices placement based on Hill Climbing (HC) and acoustic ray tracing. In hills and mountainous areas where deer damage occurs, forests account for a large proportion of the total area compared with the residential living environment. Therefore, the proposed system optimizes the placement of deer repellent devices in their living environment.

The structure of the paper is as follows. In Sect. 2, we present the proposed system. In Sect. 3, we discuss the simulation results. Finally, conclusions and future work are given in Sect. 4.

Table 1. Hearing ranges of white-tailed deer and humans.

White-tailed Deer		Human	
Frequency (kHz)	Threshold T(dB SPL)	Frequency (kHz)	Threshold T(dB SPL)
–	–	0.004	101
–	–	0.008	94
–	–	0.016	82
0.031	96.5	0.032	58
0.063	78	0.063	36
0.0125	56.5	0.0125	17
0.25	25	0.025	10
0.5	21.25	0.05	10
1	7.5	1	−4
2	14.5	2	−10
4	1.75	4	−10
8	−3	8	9
16	11.25	16	26
–	–	18	71
–	–	20	91
32	23.5	–	–
45	43	–	–
56	64	–	–
64	93	–	–

2   Proposed System

The proposed system extracts deer living areas from the target district and decides the placement of deer repellent devices that have the maximum Size of Giant Component (SGC) and the maximum range of sound that is effective in repelling deer. A sound wave propagation analysis is performed to evaluate the sound pressure in each space to derive the optimum placement of the deer repellent devices.

2.1   Hill Climbing Based Deer Repellent Device Placement Optimization

The proposed system decides the placement of deer repellent devices that maximize SGC and Area of Deer Repellent Sound ($ADRS$) [14,15]. The proposed system discretizes the simulation space and if the Euclidean distance between the center coordinates of the discretized space and the deer repellent device is within the reach of effective repellent sound, then this discretized space is considered a space where deer repellent is possible. The objective function $ADRS$ is the total number of discretized spaces where deer avoidance is possible.

In the placement optimization process, one deer repellent device is randomly selected and moved within an arbitrary distance from the coordinates of the deer repellent device within the simulation range. When the SGC is maximum after the deer repellent device is moved, the $ADRS$ is evaluated. If the SGC is not maximum, then return the moved deer repellent devices to the base coordinates. If the $ADRS$ is increasing, update the solution. If the $ADRS$ is not increased, then return the deer repellent to its pre-movement coordinates. These processes are repeated until an arbitrary number of times and the solution with the largest $ADRS$ is decided as the placement for deer repellent device.

2.2   Acoustic Ray Tracing Based Deer Repellent Device Placement Optimization

The proposed system evaluates the sound pressure distribution in the simulation area by acoustic wave propagation analysis based on acoustic ray tracing and maximizes the Average Sound Pressure of Simulation Space ($ASPSS$), which is the average value of sound pressure in the discretized simulation space [16,17].

The acoustic ray tracing method is based on the geometric-optic approximation and the geometric optic diffraction theory. It considers sound, which is a wave motion, as a ray, and performs sound wave propagation analysis by sequentially calculating the position and sound pressure of the ray. This method simulates sound pressure attenuation and the phenomena of reflection and diffraction, which are characteristics of sound waves.

In the proposed system, the coordinates of the ground surface are decided from Digital Elevation Model (DEM) of the area where deer damage has been reported and set as the medium in the simulation environment. After the material of the ground surface is defined, the sound pressure is absorbed based on the sound absorption coefficient of the material. In the acoustic ray tracing of the proposed system, a ray is ejected considering the coordinates of the deer repellent as the sound source, then is calculated the sound pressure distribution in the simulation space. When a ray collides with the edge of the simulation domain, the calculation for that ray is terminated and the next ray is calculated. The sound pressure distribution in the simulation space is derived by adding the sound pressure of the rays that pass through each discretization space.

In the proposed system, the initial solution is the placement of deer repellent devices decided based on the HC. One deer repellent device is randomly selected and moved randomly within an arbitrary distance from the coordinates of the deer repellent device. When the SGC is not maximum, the coordinates are returned to the coordinates before the move. If the SGC is maximal, then the $ASPSS$ is evaluated by acoustic ray tracing. In the case when $ASPSS$ increases, update the placement of the deer repellent device after the move as a new solution. These processes are repeated until an arbitrary number of times and the solution with the maximum $ASPSS$ is the optimal solution.

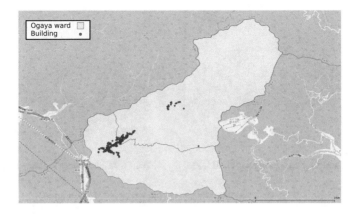

Fig. 1. Area for placement of deer repellent device.

3   Simulation Results

In the simulation, the area for deer repellent device placement is considered Ogaya Ward, Nishiawakura Village, Okayama Prefecture, Japan, which is a hilly mid-mountain area where deer damage has been reported. Deer repellent devices need to be placed in the vicinity of deer living areas because the deer damage occurs frequently in hilly and mountainous areas.

The proposed system obtains map data of the area around Ogaya Ward from OpenStreetMap (OSM). The map data of OSM has labels assigned to each polygon, such as forests, rivers and buildings. From the obtained OSM map data, polygons of buildings within the boundaries of the ward are extracted using QGIS. The area around the most dense concentration of buildings is decided to be the installation area of the deer repellent device. The attenuation due to air absorption of the ray is calculated based on the ISO 9613-1 standard [18].

Figure 1 shows the results of extracting buildings from OSM in Ogaya Ward, Nishiawakura Village, Okayama Prefecture, Japan, where deer damage has been reported. While Fig. 2 shows the place where the deer repellent devices will be installed. In Fig. 2(a) is shown the dense builings area in Ogaya Ward DEM and Fig. 2(b) shows the DEM in the building dense area. The proposed system uses these DEM for the placement optimization of deer repellent devices.

Table 2 shows the parameter settings of HC-based deer repellent device placement optimization. While Table 3 shows the parameter settings for the acoustic ray tracing based deer repellent placement optimization. Figure 3 shows the results of HC-based deer repellent device placement optimization. While Fig. 4 shows the results of placement optimization based on acoustic ray tracing. It can be seen that the SGC is always maximum 100 [$unit$]. Also, $ADRA$ increased from 1331 to 6299. The $ASPSS$ is 17.78 [dB] and it shows that the range of deer repellent sound is increased.

Figure 5 shows the visualization results of the initial placement of the deer repellent device. Figure 5(a) shows two demensional area and Fig. 5(b) shows

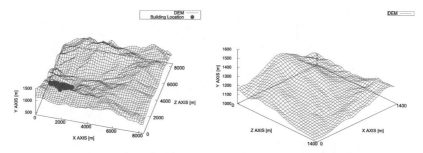

(a) Dense building area in Ogaya Ward DEM.   (b) DEM in densely building areas.

Fig. 2. Deer repellent device placement area.

Table 2. Settings parameter of HC-based deer repellent device placement optimization.

Parameter Names	Parameter Values
$Width \times Depth \times Height \ [m] \times [m] \times [m]$	$1400.0 \times 1400.0 \times 1600.0$
Number of Deer Repellent Device $[unit]$	100
Radius of Communication Range $[m]$	100
Ultrasonic Radius $[m]$	120
Mesh Size of DEM $[m]$	40
Number of Iterations $[times]$	25000

Table 3. Settings parameter of acoustic ray tracing based deer repellent device placement optimization.

Parameter Names	Parameter Values
$Width \times Height \ [m] \times [m]$	1400.0×1400.0
Number of Deer Repellent Device $[unit]$	100
Number of Rays $[unit]$	100
Radius of Communication Range $[m]$	100
Deer Repellent Sound Frequencies $[Hz]$	4000
Mesh Size of DEM $[m]$	80
Number of Iterations $[times]$	200

three dimensional area. Figure 6 shows the visualization results of the optimized placement of the deer repellent based on HC-based method. While Fig. 6(a) shows the visualization results of two dimensional area and Fig. 6(b) the visualization results of three dimensional area. Figures 5 and 6 show that the placement of deer repellent devices is spread out. Figure 7 shows the visualization results of optimized placement of deer repellent devices based on acoustic ray tracing. While Fig. 7(a) shows the visualization results for two dimensional area and Fig. 7(b) for three dimensional area.

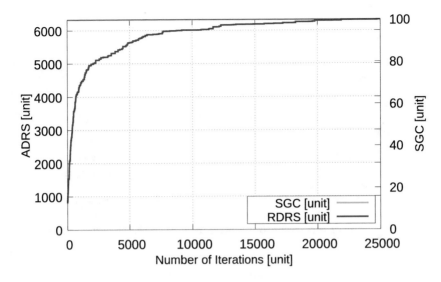

Fig. 3. ADRS and SGC vs. number of iterations.

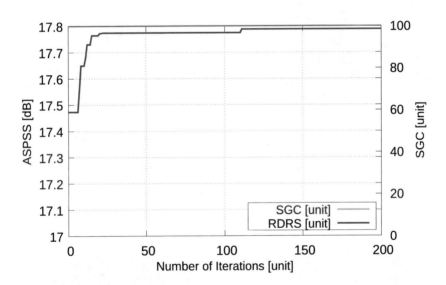

Fig. 4. ASPSS and SGC vs. number of iterations.

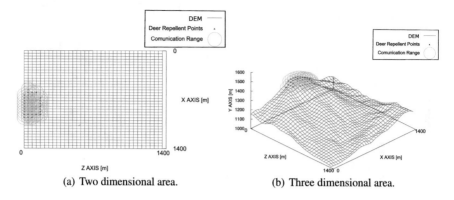

Fig. 5. Visulalization result of initial placement of deer repellent device.

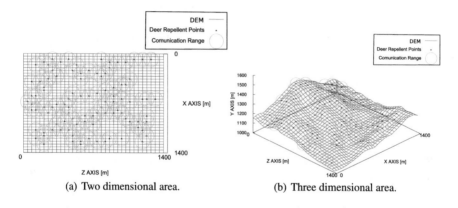

Fig. 6. Visulalization results of *HC*-based deer repellent device placement.

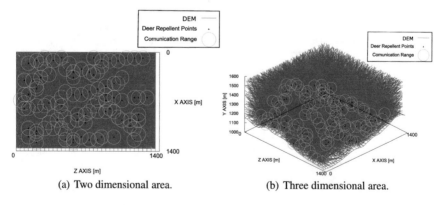

Fig. 7. Visulalization results of acoustic ray tracing based deer repellent device placement.

4   Conclusions

In this paper, we proposed an Optimization System for placement of deer repellent devices to prevent deer damage based on HC and acoustic ray tracing. Simulation results have shown that the proposed system was able to optimize the placement of deer repellent devices by maximizing network connectivity and deer repellent sound range.

In the future, we would like to improve the proposed system and carry out extensive simulations.

Acknowledgement. This work was supported by JSPS KAKENHI Grant Number JP20K19793.

References

1. Shimura, S., et al.: Development of an acoustic deterrent to prevent deer-train collisions. Q. Report RTRI **59**(3), 207–211 (2018)
2. Honda, T.: A sound deterrent prevented deer intrusions at the intersection of a river and fence. Mammal Study **44**(4), 269–274 (2019)
3. Matsuzaki, H.: Acoustic measurement of deer alert whistles and investigation of sika deer reaction to alert sounds played back through loudspeaker. Wildlife Hum. Soc. **9**, 75–85 (2021)
4. Heffner, H., et al.: The behavioral audiogram of whitetail deer (Odocoileus virginianus). J. Acoust. Soc. Am. **127**(3), 111–114 (2010)
5. Jackson, L.L., et al.: Free-field audiogram of the Japanese macaque (Macaca fuscata). J. Acoust. Soc. Am. **106**(5), 3017–3023 (1999)
6. Roubaiey, A., et al.: Reliable middleware for wireless sensor-actuator networks. IEEE Access **7**, 14099–14111 (2019)
7. Akyildiz, I.F., et al.: Wireless mesh networks: a survey. Comput. Netw. **47**(4), 445–487 (2005)
8. Sabrine, K., et. al.: Precision irrigation: an IoT-enabled wireless sensor network for smart irrigation systems. In: Women in Precision Agriculture: Technological Breakthroughs, Challenges and Aspirations for a Prosperous and Sustainable Future, pp. 107-129 (2021)
9. Saleem, R., et. al.: Industrial wireless sensor and actuator networks in industry 4.0: exploring requirements, protocols, and challenges - A MAC survey. Int. J. Commun. Syst. **32**(15), e4074 (2019)
10. Asada, S., et. al.: A simulated annealing based simulation system for optimization of wild deer damage prevention devices. In: International Conference on Broadband and Wireless Computing, Communication and Applications, pp. 38-44 (2019)
11. Harris, C.M.: Absorption of sound in air versus humidity and temperature. J. Acoust. Soc. Am. **40**(1), 148–159 (1966)
12. Seddeq, H.S.: Factors influencing acoustic performance of sound absorptive materials. Aust. J. Basic Appl. Sci. **3**(4), 4610–4617 (2009)
13. Nagai, Y., et. al.: A CCM, SA and FDTD based mesh router placement optimization in WMN. In: Conference on Complex, Intelligent, and Software Intensive Systems, pp. 48-58 (2023)

14. Nagai, Y., et. al.: A wireless sensor network testbed for monitoring a water reservoir tank: experimental results of delay and temperature prediction by LSTM. In: International Conference on Network-Based Information Systems, pp. 392-401 (2022)
15. Nagai, Y., et. al.: Wireless visual sensor node placement optimization considering different distributions of events. In: International Conference on Broadband and Wireless Computing, Communication and Applications, pp. 312-322 (2023)
16. K. M. Li, et. al.,"An improved ray-tracing algorithm for predicting sound propagation outdoors", The Journal of the Acoustical Society of America, Vol. 104, No. 4, pp. 2077-2083, 1998
17. Hou, Q., et. al.: Dynamic modeling of traffic noise in both indoor and outdoor environments by using a ray tracing method. Build. Environ. 121, 225–237 (2017)
18. Economou, P., et. al.: Accuracy of wave based calculation methods compared to ISO 9613-2. In: Proceedings of Noise-Con 2014 (2014)

Real-Time Detection of Network Exploration Behavior: A Method Based on Feature Extraction and Half-Space Trees Algorithm

Peixin Cong and Baojiang Cui[(✉)]

School of Cyberspace Security, Beijing University of Posts and Telecommunications, Haidian District, Xitucheng Road No. 10, Beijing, China
{TonySmith_2222,cuibj}@bupt.edu.cn

Abstract. In the context of an increasingly severe network security landscape, attackers often employ novel techniques and social engineering attacks to bypass network boundaries, and collect internal network information for lateral attacks after successful penetration. To timely identify and expel attackers, this paper introduces a stream-based detection method for analyzing online traffic. This method employs a custom-designed streaming algorithm and Half-Space Trees for anomaly detection. Experimental tests on the LanL dataset demonstrate that this method can effectively discover day-to-day variations in attack postures and potential attackers, thus proving its effectiveness.

1 Introduction

The MITRE ATT&CK framework [12] underscores the importance of internal network information discovery and mining. Attackers leverage this for obtaining target network information and locating optimum targets for continued attack. This step becomes critical for attackers wishing to control the entire network after securing one or several endpoints. However, attackers' exploration for valuable machines in unfamiliar networks may expose them, creating a divergence between controlled and normal hosts.

Existing methods for discovering internal network scanning by attackers [6, 9, 15] are computationally intensive and time-consuming, escalating the impact incurred during penetration. Additionally, previous research [3,5,14] necessitate more than a five-tuple information, such as login logs, which may be unattainable in certain scenarios.

This paper presents a real-time anomaly detection method, the 'Feature-based Half-space Trees' (FHT), based on feature extraction and the Half-space Trees algorithm [13]. FHT can detect high-confidence internal network exploration behavior online, revealing it in early stages, and thereby mitigating the amount of subsequent information obtained by the attacker.

The rest of this paper is organized as follows: Section two reviews related work; section three describes our method framework; section four details the procedure to extract features based on the five-tuple and their combination with the river framework and half-space trees anomaly detection algorithm; section five presents the experimental data; and finally, section six concludes the paper and discusses future research directions.

2   Related Work

Over the past decade, research utilizing machine learning to tackle challenges in developing autonomous intrusion systems for different communication networks has been proposed. For example, Aksu et al. put forward a model using deep learning and Support Vector Machine (SVM) algorithms to detect port scanning attacks [1]. Their model, with an activation function based on ReLU, changes performance with varying neurons and hidden layers. They selected the optimum neurons based on pattern accuracy without a feature selection algorithm for SVM. Their research primarily focuses on the latest CICIDS2017 dataset, and results indicate that deep learning outperforms SVM significantly.

In terms of online learning, OnCAD, an online clustering algorithm for anomaly detection, was proposed by Chenaghlou, Moshtaghi, Leckie, and Salehi in 2018 [4]. Their method uses Gaussian clustering and divides data streams into time windows. It initially generates clusters of standard network behavior, evaluating new samples against existing clusters, identifying them as potential anomalies or new clusters. Afterward, they utilize DBScan with samples from one window to discover new clusters and anomalies. Their algorithm outperforms online K-means and Adaptive Resonance Theory (ART-2), albeit with a longer runtime. Despite better detection performance on synthetic and public datasets, increased execution time results in detection delays, potentially creating security vulnerabilities.

3   Modeling with Feature-Based Half-Space Trees

Currently, the modes of network connections can be generalized as follows: In Fig. 1, we can see that 'b' serves as the server, while 'a', 'c', and 'd' are requesting access to 'b'. Conversely, in Fig. 2, 'b' serves as the client trying to access 'a', 'c', and 'd'. These two modes are the typical access modes of the Internet.

Within internal networks, the operators of network members usually have additional information. This means that regular network members under human guidance typically aim to access specific services rather than aimlessly and frequently attempting to explore other members' services in the network. This characteristic could potentially serve as a key to distinguish between normal and abnormal behaviors.

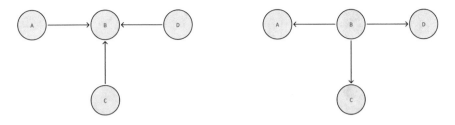

Fig. 1. Server mode.   **Fig. 2.** Client mode.

While it might seem that the problem could be solved by detecting the packet sending rate of source IPs within a unit of time and their deduplicated exploration range, this approach may inadvertently harm systems used for bulk control of terminals in the internal network, such as operation maintenance systems and quality monitoring systems. These systems, during operation, would access specific zones of the network rapidly and extensively, which is an expected behavior and not a malicious one. Additionally, if an attacker controls multiple terminals within the internal network and deliberately controls the rate at which each IP explores the internal network by distributing network probing tasks, they can circumvent detection of packet sending rate or exploration range targeted at a single IP. Hence, we need a new approach.

Facing this situation, we extract the following parameters from the traffic $K_v = \{\alpha(v), \beta(v)\}$. The overall architecture of this method is presented in Fig. 3. It consists of three parts: DataWasher, FeatureGenerator, and AnomalyDetector. The traffic from security devices such as firewall needs to be processed by DataWasher first.

When defining the data requirements for the detection method, we stipulate that only connection records from the client to the server should be inputted. All response records from the server to the client should be filtered out to prevent interference with Feature Generator's understanding of the traffic, thus affecting feature generation. The Feature Generator will read the five-tuple information processed by DataWasher and generate corresponding features, then output both the five-tuple information and feature information to Anomaly Detector. The Anomaly Detector will combine the five-tuple information with the features generated by the Feature Generator, and then input them into the Half-Space Trees model under the river framework. If the model's predicted probability result exceeds a threshold, it will trigger an alarm and provide the corresponding five-tuple.

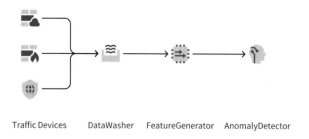

Traffic Devices   DataWasher   FeatureGenerator   AnomalyDetector

Fig. 3. FHT model.

The following sections will elaborate on these three parts in detail.

3.1   DataWasher

In the context of security software or devices, such as OpenSense and Suricata, traffic typically exists in the form of TCP state packet records, noting SYN, ACK, FIN, etc. Under the connection session between the server and the client, the security device will record traffic packets on both sides. However, due to performance reasons, packet pairing on both sides may be missing, often seeing only the server-side IP and server port as the source IP and source port, but not the client-side IP and client port. This situation may result in the subsequent phase of FeatureGenerator being unable to differentiate between the roles of the server and the client, thus necessitating the data-cleansing work of DataWasher.

We define the traffic dataset S as:

$$S = \{Time, Sip, Sport, Dip, Dport\}$$

representing time information, source IP, source port, destination IP, and destination port, respectively. Typically, the dataset includes client-to-server traffic and the server's corresponding responses to the client. However, the dataset contains instances where the client's IP and port do not pair with the server's IP and port. Therefore, we need to filter the dataset, obtaining only the IPs and ports of the client and server, and outputting a new dataset S', defined as:

$$S' = \{Time, ClientIp, ClientPort, ServerIp, ServerPort\}$$

When DataWasher encounters a new five-tuple record R in dataset S, it first checks whether the destination IP and port have previously appeared in C. If they have, it converts the record into the new format and adds it to S', and removes it from C. If they haven't, it adds the record to C.

Algorithm 1. DataWasher

1: $S' \leftarrow \emptyset$
2: $C \leftarrow \emptyset$ ▷ Initialize a collection C that stores destination IP and destination port pairs that have appeared
3: **for** each *record R* in *S* **do**
4:    **if** $\exists(R.Dip, R.Dport) \in C$ **then**
5:       $new\_record \leftarrow transformRecord(record)$
6:       $S' \leftarrow S' \cup new\_record$
7:       Remove $(R.Dip, R.Dport)$ from C
8:    **else**
9:       Add $(R.Dip, R.Dport)$ to C
10:    **end if**
11: **end for**
12: **procedure** $transformRecord(record)$
13:    $new\_record \leftarrow \{record.Time, record.Sip, record.Sport, record.Dip, record.Dport\}$
14:    **return** new_record
15: **end procedure**

3.2 FeatureGenerator

Within a corporate intranet, network traffic and member communications are structured, displaying distinct temporal patterns. Activities like employees accessing internal systems or interacting with production operation machines generate stable inter-member relationships that maintain temporal consistency. Hence, normal and orderly parts are to be filtered out. Let us consider an intranet G with n nodes denoted as $V = \{v_1, v_2, \ldots, v_k\}$. The network connection graph at time $t1$ is represented by $G(t1) = (V, E(t1))$, where $E(t1) = \{e_1, e_2, \ldots, e_n\}$ describes the connections at $t1$. For a disjoint time period $t2$, $E(t2) = \{e'_1, e'_2, \ldots, e'_m\}$ describes the connections at $t2$, forming the graph $G(t2) = (V, E(t2))$.

Define $D(v, t2, t1)$ as the differential edge count for a node v between $t1$ and $t2$:

$$D(v, t2, t1) = |E(v, t2)| - |E(v, t2) \cap E(v, t1)|$$

Subsequently, the aggregate differential edge count for all nodes is:

$$D'(V, t2, t1) = \sum_{v \in V} D(v, t2, t1)$$

This count, $D'(V, t2, t1)$, is expected to be considerably smaller than the size of the edge intersection $|E_{t1} \cap E_{t2}|$, reflecting stable temporal linkages.

For a regular network participant v_n and an anomaly v_a, their respective extra edge counts are $D(v_n, t2, t1)$ and $D(v_a, t2, t1)$. When v_a initiates intranet probing, $D(v_a, t2, t1)$ greatly exceeds $D(v_n, t2, t1)$. This behavior hints at anomaly presence through surplus edge generation. In scenarios where an attacker controls many terminals, the α characteristic of compromised terminals may resemble that of legitimate members, complicating distinction. Within a directed graph E and for a node v, let $N(v)$ represent its neighboring nodes.

The differential edge count for neighbors of v from $t1$ to $t2$ is given by:

$$F(v_n, t2, t1) = \sum_{u \in N(v)} D(u, t2, t1)$$

Here, a suspicious node v_t and a normal node v_n might satisfy:

$$D(v_t, t2, t1) \approx D(v_n, t2, t1)$$

Yet, the neighbor differential $F(v_n, t2, t1)$ will be significantly higher than $D(v_n, t2, t1)$, marking v_t as a potential risk node for alerts.

Considering DataWasher relays streaming data, if feature extraction for each record occurs, δt approaches 0. Consequently, $\alpha(v) = D(v, t2, t1)$ and $\beta(v) = F(v, t2, t1)$, forming the feature set $K_v = \{\alpha(v), \beta(v)\}$.

Upon receipt of traffic quintuple logs from DataWasher, FeatureGenerator hashes the SrcIp, DstIp, and DstPort pairs, stores them in a hash table, and maintains the relationships in memory. It then continuously updates the α and β features of SrcIp using the hash table.

Algorithm 2. FeatureGenerator Time-Window Algorithm

     $M \leftarrow$ empty dictionary
2: $K \leftarrow$ empty dictionary
     for each R in S **do**
4:       Extract Sip, $Sport$, Dip, $Dport$ from R
        $hash\_sip \leftarrow hash(Sip)$
6:       $hash\_dip\_dport \leftarrow hash(Dip + str(Dport))$
        if $hash\_dip\_dport$ not in $M[hash\_sip]$ **then**
8:          Append $hash\_dip\_dport$ to $M[hash\_sip]$
            $P \leftarrow$ value of P in K (if exists) or $P \leftarrow 1$
10:        $F \leftarrow$ value of F in K (if exists) or $F \leftarrow 0$
            $F \leftarrow F + len(M[hash\_dip\_dport])$
12:       **else**
            $P \leftarrow$ value of P in K (if exists) or 0
14:        $F \leftarrow$ value of F in K (if exists) or 0
        end if
16:       $K[hash\_sip] \leftarrow \{P, F\}$
     end for
18: **return** K

After adding records, the corresponding SrcIp and DstIp+DstPort mapping table will occupy more memory over time, which is not conducive to long-term maintenance. Therefore, the sliding window method is needed to update the mapping data to achieve stable long-term operation. As shown in Fig. 1, let the capacity of each item be 24 h, the window size be 3, and each item has a M and K for storing mapping relationships. The M and K corresponding to items outside the window capacity will be cleared. Each time a new record

$R = \{$Time,SrcIp,SrcPort,DstIp,DstPort$\}$ is added, first check whether the M of other items in the window has a SrcIp and DstIp+DstPort mapping. If it exists, check whether the current item's M has a SrcIp and DstIp+DstPort mapping. If the current item's M does not have a mapping, add the mapping to the current item's M. If the other item's M does not have the corresponding SrcIp and DstIp+DstPort mapping, then add the mapping to the current item's M and update K.

Algorithm 3. Sliding Window Method for Updating Mapping Data

    $windowSize \leftarrow 3$
    $itemCapacity \leftarrow 24$ (h)
3: $M \leftarrow$ empty mapping storage for each item
    $K \leftarrow$ empty mapping storage for each item
    for each new record R **do**
6:      $R \leftarrow \{Time, SrcIp, SrcPort, DstIp, DstPort\}$
       for each item i in window **do**
         if item i is outside the window capacity **then**
9:           delete mappings in M_i and K_i
         end if
         if there is a mapping in M_i for $SrcIp$ and $DstIp + DstPort$ **then**
12:          **if** there is no such mapping in $M_{current}$ **then**
             add mapping to $M_{current}$
          end if
15:        **else**
          add mapping to $M_{current}$ and update $K_{current}$
         end if
18:      **end for**
    end for

After the feature set $K_v = \{\alpha(v), \beta(v)\}$ is calculated by the above time-window-processed algorithm, the corresponding five-tuple S record will be combined with the feature set K_v and sent to AnomalyDetector for anomaly detection.

3.3   AnomalyDetector

Machine Learning (ML) has found extensive applications in network security, offering advanced solutions to complex issues. However, evaluating these solutions is challenging due to unique issues in this field, such as concept drift. This term refers to changes in data distribution over time, allowing malicious actors to create new threats that existing models might fail to recognize.

We address concept drift using online machine learning, also known as incremental or stream learning. This paradigm enables real-time learning from new data during model training, without waiting for extensive data accumulation. Such real-time learning is advantageous in scenarios demanding swift responses and in adapting to changes in data distribution, crucial in contexts like financial trading and network security where data patterns can evolve.

River, a Python tool designed for streaming data, supports online learning and provides tools for executing online machine learning on streaming data [11]. It allows learning from and modifying the model in continuously arriving data. As each network environment is unique, collecting substantial historical data sets of intranet environments is challenging, necessitating unsupervised anomaly detection based on online learning.

The HalfSpaceTrees (HST) algorithm, proposed by Tan, is a tree-based data structure designed for anomaly detection in multidimensional data streams [13]. The following section will explicate the underlying mathematical steps of the HST algorithm.

Given a data space $X \subset \mathbb{R}^d$ and a depth limit D.

1. Select an attribute q, where $q \in \{1, \ldots, d\}$, and calculate the split point $p = \frac{1}{2}(\max_{x \in X} x_q + \min_{x \in X} x_q)$, where x_q is the qth element of vector x.
2. Based on the calculated attribute q and split point p, the space X is split into two subspaces $X_L = \{x \in X | x_q \leq p\}$ and $X_R = \{x \in X | x_q > p\}$.
3. If the current depth equals the maximum depth D, then create a leaf node and save the statistic information of the current space. Otherwise, apply the above process recursively on the subspaces X_L and X_R to create subtrees.

The mathematical formulas are as follows:

- For internal nodes: $N = \{L, R, q, p\}$, where L and R are the left and right subtrees, q is the split attribute, and p is the split point.
- For leaf nodes: $L = \{n, \mu, \sigma^2\}$, where n is the number of samples, μ is the sample mean, and σ^2 is the sample variance.

4   Experiments on LanL Dataset

The Comprehensive Multi-Source Cyber-Security Events of LANL is a dataset containing a large number of network security events from Los Alamos National Laboratory (LANL) [7,8]. This dataset includes various network events collected from the computer network of Los Alamos National Laboratory over a continuous period of 18 months, including network traffic, authorized and unauthorized login events, file uploads and downloads, device connections and disconnections, and other various network activities. The events in the dataset include the start time, end time, event type, event source, target, and other information related to the event.

Since the network traffic records in the dataset are bidirectional, they should be fully recorded in both directions, but there are cases where the bidirectional records in the dataset are incomplete, making it impossible for the model to accurately analyze the client and server in the network. Therefore, it is necessary to process the dataset and perform data cleaning first. In Fig. 4 is shown the related information of the dataset cleaned by DataWasher.

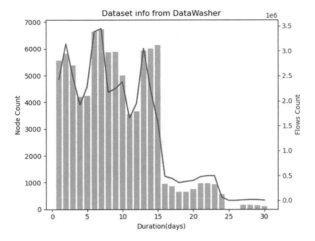

Fig. 4. Dataset info from DataWasher.

We have identified indicators reflecting the interception effect: we define the α interception rate of node v as P_α, the α interception quantity as $X_\alpha(v)$, and G as the set of suspicious nodes with length C. We have identified two indices:

The average interception quantity reflecting the α feature is calculated by Eq. (1).

$$P_\alpha = \frac{1}{C} \sum_{v \in G} X_\alpha(v) \tag{1}$$

While, the interception rate reflecting the β feature is calculated by Eq. (2).

$$P_\beta = \frac{1}{C} \sum_{v \in G} X_\beta(v) \tag{2}$$

We perform one-hot encoding on the SrcIp field. The experimental model is shown in Fig. 5.

In previous research [2,10], OneClassSVM was explored as an effective tool for network attack detection. As an unsupervised learning method based on support vector machines, OneClassSVM aims to identify and distinguish normal data from anomalies, and it does so by constructing a decision boundary for classifying new samples. Originally designed to deal with single-class data, OneClassSVM exhibits robustness when handling imbalanced datasets, especially in the field of network security.

To further evaluate the performance of HalfSpaceTrees, this study uses OneClassSVM for comparison. HalfSpaceTrees is an efficient online anomaly detection method that uses multiple decision trees to quickly adapt to changes in data flow and is highly sensitive to minute changes in data distribution.

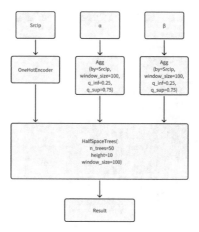

Fig. 5. Experimental model.

In the experiment, we used an AMD Ryzen 9 3900XT CPU, Ubuntu 20.04, Python 3.8.10 for calculations. After preliminary data screening, we used a length with a step size of 50 as the input. We perform one-hot encoding on the SrcIp field. The experimental results are shown in Fig. 6.

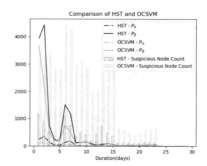

Fig. 6. Comparison of HST and OCSVM algorithm.

The experimental results indicate that although OneClassSVM exceeds Half-SpaceTrees in the number of anomalies detected, it performs inadequately in interception rates P_α and P_β, resulting in a higher false positive rate and inefficient attack blocking. In contrast, HalfSpaceTrees surpass OneClassSVM on both metrics, demonstrating a more balanced and effective detection capability.

5   Conclusions

In this paper, we propose a new method for analyzing online traffic and discovering intranet exploration behavior within it. We introduce an algorithm that the online traffic analysis system uses to output client-to-server connection information after matching the source IP and port with the destination IP and port, allowing the model to accurately understand the meaning of the traffic. Next, we introduce two key parameters α and β extracted from the traffic. These two key parameters reflect the new connection relationships of the source IP in the network traffic directed graph, and can effectively measure the extent to which the source IP collects information about intranet IP survival and port opening. We designed a streaming extraction algorithm with a complexity of $O(n)$ that can efficiently extract features in actual enterprise-level network traffic. Then, we implemented an online learning method for detection in machine learning using the HalfSpaceTree algorithm and the River framework. In simulating real-time traffic input to the model in the public dataset, we found several suspicious characters trying to obtain intranet information. In the future, we plan to improve the speed of anomaly detection, reduce the false alarm rate, and generate meaningful intranet exploration datasets for anomaly detection training.

References

1. Aksu, D., Aydin, M.A.: Detecting port scan attempts with comparative analysis of deep learning and support vector machine algorithms. In: 2018 International Congress on Big Data, Deep Learning and Fighting Cyber Terrorism (IBIGDELFT), pp. 77–80. IEEE (2018)
2. Binbusayyis, A., Vaiyapuri, T.: Unsupervised deep learning approach for network intrusion detection combining convolutional autoencoder and one-class SVM. Appl. Intell. **51**(10), 7094–7108 (2021). https://doi.org/10.1007/s10489-021-02205-9
3. Bowman, B., Laprade, C., Ji, Y., Huang, H.H.: Detecting lateral movement in enterprise computer networks with unsupervised graph AI. In: Proceedings of the 23rd International Symposium on Research in Attacks, Intrusions and Defenses, RAID 2020, pp. 257–268 (2020)
4. Chenaghlou, M., Moshtaghi, M., Leckie, C., Salehi, M.: Online clustering for evolving data streams with online anomaly detection. In: Phung, D., Tseng, V.S., Webb, G.I., Ho, B., Ganji, M., Rashidi, L. (eds.) PAKDD 2018, Part II 22. LNCS (LNAI), vol. 10938, pp. 508–521. Springer, Cham (2018). https://doi.org/10.1007/978-3-319-93037-4_40
5. Dong, C., Yang, J., Liu, S., Wang, Z., Liu, Y., Lu, Z.: C-BEDIM and S-BEDIM: lateral movement detection in enterprise network through behavior deviation measurement. Comput. Secur. **130**, 103267 (2023). https://doi.org/10.1016/j.cose.2023.103267
6. Fang, Y., Wang, C., Fang, Z., Huang, C.: LMTracker: lateral movement path detection based on heterogeneous graph embedding. Neurocomputing **474**, 37–47 (2022). https://doi.org/10.1016/j.neucom.2021.12.026
7. Kent, A.D.: Comprehensive, Multi-Source Cyber-Security Events. Los Alamos National Laboratory (2015). https://doi.org/10.17021/1179829
8. Kent, A.D.: Cybersecurity data sources for dynamic network research. In: Dynamic Networks in Cybersecurity. Imperial College Press (2015)
9. Kinable, J.: Detection of network scan attacks using flow data. In: 9th Twente Student Conference on IT, 23 June 2008 (2008). http://www.utwente.nl/ewi/dacs/assignments/completed/bachelor/reports/2008-kinable.pdf

10. Miao, X., Liu, Y., Zhao, H., Li, C.: Distributed online one-class support vector machine for anomaly detection over networks. IEEE Trans. Cybern. **49**(4), 1475–1488 (2018)
11. Montiel, J., et al.: River: machine learning for streaming data in Python. J. Mach. Learn. Res. **22**(1), 4945–4952 (2021)
12. Strom, B.E., Applebaum, A., Miller, D.P., Nickels, K.C., Pennington, A.G., Thomas, C.B.: MITRE ATT&CK: design and philosophy. Technical report. The MITRE Corporation (2018)
13. Tan, S.C., Ting, K.M., Liu, T.F.: Fast anomaly detection for streaming data. In: IJCAI International Joint Conference on Artificial Intelligence, pp. 1511–1516 (2011). https://doi.org/10.5591/978-1-57735-516-8/IJCAI11-254
14. Tian, Z., et al.: Real-time lateral movement detection based on evidence reasoning network for edge computing environment. IEEE Trans. Ind. Inform. **15**(7), 4285–4294 (2019). https://doi.org/10.1109/TII.2019.2907754
15. Viet, H.N., Trang, L.L.T., Nguyen Van, Q., Nathan, S.: Using deep learning model for network scanning detection. In: ACM International Conference Proceeding Series, pp. 117–121 (2018). https://doi.org/10.1145/3233347.3233379

Star Point Target Extraction with High Noise Background Based on Morphology

Zhen-zhen Li[✉]

Officers College of PAP, Chengdu, Sichuan, China
524495155@qq.com

Abstract. Star sensor positioning accuracy is the most important performance index of Star Sensor. However, noise is also an important factor affecting the positioning accuracy of star sensor. In order to extract the small star-point target effectively and improve the efficiency of star-image recognition and count accuracy of satellite attitude, a morphological star-point extraction method is adopted in this paper. We can get an estimate of the image background based on the characteristics of mathematical morphology operation. By using the Top-Hat transform of gray-scale morphology to cancel the background, we can get images with targets and high-frequency noise. In addition, in this paper, we also use adaptive threshold method to determine the actual image threshold. The experimental results [1] show that this method can filter background noise very well, and it is helpful to select the threshold value and determine the target brightness. The object luminance determined by this method is better than that obtained by the traditional method. It is proved that the difference between the centroid coordinates and the true values is no more than 0.26 pixel units.

1 Introduction

With the development of aerospace technology, As a kind of high precision attitude measurement equipment, star sensor is more and more widely used in satellite and other spacecraft. However, because the star-sensitive carrier is an airtight cabin, the star-sensitive work is limited by time and space. At different times, through the front glass window when observing, star selection by the Sun, the earth and other bright celestial bodies, therefore, the working time of Star sensor is limited. Spatially, when the evasive angle is less than the design value, stray light will cause the background of the image to become large and uneven. Stars in the CCD plane generally appear as spots on a darker background. The diameter of the spot is generally 3–5 pixels. Because the light intensity of the star is not very prominent relative to the background light, the background information must be filtered out as noise for the recognition of the star [2].

In order to extract the star point target effectively and improve the star point position precision, in this paper, a non-linear background estimation method based on the open and closed operation of gray morphology is adopted.

In this paper, a non-linear background estimation method based on the open and closed operation of gray morphology and a background-to-small target detection method based on the Top-Hat [3] transform of gray morphology are adopted. At the same time, the

L. Barolli (Ed.): EIDWT 2024, LNDECT 193, pp. 567–574, 2024.
https://doi.org/10.1007/978-3-031-53555-0_54

method of small target detection based on gray-scale morphological Top-Hat transform [4] is adopted.

2 Point Target Extraction Algorithm Based on Morphology

Against a background of noise, a star image can be represented as:

$$I(m, n) = I_t(m, n) + n(m, n) \tag{1}$$

Formula (1), $I_t(m, n)$ is the target image, $n(m, n)$ is Background noise, background noise includes strong stray light and system noise.

Figure 1 the original image in the background of stray light. Figure 2 shows the target in complicated background. Noise mainly comes from sky background noise, molecular noise, electronic noise and so on [5]. This paper discusses the background noise in the sky. It is clear from the picture that the signal is buried in the background. At present, the common background suppression methods are mainly based on the distribution characteristics of target and background. However, the relevance between the objective and the context is not taken into account. In many cases the signal can not be completely extracted, so it is necessary to find a new way to solve this problem.

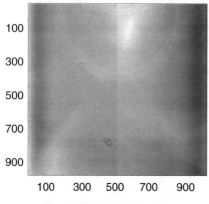

Fig. 1. The original image

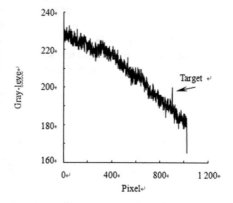

Fig. 2. The target in complicated background

In order to effectively extract the starry target in the background of stray light, we adopt the mathematical morphology method to process the image. In morphology, operations are used to deal small bright details, At the same time, can relatively keep the overall grayscale and large bright areas unchanged.

Because the target, noise and the edge of the background are the high-frequency parts of the image, we can use morphology to estimate the background, and then perform image cancellation to detect the target.

2.1 Basic Operations of Morphology

Mathematical morphology is an important discipline in image analysis. And it is widely used in image analysis. In morphology, open operation can remove small bright details.

At the same time, open operation can relatively keep the overall gray level and larger bright areas unchanged. The edges of target, noise and background appear as the high frequency part of the image. Therefore, we can use morphology operation to estimate the background, and then perform image cancellation operation to detect the target. f(x, y) is the coordinate of some point in the image. $b(i, j)$ is a structural element. Gray-scale morphological corrosion operations are defined as:

$$(f \ominus b)(x, y) = min\{f(x + i, y + j) - b(i, j)|(x + i, y + j) \in D_f; (i, j) \in D_b\} \quad (2)$$

The expansion operation of gray-scale morphology is defined as:

$$(f \oplus b)(x, y) = max\{f(x - i, y - j) - b(i, j)|(x - i, y - j) \in D_f; (i, j) \in D_b\} \quad (3)$$

Formula (2) and formula (3), D_f and D_b are the defined fields of f and b, x, y, i, j are the image and structure elements of the coordinates of a point. $\ominus, \oplus, \circ$ are corrosion, expansion, and opening operation, respectively. Firstly opening operation is carried out by corrosion operation, and then by expansion operation. Formula (4) is the expression of opening operation:

$$f \circ b = (f \ominus b) \oplus b \quad (4)$$

2.2 Background Estimation Filtering Based on Top-Hat and Star-Point Target Extraction

The definition of the Top-Hat transformation operator is:

$$TOPHAT(f) = f - f \circ g \quad (5)$$

Formula (5), g is the structural element, Because the structure element always moves under the signal in the open operation, the result of the open operation will never exceed the original signal. $TOPHAT(f)$ is non-negative [6]. Structural elements can be any 3-D structure, the more commonly used are conical, cylindrical, hemispherical or parabolic structures. Structure element sizes are always odd, so that the center of the structure corresponds to exactly one pixel. The relationship between the parts of an image can be examined by moving the structural elements in the image. The pixel of star-point target is 3–5 units, and it can use 5×5 structural elements [7]. Different structural elements have different effect on image denoising.

Figure 3(a) can remove square imperfections. Figure 3(b) can remove diamond-shaped imperfections. Figure 3(c) can remove cross imperfections. The defects in the actual processed image may be a combination of the three, so the three structural elements are used together in this paper. Firstly, the star image is opened, then the result is calculated by logic or, and then the estimated background is subtracted to get the image containing target. Formula (6) express the idea.

$$TOPHAT(f) = max\{(f \circ a), (f \circ b), (f \circ c)\} \quad (6)$$

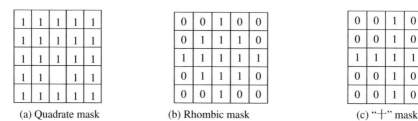

Fig. 3. Three morphological structuring elements

2.3 Calculation of Star Brightness

Because the open operation in morphology can eliminate the isolated scattered points, burrs and bridges on the star image [8]. Combined with the geometric shape of the star points on the star image, Firstly, the background of the star map is estimated by the open operation. Secondly, the brightness of the target star point can be obtained by subtracting the estimated background from the original star map.

Figure 4 shows the grayscale values for the target points (320:322, 3491:393). Figure 4(a) denotes the true value of the target point. Figure 4(b) is the target gray value obtained by this method. Figure 4(c)–(f) is the average value of the pixel gray value in the open window after using different windows. The original star image subtracts the average value to get the gray value of the target point.

4	26	8
2019	26	19
11	19	9

(a) The ture value of star

6	25	5
17	30	20
12	24	14

(b) Background estimation

4.3	24.2	12.9
17.9	29.1	21.2
13.5	18.3	12.9

(c) Window 5×5

4	22	7
17	27	16
12	21	14

(d) Window 23×23

4.5	23.5	6.9
18.0	28.42	18.2
13.7	20.2	12.9

(e) Window 33×33

4.5	24.4	7.4
17.9	29.4	17.4
14.0	20.4	12.4

(f) Window 43×43

Fig. 4. The comparison of target brightness

2.4 Threshold Processing

After the image is processed by morphological method, the background is greatly suppressed. Most of the pixels are concentrated in the low gray area [7], only the target and a small amount of noise are distributed in the high brightness area. Adaptive threshold method [3] is used to find the target point. Threshold $T = Mean + k\sigma$

$$f(i,j) = \begin{cases} f(i,j) \text{ if } f(i,j) > T \\ 0, \quad else \end{cases} \tag{7}$$

"Mean" and "σ" are the Mean and variance of the filter output. "K" is the constant. Whether the threshold is suitable or not directly affects the false alarm probability of the target [9], according to experience "k" generally takes 5–15 [10]. Formula (7) is the threshold processing expression.

2.5 Centroid Extraction

A star-point target takes 3–5 pixels, so we need to find the center of the target-the center of mass [11]. Centroid is the gray center of the target image. Hypothetically in an M × N image, the gray value of the target at position (i, j) is f(i, j). The centroid (x_c, y_x) expression [8] for the target is:

$$x_c = \frac{\sum_{i=1}^{M} \sum_{j=1}^{N} x_{ij} f(i,j)}{\sum_{i=1}^{M} \sum_{j=1}^{N} f(i,j)} \quad y_c = \frac{\sum_{i=1}^{M} \sum_{j=1}^{N} y_{ij} f(i,j)}{\sum_{i=1}^{M} \sum_{j=1}^{N} f(i,j)} \quad (8)$$

Formula (8), x_{ij}, y_{ij} represent the horizontal and vertical coordinates of the target at (i, j), respectively.

3   Test Results and Analysis

We chose Fig. 1 as the original diagram of the experiment. In Fig. 3, the three structures are used to process the image, and the three morphological mechanism elements are combined to process the image. The results are shown in Table 1:

Table 1. The results were compared between the three structural elements and the combined elements

	Maximum	Mean	Standard deviation	SNR
The article methods	28	1.4012	1.6915	15.3215
Quadrate mask	31	2.5329	2.1018	13.4908
Rhombic mask	29	1.7180	1.9300	14.3200
"+" mask	28	1.8960	1.8875	14.0111

By comparing the image gray mean value after processing, Three structural element methods are used to process the image, and the results show that the mean value is minimum, the standard deviation is minimum, and the signal-to-noise ratio is maximum. The above results show that the method of image processing with three structures can remove the noise background well. In addition, the experiment compared the variance before and after image processing. The result is shown in Fig. 5. The signal-to-noise ratio of the image can be increased up to 10 times of the original image by using the morphological background estimation method.

In addition, the experiment also compared the extracted star-point brightness with the traditional window-opening method. The traditional method is to open a window of

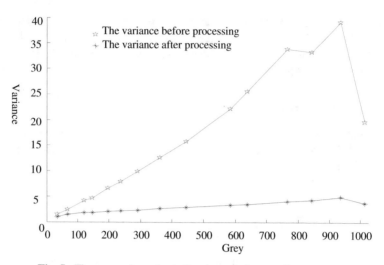

Fig. 5. The comparison chart of variance before an after treatment

N × N at the star-point target and get the average value in the window. In Table 2 the deviation expression [8] is: $\delta = (x - x_{true})/x_{true}$. The window gray value is the sum of the gray values of the pixels where the target point is located. We can see that the error of star luminance obtained by morphological background estimation is smaller than that obtained by opening 5 × 5 and 43 × 43 windows.

Table 2. The comparison of grey error between estimation background method and opening window

	Background estimation	Window 5 × 5	Window 23 × 23	Window 33 × 33	Window 43 × 43
Window PN	146	151.8	143	146.8	148.6
Deviation (%)	6.50	9.17	3.52	6.28	7.58

The centroid coordinates of star points are extracted by Formula (7). Figure 6 shows the deviation of centroid coordinates from true value obtained by background estimation. The Red Star represents the deviation of the center of mass on the x-axis. The black circle represents the deviation of the center of mass on the Y-axis. Figure 6 shows that the centroid error obtained by the morphological background estimation method is very small, not exceed 0.26 pixels. The deviation between the experimental centroid coordinates and the true values is less than 0.26 pixels.

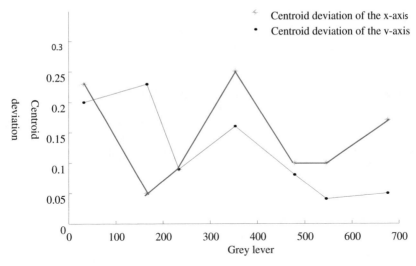

Fig. 6. The deviation between true value of centroid coordinates and extracted

4 Conclusion

In this paper, the method of star-point target signal extraction based on morphology theory is adopted, and the operation characteristics of morphology operation are introduced. The combination of three structures is used to extract the target points, and the method is verified by practice. The experimental results show that the method can effectively [8] remove the noise and improve the signal-to-noise ratio by up to 10 times. In the experiment, The target brightness is obtained by adaptive threshold processing, which is less error than traditional window-opening method. In the aspect of centroid extraction, the centroid method [12] is used to get the coordinate error and the true value error less than 0.26 pixel units.

References

1. Wang, S., Yang, S.: Technique of IR background suppression based on spatial matched filter. Appl. Sci. Technol. **33**(1), 26–28 (2006)
2. Ma, Z., Wang, J., et al.: Point target detection based on image fusion of two infrared bands. Laser Infrared **33**(3), 228–230 (2003)
3. Xu, B., Zheng, Y., et al.: The improvement of star target region extraction algorithm for star centroid. Acta Geodaeticaet Cartographica Sinica **5**, 760–768 (2023)
4. Ning, M.: The improvement of star target region extraction algorithm for star centroid. Telecommun. Eng. **4**, 499–504 (2023)
5. Maragos, P.: Differential morphology and image processing. IEEE Trans. Image Process. **5**(6), 922–937 (1996)
6. Li, X., Hao, Z., et al.: The research on the method of the star's position determination of the star sensor. J. Electron Dev. **27**(4), 571–574 (2004)
7. Wang, H., Luo, C., et al.: Algorithm for star extraction based on self-adaptive background prediction. Opt. Tech. **35**(3), 413–414 (2009)

8. Zhang, Y., Zhang, J., Wang, D., Chen, C.: Infrared small target detection based on morphology and wavelet transform. In: 2011 2nd International Conference on Artificial Intelligence, Management Science and Electronic Commerce (AIMSEC) (2011)

9. Wang, Y.: The Feature Extraction and Recognition for Small Target under Complex Background, pp. 9–12. National University of Defense Technology, Changsha (2007)

10. Zou, J., Jiang, W., Chen, Z.: Target extraction method of low SNR image sequences based on morphology. Optoelectron. Technol. Inf. **18**(3), 67–71 (2005)

11. Tian, J., Ou-yang, H., et al.: Method of star acquisition from star image. J. Huazhong Univ. Sci. Technol. **33**(4), 38–40 (2005)

12. Li, C., Xie, H., et al.: Centroiding algorithm for high-accuracy star tracker. Opto-Electron. Eng. **33**(2), 41–44 (2006)

Toward Detection of Fake News Using Sentiment Analysis for Albanian News Articles

Besjana Muraku[1(✉)], Lu Xiao[1], and Elinda Kajo Meçe[2]

[1] School of Information Studies, Syracuse University, Syracuse, USA
{bmuraku,lxiao04}@syr.edu
[2] Faculty of Information Technology, Polytechnic University of Tirana, Tirana, Albania
ekajo@fti.edu.al

Abstract. The public's concern of fake news has grown due to its potential to challenge social cohesion, and foster mistrust of the government and society in general. Recent research has revealed a concerning trend in which a sizable portion of the population has an inability to distinguish between authentic and non-authentic news, moreover in Western Balkans 75% of the population tend to believe in fake news. Writers of fake news often employ certain strategies in their articles that, from a sentiment perspective, exhibit a higher degree of polarity in comparison to authentic news articles. The rapid growth of online communities in Albania have given rise to fake news risks, however there is a limited body of research on the subject matter. In this research we will leverage the sentiment analysis techniques through feature engineering to identify the characteristics of sentiment polarity of fake news in Albanian articles.

1 Introduction

The amount of time individuals spends engaging with online activities has shown a rapid increase over the course of recent years. The rapid growth of internet connectivity and the expansion of digital platforms have significantly enhanced the accessibility and efficiency of information dissemination. Social media platforms play a significant role in the spread of information, but it is necessary to also acknowledge the contribution of news websites in this regard. Newman et al., [1] showed that users that visit social media and online platforms for news and information have increased significantly through years, and this number has projected to grow. The term "fake news" commonly refers to fraudulent or misleading information that is disseminated through many news outlets, including traditional or internet media and social networks [2]. During the 2016 US presidential elections, Facebook emerged as a significant disseminator of false information due to weaknesses in its algorithmic design [3]. This phenomenon exerted a considerable influence on individuals' voting preferences and profoundly impacted the election outcome. McGonagle [4] calls it "post-truth" area due to the increase popularity and attention of the fake news. Botha and Pieterse [5] discuss the emergence of deep-fakes, which are AI-generated fake videos that can further contribute to the spread of fake news. The role of sentiment expression has a significant impact in the dissemination of fake news. To boost the dissemination of news, it is necessary that headlines exhibit a tendency to catch the reader's attention and trigger an emotional response.

L. Barolli (Ed.): EIDWT 2024, LNDECT 193, pp. 575–585, 2024.
https://doi.org/10.1007/978-3-031-53555-0_55

The Stamatoukou [6] study shows that disinformation is a significant issue also in the Western Balkans, posing challenges to EU credibility, COVID-19-related disinformation, and elections. The findings of the study show that over 75% of the population in the Balkans believe in one or more conspiracy theories spread through disinformation. Despite being an ongoing problem, little study has been done on solutions specific to the Albanian setting.

The Albanian language belongs to the Indo-European (IE) language family and is classified as an independent branch within this family. It possesses unique characteristics that encompass several aspects, including morphology and lexicon [7]. In instances of this nature, it is necessary to do research on language-specific solutions as opposed to more generalized alternatives. Canhasi et al., [8] and Hoti et al., [9] both focus on the development and evaluation of machine learning methods for fake news detection in the Albanian language.

The use of sentiment analysis in the identification of false news is a significant subject of interest inside current research efforts. The research Ibrahim [10] provides a survey of sentiment analysis and machine learning methodologies in fake news detection, emphasizing the need for automated methods due to the difficulty of manual verification. Alonso et al., [11] discusses the use of sentiment analysis in fake news detection, considering the impact of fake news on social well-being and the need for automatic systems. Multi-EmoBERT, a multi-label emotion recognition tool, was undertaken by Li and Xiao [12] with the aim of investigating the connection between the emotional content of articles and the stance that they take in datasets including fake news. Ding et al., [13] and Bhutani et al. [14] explore sentiment analysis of real and fake news on social media, demonstrating how social media sentiment can improve the accuracy of fake news identification. However, in the context of Albanian language, no previous research has been conducted on the sentiment analysis of fake news.

In this paper, we first attempt to improve the state of the art for sentiment classification in existing dataset using feature engineering and then investigate the variations in sentiment among fake articles written in the Albanian language. The paper is structured in the following sections: Sect. 2 we provide a literature review for sentiment analysis and fake news. Section 3 describes the sentiment classification methodology. Section 4, phase one we evaluate four models for sentiment classification on two different datasets, phase two we predict sentiment on fake news dataset. In Sect. 5 we discuss our findings and state the limitations. Finally, the conclusion is presented in Sect. 6.

2   Literature Review

Sentiment analysis classification models are traditionally based on feature engineering. This is a common practice of retrieving meaningful features from the data itself. Turner [15] introduces the idea of a feature-oriented approach in software development, emphasizing the benefits of explicitly treating features throughout the software life cycle. Khurana [16] proposes a framework for automating feature engineering using reinforcement learning, aiming to reduce the human effort and cost involved in the process. Ghosh and Sanyal [17] explores the efficacy of combined feature selection techniques on machine learning classifiers for sentiment analysis. Zou et al., [18] train syntax features on movie reviews as data with machine learning methods.

The more recent literature of sentiment analysis is focused on using deep learning models such as convolutional neural network or recurrent neural network. These models are highly dependent on word embeddings such as BERT, ELMo, to predict with high accuracy sentiment polarity. These models are pre-trained on extensive text corpora and generally entail unsupervised tasks. Various derivations of these models have been identified in subsequent researches, including the incorporation of multilingual capabilities, such as mBERT. The performance of multilingual BERT in several languages remains uncertain, given its coverage of just 2.6% of languages within the real-life group.

Catelli et al., [19] compares the efficacy of a deep neural network-based language model, BERTBase Italian XXL, with a lexicon-based system, NooJ, for sentiment analysis in Italian language. Existing scientific literature generally indicates that language models, particularly those based on deep neural networks like BERT, exhibit superior performance. However, this study raises important questions regarding the effectiveness and potential enhancements of these solutions when compared to lexicon-based approaches. This situation is commonly encountered when working with languages other than English or Chinese.

Numerous research studies have been conducted on the identification of fake information through the use of sentiment analysis approaches. The study of Iwedi et al., [20] employs the technique of Information Fusion to extract authentic news data and introduces a set of 39 original text features, encompassing aspects such as sentiment analysis, linguistic characteristics, and named entity-based attributes. The aforementioned features exhibit a detection capability for COVID-19-related misinformation with an accuracy rate of 86.12%, representing a notable improvement of 20% compared to the initial set of features.

The study of Subramanian et al. [21] encompassed a wide range of approaches and datasets, emphasizing the notable progress achieved in hate speech domain and sentiment analysis. Another study proposes a deep framework for detecting fake news, incorporating various news features and an adversarial mechanism to maintain semantic relevance [22]. The authors employ statistical analysis to validate the effectiveness of user sentiment, and they leverage the sentiment polarities of users to make it easier to identify fake news.

Interestingly, the research of Canhasi et al., [8] on Albanian fake news detection explores key aspects, including optimal feature selection, machine learning methods, text size impact, and the influence of news sources. Their tests revealed no significant difference between engineered and language model-derived features and evaluated that the highest performing machine learning models were XGBoost, LR, SVM, and RF with the highest ratio for a fake news correctly identified of 91.1%. The features examined in this research consist of: number of characters and words, part-of-speech related features, vocabulary complexity and psycholinguistic features. Among these, the most significant factor influencing successful fake news classification was identified as the news source and associated writing style. The potential of sentiment features in the news content was not explored.

Given these findings of our literature review, we apply feature engineering approach to improve state-of-the-art for sentiment classification and investigate the use of sentiment polarity features in fake news for Albanian language context.

3 Methodology

The research can be considered in two phases. In the first phase we identify the features for Albanian language that counts as labels for sentiment analysis. Our objective is to determine the most optimal features for sentiment analysis when utilized in alongside a classification model. The dataset classified by sentiment is initially subjected to pre-processing. In this step, the text undergoes tokenization, and all terms that are non-alpha-numeric in nature are eliminated from the dataset. Capitalized words serve as indicators of sentiment for this reason they are not converted into lower case. In the subsequent stage, referred to as feature extraction, the tokens are ranked according to their respective frequencies. The top 2000 most frequent are selected from the obtained results, which will then be used into the machine learning models. Various methodologies are employed in the process of picking the initial 2000 characteristics. In the initial iteration, the bag of words is obtained without undergoing any form of filtration. In the subsequent iteration, the elimination of stop words is executed, and in the ultimate iteration, the subsequent word after a negation is annotated. The evaluation of the best performing model is done considering the accuracy, recall, precision and f1 results.

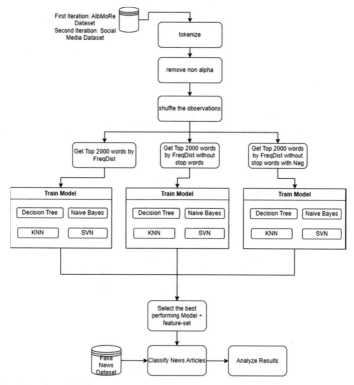

Fig. 1. Flowchart for selecting the sentiment classification model and using it for sentiment analysis on fake news dataset

In response to the linguistic characteristics of the Albanian language, specifically the prevalence of various dialects and the informal nature of language used in social networks, the stop word list underwent multiple revisions. Notably, the inclusion of the letter 'ë' which is frequently employed in Albanian, posed a challenge, also due to its limited accessibility on standard keyboards. Consequently, common words such as 'për' equivalent to 'for' in English, were frequently observed in the alternate form 'per' due to the practical constraints associated with typing the character 'ë'. This necessitated iterative adjustments to the stop word list to accurately account for the linguistic nuances present in the Albanian text data.

In the second phase, the most effective combination of feature-set selection and classification model is employed to analyze the dataset of fake news articles. Each data point in the fake news dataset corresponds to one article. Our objective is to extract each sentence from these articles and predict its sentiment. After predicting sentiment on all sentences in each news article, we examine the differences between these two aforementioned groups for each sentiment category. Figure 1 provides the flowchart of the overall implementation.

4   Results

4.1   Phase I: Training of Sentiment Analysis Models

We use two datasets in the first phase. AlbMoRe (Albanian movie review) is a dataset of movies reviews developed by Cano E. [23]. The author has translated movies reviews from IMBD from English to Albanian and have manually labeled sentiment polarity of each review. The overall dataset consists of 800 reviews. The corpus is balanced, comprising 400 positive and 400 negative evaluations pertaining to 67 distinct movies spanning various genres [23]. The other existing and public dataset we use is from social media data and manually annotated by Kadiu et al., [24]. The dataset utilized in this work consists of comments that were gathered from the official Facebook page of the NIPHK Institute during a span of six months, from March 12 to August 31, 2020. The sentiment of each an observation in the dataset is evaluated by three academics. Comments are categorized based on their sentiment: neutral comments are labeled as 0, positive comments are labeled as 1, and negative comments are labeled as 2. The dataset has been adjusted for class balance.

Table 1. Results from AlbMoRe dataset sentiment classification training

Top frequent 2000 words				
	Accuracy	Precision	Recall	F1
Naïve Bayes	0.949	0.950	0.949	0.949
Decision Tree	0.839	0.847	0.837	0.841
SVM	0.925	0.926	0.925	0.924

(continued)

Table 1. (*continued*)

Top frequent 2000 words

	Accuracy	Precision	Recall	F1
K-NN	0.818	0.821	0.819	0.818

Top 2000 features without stop words

	Accuracy	Precision	Recall	F1
Naïve Bayes	0.950	0.932	0.965	0.948
Decision Tree	0.854	0.860	0.852	0.860
SVM	0.931	0.933	0.932	0.931
K-NN	0.863	0.865	0.863	0.862

Top 2000 features without stop words with negation words

	Accuracy	Precision	Recall	F1
Naïve Bayes	0.944	0.992	0.980	0.947
Decision Tree	0.820	0.827	0.830	0.831
SVM	0.925	0.925	0.925	0.925
K-NN	0.845	0.849	0.847	0.844

Table 1 shows the results from the first dataset analyzed, AlbMoRe. For this dataset we tested three different feature sets and four different classification models. The best performing model in combination with feature set is Naïve Bayes and the list of top 2000 most frequent words. The results outperformed the baseline of the [23] on which the best performing model was Support Vector Machine with an accuracy of 0.946.

In Table 2 are shown the results from the second dataset analyzed, the social media dataset. Also, in this case, we tested three different feature sets and five different classification models, including the Multinomial Naïve Bayes. The results from this dataset are scientifically lower performing comparing to the first case. The best performing model in combination with feature set, for social media dataset, is Naïve Bayes and the list of top 2000 most frequent words without stop words and negation words with an accuracy of 64.3%. The precision of for this model is 64.3%, the recall of 64.2% and F1 score of 64.2%.

Table 2. Sentiment Analysis of Social Network Dataset and top frequent words

Top frequent 2000 words

	Accuracy	Precision	Recall	F1
Naïve Bayes	0.622	0.621	0.622	0.622
MultinomialNB	0.640	0.642	0.641	0.640
Decision Tree	0.535	0.528	0.533	0.535

(*continued*)

Table 2. (*continued*)

Top frequent 2000 words				
	Accuracy	Precision	Recall	F1
SVM	0.630	0.644	0.630	0.633
K-NN	0.480	0.482	0.479	0.475
Top 2000 features without stop words				
Naïve Bayes	0.633	0.633	0.633	0.633
MultinomialNB	0.642	0.646	0.640	0.640
Decision Tree	0.583	0.590	0.587	0.587
SVM	0.626	0.659	0.626	0.630
K-NN	0.478	0.486	0.478	0.476
Top 2000 features without stop words with negation words				
Naïve Bayes	0.643	0.643	0.642	0.642
MultinomialNB	0.635	0.639	0.636	0.631
Decision Tree	0.581	0.588	0.589	0.580
SVM	0.626	0.659	0.626	0.630
K-NN	0.476	0.485	0.476	0.470

4.2 Phase II: Sentiment Analysis in Fake News Dataset

In the second phase, we employ the Naïve Bayes model trained using the AlbMoRe dataset, which exhibited better performance compared to other machine learning models, on categorizing the sentiments of news items, to the Albanian fake news corpus [8]. The news content in this dataset is collected from various domains, including news agencies, and manually searched for fake articles on Facebook. The dataset contains 3996 news articles, from which 1998 are labeled as False news, 1997 are labeled as True news and one observation is incorrectly labeled and is discarded from before processing.

We investigate the relationship among sentiment and news type: fake/true by conducting a Chi-square test. Initially we define the article sentiment which is calculated by investigating the sentiment for each sentence on the article. The sentence sentiment is predicted using the Naïve Bayes model trained with AlbMoRe dataset, and for each article we calculate total positive and negative sentences. The sentiment on article level is defined as: 0 – if the article has more negative sentences, 1 – article has equal number of positive and negative and 2 – if there are more positive sentences. [8] categorizes news as false – being a not-fake news and true – being a fake news. We conduct a Chi-square test between article sentiment and its type.

In Table 3 we present the observed and expected in parenthesis values along with the p-value. The chi-squared test results suggest a significant association between the authenticity of news (fake or true) and its sentiment classification (negative, neutral, positive). The observed frequencies, such as more negative sentiments for false news and more neutral sentiments for true news, differ notably from the expected frequencies under the assumption of no association. With a low p-value of 0.000355, the likelihood that these differences occurred by chance is very small, indicating a statistically significant relationship between news authenticity and sentiment. For this dataset, true news, are more likely to have negative and positive sentiment, and less likely to have neutral sentiment, while fake news is more likely to have more neutral news.

Table 3. Chi-square test for veracity and sentiment categories

	Negative (0)	Neutral (1)	Positive (2)
Fake News:False	897(872.88)	293(340.2)	801(777.9)
Fake News:True	840(864)	384(336)	747(770)
		p = 0.000355	

Furthermore, interested in whether the sentiment feature of news content contributes to fake news detection, we conduct a comprehensive assessment of a machine learning model employing the Word2Vec technique. Initially, our analysis involves predictive modeling without the consideration of sentiment results. Subsequently, we augment our investigation by incorporating sentiment into the prediction process. This sequential approach enables us to evaluate the model's performance under varying conditions and provides a nuanced understanding of its efficacy with and without sentiment considerations. Given the favorable performance of XGBoost among the models assessed in reference [8], our experimental framework adopts this model. The outcomes, shown in Table 4, demonstrate an enhancement in precision for the identification of news categorized as true (thus being fake news), rising from 0.82 to 0.86, at the cost of a lower recall. This result suggests that while the sentiment polarity of an article is not a strong indicator of its fakeness when it is considered in deciding whether the questioned article is fake it adds precision.

Table 4. Comparison of baseline model and new model including sentiment as parameter

Model	Accuracy	Precision	Recall	F1
Baseline model XGBoost	0.81			
False		0.81	0.86	0.83
True		0.82	0.76	0.79
XGBoost with Sentiment	0.80			
False		0.75	0.87	0.81
True		0.86	0.73	0.79

5 Discussion and Limitation

In this study, we analyzed the sentiments in news articles and investigated its effect in predicting their tunefulness. We improved the baseline model [23] for sentiment prediction. We found out that for the [8] dataset the true news articles are more likely to have negative sentiment and positive sentiment, and less likely to have neutral sentiment. This is an interesting outcome that will be investigated further to understand the nature of the dataset or whether fake news is crafted with a tone that is neither overtly positive nor extremely negative. The next step which regarded including sentiment as an attribute for predicting the fake news we noticed an improvement of the precision of the baseline model of XGBoost by 0.4% for the fake news identification. This result echoes the trends identified by the studies listed in the literature review, on improving the performance of fake news identification including the sentiment. Our study lays the groundwork for future research on sentiment analysis in fake news. Further studies could be expanded by examining a broader range of emotions. Moreover, as the Deep Learning algorithms have confirmed superiority in many areas including fake news detection, new studies can explore them for the Albanian language context.

Nonetheless, the constraints of this research must be recognized. Due to the small sample size, the dataset might not have comprehensively represented the diversity of all relevant domains. Thus, it is imperative to show discretion when interpreting the results of this research, as they might not fully reflect entirely population or all possible domains.

Another noteworthy limitation of this study is to the linguistic diversity seen within the Albanian language. The Albanian language demonstrates a wide range of linguistic nuances, including dialectal variations. The integration of diverse linguistic variances into a unified natural language processing model is a challenge. Although attempts were made to include a diverse sample, it is possible that the complications originating from the variety of the Albanian language were not completely accounted for. Therefore, the extent to which the natural language processing model can be applied to all linguistic variants within the Albanian language may be constrained.

6 Conclusion

The study of fake news continues to be a significant subject of investigation. Numerous ideas employ various qualities that might enhance the accuracy of fake news detection. This article starts with building a model for sentiment analysis using the process of feature engineering. The models were evaluated on two distinct datasets. The first dataset exhibits a polarity categorization, comprising two distinct states: positive and negative. The second dataset is categorized into three distinct states, namely neutral, positive, and negative. The traits were derived by analyzing word frequency and subsequently choosing the most prominent 2000 attributes. Subsequently, many models were subjected to testing using distinct feature-sets. The optimal model, together with its associated feature set, was chosen for each dataset. The best performing model was then employed to categorize each sentence of the news articles and then calculate the article sentiment. Next, we proceed to examine the polarity of sentiments in both authentic and fake news articles.

The results given in this study signify the beginning of a more extensive investigation into the detection of news veracity. It is expected that future research will expand upon these first findings and investigate further in the areas of emotion expression and news veracity.

References

1. Newman, N., Fletcher, R., Kalogeropoulos, A., Levy, D., Nielsen, R.K.: Reuters Institute Digital News Report 2018, Rochester, NY, 14 June 2018. https://papers.ssrn.com/abstract= 3245355. Accessed 08 Nov 2023
2. Kula, S., Choraś, M., Kozik, R., Ksieniewicz, P., Woźniak, M.: Sentiment analysis for fake news detection by means of neural networks. In: Krzhizhanovskaya, V.V., et al. (eds.) ICCS 2020. LNCS, vol. 12140, pp. 653–666. Springer, Cham (2020). https://doi.org/10.1007/978-3-030-50423-6_49
3. Zannettou, S., Sirivianos, M., Blackburn, J., Kourtellis, N.: The web of false information: rumors, fake news, hoaxes, clickbait, and various other shenanigans. J. Data Inf. Qual. **11**(3), 1–37 (2019). https://doi.org/10.1145/3309699
4. McGonagle, T.: 'Fake news': false fears or real concerns? Netherlands Q. Hum. Rights **35**(4), 203–209 (2017). https://doi.org/10.1177/0924051917738685
5. Botha, J., Pieterse, H.: Fake News and Deepfakes: A Dangerous Threat for 21st Century Information Security, March 2020
6. Stamatoukou, E.: BIRN Albania and SCiDEV Present Studies on Disinformation, Propaganda and Fake News. BIRN. https://birn.eu.com/news-and-events/birn-albania-and-scidev-present-studies-on-disinformation-propaganda-and-fake-news/. Accessed 08 Nov 2023
7. Kastrati, M., Biba, M.: Natural language processing for Albanian: a state-of-the-art survey. Int. J. Electr. Comput. Eng. (IJECE) **12**(6), 6432 (2022). https://doi.org/10.11591/ijece.v12i6.pp6432-6439
8. Canhasi, E., Shijaku, R., Berisha, E.: Albanian fake news detection. ACM Trans. Asian Low-Resour. Lang. Inf. Process. **21**(5), 1–24 (2022). https://doi.org/10.1145/3487288
9. Hoti, A.H., Hoti, M.H., Hoti, H., Salihu, A.: Identifying Fake News written on Albanian language in social media using Naïve Bayes, SVM, Logistic Regression, Decision Tree and Random Forest algorithms. In: 2022 11th Mediterranean Conference on Embedded Computing (MECO), June 2022, pp. 1–6 (2022). https://doi.org/10.1109/MECO55406.2022.9797147
10. Ibrahim, M.: Sentiment analysis for fake news detection in online media networks: a survey, fusion techniques and quality metrics. Neutrosophic Inf. Fus. **1**(2), 44–68 (2023). https://doi.org/10.54216/NIF.010205
11. Alonso, M.A., Vilares, D., Gómez-Rodríguez, C., Vilares, J.: Sentiment analysis for fake news detection. Electronics **10**(11), 1348 (2021). https://doi.org/10.3390/electronics10111348
12. Li, J., Xiao, L.: Multi-emotion recognition using multi-EmoBERT and emotion analysis in fake news. In: Proceedings of the 15th ACM Web Science Conference 2023, WebSci 2023, April 2023, pp. 128–135. Association for Computing Machinery, New York (2023). https://doi.org/10.1145/3578503.3583595
13. Ding, L., Ding, L., Sinnott, R.O.: Fake news classification of social media through sentiment analysis. In: Nepal, S., Cao, W., Nasridinov, A., Bhuiyan, M.D.Z.A., Guo, X., Zhang, L.-J. (eds.) BIGDATA 2020. LNCS, vol. 12402, pp. 52–67. Springer, Cham (2020). https://doi.org/10.1007/978-3-030-59612-5_5
14. Bhutani, B., Rastogi, N., Sehgal, P., Purwar, A.: Fake news detection using sentiment analysis. In: 2019 Twelfth International Conference on Contemporary Computing (IC3), August 2019, pp. 1–5 (2019). https://doi.org/10.1109/IC3.2019.8844880

15. Turner, C.R., Fuggetta, A., Lavazza, L., Wolf, A.L.: Feature engineering [software development]. In: Proceedings Ninth International Workshop on Software Specification and Design, April 1998, pp. 162–164 (1998). https://doi.org/10.1109/IWSSD.1998.667935

16. Khurana, U., Samulowitz, H., Turaga, D.: Feature engineering for predictive modeling using reinforcement learning. In: Proceedings of the AAAI Conference on Artificial Intelligence, vol. 32, no. 1, Art. no. 1, April 2018. https://doi.org/10.1609/aaai.v32i1.11678

17. Ghosh, M., Sanyal, G.: Analysing sentiments based on multi feature combination with supervised learning. IJDMMM 11(4), 391 (2019). https://doi.org/10.1504/IJDMMM.2019.102728

18. Zou, H., Tang, X., Xie, B., Liu, B.: Sentiment classification using machine learning techniques with syntax features. In: 2015 International Conference on Computational Science and Computational Intelligence (CSCI), December 2015, pp. 175–179 (2015). https://doi.org/10.1109/CSCI.2015.44

19. Catelli, R., Pelosi, S., Esposito, M.: Lexicon-based vs. BERT-based sentiment analysis: a comparative study in Italian. Electronics 11(3), 374 (2022). https://doi.org/10.3390/electronics11030374

20. Iwendi, C., Mohan, S., Khan, S., Ibeke, E., Ahmadian, A., Ciano, T.: Covid-19 fake news sentiment analysis. Comput. Electr. Eng. 101, 107967 (2022). https://doi.org/10.1016/j.compeleceng.2022.107967

21. Subramanian, M., Sathiskumar, V.E., Deepalakshmi, G., Cho, J., Manikandan, G.: A survey on hate speech detection and sentiment analysis using machine learning and deep learning models. Alexandria Eng. J. 80, 110–121 (2023). https://doi.org/10.1016/j.aej.2023.08.038

22. Cui, L., Wang, S., Lee, D.: SAME: sentiment-aware multi-modal embedding for detecting fake news. In: Proceedings of the 2019 IEEE/ACM International Conference on Advances in Social Networks Analysis and Mining, ASONAM 2019, January 2020, pp. 41–48. Association for Computing Machinery, New York (2020). https://doi.org/10.1145/3341161.3342894

23. Çano, E.: AlbMoRe: a corpus of movie reviews for sentiment analysis in Albanian. arXiv, 14 June 2023. https://doi.org/10.48550/arXiv.2306.08526

24. Kadriu, F., Murtezaj, D., Gashi, F., Ahmedi, L., Kurti, A., Kastrati, Z.: Human-annotated dataset for social media sentiment analysis for Albanian language. Data Brief 43, 108436 (2022). https://doi.org/10.1016/j.dib.2022.108436

Proposal of My Hazard Map System for Disaster Prevention Education

Toshiya Morita and Tomoyuki Ishida[✉]

Fukuoka Institute of Technology, Fukuoka, Fukuoka 811-0295, Japan
s20b1051@bene.fit.ac.jp, t-ishida@fit.ac.jp

Abstract. In this study, a "My Hazard Map System" is developed that can visualize dangerous places in the event of a disaster on school routes. This system, which is primarily designed for the younger generation, especially elementary school students, requires the students take photographs of places or objects on their route to school that could be dangerous in the event of a disaster. They can take pictures of objects, such as block walls, telephone poles, and vending machines, and send them to a server. As the images sent to the server include data on their latitude and longitude, the images are mapped to Web-GIS according to the location information. This allows elementary school students to create their own digital "My Hazard Map."

1   Introduction

Compared with other countries, Japan is prone to natural disasters, such as typhoons, heavy rain and snow, landslides, floods, earthquakes, tsunamis, and volcanic eruptions [1]. In recent years, various natural disasters that caused severe damage have occurred throughout Japan, including the Great East Japan Earthquake of 2011, the Kumamoto Earthquake of 2016, the heavy rains of July 2018, and the East Japan Typhoon of 2019. Therefore, for the Japanese people, natural disasters are inseparable natural phenomena that are part of their lives. In Japan, where there are many natural disasters, "disaster prevention and evacuation training" has been established as a realm in which children can learn how to protect themselves from disasters. Among these, disaster prevention and evacuation trainings held in many schools include "evacuation trainings that simulate an earthquake." However, this such a training assumes that an earthquake occurs while students are at school; this does not address the disaster risks present while they are commuting to and from school." Because we do not know when or where a natural disaster will strike, there must be a way to cover areas outside of school as well.

A "hazard map" is a medium that enables people to understand the dangers of disasters. Hazard maps "display the locations of disaster prevention-related facilities, such as predicted disaster areas, evacuation sites, and evacuation routes, for the purpose of reducing damage caused by natural disasters and disaster prevention measures." Local governments typically distribute these maps to residents in paper form or publish it on their websites [2]. However, hazard maps have the problem of low awareness.

Therefore, in the present study, we developed a "My Hazard Map System" that allows young people, especially elementary school students, to visualize dangerous places in the

L. Barolli (Ed.): EIDWT 2024, LNDECT 193, pp. 586–596, 2024.
https://doi.org/10.1007/978-3-031-53555-0_56

event of a disaster on their own routes to school. This system allows elementary school students to take photographs of places on their way to school, which could be dangerous in the event of an earthquake. In this system, they take images of block walls, telephone poles, and vending machines, among others, and send them to a server. Because the images sent to the server include data on their latitude and longitude, the images can be mapped to Web-GIS according to the location information. This allows elementary school students to create their own digital "My Hazard Map."

2 Related Works

Okazaki et al. [3] developed a system that supports the creation of hazard maps by local residents themselves. Their proposed system, which targets local cities with historic townscapes, residents register information on points that could be dangerous in the event of a disaster, which then creates a hazard map by displaying the information on a map. This system realizes the sharing of detailed local information through posting by local residents themselves.

Tajima et al. [4] developed a disaster prevention map creation support system that uses smartphones with a GPS function to easily register comments, photographs icons, etc. at the current location on a map while walking around the town. Using their smartphone's GPS and camera functions, users can easily register information, such as comments, photos, and icons, at their current location on a map. Furthermore, they can use the chat function for sharing purposes and the editing function.

Tsuji et al. [5] developed a system to instruct children on what to do during disasters. They created a system in which a simulated computer environment allows parents to plan and prepare guidelines for their children on their way to and from school in the event of a disaster. Using the system, parents can register information regarding certain dangers and behavioral guidelines as their children travel through their routes to school, thus creating safe evacuation route maps for their children. When a child activates the system during a disaster, the information entered by the parent is displayed on a map, allowing the child to act quickly and effectively when disasters strike.

Okabe et al. [6] developed a disaster prevention map system using information technology (IT) that allows for the easy updating of information in response to the problem of disaster prevention maps published by local governments being poorly understood. This system provides two types of information registration screens: a basic and a detailed information registration screen, which make it possible to register more detailed information, such as the congestion status of evacuation centers and hospitals.

Kometani et al. [7] developed the crime prevention walking app called "Walking Miimai" as an approach to creating a town wherein people can live safely and with peace of mind. This system is equipped with two functions: a reporting and sharing function for safety/danger points. Here, users can take photos and register their safety/danger judgments and reasons for such judgments in the system. The GPS function allows registered reports with longitude and latitude data to be visualized on the regional safety map.

Enokida et al. [8] developed the web GIS system called the "Agara Map," which consistently supports the creation of disaster prevention maps while walking around town.

This system mainly implements several functions, including disaster prevention information and word-of-mouth registration, hazard map display, current location display, and altitude display functions, to achieve consistent support from collected information generated by users walking around town. Using such information, disaster prevention maps are created.

Hayakawa et al. [9] developed a community safety map system in which children can input information on places where they think crimes are likely to occur. The system also allows them to share the information encoded by the children. This system is equipped with functions for creating, viewing, editing, and printing regional safety maps. It also uses WebSocket technology to provide a collaborative work function that allows users to create maps while sharing real-time editing information.

3 Research Objective

In this study, we develop a "My Hazard Map System" designed to allow elementary school students to visualize dangerous places in the event of a disaster on their own school routes. In the proposed system, we introduce a method of mapping on Web-GIS from the latitude and longitude information attached to photographs when registering information about specific dangerous places. Users can register on the system's server-specific photographs of dangerous places along their school routes, as well as disaster-type icons, comments, and information about the dangerous spots identified. When registering disaster-type icons, we provide a function that allows users to understand at a glance what kind of disaster is considered dangerous by linking the disaster type to the photograph of the dangerous place. In addition, when registering comments on dangerous place information, users are provided with the ability to register comments related to the photographs of dangerous places and edit comments after posting them. Furthermore, by preparing a "disaster prevention site" that provides information on disaster prevention, the system can provide users with a function that gives easy-to-understand explanations about disaster prevention and what actions to take. With this proposed system, even elementary school students can easily create their own hazard maps that would allow them to visualize disaster risks in their own areas or school routes. Moreover, this proposed system aims to improve disaster prevention awareness by helping elementary school students develop appropriate behavior and decision-making skills during disasters.

4 My Hazard Map System Configuration and Architecture

The configuration of this system is shown in Fig. 1, and the system architecture is shown in Fig. 2. This system consists of a registration/browsing agent that registers photographs of dangerous places, disaster-type icons, and disaster information along with location information, and browses the registered information on Web-GIS or a list of dangerous places; a management server, which stores registered information about dangerous places in the database server and reflects it on Web-GIS; a database server that stores photographs of dangerous locations, latitude and longitude information obtained from the photographs, disaster-type icons, and disaster information, among others; and a map information provision API that visualizes registered information about dangerous places on Web-GIS).

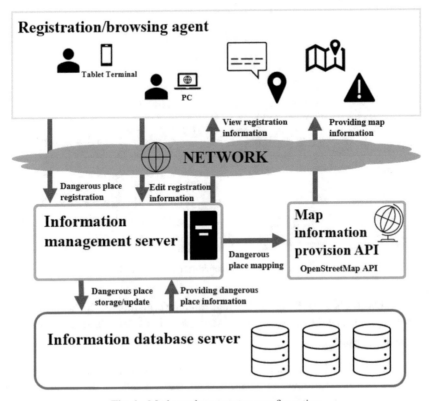

Fig. 1. My hazard map system configuration.

- Registration/Browsing Agent

The registration/browsing agent should be operated by elementary school students as part of their disaster prevention education. Students are asked to take photographs of places on their own school routes that are considered at risk in the event of a disaster. They then select the photographs they have taken and then input the latitude and longitude information automatically obtained from the photographs, the disaster-type icons, disaster information, and so on, and send them to the information management server. The data sent to the server are then stored in the database server. The data registered by the registration/browsing agent are reflected on Web-GIS, allowing the students to browse through them via a generated personal hazard map that visualizes dangerous places on their school routes. When a user selects a disaster-type icon that represents registered information on Web-GIS, the registered photos, disaster information, comments, etc. are displayed on a pop-up screen. In addition, as an auxiliary function, it is also possible to map the open data provided by the Geospatial Information Authority

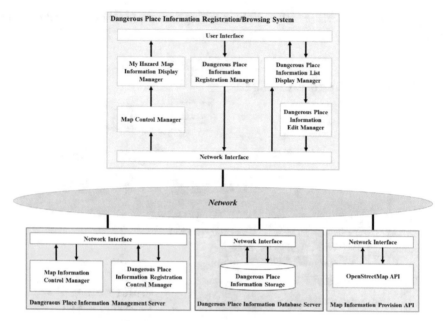

Fig. 2. My hazard map system architecture.

of Japan on Web-GIS. Furthermore, the registered information can be browsed in a list format that provides links to pages where users can edit or delete registered information.

- Information Management Server

The dangerous place information management server stores the registered and edited data by the registration/browsing agent. The server also reflects the dangerous place information on the Web-GIS and in the dangerous place list.

- Information Database Server

The dangerous place information database server stores photographs of dangerous places registered by the registration/browsing agent, along with the latitude and longitude data obtained from the photographs, disaster-type icons, disaster information, and so on.

- Map Information Provision API

The map information provision API uses the OpenStreetMap API [10] to visualize the dangerous place information registered by the dangerous place information registration/browsing agent on Web-GIS.

5 My Hazard Map System

5.1 Dangerous Place Information Registration Screen

When a user clicks "Register Dangerous Place Information" from the menu bar, the screen transitions to the dangerous place information registration screen shown in Fig. 3. They can then register information using the following steps:

STEP 1. After turning on the location information on tablet device, take a photo of the dangerous place using the camera application.

STEP 2. Upon clicking the "Select Photo" button, choose a photo of a dangerous place, and the latitude and longitude information attached to the selected photo will be displayed.

STEP 3. The user selects the icon corresponding to the dangerous place information. This system provides five icons: flood, earthquake, landslide, evacuation center, and danger (other).

STEP 4. Registration is completed when the user clicks the upload button.

Figure 4 shows an example of how to register information about dangerous places identified.

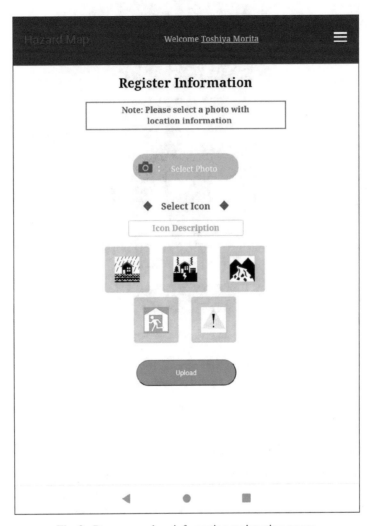

Fig. 3. Dangerous place information registration screen.

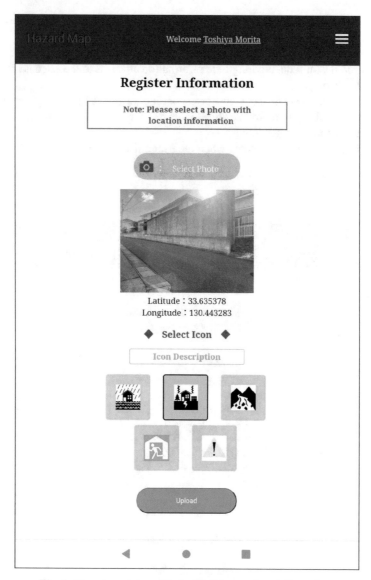

Fig. 4. How to register a piece of dangerous place information.

5.2 Map Screen After Registering Dangerous Place Information

When a user clicks on the icon on the map, the registered dangerous place information (disaster prevention information, latitude/longitude information, dangerous place image, and registration date and time) is displayed in a pop-up screen, as shown in Fig. 5.

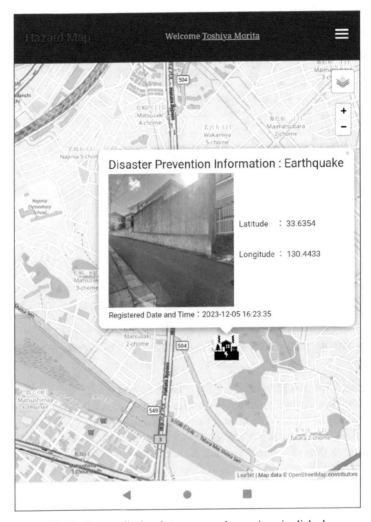

Fig. 5. Pop-up display that appears when an icon is clicked.

5.3 Registration Information Viewing Screen (List Format)

On the registered information viewing screen, images of registered dangerous places, icons, latitude and longitude data, disaster prevention information, comments, and registration date and time are displayed in a list format as shown in Fig. 6.

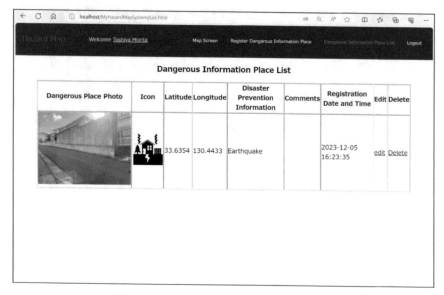

Fig. 6. Registered information viewing screen.

5.4 Map Auxiliary Function (Superimposed Display of Overlapping Hazard Maps)

We implemented the superimposed display of the overlapping hazard maps as a map auxiliary function. In particular, we used open data called "Overlapping Hazard Maps" [11], which are provided by the Ministry of Land, Infrastructure, Transport and Tourism of Japan. Here, users can select checkboxes for flood inundation risk area, flood inundation risk area (high wave or tsunami), flooding duration, debris flow risk area, steep slope, landslide risk area, and collapse risk area. As shown in Fig. 7, it is possible to display a hazard map superimposed on the actual map.

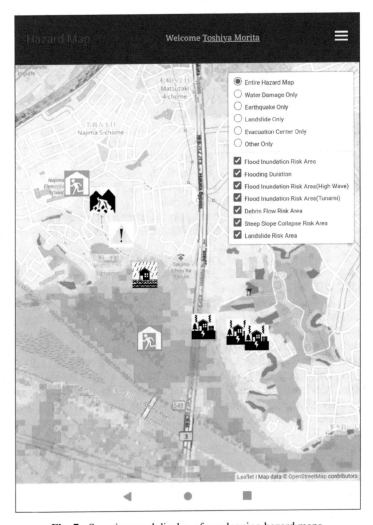

Fig. 7. Superimposed display of overlapping hazard maps.

6   Conclusion

A hazard map can be used to reduce damage caused by natural disasters and to introduce disaster prevention measures. However, hazard maps are not widely used, especially among the younger generation. Therefore, in this study, we developed the "My Hazard Map System" for elementary school students. The proposed system allows them to automatically create their own hazard maps by sending photographs of dangerous places on their way to school to the system server. By creating their own digitized hazard maps, elementary school students can learn about dangerous places on their way to school and know what to do in the event of a disaster.

References

1. Japan Institute of Country-ology and Engineering: Japan, a country with many natural disasters (2023). https://www.jice.or.jp/knowledge/japan/commentary09. Accessed Nov 2023
2. Geospatial Information Authority of Japan: Hazard Map (2023). https://www.gsi.go.jp/hokkaido/bousai-hazard-hazard.htm. Accessed Nov 2023
3. Okazaki, Y., Mori, S., Mishima, N.: Development of a regional hazard map generation support system based on the residents' questionnaire survey in a historic local town. In: Proceedings of the 40th National Convention of JSiSE, pp.75–76 (2015). (in Japanese)
4. Tajima, S., Murakami, Y., Uchida, O., Kajita, Y.: The development of a creation system for a disaster-avoidance map using a smartphone and the evaluation of educational practice. Japan J. Educ. Technol. **41**, 85–88 (2017). (in Japanese)
5. Tsuji, K., Nakatani, Y.: Parents support system of disaster planning for their children on the school route. In: Proceedings of the 73rd National Convention of IPSJ, pp. 4-657–4-658 (2011). (in Japanese)
6. Okabe, Y., Tanaka, Y, Notomi, K.: Prototyping of disaster prevention map system for the purposes of information sharing. In: Proceedings of the 74th National Convention of IPSJ, pp. 3-535–3-536 (2012). (in Japanese)
7. Kometani, Y., Isono, T., Yabe, T., Okubo, T., Takeshita, Y., Yaegashi, R.: Development of "Aruite Mi Mai," a walking application for crime prevention for safe and secure city development. JSiSE Res. Rep. **33**(7), 101–108 (2019). (in Japanese)
8. Enokida, S., Fukushima, T., Yoshino, T., Sugimoto, K., Egusa, N.: Proposal of town-walk type disaster-preparedness map making support system. In: Proceedings of the Multimedia, Distributed, Cooperative, and Mobile Symposium 2017, pp. 1721–1732 (2017). (in Japanese)
9. Hayakawa, T., Matsuda, K., Ito, T.: Prototype of community safety map system that can collaboratively edit using OpenStreetMap. IPSJ J. **59**(3), 1095–1105 (2018). (in Japanese)
10. OpenStreetMap: OpenStreetMap API (2022). https://wiki.openstreetmap.org/wiki/API. Accessed Nov 2023
11. Ministry of Land, Infrastructure, Transport and Tourism of Japan: Overlapping Hazard Maps (2023). https://disaportal.gsi.go.jp/maps/?ll=35.371135,138.735352&z=5&base=pale&vs=c1j0l0u0t0h0z0. Accessed Nov 2023

Automated Storytelling Technologies for Cultural Heritage

Luigi Colucci Cante[1], Beniamino Di Martino[1,2,3], Mariangela Graziano[1],
Dario Branco[1(✉)], and Gennaro Junior Pezzullo[4]

[1] Department of Engineering, University of Campania L. Vanvitelli, Aversa, Italy
{luigi.coluccicante,beniamino.dimartino,mariangela.graziano,
dario.branco}@unicampania.it
[2] Department of Computer Science and Information Engineering, Asia University,
Taichung, Taiwan
[3] Department of Computer Science, University of Vienna, Vienna, Austria
[4] University of Rome Campus Bio-Medico, Rome, Italy
gennaro.pezzullo@unicampus.it

Abstract. The awareness of the importance of communicating cultural heritage through innovative and automated means is steadily increasing. This awareness is fueled by the need to promote cultural heritage knowledge in accessible and engaging ways for an increasingly digital audience. This article aims to examine in detail the various methodologies used in the creation of automated storytelling for historical cultural heritage communication. In particular, attention is given to models based on large language models, such as those derived from advanced neural networks, which demonstrate significant potential in generating rich and engaging narratives. Another key aspect of the analysis concerns the use of chatbots in the context of automated storytelling. The possibilities offered by interactive conversation managed by virtual agents are explored, which can enhance the user experience through personalized dialogues and contextualized information. Additionally, the article focuses on various tools dedicated to automatic storytelling writing, evaluating their effectiveness and versatility in the specific context of historical cultural heritage. These tools are essential to facilitate the creative process and ensure narrative coherence. A significant part of the analysis addresses automated storytelling models based on processes, exploring how the sequence of events and concepts can be managed automatically to construct meaningful and relevant stories from a historical perspective. Furthermore, applications of ontologies and other tools are examined with the aim of improving the structure and understanding of narratives, ensuring effective linkage between different elements of historical cultural heritage. The overall goal of the article is to provide a comprehensive view of the current landscape of automated storytelling techniques, highlighting the still-open challenges and potential future development directions in this field.

1 Introduction

The introduction to automated storytelling technologies for cultural heritage unfolds as a captivating journey into the synergistic integration of various innovations, all aimed at

L. Barolli (Ed.): EIDWT 2024, LNDECT 193, pp. 597–606, 2024.
https://doi.org/10.1007/978-3-031-53555-0_57

enhancing and sharing the stories that enrich our historical and cultural heritage. Large Language Models (LLM) stand as fundamental pillars, capable of generating engaging and meticulously contextualized narratives through their advanced understanding of language. Concurrently, models based on semantics add layers of meaning and detail, enriching the narrative plot with a deeper comprehension of content. The ecosystem of automatic storytelling technologies also embraces the power of chatbots, intrinsically linked to advanced natural language models. These chatbots not only facilitate direct interaction with the audience, answering questions and providing detailed information, but they can also adapt to the context of cultural heritage, creating engaging and personalized conversations. In parallel, tools for automatic composition and writing emerge as allies in crafting unique narratives, allowing targeted customization and the formulation of stories that intricately weave with the fabric of cultural heritage. This convergence of technologies opens new perspectives, transforming automated storytelling into an art form that not only preserves and transmits our cultural heritage but does so in an accessible, engaging manner intrinsically linked to the expectations and preferences of the audience. Within this amalgamation of advanced models, semantics, chatbots, and composition tools, emerges a landscape of automated storytelling that harmoniously blends with the complexity of cultural narratives, promoting interactivity and active participation in this captivating exploration of our shared past.

2   Models

2.1   Models Based on Large Language Models

The complexity of LLM has revolutionized automated storytelling, particularly in the context of cultural heritage communication and valorization. Models rooted in LLM play a crucial role in advancing automated storytelling for cultural heritage. Underpinning technologies like Natural Language Processing (NLP) empower models to comprehend implicit meanings within texts, extracting pertinent information from cultural heritage to enhance narratives with nuanced details and historical contexts. Simultaneously, Machine Learning (ML) optimizes models, enhancing their adaptability to cater dynamically to user preferences and specific contexts. The convergence of these technologies provides a robust platform for deploying advanced solutions, enabling the generation of storytelling that go beyond mere historical facts, creating an immersive experience through engaging and personalized stories. Additionally, LLM-based models integrate automatic translation functionalities, facilitating global appreciation of cultural heritage. Various transformer-based models, such as GPT, BERT, XLNet, and RoBERTa, play a crucial role in automated storytelling and cultural heritage enhancement, offering unique features for diverse applications. These models contribute to improved user experiences and the preservation of historical and cultural richness. To compare their effectiveness in cultural heritage storytelling, considerations include creativity, historical context understanding, computational efficiency, and adaptability to cultural sector needs. **GPT (Generative Pre-trained Transformer)** [6], a pre-trained transformer model, excels in creative text generation, capturing nuances in natural language. Despite requiring significant computational resources, its pre-training simplifies cultural heritage application development, providing adaptability across tasks [18].

BERT (Bidirectional Encoder Representations from Transformers) [17], known for bidirectional context understanding, is efficient for natural language processing and context-dependent tasks. It is more focused on context understanding than creative generation, making it suitable for creating stories based on historical information with fewer resource requirements. **XLNet (Transformer-XL Network)** [19], integrating bidirectional and long-term dependency capabilities, excels in understanding longer sequences, making it suitable for more detailed narratives. Despite demanding computational resources, its advantage lies in managing complex stories. **RoBERTa (Robustly optimized BERT approach)** [12], an optimized BERT implementation, maintains language understanding capabilities with a focus on efficiency. Like BERT, it is suitable for context understanding and story generation based on existing data, optimized for resource usage. In conclusion, the optimal model selection for automated storytelling in cultural heritage depends on specific application requirements. GPT showcase creativity and adaptability, while BERT and RoBERTa excel in historical context precision. XLNet's capacity for longer sequences suits comprehensive narratives. The decision should carefully balance creativity and historical accuracy, considering available computational resources and project requirements.

2.2 Chatbot

Interactive storytelling, in cultural heritage settings, is strongly supported by the use of chatbot technology. A chatbot is a conversational agent that, trained appropriately, provides the user with virtual guidance toward a structured, engaging, and coherent narrative in cultural contexts. Their strong development is certainly due to their clear ability to adapt the proposed content based on the conversations held with users. Chatbots based on **Google Dialogflow** or **RASA** demonstrate how, using NLP techniques, it is possible to create compelling narratives with valid and reliable content. Typically, chatbots rely on a series of intents, stories, and actions. An intent corresponds to a precise user request, and each intent, in the chatbot's knowledge base, corresponds to a set of predetermined sentences (and their variants). Intent recognition is a key component, enabling the chatbot to understand the user's goals or requests. For instance, if a user shows interest in a particular artifact, the chatbot identifies the intent and adjusts the narrative accordingly. In the context of chatbots and stories, we are referring to pre-defined narrative sequences that the chatbot follows based on the user's messages, typically a series of recognized intents. These stories are crafted to lead the user through a meaningful and purposeful conversation [1]. With the ability to define paths that deviate from the main story, conversational agents can modify the narrative based on choices made by users, providing a personalized and interactive experience. Finally, actions represent operations that the conversational agent can perform as a result of the user's recognition of intent that involves more than simply sending a message. Actions can include executing code, calling APIs, interacting with third-party components, etc. Consider a conversational agent developed for guiding users through a museum tour. When a user expresses interest in a specific artwork, the chatbot can initiate a detailed narrative, covering aspects like the author, material, and artistic influences. During the interaction, using actions, the conversational agent can prompt other system components to play relevant audio or video files, enhancing the user's understanding of the artwork.

Even with their great potential, the use of chatbots also presents challenges. Several weaknesses of this technology can range from potential limitations in user comprehension of input to the possible lack of context if the chatbot is not perfectly trained to the current technological limitation that mandates the use of dyadic conversations only [2]. Despite this, several advances in NLP and ML are underway, offering promising avenues for overcoming these limitations. As an example, the integration of traditional chatbots with emerging technologies such as LLM can contribute to the creation of more sophisticated and robust conversational agents capable of responding consistently even to user requests that perhaps do not sufficiently match any of the chatbot's intent, improving the overall user experience.

2.3   Semantic and Process-Based Models

Several models of storytelling representation based on semantics are available in the literature. In particular, with the use of ontologies it is possible to provide better logical structuring of the story and greater interoperability of information, enabling better management and generation of content for storytelling. The study [4] introduces a storytelling ontology developed in Ontology Web Language (OWL), drawing inspiration from existing models like the **Rhetorical Structure Theory Ontology** [14] and the **Common Narrative Model Ontology** [5]. Furthermore, [4] presents an overview of utilizing process-based models to depict a story. Other works, such as [11, 13], contribute by presenting an ontology for narrative representation, accompanied by tools for constructing narratives based on the ontology and querying stories through a SPARQL querying system. Meanwhile, [7, 8] outline a methodology that combines the process-based and semantic approaches, enabling the semantic annotation of structural elements in a process model (e.g., scenes, events, or actions) with domain concepts from an OWL ontology.

3   Tools for Automated Storytelling

3.1   Story Composition Tools

To develop automated storytelling systems, it is frequently essential to establish a structured, cohesive, and coherent representation of the narrative. Utilizing story composition tools can significantly aid scriptwriters in the creation process of storytelling. This section offers an overview of the cutting-edge story composition tools designed to assist scriptwriters, with a particular focus on those that provide a user-friendly graphical interface for crafting stories. The work [9] presents an web-based tool named **Egon.io**[1], which is a lightweigh tool that allows one to create models of domain storytelling to represent through graphical notation the functional requirements of a software system. Domain Storytelling is a collaborative, visual, and agile way to build domain-driven software. Egon.io is an open-source online tool for visualizing domain stories, that is an extension version of *BPMN.io*[2] that provide the possibility through a similar formalism to represent not only processes, but any graphical element. In fact, the tool allows

---

[1] https://egon.io/.
[2] https://bpmn.io/.

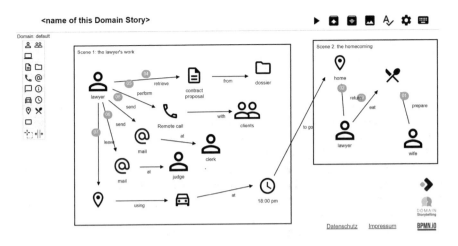

Fig. 1. Domain story modeled with the Egon.io Tool.

the user to compose a story using the elements in a graphical palette, which can be customized according to the user's preferences. Upon completion of creation, the tool also offers a very simple story "run" mode. Figure 1 shows an example of a domain story modeled with the Egon.io tool. As can be seen from Fig. 1, Egon.io allows to explicitly model scenes and scene changes, which usually occur when a change of location or time occurs, or when new characters enter the scene. The work [3] describes an open-source tool for telling interactive and nonlinear stories named **Twine**[3]. Twine enables users to build projects by incorporating pages or scenes that are interconnected through links, dictating the storyline's progression. Each scene is customizable with text, images, videos, or various multimedia elements. Users have the flexibility to insert

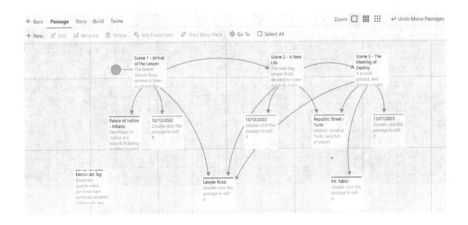

Fig. 2. Story modeled with the Twine Tool.

---

[3] https://twinery.org/.

decision points within scenes, altering the story's direction based on the choices made by the audience. Twine then generates an HTML file, which can be shared or published online, facilitating easy dissemination of the project to a broad audience. Twine is supported by a large community and the *"IFTF (Interactive Fiction Technology Foundation)*[4]", a nonprofit organization that provides infrastructure support for the interactive fiction community. Figure 2 shows an example of a story, consisting of three scenes, modeled using the Twine tool. Through the use of tags, Twine enables the definition of custom types of story elements. Several tools are employed in game design to facilitate story modeling, and among them, **Arcweave**[5] tool is one of the most widely used. As described by work [16], Arcweave is a cloud-based application designed for creating and prototyping games directly within a web browser. The game's storyline and logic are visualized through diagrams across multiple boards, connecting nodes and conditions. Subsequently, users can test the game prototype in play mode. Arcweave seeks to close the gap between writers and developers by implementing logic while maintaining a user-friendly interface for writers. Additionally, it facilitates real-time collaboration for teams of two or more. Figure 3 shows an example of a story modeled using the Arcweave tool. One of the most commonly used tools for narrative design of a video game is **Articy**[6], described in work [10]. Articy is a versatile software application tailored for creating and organizing diverse narrative content, including screenplays, video game stories, drama scripts, and novel writing. Its intuitive user interface empowers creators to structure their narrative seamlessly, managing plot, character development, and environment creation efficiently. Notably, Articy automates the generation of reference and technical documentation for development teams, facilitating clear communication. Fostering real-time collaboration, Articy ensures easy synchronization of projects among creatives. It allows users to define custom palettes for actors, locations, and scene objects, providing a high degree of personalization. The tool also supports the

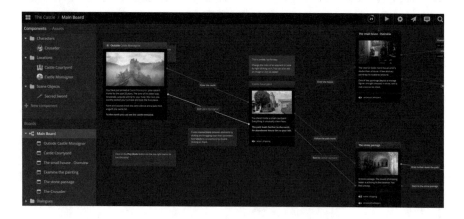

Fig. 3. Story modeled with the ArcWeave Tool.

---

[4] https://iftechfoundation.org/.

[5] https://arcweave.com/.

[6] https://www.articy.com/en/.

creation of custom maps for locations, enabling users to specify the positions of scene objects and actors throughout the story. Figure 4 shows an example of a story modeled using Articy. Table 1 presents a comparison of the features offered by the previously described tools.

3.2 Automatic Writing Tools

The strong demand for this type of tool has led to numerous efforts for the development of such tools both in the academic field and in industry. Among the main factors that support this process we must certainly mention: efficiency, as these tools allow the production of even large quantities of written content in a relatively short time. Furthermore, they are very versatile technologies capable of adapting to any context, from technical and formal writing to more creative and confidential writing. Finally, a final factor that has allowed the spread of this technology is accessibility: these tools are increasingly integrated into user-friendly platforms and consequently are easily accessible even to users who do not have a technical-scientific training. However, despite these advances, the development of such tools still faces numerous challenges, not only in the technical field but also from an ethical point of view. Certainly, among the most obvious problems is the lack of precision or coherence, especially in specialized contexts. Furthermore, there are also quite a few concerns regarding bias and privacy [15] and, last but not least, it should be emphasized that the use of such tools can reduce creativity and critical thinking in the long term. Currently, the market has many tools, some open-source, others with advanced functions accessible only for a fee. In order to illustrate the characteristics of some of the most used tools, a comparative analysis summarized in Table 2 was carried out.

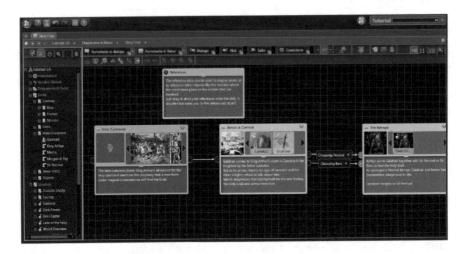

Fig. 4. Story modeled with the Articy Tool.

Table 1. Comparison of Story Composition Tools

Feature	Egon	Twine	Arcweave	Articy
Creating Custom Story Elements		X	X	X
Modelling Story Flow	X	X	X	X
Representation of Times	X	X	X	X
Representation of Locations	X	X	X	X
Representation of Scene Objects	X	X	X	X
Play Mode	X	X	X	X
Creating Palettes of Actors and Locations			X	X
Real-time collaboration for teams			X	X
Define World's Map and the Pathways of History				X
Mapping Actors and Scene Objects to Locations during Story				X

Table 2. Comparison of capabilities among various Automatic Writing Tools.

Capability	Smodin	ChatGPT	Grammarly	DeepL	Bard AI	QuillBot
Question Answering	X	X			X	
Text Generation	X	X	X		X	X
Translation	X	X	X	X	X	X
Text Re-elaboration	X	X	X	X	X	X
Grammatical Correction	X	X	X	X	X	X
Plagiarism Check	X		X			X
Text Summarization	X	X			X	X
Stylistic Suggestions	X	X	X	X	X	X

The comparison between the different writing platforms highlights a constant trend which is the continuous increase in the set of functions available for each tool. However, this homogeneity in the provision of these functions does not make the tools equivalent since factors such as accuracy, precision or recall stand out as critical in evaluating the effectiveness of these tools. While all platforms adhere to a high standard of service, there are notable distinctions and specialized areas of expertise in managing tasks.

4   Conclusion and Future Works

In conclusion, the importance of communicating cultural heritage through innovative and automated means has gained significant attention due to the increasing demand for accessible and engaging ways to promote cultural heritage knowledge for a digital audience. This article has explored various methodologies used in the creation of automated storytelling for historical cultural heritage communication, with a focus on models based on LLM and chatbots, automatic storytelling writing tools, semantic-based and process-based models for storytelling and story composition tools. As technology continues to advance, it is imperative that we continue to explore new and innovative ways to promote cultural heritage knowledge, ensuring that it remains accessible and engaging for future generations.

Acknowledgements. The work described in this paper has been supported by the research project RASTA: Realtà Aumentata e Story-Telling Automatizzato per la valorizzazione di Beni Culturali ed Itinerari; Italian MUR PON Proj. ARS01 00540.

References

1. Amato, A., Aversa, R., Branco, D., Venticinque, S., Renda, G., Mataluna, S.: Porting of semantically annotated and geo-located images to an interoperability framework. In: Barolli, L. (eds.) Complex, Intelligent and Software Intensive Systems. CISIS 2022. LNNS, vol. 497, pp. 508–516. Springer, Cham (2022). https://doi.org/10.1007/978-3-031-08812-4_49
2. Ambrisi, A., et al.: Intelligent agents for diffused cyber-physical museums. In: Camacho, D., Rosaci, D., Sarné, G.M.L., Versaci, M. (eds.) Intelligent Distributed Computing XIV. IDC 2021. Studies in Computational Intelligence, vol. 1026, pp 285–295. Springer, Cham (2022). https://doi.org/10.1007/978-3-030-96627-0_26
3. Carlon, M.K.J., et al.: Educational nonlinear stories with twine. In: Proceedings of the Ninth ACM Conference on Learning@ Scale, pp. 248–251 (2022)
4. Colucci Cante, L., Di Martino, B., Graziano, M.: A comparative analysis of formal story-telling representation models. In: Barolli, L. (ed.) Complex, Intelligent and Software Intensive Systems. CISIS 2023. LNDECT, vol. 176, pp. 327–336. Springer, Cham (2023). https://doi.org/10.1007/978-3-031-35734-3_33
5. Concepción, E., Gervás, P., Méndez, G.: A common model for representing stories in automatic storytelling. In: 6th International Workshop on Computational Creativity, Concept Invention, and General Intelligence. C3GI, Madrid. Spain vol. 12, no. 2017, 2017 (2017)
6. Davis, G., Grierson, M., et al.: Investigating attitudes of professional writers to GPT text generation AI based creative support tools (2022)
7. Di Martino, B., Colucci Cante, L., Esposito, A., Graziano, M.: A tool for the semantic annotation, validation and optimization of business process models. Softw. Pract. Exp. 53(5), 1174–1195 (2023)
8. Di Martino, B., Graziano, M., Colucci Cante, L., Esposito, A., Epifania, M.: Application of business process semantic annotation techniques to perform pattern recognition activities applied to the generalized civic access. In: Barolli, L. (ed.) Complex, Intelligent and Software Intensive Systems. CISIS 2022, LNNS, vol. 497, pp. 404–413. Springer, Cham (2022). https://doi.org/10.1007/978-3-031-08812-4_39
9. Hofer, S., Schwentner, H.: Domain Storytelling: A Collaborative, Visual, and Agile Way to Build Domain-Driven Software. Addison-Wesley Professional, Boston (2021)
10. Horst, R., Lanvers, M., Dörner, R.: Quest-centric authoring of stories, quests, and dialogues for computer game modifications. In: Paljic, A., Ziat, M., Bouatouch, K., (eds.), Proceedings of the 17th International Joint Conference on Computer Vision, Imaging and Computer Graphics Theory and Applications, VISIGRAPP 2022, Volume 2: HUCAPP, Online Streaming, 6–8 February 2022, pp. 217–224. SCITEPRESS (2022)
11. Lenzi, V.B., Marcelloni, F., Meghini, D.C., Doerr, D.M., Luise, M.: An ontology for narratives, Franco Niccolucci (2017)
12. Liu, Y., et al.: Roberta: a robustly optimized bert pretraining approach. arXiv preprint arXiv:1907.11692 (2019)

13. Meghini, C., Bartalesi, V., Metilli, D.: Representing narratives in digital libraries: the narrative ontology. Semant. Web **12**(2), 241–264 (2021)

14. Nakasone, A., Ishizuka, M.: Storytelling ontology model using RST. In: 2006 IEEE/WIC/ACM International Conference on Intelligent Agent Technology, pp. 163–169 (2006)

15. Nelson, G.S.: Bias in artificial intelligence. North Carolina Med. J. **80**(4), 220–222 (2019)

16. Nowak, L., Grabska-Gradzińska, I., Palacz, W., Grabska, E., Guzik, M.: Tool for game plot line visualization for designers, testers and players. In: Luo, Y. (ed.) Cooperative Design, Visualization, and Engineering. CDVE 2023. LNCS, vol. 14166, pp. 85-93. Springer, Cham (2023). https://doi.org/10.1007/978-3-031-43815-8_8

17. Rothman, D.: Transformers for Natural Language Processing: Build Innovative Deep Neural Network Architectures for NLP with Python, PyTorch, TensorFlow. RoBERTa, and more. Packt Publishing Ltd, BERT, Birmingham (2021)

18. Yang, X., Tiddi, I.: Creative storytelling with language models and knowledge graphs. In: CIKM (Workshops) (2020)

19. Yang, Z., et al.: Xlnet: Generalized autoregressive pretraining for language understanding. In: Advances in Neural Information Processing Systems, vol. 32 (2019)

Support for Automated Story Telling Using Natural Language Processing Techniques Aimed at Recognizing Narrative Elements

Beniamino Di Martino[1,2,3], Gennaro Junior Pezzullo[1,4(✉)], and Emanuela Grassia[1]

[1] Department of Engineering, University of Campania "Luigi Vanvitelli", Caserta, Italy
beniamino.dimartino@unicampania.it,
emanuela.grassia@studenti.unicampania.it
[2] Department of Computer Science and Information Engineering, Asia University, Taichung City, Taiwan
[3] Department of Computer Science, University of Vienna, Vienna, Austria
[4] Department of Engineering, University of Rome "Campus Bio-Medico", Rome, Italy
gennaro.pezzullo@unicampus.it

Abstract. Exploring the cutting-edge field of artificial intelligence, this work delves into how this emerging technology can be applied to the field of automated storytelling. The objective is to identify a clear and well-defined methodology capable of analyzing a text, recognizing the narrative elements within the story and, based on this, structuring the narrative into sequences. All these processes are to be carried out in an automated manner. Subsequently, based on this methodology, an application will be implemented that introduces a new algorithm that integrates natural language processing and unsupervised machine learning techniques. Finally, through a series of experiments on historical texts and the use of metrics, the effectiveness and potential of the proposed model will be demonstrated.

1 Introduction

According to what is reported by the Treccani Encyclopedia, storytelling is the art of writing or telling stories capturing the attention and interest of the public and is a discipline that uses the principles of rhetoric and narratology. When we talk about narratology we are referring to the part of literary theory that deals with the processes of representation and narrative communication. Although narratology was born as a particular form of analysis of the written text, starting from the end of the 1960s the study of the elements of narration has also been extended towards other expressive forms: from comics to theatre, from music to cinema. Within the structure of a narrative text different phases can be identified: background, initial situation, breakdown of balance, evolution of the story or development, spanning, resolution and conclusion. Although a narrative text presents this structure, there is an element that is fundamental to detect within it and this is the sequence [8]. The sequence is a segment that presents a certain autonomy of content, a sequence is each portion of the narrative text that differs from the previous or subsequent one for the introduction of new elements in the narrative.

L. Barolli (Ed.): EIDWT 2024, LNDECT 193, pp. 607–616, 2024.
https://doi.org/10.1007/978-3-031-53555-0_58

Over the years, especially with the arrival of machine learning and more generally artificial intelligence, in many sectors there is a desire to "automate" increasingly complex operations that require underlying reasoning that is not limited to a simple case-based algorithm [10]. In this context, the technology that lends itself to this type of operation is Natural Language Processing (NLP) which refers to the automatic processing of information written or spoken in natural language. This involves processing natural language datasets using rule-based or probabilistic machine learning techniques. When we talk about natural language it refers to the language that human beings commonly use every day to exchange information with the aim of creating a bridge between it and the language of computers [3].

2 State of the Art

2.1 Storytelling

Generally, storytelling is present in everyday life: at the table with the family to share experiences; through fairy tales to teach values; through journalism to communicate important events, and in entertainment films, and computer games for fun. Stories also motivate people to learn, which is why they form the backbone of training scenarios and case studies at school or at work. Therefore it is a methodology that uses narration as a means created in the mind to frame events in reality, explain them according to a logic that makes sense and portray them through words, images, sounds, videos. To date, storytelling plays a fundamental role because it is the simplest way we have to communicate, transmit information and knowledge but also to persuade and "format people". It is seen as a collaborative activity in which there is a narrator and a listener and the human brain understands the narrated stories much more easily rather than the logical-mathematical processes. Much of this content is communicated using natural language by creating narratives that refer to stories involving multiple actors and events, which occur in different places according to a temporal sequence. The composition of a narrative is given by the entities that participate in the story, by the events of which they are agents and by the temporal data that define the sequence of narrated events. In this sense, as stated by Toolan [12], a narrative can be seen as a sequence of nonrandomly connected events, a sequence of interconnected facts observed over a period of time involving basic elements such as organizations, people , places or time. There is a variety of existing research and narrative is applied in the most disparate fields and to different types of data ranging from images and videos to written texts. In this case the focus is on techniques that work with narrative texts, identifying different strands such as:

- **Automatic story generation**: machines create and tell stories with attempts since 1843 [2].
- **Extraction of the narrative structure**: the idea is to identify all the elements that constitute the narrative text [5].
- **Understanding of story**: important work has been done to teach machines how to understand and answer questions using unstructured text [6].
- **Classification of narrative text**: the idea is to find out which category the narrative text belongs to in order to have a clear idea of the type of narrative text [11].

2.2 Natural Language Processing

NLP had an enormous development over the years as everything that surrounds us represents an infinite source of data and this is why we want to equip machines with linguistic knowledge in order to design computer systems that can help humans in the classics linguistic tasks such as translation and document and/or knowledge management. Another objective is to create computer systems capable of automatically extracting useful information from texts or other media that use natural language [7]. A system for the analysis of natural language is made up of various blocks such as first a phoneme recognition system which works in parallel with a character recognition system as it allows both written and oral texts as input; a macro-block that performs lexical analysis; a macro-block that performs syntactic analysis and finally a macro-block that performs semantic analysis. The process of converting data into something that a computer can analyze is called **preprocessing**. The preprocessing activity consists of the following steps:

- **Tokenization**: subdivision of the text into single words called tokens.
- **Removal of stopwords**: removal of meaningless words also called empty words, i.e. clauses, conjunctions, pronouns, commonly used verbs.
- **Removing punctuation, numbers and spaces**
- **Stemming**: reduction of the word to the basic form called stem (root).
- **Lemmatization**: reduction of the word from the inflected form to the canonical form called lemma.
- **Normalization**: Reduce words from uppercase to lowercase.

Part of Speech Tagging (POS) [13] is also part of NLP, which is a process of identifying the part of speech of each word within a text. **Named Entity Recognition (NER)** [4] or the process of identifying the class to which a word belongs within a text document. NER involves the detection and categorization of named entities, a term used to refer to keywords in a text such as people, places, companies, events, products and many others.

3 Objective

Following the analysis of what has been carried out up to now in the research field, the main objective of this work has been the analysis of narrative texts aimed at identifying the main characterizing elements using Machine Learning and Natural Language Processing techniques in order to develop an algorithm that would allow an automated subdivision of a text into sequences on the basis of a multi-objective function that reflects what was identified in the state-of-the-art analysis. The sequence has well-defined boundaries and the rule that describes the transition from one sequence to another is very clear. To achieve the objective set, it was possible to organize this project into sub-objectives to be completed:

1. **Pre-processing and data search**: A preliminary search was conducted for the data to be analyzed, and subsequently, the preprocessing techniques cited in the state-of-the-art were applied.

2. **Named Entity Recognition (NER)**: it was necessary to perform an initial analysis of the text by filtering the entities representing some elements of the narrative text.
3. **Analysis of the structure of the dates**: following the identification of the "date" entity it was necessary to reconstruct the structure of this date in order to be able to understand if there were multiple instances that referred to the same time interval .
4. **POS (Part-of-Speech) analysis and identification of pseudonyms**: following the identification of the "people" entity, it became necessary to develop a technique that would allow the clustering of different names referring to the same actor with the aim of creating as many clusters as there are identified actors and to insert within the same group all the pseudonyms.
5. **Identification of the sequence**: by exploiting the characteristics that establish the transition from one sequence to another, a rules approach was used in order to identify the structure of the narrative text in terms of sequence.
6. **Validation of results**: after defining the evaluation metrics used, they will be applied to the results obtained in order to obtain an estimate of the correctness of the algorithm created.

4   Methodology

In order to achieve the objectives listed above, a clear methodology has been defined and shown in Fig. 1 which will be described in this chapter.

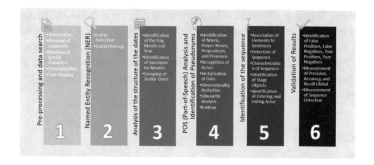

Fig. 1. Description of the methodology applied to achieve the objectives

Pre-processing and data search was preliminarily carried out to be subjected to the extraction of the narrative elements. Due to the lack of datasets in Italian labeled with the elements of the narrative text, it was decided to work directly on the texts provided by the RASTA project. Following the choice of data, these were pre-processed through: tokenization activities, removal of stopwords and special characters, lemmatization and text cleaning.

A passage between one sequence and another is identified by the change of actors, by the change of place, by the temporal change. Based on this, in **Named Entity Recognition (NER)** the entities extracted from the text are: actors, places, dates and times. Once the narrative text has been provided by the user, the entities defined by Named Entity Recognition are recognized. But the entities needed for the objective are only a subset of those automatically recognized, so only those previously mentioned are filtered and extracted. The extraction occurs through the respective labels and once filtered, they are set aside to achieve subsequent objectives.

In the **analysis of the structure of the dates** all the instances highlighted by the "DATE" label recognized in the previous objective stored in a text file were used. An analysis is considered necessary as it was rather restrictive to validate two dates as identical if the strings that identify them correspond. This phase deals with minimizing errors related to the comparison of similar dates identified in the text with different strings or due to spelling errors. In detail, in this phase information such as *day*, *month* and *year* are extracted where present. Furthermore, for the *month* category, any synonyms used in the text are also identified. Finally, dates with similar characteristics were included in the same set in order to evaluate them as referring to the same period of time.

In the context of historical texts or texts that refer to real people, it is common to refer to the same person in a different way, for example, specifying only the first name, only the surname or full name or mentioning that person with the title that he or she holds or with a diminutive or even with a nickname. This represents a limit for the comparison since it does not allow using a simple textual comparison to identify the same actors but requires a preliminary phase of **POS (Part-of-Speech) analysis and identification of pseudonyms** used in the text to refer to each participating actor. For the recognition of the "Actor" entities, in addition to making use of the Named Entity Recognition labels, the labels that identify the Parts of Speech were used. To cluster the data, this textual information must necessarily be translated into numerical format and this is why this data was vectorized using the *"Term Frequency - Inverse Document Frequency"* [1]. In order to eliminate redundant and/or useless data and to obtain a graphical display, it is necessary to carry out a dimensionality reduction where the number of numerical features is greater than two. But quite important information is the number of clusters into which to divide the starting data obtained using *Silhouette Analysis* [9]. Finally, the clustering was obtained by exploiting the *K-Means* technique. However, the results obtained were validated through human intervention. To **identify the sequence**, all the elements capable of defining it through the previous objectives were identified. At this point, each recognized entity is associated with the respective sentence to which it belongs. Afterwards, by examining the various entities phrase by phrase, any variations are discovered and consequently a passage between one sequence and another. An initial comparison is carried out on subsequent sentences but this work is subsequently completed by a comparison between groups of non-consecutive sentences in the text in order to identify a continuation of the same sequence in portions of text that are not necessarily successive. Following the identification of the sequence, a characterization of the elements that define it was carried out including *scene objects* and the characterization of the actors as *entering* and *leaving*.

5   Implementation and Testing

To accomplish this objective, *Python* was used as the programming language and the main Python libraries used were: *Numpy*, *Pandas* , *Scikit-learn*, *NLTK* and *Spacy*. Instead, *Google Colab* was used to implement the project.

In the **pre-processing and data search** phase, the data was pre-processed using libraries such as NLTK and Spacy.

For **Named Entity Recognition**, a Spacy model optimized for the NER activity was used, whose name is *it_nerIta_trf*, based on a pipeline composed by the *hseBert-it-cased* transformer. This model gives the possibility of recognizing eighteen different entities: PER, NORP, ORG, GPE, LOC, DATE, MONEY, FAC, PRODUCT, EVENT, WORK_OF_ART, LAW, LANGUAGE, TIME, PERCENT, QUANTITY, ORDINAL, CARDINAL and MISC. In this phase, the text provided as input by the user is processed in the form of a text file whose name is requested on the command line and opened using the *open* function by specifying three parameters: file name, *newline* and the encoding parameter. Instead, reading occurred via the *read* function by storing its textual content in a variable. After reading the input, the Spacy model was applied on the previous variable, returning an object of type *Doc* as output. It was possible to perform the entity recognition activity using the *ents* attribute of the object Doc. In the next phase, a filtering of the entities was carried out by selecting only the entities of interest for the objective to be completed whose labels were specified in an appropriate variable. The recognized entities were written to the respective file and stored in variables to be processed further later.

In the **analysis of the date structure** phase, a text file was read and the contents were extracted in a list using the *split* function. The extracted information was stored within a DataFrame initially made up of only two columns:"words" containing the elements read, "cleaned" containing the elements subjected to the pre-processing function. Subsequently, the nlp function (Spacy module) was applied to each individual "date" element in order to check whether the token of the "date" element corresponded to a day, a month or a year and in the case of the month the cluster index to which it belongs has been associated. It was also checked whether synonyms resulting from possible spelling errors had been used for the month. To do this, the *levehenstein_distance* function of the *jellyfish* library was used with the aim of finding out whether the number of different characters between the element under examination and the months of the year had been less than or equal to 1. If this had been the case, that element would have fallen into the "month" category and a cluster would have been defined consisting of the month and the corresponding synonymous identified. However, if this further verification had failed, he would have been included in the "other" list because he did not belong to any category. To group similar dates, it was first checked whether the value associated with the day, month and year was equal to the "undefined" string, because in this case it would not be a date to be grouped with others but would was a date belonging to a set consisting only of itself. For all other values to be examined the following check was performed: *if index i not examined value and index j not examined value and cluster identifier month i == cluster identifier month j and (day i == day j or day i == "undefined" or day j == "undefined") and (year i == year j or year i == "undefined" or year j == "undefined")*. In the end, the final DataFrame contained the following information: day, month, month cluster, and year of the given instance, other words, and the identifier of the cluster to which that instance belongs.

In the **POS (Part-of-Speech) analysis and identification of pseudonyms** phase, the actors were preliminarily recognized through the Named Entity Recognition activities of the optimized Spacy model and Part Of Speech Tagging of the classic Spacy model. In the second case, rules were defined that provided for inserting into this category all the instances in the text consisting of **NOUN** followed by **ADP** followed by **PROPN** or the instances constituted belonging to the category **NOUN** or **PROPN** as long as the respective strings fell within specific appropriately defined lists. The first thing that was done was the creation of a DataFrame containing three columns: "words" containing the values of the list, "cleaned" containing the values of the list cleaned thanks to the application of the pre-processing function, "lower" containing the values of the list in lowercase. For the data vectorization phase, *TfidVectorizer* was used which applies the TF-IDF technique on textual data. Subsequently, a dimensionality reduction was performed where the number of features was greater than two using *Principal Component Analysis* and a Silhouette Analysis. For the clustering execution, the value that maximizes the silhouette coefficient was taken into consideration as the number of clusters to be fed to the *KMeans* function of scikit-learn. Although the results obtained were partially correct, they required human intervention carried out through element-by-element comparison of each cluster, merging similar clusters and eliminating incorrect clusters. Once the validation was completed, the information regarding the cluster to which it belongs was inserted in the "new cluster" column of the DataFrame declared at the start.

In the **identification of the sequence** phase, an association of the entities with the respective sentences was carried out as the reasoning used to identify the sequence involves checking whether there have been any changes in moving from the current sentence to the next one. To identify the change of sequence it was enough to satisfy one of the following four conditions: *if (cluster set of the sequence != cluster set of the next sentence and cluster set of the next sentence != empty) or not all locations of the next sentence in location of the sequence or not all cluster of the dates of the next sentence in cluster date of the sequence or (next sentence time != empty and current time != next sentence time).* Based on the satisfied condition, the respective element of the "sequences_motivation" dictionary was set. In the event that the motivation had been the change of actors, the incoming actors would have been identified who are the uncommon ones between those of the set of sentences constituting the sequence and those of the following sentence. The remaining information set to carry out the characterization of the sequence was: the portion of text; the identifiers of the belonging sentences; the list of participating actors; the identifiers of the corresponding clusters containing the actors; the pseudonyms of the participating actors; places; date; the corresponding cluster identifier containing the date; time; the scene object. To identify the exiting actors, some additional processing was carried out which involves the recognition of verbs that could testify to the exit of a character such as verbs of movement or the verb "to die". For each verb identified, it was checked that the subject of that verb was not present in the following sentence. Finally, to complete the identification of the sequence, it was checked whether there were multiple sequences that had the same characteristics (actors, places, dates and times) and in case of a positive outcome, the user was asked if those two sequences should be merged in a single sequence. If the user's response had been in favor of the union, the second sequence would have been eliminated and the information from the new one would have also been inserted in the first, in particular the new sentences and their relative indices.

6   Validation

In the evaluation of algorithms, particularly in the context of classification tasks, several metrics are employed. These include the four fundamental outcomes: True Positive (TP), True Negative (TN), False Positive (FP), False Negative (FN). based on this it' s possible to mathematically define accuracy, precision and recall in the following ways:

$$\text{Accuracy} = \frac{TP+TN}{TP+TN+FP+FN} \quad \text{Precision} = \frac{TP}{TP+FP} \quad \text{Recall} = \frac{TP}{TP+FN}$$

The values obtained for the three selected metrics were the following:

- Global Accuracy = 0.82
- Global Precision= 0.85
- Global Recall = 0.83

These values were obtained by considering the four entities that discriminate the passage from one sequence to the next. Taking into account these values obtained it was possible to globally evaluate the recognizing of the sequences and it can be stated that this value is equal to 0.82. Furthermore, to have a clearer vision of what may be the most difficult elements to identify compared to others, in addition to a general measurement, more specific vertical measurements were carried out on the individual elements. Below we present a table for each measurement showing how the various parameters were measured Table 1. These are just some of the measurements that were made to evaluate the algorithm. But already by observing these we can draw some important conclusions. In particular, we can note that the dates and times return much more reliable results than places and actors. This result was expected as the latter ones, due to the presence of nicknames, pseudonyms and much more, are more difficult to identify than a string that refers to date or time. Furthermore, in many cases the accuracy and precision of the actors decrease. This is due to the higher presence of false positives. The reason for this incorrect classification is often caused by the fact that in some texts the subject is often not a person but an object.

Table 1. A comparison between different written texts of precision accuracy and recall regarding the identification of actors, dates, times and places

Entità	Text1			Text2			Text3		
	Accuracy	Precision	Recall	Accuracy	Precision	Recall	Accuracy	Precision	Recall
Attore	0.85	0.96	0.88	0.58	0.69	0.79	0.71	0.83	0.82
Luoghi	0.84	0.85	0.88	0.86	0.75	1.00	0.82	0.68	0.92
Date	1.00	1.00	1.00	0.94	0.94	0.94	0.98	0.97	1.00
Tempo	1.00	1.00	1.00	0.89	0.33	1.00	0.98	0.67	1.00

7   Conclusion

In this work a research phase was conducted regarding automated storytelling and the sectors in which it fits. Following this study, the path that was decided to pursue was that of identifying the elements constituting narrative texts by ML and NLP technique. Starting from the texts, a process of extracting the information characterizing each narrative text was conducted, focusing particular attention on the "sequence" element of which all the objects that constitute it were identified. In the last phase of the study, a validation of the results was conducted in order to obtain a measure of the performance of the developed algorithm. Starting from this work, a future development could certainly be to define a graphic methodology and develop an application capable of visualizing these individual elements through a logical graph. Furthermore, the recognition phase of the elements and specifically of the actors could be improved by carrying out a logical analysis of the text which was not performed in this work. A final consideration is that the algorithm could be extended include a more comprehensive temporal analysis that allows the sequences to be placed respecting the temporal order.

Acknowledgements. The work described in this paper has been supported by the research project RASTA: Realtà Aumentata e Story-Telling Automatizzato per la valorizzazione di Beni Culturali ed Itinerari; Italian MUR PON Proj. ARS01 00540.

Gennaro Junior Pezzullo is a PhD student enrolled in the National PhD in Artificial Intelligence, XXXVII cycle, course on Health and life sciences, organized by "Università Campus Bio-Medico di Roma".

References

1. Al-Obaydy, W.N.I., Hashim, H.A., Najm, Y.A., Jalal, A.A.: Document classification using term frequency-inverse document frequency and k-means clustering. Indonesian J. Electr. Eng. Comput. Sci. **27**(3), 1517–1524 (2022)
2. Alabdulkarim, A., Li, S., Peng, X.: Automatic story generation: challenges and attempts. arXiv preprint: arXiv:2102.12634 (2021)
3. Jones, K.S.: Natural language processing: a historical review. In: Zampolli, A., Calzolari, N., Palmer, M. (eds.) Current Issues in Computational Linguistics: In Honour of Don Walker. Linguistica Computazionale, vol. 9, pp. 3–16. Springer, Dordrecht (1994). https://doi.org/10.1007/978-0-585-35958-8_1
4. Li, J., Sun, A., Han, J., Li, C.: A survey on deep learning for named entity recognition. IEEE Trans. Knowl. Data Eng. **34**(1), 50–70 (2020)
5. Metilli, D., Bartalesi, V., Meghini, C.: Steps towards a system to extract formal narratives from text. In: Text2Story@ ECIR, pp. 53–61 (2019)
6. Mostafazadeh, N., et al.: A corpus and cloze evaluation for deeper understanding of commonsense stories. In: Proceedings of the 2016 Conference of the North American Chapter of the Association for Computational Linguistics: Human Language Technologies, pp. 839–849 (2016)
7. Nadkarni, P.M., Ohno-Machado, L., Chapman, W.W.: Natural language processing: an introduction. J. Am. Med. Inf. Assoc. **18**(5), 544–551 (2011)
8. Panebianco, B., Seminara, S., Gineprini, M.: LetterAutori. Zanichelli, Bologna (2011)

9. Shahapure, K.R., Nicholas, C.: Cluster quality analysis using silhouette score. In: 2020 IEEE 7th International Conference on Data Science and Advanced Analytics (DSAA), pp. 747–748. IEEE (2020)

10. Shinde, P.P., Shah, S.: A review of machine learning and deep learning applications. In: 2018 Fourth International Conference on Computing Communication Control and Automation (ICCUBEA), pp. 1–6. IEEE (2018)

11. Tikhonov, A., Samenko, I., Yamshchikov, I.P.: StoryDB: broad multi-language narrative dataset. arXiv preprint: arXiv:2109.14396 (2021)

12. Toolan, M.J.: Narrative: A Critical Linguistic Introduction. Routledge, Milton Park (2013)

13. Atro, V.: Part-of-Speech Tagging, vol. 219. The Oxford Handbook of Computational Linguistics (2003)

Travel Air IQ: A Tool for Air Quality-Aware Tourists

Divya Pragna Mulla[1]($\boxtimes$), Antonella Calo[1], and Antonella Longo[1,2]

[1] Department of Engineering for Innovation, University of Salento, Lecce, Italy
{divyapragna.mulla,antonella.calo,antonella.longo}@unisalento.it
[2] Italian Research Center on High-Performance Computing, Big Data and Quantum Computing (ICSC), Bologna, Italy

Abstract. This article introduces "Travel Air IQ," a web solution designed for European tourists, integrating advanced Decision Support Systems and real-time air quality data. Beyond conventional platforms, it serves tourists, aids tourism departments, and supports public administration. The study explores Travel Air IQ's capabilities, highlighting its potential to empower tourists, assist in resource management, and enable proactive responses to challenges. By leveraging data and computer science principles, it addresses varied needs, enhancing the tourist experience and efficiently managing resources. The integration of advanced systems contributes to sustainability and adaptability in tourism practices. This research aligns with the conference's focus on advancing internet, data, and web technologies, showcasing how these innovations reshape tourism towards comprehensive, user-centric solutions.

1 Introduction

In the era of data-driven decision-making, the intersection of tourism and air quality management presents a compelling space for innovation and technological advancement. This article introduces "Travel Air IQ," a sophisticated web solution meticulously engineered for European tourists, with a primary focus on integrating advanced Decision Support Systems (DSS) and real-time air quality data. Through the lens of data science, this platform aims to redefine the tourist experience, empower travellers with informed choices, and contribute to sustainable tourism practices.

Contemporary tourism is increasingly shaped by the fusion of data science principles with real-time analytics, offering novel insights and personalized experiences. The conventional approach to air quality management for tourists often falls short of leveraging the full potential of data science methodologies. "Travel Air IQ" represents a paradigm shift, where data science techniques are harnessed to provide not only real-time air quality information but also predictive analytics to anticipate and mitigate potential challenges.

The utilization of Decision Support Systems in the platform draws inspiration from recent advancements in data science applications for environmental monitoring and management [1]. By incorporating machine learning algorithms, "Travel Air IQ" not

L. Barolli (Ed.): EIDWT 2024, LNDECT 193, pp. 617–628, 2024.
https://doi.org/10.1007/978-3-031-53555-0_59

only informs tourists about existing air quality conditions but also adapts to changing circumstances, offering dynamic recommendations based on predictive modelling.

Moreover, the platform aligns with the broader landscape of technological interventions in tourism, as evidenced on the comprehensive integration of technology to enhance the tourist experience [2]. "Travel Air IQ" emerges as a pioneering solution that not only aids individual travellers but also serves as a valuable tool for tourism departments and public administration by employing data-driven insights for effective resource management.

In this context, it is important to highlight the significance of open data and the fairness of data in the development of this platform. Open Data plays a crucial role in ensuring transparency, fostering innovation, and enabling collaboration in the creation of products such as this one. By incorporating open data, the platform can enhance the accessibility of air quality information for tourists and contribute to the broader goal of promoting sustainable tourism practices.

Furthermore, attention to fairness of data is essential to ensure that the sensitive needs of tourists, such as those with respiratory conditions, are respected. Research has shown that the fairness of data, particularly in the context of air quality information, is vital for addressing potential disparities and safeguarding the well-being of vulnerable tourist populations [17]. Therefore, the integration of fairness aware data processes in "Travel Air IQ" is imperative to uphold ethical standards and provide equitable support to all tourists. This would make the platform's environment more responsible and inclusive.

This research contributes to the evolving narrative of technology-driven solutions in tourism, showcasing the transformative potential of data science in shaping user-centric and sustainable practices. By delving into the intricate relationship between tourism and air quality through the lens of data science, "Travel Air IQ" sets the stage for a new era of intelligent, adaptive, and informed travel experiences. The State of Art section of this paper reviews the existing literature on the real time air quality data use and its incorporation in tourism decision making. This is followed by a section on the presentation of the adopted Method and Materials, which covers research areas, data sources, research methods, data preparation and validation. Next section of Results and Discussion then highlights the findings and discuss the main features. The final section is devoted to conclusions and ideas for future research.

2 State of Art

Recent years have witnessed a surge in the development of decision support systems and technology-driven solutions aimed at enhancing various aspects of our lives. In the context of air quality management, several platforms provide real-time data and insights. However, the existing systems often lack a comprehensive approach that caters specifically to the needs of tourists. "Travel Air IQ" addresses this gap by offering a holistic solution designed to meet the unique requirements of European travellers.

The landscape of air quality management is undergoing a paradigm shift, driven by advancements in data science and technology. Within this context, Travel Air IQ emerges as a groundbreaking web-based application, amalgamating cutting-edge Decision Support Systems (DSS) with real-time air quality data. As we delve into the current state of the art, each referenced work serves as a guiding beacon, illuminating the transformative potential of Travel Air IQ while shedding light on notable research gaps that the application aims to address.

The foundational work of Smith, J. et al. outlines the historical evolution of Decision Support Systems (DSS) in air quality management [1]. While acknowledging the progress made, a notable research gap exists in the lack of personalized approaches to decision-making. Travel Air IQ aims to bridge this gap by introducing a user-centric model, tailoring air quality predictions based on individual preferences and lifestyles. The application seeks to fill the void in personalized decision support within the realm of air quality management. Studies showed the pioneering research on predicting the Air Quality Index through deep learning techniques showcases the potential of advanced algorithms [2]. However, a critical research gap lies in the adaptability of predictive models to dynamic air quality conditions. Travel Air IQ steps into this gap, leveraging sophisticated deep learning methodologies to not only enhance accuracy but also ensure responsiveness to rapidly changing environmental factors.

The exploration of personalized travel recommender systems introduces a novel perspective that directly influences Travel Air IQ's user-centric design [3]. The research gap addressed here lies in the absence of personalized decision support systems in the context of air quality management. By incorporating individual preferences into the decision-making process, Travel Air IQ seeks to fill this void, providing tailored air quality predictions for users with diverse lifestyles. The spatial-temporal analysis in Beijing underscores the importance of geographical context in air quality management [4]. However, a notable research gap persists in the lack of localized and contextually relevant air quality predictions on a broader scale. Travel Air IQ aims to fill this gap by integrating Geographic Information Systems (GIS) principles, ensuring that predictions are not only accurate but also highly relevant to the user's specific location.

The exploration of big data in tourism sets the stage for Travel Air IQ's reliance on extensive datasets [5]. A critical research gap lies in the underutilization of big data for nuanced insights into air quality. Travel Air IQ addresses this gap by harnessing vast datasets, providing a comprehensive understanding of air quality and contributing to the broader trend of integrating big data into environmental research. The overview on machine learning applications in environmental sciences highlights the potential of advanced algorithms [6]. However, a research gap exists in the application of machine learning specifically for air quality predictions. Travel Air IQ addresses this gap by leveraging machine learning algorithms for robust predictive modelling, contributing to the evolving landscape of environmental research.

The review of IoT-based air quality monitoring systems underscores the significance of real-time data collection [7]. Yet, a notable research gap exists in the lack of comprehensive and user-friendly applications that harness IoT for personalized air quality insights. Travel Air IQ steps into this gap, offering a dynamic platform that continuously updates and adapts to changing environmental conditions. A comprehensive review on data-driven approaches establishes the foundation for Travel Air IQ's methodology [8]. The research gap addressed here pertains to the limited integration of data-driven insights into decision support systems. Travel Air IQ seeks to fill this gap by prioritizing data-driven approaches, ensuring that predictions are not only accurate but also dynamically responsive to evolving air quality scenarios.

Some interesting exploration of GIS and machine learning synergy guides Travel Air IQ's interdisciplinary approach [9]. However, a research gap exists in the underexplored intersection of spatial analysis and machine learning for air quality predictions. Travel Air IQ fills this void by seamlessly integrating GIS principles with advanced algorithms, ensuring precise and contextually aware predictions. The review of web- based Decision Support Systems (DSS) in environmental management influences Travel Air IQ's user-centric design [10]. The research gap addressed here is the lack of accessible and user-friendly platforms for decision support in air quality management. Travel Air IQ steps into this gap, prioritizing accessibility and a seamless user interface for effective decision-making.

Some valuable insights into personalized DSS directly align with Travel Air IQ's philosophy of tailoring air quality predictions based on individual preferences [11]. The research gap addressed is the absence of personalized decision support systems catering specifically to air quality management. Travel Air IQ fills this gap by providing a platform that adapts to individual user contexts, enhancing the relevance of air quality predictions. The exploration of IoT-enabled air quality monitoring resonates with Travel Air IQ's real-time data collection approach [12]. However, a research gap exists in the limited availability of applications that comprehensively integrate IoT for continuous updates and accuracy. Travel Air IQ addresses this gap, offering a dynamic and responsive platform that leverages IoT for personalized air quality insights.

The emphasis on enhancing DSS with real-time data aligns seamlessly with Travel Air IQ's commitment to up-to-date information [13]. The research gap addressed here pertains to the limited availability of decision support systems that dynamically adapt to real-time environmental conditions. Travel Air IQ fills this void, ensuring timely and relevant air quality predictions through continuous updates. The exploration of machine learning approaches in air quality forecasting directly influences Travel Air IQ's integration of advanced algorithms [14].

The research gap addressed here lies in the limited application of machine learning specifically for air quality predictions. Travel Air IQ fills this void by leveraging machine learning for accurate and personalized predictions, contributing to the advancement of environmental forecasting. The review on urban air quality monitoring with sensor networks resonates with Travel Air IQ's networked approach [15]. The research gap addressed here is the limited utilization of comprehensive sensor networks for localized air quality insights. Travel Air IQ steps into this gap, offering a networked approach that ensures precise and localized predictions, contributing to a more granular understanding of air quality.

The integration of advanced Decision Support Systems (DSS) with real-time air quality data distinguishes "Travel Air IQ" from conventional platforms. This approach allows tourists to access up-to-date information about air quality in their chosen destinations, enabling them to make informed decisions for their well-being. Moreover, the system aids tourism departments in efficiently managing resources, as it provides a valuable tool for monitoring and responding to air quality challenges in real time. This research builds upon the synergy of data and computer science principles, utilizing cutting-edge technologies to address diverse needs within the realm of tourism. By leveraging the power of data analytics, machine learning, and real-time monitoring, "Travel Air IQ" contributes not only to enhancing the tourist experience but also to the sustainability and adaptability of tourism practices.

3 Travel Air IQ: Tool Design

The flow chart of the Web application development is shown in Fig. 1. The architectural design flow for TRAVEL AIR revolves around creating a user-centric and efficient system. The user interface (UI) is designed with the objective of providing tourists with a friendly platform to access air quality information and receive personalized recommendations. Key components include an interactive map displaying air quality levels, user preference input forms, and a recommendation display panel. Frontend development involves translating UI designs into a functional application using HTML, CSS, JavaScript, and integration with a frontend framework for dynamic content.

User preferences handling captures and processes traveler preferences, incorporating components such as user profile management and preference forms for travel styles, activities, and accommodation. Real-time data collection gathers up-to-date air quality information through API integration with monitoring stations, satellite data retrieval mechanisms, and ground sensor network data acquisition. Data processing and preprocessing involve cleaning, augmenting, and enriching raw data using algorithms and quality control mechanisms. The Decision Support System (DSS) integrates the Extra Trees Regressor algorithm for personalized air quality predictions, with components including the Extra Trees model, continuous model updating, and integration with user preferences.

Backend development focuses on creating server-side logic for data processing, model integration, and communication with the frontend, utilizing server-side scripting and integration with databases. Security measures are implemented to protect user data and system integrity, incorporating encryption, authentication, authorization mechanisms, and regular security audits. Scalability considerations involve designing the system to handle varying loads and user demands through load balancing and cloud-based infrastructure. Documentation is maintained for system understanding and future development, including API documentation, code comments, inline documentation, and system architecture diagrams. Testing and quality assurance ensure system reliability, functionality, and user experience through unit testing, integration testing, and user acceptance testing. Deployment involves launching the TRAVEL AIR application for public access, deploying components on hosting servers, and monitoring system performance. Continuous improvement is emphasized, with mechanisms for user feedback collection and regular updates to enhance recommendations and system responsiveness. This comprehensive design ensures a seamless and reliable TRAVEL AIR system, addressing various aspects of user interaction, data processing, security, scalability, and ongoing refinement.

- User-Centric UI Design with an objective to design a user-friendly interface for tourists to access air quality information and receive personalized recommendations

Fig. 1. Flow chart of the Web Application development.

with the components of Interactive maps of air quality, user preference input forms, recommendation display panel where the following technologies are incorporated HTML, frontend framework for dynamic content. In this study Wix website platform has been used for the development of the application has been customized based on the need of the application.

- User Preferences Handling helps in capturing and processing travelers preferences by using user profile management, preference forms for travel styles, activities, and accommodation.
- Design: Real-time data collection is a dynamic process that involves gathering the most current and relevant information available at the time of retrieval. In the context of the described study, real-time data collection is employed to capture up- to-date air quality information. The data sources for this study primarily include ARPA PUGLIA, an organization presumably associated with air quality monitoring.

The following components contribute to the real-time data collection process:

- API Integration with Monitoring Stations, uses APIs to seamlessly integrate with monitoring stations for direct communication and data exchange. By ensuring constant reception of live updates and real-time air quality measurements from ARPA PUGLIA-operated stations.
- Satellite Data Retrieval Mechanisms are done by obtaining information from Earth-orbiting satellites to complement ground-level measurements. Enhancing air quality information comprehensiveness is done by incorporating satellite data retrieval mechanisms for insights across a wider geographical scope.
- Ground Sensor Network Data Acquisition is achieved by deploying distributed ground sensors strategically to monitor specific parameters like air quality.
- Collect real-time data from ground sensors to contribute valuable, localized insights to the overall air quality dataset.

The datasets used for the air quality data are specifically collected from the ARPA PUGLIA website for a single day. This implies that the study focuses on a snapshot of air quality conditions for a specific time period, providing a detailed and time-sensitive perspective. The utilization of datasets from ARPA PUGLIA's website underscores the reliance on authoritative and reputable sources for acquiring air quality information.

- Data is prepared by utilizing Extra Trees Regressor Algorithm for regression tasks, specifically tailored for personalized air quality predictions. Focused on creating personalized predictions by adapting the model to individual or specific locations, considering unique variables like local pollution sources and weather patterns.

- Data Cleaning is done by ensuring cleanliness of datasets by addressing missing values, outliers, and standardizing formats and they are thoroughly cleaned to enhance the reliability of the model for personalized air quality predictions. Data Processing is made by leveraging Jupyter Notebook IDE for an interactive and collaborative computing environment with the help of Python programming language within Jupyter Notebook for processing and analyzing cleaned datasets using the Extra Trees Regressor algorithm. Finally, Decision Making is achieved by developing a predictive model using the Extra Trees Regressor algorithm trained on historical data.
- Utilize model predictions for decision-making, providing valuable insights into future air quality conditions.
- Decision Support System (DSS) is used for the provision of personalized air quality predictions with incorporation of Extra Trees model and followed by continuous model updating, integration with user preferences. In this study personalized predictive Decision Support System is incorporated to assist with objectives of the application.
- Backend Development is created by using server-side logic for data processing, model integration, and communication with the frontend with the help of Server- side scripting, integration with databases. Suggestion System for advising the tourists to choose a desirable suggestion is by comparing the real-time air quality with the predefined threshold. If it is below the threshold, trigger the system to suggest an alternative city and this aspect is still under development and will be incorporated in the next steps of the research.
- Security Measures are carefully taken into consideration for the user data protection and system integrity by using encryption, authentication, authorization mechanisms have to be developed.
- Testing and Quality Assurance is conducted to ensure system reliability, functionality, and user experience by Unit testing, integration testing, user acceptance testing has to be developed.
- Deployment, the last stage involves launching the TRAVEL AIR application for public access to assess the deployment components on hosting servers, monitor system performance.
- Continuous Improvement, after the launch of application to emphasize on the refinement we require the User feedback collection, regular updates to enhance recommendations and system responsiveness.

This comprehensive design methodology ensures a seamless and reliable TRAVEL AIR system, addressing various aspects of user interaction, data processing, security, scalability, and continuous improvement.

4   Results and Discussion

The website upon launching has the possibility of providing multifaceted data driven decision making assistance to the tourists. The validation of TRAVELAIR IQ website features involves confirming the accuracy, reliability, and effectiveness of its functionalities using datasets provided by ARPA PUGLIA. This process ensures that the website's capabilities align with the real-world air quality data collected by ARPA PUGLIA, a recognized authority in environmental monitoring. The website is multi page application where the Home Page (as shown in Fig. 2) provides the glimpse of the air quality in the given region and asks for the registration/login for accessing other features.

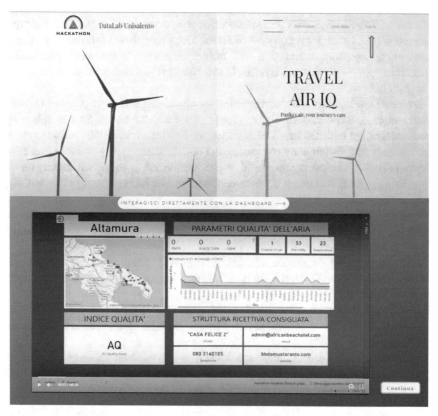

Fig. 2. Home Page of the TRAVEL AIR IQ website.

The login page is shown in Fig. 3. Upon registration, the user is given access to check for the current region Dashboard as shown in Fig. 4 and also an option (Consigliami Una Citta) for asking the website suggestion. Upon clicking this the user will be directed to the next page showing the selected city with good air quality in our case it was Monte Sant Angelo for better touristic experience depending upon the air quality (see Fig. 5).

Iscriviti per utilizzare tutte le funzionalità

Registrazione

Nome   Cognome

Email *

☐ Accetto termini e condizioni

Registrati

Fig. 3. Login Page.

Fig. 4. Dashboard

Furthermore, suggestions are made for activities other than visiting such as restaurants, historical places etc. The suggestion system is under development for the full functionality of the website. The introduction of Travel AirIQ represents a significant advancement in the realm of travel technology, particularly in the context of European tourism. The strategic integration of advanced Decision Support Systems (DSS) and real-time air quality data distinguishes Travel AirIQ as a one-stop web solution designed to enhance the holistic decision-making process for tourists exploring Europe.

One notable aspect of Travel AirIQ is its triple-purpose functionality. It not only serves the immediate needs of tourists but also supports tourism departments and facilitates public administration. This multifaceted approach is crucial in addressing the complex dynamics of the tourism industry. For tourists, the platform provides an invaluable

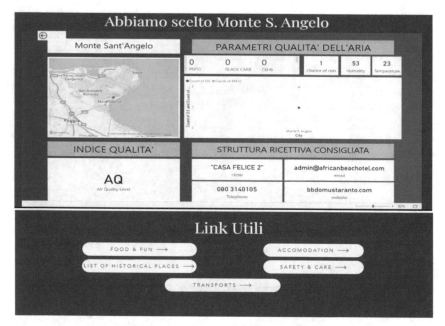

Fig. 5. Suggestion Page with other features.

tool for making informed choices, considering factors beyond traditional travel consid-erations, such as real-time air quality. The support extended to tourism departments is evident in the platform's capabilities for resource management and strategic planning. By leveraging cutting-edge data science principles, Travel AirIQ enables authorities to adaptively respond to emerging challenges. This aspect is particularly crucial in an industry susceptible to dynamic factors, such as weather conditions, events, and environ-mental changes. Furthermore, the integration of advanced DSS and real-time data not only enhances the tourist experience but also contributes to the efficient management of tourism resources. This aligns with the overarching theme of sustainability and adapt-ability in tourism practices. By proactively addressing challenges through technology, Travel AirIQ signifies a shift towards a more resilient and responsive tourism ecosystem.

In the broader context of the conference's focus on advancing internet, data, and web technologies, Travel AirIQ serves as a noteworthy case study. It exemplifies how leveraging these technologies can result in a comprehensive, user-centric solution that reshapes conventional tourism practices. This discussion underscores the transformative potential of Travel AirIQ in fostering sustainability, adaptability, and an enhanced overall experience for tourists, tourism authorities, and public governance alike.

5   Conclusion and Future Research

In conclusion, TRAVEL AIR IQ transforms tourist experiences in Puglia by seam-lessly merging real-time air quality data with a user-friendly Decision Support System. The collaborative efforts with local authorities ensure a reliable platform, empowering travelers with the insights needed for informed and health-conscious travel decisions. Looking ahead, ongoing enhancements to the Decision Support System, integration of advanced machine learning algorithms, and a real-time user feedback loop are key priorities. Exploring partnerships for expanded data sources, embracing augmented reality features, and extending collaborations to other regions will solidify TRAVEL AIR IQ's position as a dynamic and indispensable tool for sustainable and informed tourism worldwide.

Acknowledgement. This work is part of the research activity developed by the authors within the framework of the Italian Research Center on High Performance Computing, Big Data and Quantum Computing (ICSC) funded by EU – NextGenerationEU (PNRR-HPC, CUP:C83C22 000560007) and the "PNRR CN00000023 - PNRR – M4C2 Inv. 1.4 - MOST": SPOKE 7 "CCAM, Connected Networks and Smart Infrastructure" – WP4.

References

1. Evagelopoulos, V., Charisiou, N.D., Logothetis, M., Evagelopoulos, G., Logothetis, C.: Cloud-based decision support system for air quality management. Climate **10**(3), 39 (2022). https://doi.org/10.3390/cli10030039
2. Mc Grath, S., Garrigan, E., Zeng, L.: Predicting air quality index using deep neural networks. In: 2021 IEEE International Conference on Electronic Technology, Communication and Information (ICETCI), pp. 341–344. IEEE (2021). https://doi.org/10.1109/ICETCI53161.2021.9563356
3. Majid, A., Chen, L., Chen, G., Mirza, H.T., Hussain, I., Woodward, J.: A context-aware personalized travel recommendation system based on geotagged social media data mining. Int. J. Geogr. Inf. Sci. **27**(4), 662–684 (2013). https://doi.org/10.1080/13658816.2012.696649
4. Ye, L., Ou, X.: Spatial-temporal analysis of daily air quality index in the Yangtze River delta region of China during 2014 and 2016. Chin. Geogr. Sci. **29**(3), 382–393 (2019). https://doi.org/10.1007/s11769-019-1036-0
5. Ardito, L., Cerchione, R., Del Vecchio, P., Raguseo, E.: Big data in smart tourism: challenges, issues and opportunities. Curr. Issues Tourism **22**(15), 1805–1809 (2019). https://doi.org/10.1080/13683500.2019.1612860
6. Zhong, S., et al.: Machine learning: new ideas and tools in environmental science and engineering. Environ. Sci. Technol. **55**(19), 12741–12754 (2021). https://doi.org/10.1021/acs.est.1c01339
7. IoT-based air quality monitoring systems for smart cities: a systematic mapping study. (n.d.). ProQuest. https://www.proquest.com/openview/0c72fbf97836137fcecb8acd1aa3c7f5/1?cbl=1686344&pq-%20origsite=gscholar&parentSessionId=kyUct6tIK%2BH7m%%202B6Ubrq%2BP1pgAVYPMyyQnfvN3qtVBXU%3D
8. Sun, Y., Haghighat, F., Fung, B.C.M.: A review of the-state-of-the- art in data-driven approaches for building energy prediction. Energy Build. **221**, 110022 (2020). https://doi.org/10.1016/j.enbuild.2020.110022
9. Ma, J., et al.: Identification of high impact factors of air quality on a national scale using big data and machine learning techniques. J. Clean. Prod. **244**, 118955 (2020). https://doi.org/10.1016/j.jclepro.2019.118955
10. Sugumaran, R., Meyer, J.C., Davis, J.: A web-based environmental decision support system (WEDSS) for environmental planning and watershed management. J. Geogr. Syst. **6**(3), 307–322 (2004). https://doi.org/10.1007/s10109-004-0137-0

11. Shambaugh, N.: Personalized decision support systems. Encycl. Artif. Intell., 1310–1315 (2009). https://www.igi-global.com/chapter/encyclopedia-artificial-intelligence/www.% 20igi-%20global.com/chapter/encyclopedia-artificial-intelligence/10409
12. Sundar Ganesh, C.S., Akshaya Prasaath, V., Arun, A., Bharath, M., Kanagasabapathy, E.: Internet of things enabled air quality monitoring system. In 2023 International Conference on Sustainable Computing and Smart Systems (ICSCSS), pp. 934–937 (2023). IEEE. https:// doi.org/10.1109/ICSCSS57650.2023.10169509
13. Liu, S., Duffy, A., Whitfield, R., Boyle, I.: Integration of decision support systems to improve decision support performance. Knowl. Inf. Syst. **22**(3), 261–286 (2010). https://doi.org/10. 1007/s10115-009-0192-4
14. Castelli, M., Clemente, F.M., Popovič, A., Silva, S., Vanneschi, L.: A machine learning approach to predict air quality in California. Complexity, e8049504 (2020). https://www.hin dawi.com/journals/complexity/2020/8049504/
15. Lo Re, G., Peri, D., Vassallo, S.D.: Urban air quality monitoring using vehicular sensor networks. In: Gaglio, S., Lo Re, G. (eds.) Advances onto the Internet of Things. Advances in Intelligent Systems and Computing, vol. 260, pp. 311–323. Springer, Cham (2014). https:// doi.org/10.1007/978-3-319-03992-3_22
16. ARPA PUGLIA. http://old.arpa.puglia.it/web/guest/qariainq2
17. Selbst, A.D., Boyd, D., Friedler, S.A., Venkatasubramanian, S., Vertesi, J.: Fairness and abstraction in sociotechnical systems. In: Proceedings of the Conference on Fairness, Accountability, and Transparency (FAT* '19), New York, NY, USA, Association for Computing Machinery, pp. 59–68 (2019) https://doi.org/10.1145/3287560.3287598

Advanced IT Technologies Applied to Archaeological Park of *Norba* (Latium, Italy)

Dario Branco[✉], Antonio Coppa, Salvatore D'Angelo, Stefania Quilici Gigli, Giuseppina Renda, and Salvatore Venticinque

Universitá della Campania "Luigi Vanvitelli", Caserta, Italy
{dario.branco,antonio.coppa,salvatore.dangelo,stefania.gigli,
giuseppina.renda,salvatore.venticinque}@unicampania.it

Abstract. This study proposes the integrated utilization of advanced information technologies to enrich conservation, research, and visitors' experience at the *Norba* Archaeological Park, located in *Latium*, Italy. Using 3D digitalization, augmented reality (AR), and geographical information systems (GIS) technologies, this work aims to transform the archaeological and historical heritage into material that can be used via smart devices, thus improving and expanding the methodologies of fruition.

1 Introduction

The application of Information and Communication Technologies (ICT) in the cultural heritage domain promoted the valorization of less-known assets and the understanding of related cultural contents [2]. The ease with which it is possible today to create quality 3D surveys at low cost has made it possible, to create online 3D galleries accessible from both computers and smartphones. For a museum, sharing its collections online means increasing its visibility, involving the online community and young generations, and making content available that can be reused to create new tools for studying and understanding the artistic heritage. This is the case, for example, of the 3D Virtual Museum[1], which with its 95 three-dimensional models is the largest national representative. The British Museum in London has an account on Sketchfab, a 3D modeling platform, which even offers the possibility of ordering a 3D printed replica of one of the uploaded models. Also based on the same platform is African Fossils, an international project with the objective of creating a large gallery of 3D models of African fossils. Hence, digital objects, such as 3D reconstructions, multimedia contents, and information services represent relevant resources to provide a complete and multi-modal fruition of cultural heritage. Nevertheless, without a guide, it may be difficult for users to fully understand the works they appreciate, and to explore selected contents following a path that is tailored to his interests. With this concern, the Smithsonian Museum[2], in addition to the common three-dimensional rotation, provides a guided tour mode, which describes the history of the object in detail with textual content and photographs. In this paper, we present how advanced techniques and original

---

[1] https://www.3d-virtualmuseum.it.

[2] https://www.si.edu/.

L. Barolli (Ed.): EIDWT 2024, LNDECT 193, pp. 629–638, 2024.
https://doi.org/10.1007/978-3-031-53555-0_60

technologies have been integrated to allow for going beyond the common way of visiting archaeological sites, natural landscapes, or traditional museums. Live experiences are enhanced by augmented and virtual reality (Hybrid Reality) technologies, with the aim of enhancing and expanding the ways of experiencing the site. In particular, intelligent conversational agents are exploited to recommend and present cultural contents, at relevant Points of Interest (PoIs) of alternative touristic itineraries. Intelligent agents are decision makers which are context aware and use multi-criteria optimization strategies to select the best content and services which match the user's profile [3]. The research activities have been developed in the context of the RASTA project (Augmented Reality and Automated Storytelling for Cultural Heritage and Tourist Routes). The RASTA project aims to create a framework that exploits the potential of integration between augmented and virtual reality technologies and with intelligent storytelling technologies for the personalization of narratives through sophisticated user recommendation and profiling systems.

2   The Archaeological Park of *Norba*

The archaeological park of *Norba* represents a pilot case study of the RASTA research project. *Norba* is located on a plateau overlooking the Pontine Plain and stands out for its spectacular position and the imposing polygonal walls that surround it, making it known since the 18th century [10]. Dionysius of Halicarnassus mentions it among the Latin cities that in the 5th century BC would have allied against Rome on the eve of the Battle of *Lake Regillus* (*Dion. Hal., V, 50, 1*). The historian Livy also recalls that in 491 BC, after the conclusion of the so-called *Foedus Cassianum*, a *nova colonia* was sent to *Norba*, with the strategic function of establishing a stronghold in the Pontine territory, presumably to counter the Volscian advance (*Liv., II, 34, 6*). During the civil wars between *Marius* and *Sulla*, *Norba*, aligning itself with *Marius*, was the last city to fall in 81 BC. An intense passage by Appian (*App., Bell. Civ., I 94, 439*) recounts the dramatic fate of the city and its inhabitants, who chose to take their own lives rather than fall into the hands of the enemy [7]. Since then, the city has remained frozen and in time, only occasionally reoccupied, miraculously escaping modern urban expansion: thus, it constitutes a privileged observatory for understanding the architecture and urban planning of a city from the middle and late Republican era [11].

The Archaeological Park of *Norba* is a creation of recent decades, the result of forward-looking political and scientific commitment to research and conservation [14]. This effort has allowed the reconstruction of the landscape and the urban reading of the city, enclosed by perfectly preserved polygonal walls that encompass an urban area of about 44 hectares over a path of more than 2.5 km. Within the walls, four gates and several small gates open; one of the main entrances to the city is through the monumental Porta Maggiore, now the entrance to the Park, where the most spectacular stretch of the walls (Fig. 1) is preserved. Inside the walls, topographic research and excavation activities [15] have allowed the recognition and delineation of an urban layout with orthogonal axes, marked by terraces in polygonal masonry that regularize the elevations (Fig. 2). Of particular relevance in urban planning is the long northwest/southeast road that runs straight through the city from the steps of the Minor Acropolis to the Porta

Fig. 1. *Norba*: section of wall and view of Porta Maggiore

Fig. 2. *Norba*: reconstructive plan of the ancient city

Serrone di Bove gate. On the road, a planning complex for orthogonal axes was set up, both uphill and downhill, and a road branched off leading to the large terrace of the Sanctuary of *Iuno* [16]. Almost in the center of the road was the monumental thermal complex. In the urban layout, 12 *domus* have been brought to light. Their plan, in most cases, recalls the canonical plan of the typical Roman house with an atrium; sometimes, they show significant restructuring signs aimed at adapting the spaces of the house, with the creation of *triclinium*, to the new convivial and social customs that spread in *Norba* in the mid-2nd century BC, following the example of Rome [12].

3   Visit Itineraries

Visit itineraries are designed as a sequence of geolocated Points of Interest (POIs), enriched with multimedia contents (texts, images, videos, audio). Itineraries start from the main entrance to the ancient city, now also the entrance to the archaeological park.

Porta Maggiore (Fig. 3), which, with its eight-meter high doorposts and the mighty bastion on the right side, represents one of the most monumental and well-preserved urban gates from the Republican era. Immediately after Porta Maggiore, you can admire the ancient paved road [6], which ascended from the gate to reach an important junction in the city's urban layout. In this section, the road took on a highly monumental appearance, given by a colonnaded portico set on the sidewalk along the entire route. Just before the portico, on the left, a dwelling has been brought to light. Along this axis, streets branched off leading southward to the main street crossing the city and the Minor Acropolis, and northward to the Major Acropolis; and it is precisely at this point that the prepared itineraries diverge. Continuing northward along a section of the paved road, one crossed a presumably residential district downstream of the Major Acropolis, then reached a large square-shaped basin completely embedded in the ground. A large cistern, constructed with polygonal masonry and coated with "cocciopesto" (a type of waterproof plaster), highlights its fundamental importance in a city lacking springs and aqueducts. The enormous capacity, estimated at around 4000 m3, underscores the crucial role of public infrastructure in a city without water sources and aqueducts. Beyond the cistern, along an uphill paved road, one reached the highest point of the city, the so-called Major Acropolis, where there was a temple dedicated to Diana, within a vast sacred area [9]. Continuing southward, however, along the porticoed road, one walks upon the so-called Second Traverse, characterized by a significant slope, leading directly to the major street crossing the city. Along this road, on the left side, three dwellings have been brought to light. Of particular interest is *Domus* IV, which exhibits the peculiarity of an atrium arranged orthogonally to the entrance, a planimetric organization reminiscent of the canonical plan of the typical Roman house with an atrium [13], like the other dwellings found in *Norba*, but above all, a significant intervention to expand the space located at the back of the atrium, to the left of the tablinum. This intervention was not insignificant; in fact, it involved an expansion beyond the original perimeter of the room, with the significant need to also redo its roof; the discovery of two bronze bed legs [13] in a corner of the room confirmed that the intervention was useful for creating a *triclinium*. The traverse leads to the major street crossing the city, which, starting from Porta Serrone di Bove, reached the steps of the Minor Acropolis,

Fig. 3. *Norba*: the entrance from Porta Maggiore

crossing the city in its entirety from west to east. To the east is the small hill on which the so-called Minor Acropolis stands, a monumental and scenic sacred complex with two temples, of which only the polygonal masonry foundations are preserved. Continuing westward, towards Porta Serrone di Bove, one reaches a junction between the Third Traverse and a transverse road leading to Porta Ninfina, the oldest urban gate with a fortified enclosure. Just beyond the intersection are the remains of a grand thermal complex. The monument consists of a series of rooms constructed in concrete with uncertain masonry, and a large oval basin, a probable water reservoir for the functioning of the facility. Beyond the thermal complex, a paved road known as the "Road to *Iuno*" [5] leads to the large terrace of the Sanctuary of *Iuno Lucina*, the final stop on our itinerary. The rectangular base of a temple, made of large blocks of polygonal masonry, is visible, along with a portico and structures connected to worship on the terrace that extends to the edge of the hill. Several votive offerings have been found in the fill of the lower terrace.

For assisted fruition of the archaeological sites of the project, 3D models and virtual reconstructions are being arranged, that improve the understanding of the archaeological remains.

Mapping and 3D scanning have been carried out on monuments and archaeological remains of the *Norba* site. A virtual 3D reconstruction of *domus* IV was designed and developed. This reconstruction is the result of a complex process, starting from the careful study and analysis of the archaeological remains, avoiding unreal reconstructions. In addition to the 3D reconstruction of the entire house, portions of it were created (walls, floors, wall decoration, roof, entrance door, furnishings inside the room, and wardrobe immediately outside the room), so that the user can interact with the entire *domus* or parts of it. We are also considering the link between the archaeological park and the Museum, where the objects found on the site are exhibited, through 3D models of the

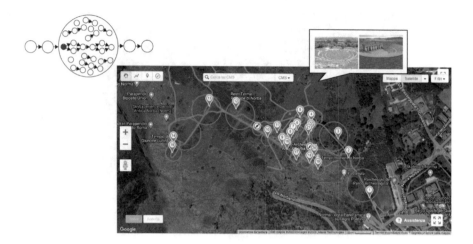

Fig. 4. Visit itinerary

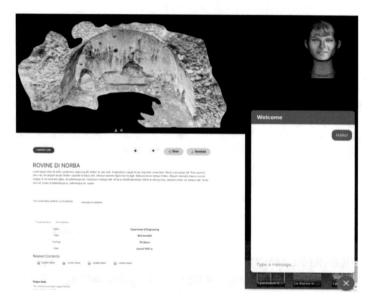

Fig. 5. 3D scan of archaeological remain

findings, that can be "virtually" relocated to their place of origin. 3D models of two feet of a *triclinium* bed found in *domus* IV have been carried out. Starting from these objects, the entire bed was subsequently reconstructed, available as a 3D object.

Fig. 6. Interactive presentation of ancient coin

4   Content Delivery and Interactive Storytelling

An example of a visit itinerary is shown in Fig. 4. Even if the physical route is statically defined, the cultural experience is dynamically built according to the user's interest, and the current context, and consistently with the visit itinerary, interactive storytelling is modeled as a *String of Pearls*[8]. The *String of Pearls* is depicted in the upper left corner of Fig. 4. In general, a string of pearls is a graph representation of a set of states in which the user experience may be. That graph is then divided into consecutive subgraphs (called pearls) and each pearl is connected to the next by a pair of nodes (one node exiting the previous pearl and one entering the next pearl) and an arc representing the transition action from the previous pearl to the next. Typically, the user can move within a pearl by traversing its various states asking for related information, accepting to enjoy a recommendation, searching and exploiting additional content, or executing specific actions. All operations do not have to have a precise temporal sequence, giving the visitor the idea of having a kind of freedom in living their cultural experience. Only upon the occurrence of specific events the user can continue along the itinerary, and thus proceeding to the next pearl and unlock another area with another set of walkable states. In our case each pearl corresponds to a PoIs of the itinerary following the static order along the defined route. Users can navigate in complete autonomy within the pearl, enjoying the tour they please. Constraints, such as a time limit within a pearl, and requirements to be met, such as the enjoyment of an amount of contents pertaining to that pearl, affect the probability to recommend to move to the next pearl. This representation and management of the tourist flow also ensures the linearity and consistency of the tourist itinerary in terms of topics covered and information preparatory to understanding the 'dependent' works. As the visitor is guided by a chat-bot, which presents the digital contents and interacts in natural language, the string of pearls guarantees that the conversation does not diverge from the storyline that describes the visit itinerary.

Fig. 7. Browsing of virtual reconstruction of a Roman *domus* in a 3Dtiles map

Let's imagine that the visitor has just arrived at the site, and uses a special app installed on their smartphone, or on a special device provided to them by the park. In a first example of interaction, the application, having located the visitor close the ruin of an architectural element, which cannot be approached for safety reasons, recommends to enjoy his digital 3D scan. In Fig. 5, we see metadata retrieved from our digital archive and the 3d Model [1]. The user can manipulate the 3D model, zoom in, look at the architectural element from any perspective. The software agent is aware about what the user is looking at, and reacts describing the current view, whenever the view is focused on a detail of particular interest, or answers to users' questions such as, "what is this?" [4].

In Fig. 6 the user accepted to receive information about a ancient coin found nearby. The software agent start to describe the model according to a well defined presentation stereotype common to money-type objects. In this case, it is the software agent itself that moves the 3D model, and referring to the current view, provides information about the obverse and the reverse of the coin. Obviously, the visitor can interrupt the presentation asking semantically related information, concerning, for example, the depicted character, and so continue to build his experience choosing cultural contents pertaining to history, rather than economics, and so on. Note that the model shown in Fig. 6 does not correspond to a Roman coin, but it has been used just for demonstration purpose.

Finally, we imagine that the visitor, enthusiastic about his last experience, chooses to share a content with his contacts. In Fig. 7, a virtual reconstruction of a *domus* is geopositioned at its place of origin, and made accessible through 3Dtiles technology, which support the visualization trough a browser. This technology allows downloading only the tiles necessary for viewing the details focused on by the user. The complexity of the model affect a comfortable navigation of the model and the understanding of details, such as the interiors that is shown in Fig. 8. Neither the navigation of the model is trivial, nor the retrieval of relevant information. However, the smart guide can illustrates the model, according a personalized storytelling as well, moving the camera and providing information about the view. Moreover, it will leave the visitors free to take back the control, when they wish.

Fig. 8. Details of virtual reconstruction of a room with Roman *triclinium*

5   Conclusion

This paper demonstrates how the augmentation of visit itineraries in archaeological or natural sites delivering digital contents leveraging augmented and virtual reality technologies allow to enhance the visitors' experience and to promote assets of less known secondary sites, increasing their visibility. The propose solution meet relevant requirements to maximize the impact of such technologies. In particular it allows for the delivery of relevant cultural contents consistently with the visit itinerary, in a personalized manner based on the user profile, leaving the user free to build their own cultural experience.

Acknowledgments. The work has been supported by ARS01_00540 - RASTA project, funded by the Italian Ministry of Research PNR 2015–2020.

Thanks to Alessandro Ciancio for the 3D virtual model of the *domus* IV of *Norba*.

References

1. Amato, A., Aversa, R., Branco, D., Venticinque, S., Renda, G., Mataluna, S.: Porting of semantically annotated and geo-located images to an interoperability framework. In: Barolli, L. (eds.) Complex, Intelligent and Software Intensive Systems. CISIS 2022. LNNS, vol 497, pp. 508–516. Springer, Cham (2022). https://doi.org/10.1007/978-3-031-08812-4_49
2. Amato, A., Di Martino, B., Scialdone, M., Venticinque, S.: Personalized recommendation of semantically annotated media contents. Stud. Comput. Intell. **511**, 261–270 (2014)
3. Amato, A., Venticinque, S.: Multiobjective optimization for brokering of multicloud service composition. ACM Trans. Internet Technol. **16**(2), 1–10 (2016)
4. Ambrisi, A., et al.: Intelligent agents for diffused cyber-physical museums. In: Camacho, D., Rosaci, D., Sarné, G.M.L., Versaci, M. (eds.) Intelligent Distributed Computing XIV. IDC 2021. Studies in Computational Intelligence, vol. 1026, pp. 285–295. Springer, Cham (2022). https://doi.org/10.1007/978-3-030-96627-0_26
5. Carfora, P., Ferrante, S., Quilici Gigli, S.: Norba : la strada per il santuario di giunone : note topografiche e indagini archeologiche, pp. pp. 151–182 (2009)

6. Carinci, F., Leopardi, A.: La strada che entra da porta maggiore: saggi di scavo negli anni 2010–2012, pp. 175–186 (2016). https://doi.org/10.1400/259593

7. Chiabá, M.: Roma e le priscae latinae coloniae. Ricerche sulla colonizzazione del Lazio dalla costituzione della repubblica alla guerra latina, 1 edn. EUT-Edizioni Università di Trieste, Trieste, Italia (2011). http://hdl.handle.net/10077/6083. Anno accademico: [Anno]

8. Finn, M.: Computer games and narrative progression. M/C J. **3**(5) (2000). https://journal.media-culture.org.au/index.php/mcjournal/article/view/1876

9. Quilici, L., Quilici Gigli, S.: Norba. la monumentalizzazione tardo repubblicana dell'acropoli maggiore. In: Atlante Tematico di Topografia Antica, vol. 7, pp. 240–246. L'Erma di Bretschneider, Roma (1998)

10. Quilici Gigli, S.: L'acropoli minore e i suoi templi. In: Atlante Tematico di Topografia Antica, vol. 12, pp. 289–322. L'Erma di Bretschneider, Roma (2003)

11. Quilici Gigli, S.: Il parco archeologico di norba. In: Lazio e Sabina. Scoperte Scavi e Ricerche, vol. 5, pp. 437–443. L'Erma di Bretschneider (2009)

12. Quilici Gigli, S.: Edilizia privata medio-tardo repubblicana: documentazione da norba. In: CIAC, vol. XVIII, pp. 67–70. Museo Nacional de Arte Romano, Mérida (2014)

13. Quilici Gigli, S.: Triclini e letti bronzei: testimonianze da norba sulla recezione culturale e materiale in ambito locale. Orizzonti **XVI**, 21–30 (2014)

14. Quilici Gigli, S.: Archeologia nel parco. In: Atlante Tematico di Topografia Antica, vol. XXI supp., pp. 9–17. L'Erma di Bretschneider, Roma (2016)

15. Quilici Gigli, S.: Topografia, scavo, indagini geofisiche, valorizzazione. esperienze integrate a norba. Rendiconti della Pontificia Accademia di Archeologia, pp. 291–334 (2021)

16. Rescigno, C.: Norba: Santuario di giunone lucina. appunti topografici. In: Atlante di Topografia Antica, vol. 12, pp. 336–351. L'Erma di Bretschneider, Roma (2003)

XR Theater: An Experience

Mario Alessandro Bochicchio[1] and Eleonora Miccoli[2(✉)]

[1] Dipartimento di Informatica, Università degli studi di Bari, Bari, Italy
mario.bochicchio@uniba.it
[2] Dipartimento di Ingegneria dell'innovazione, Università del Salento, Lecce, Italy
miccolieleonora@gmail.com

Abstract. Live theatrical performances constitutes an evolving art form that, since its earliest performances, has made extensive use of techniques and stage effects in search of effective ways to connect and engage audiences. Extended reality (XR) fits well into this search by multiplying the creative and expressive potential of performers and allowing interesting forms of interaction between them and the audience. This is the context of the RASTA project for the promotion of cultural heritage and the development of tourism in Italy, which includes, among other actions, the realization of extended reality theater performances in Apulia. This article reports the preliminary results of the requirements definition phase of the software system proposed by RASTA for the creation and staging of the extended reality theater productions envisaged by the project.

1 Motivations and Background

The work described in this article is about a research project funded by the Italian Ministry of University and Research within the "PON 12 aree" framework, i.e. the National Program on Industrial Research and Experimental Development in the 12 Areas of Smart Specialization. The project name is "RASTA - Augmented Reality and Automated Storytelling for the Promotion of Cultural Heritage and Tourist Itineraries", and the area of smart specialization is "Cultural Heritage".

The objective of RASTA is to define a technological framework for the promotion of cultural heritage and tourism in Italy, also through extended reality, intelligent storytelling technologies for the personalization of the narrative, and sophisticated user recommendation and profiling systems.

In this context, the work described in this paper is about the preliminary results produced by the research unit of the University of Salento, which is working on "Tourist itineraries in Apulia." The focus of our Research Unit is on the production, in extended reality, of three theatrical events about the history of three iconic Apulian medieval castles. Here we describe the results of the requirement analysis in terms of use cases and sequence diagrams of the subsystem we are creating to produce the extended reality theatrical events.

L. Barolli (Ed.): EIDWT 2024, LNDECT 193, pp. 639–648, 2024.
https://doi.org/10.1007/978-3-031-53555-0_61

2 Related Works

Since its inception, theater has paid great attention to stage technologies to bring different environments to the stage or to create effects such as the intervention of deities in human affairs (deus ex machina). Subsequently, developments in stage machinery and lighting techniques based on gas and then electric lamps have profoundly influenced the way theater and all performing arts are done. In this sense, the use of digital technologies to generate immersive environments, virtual objects and characters, new forms of interaction between actors, props and audiences, and acoustic effects in theater stands as a sign of continuity and is the subject of experimentation and widespread use by numerous groups in many theatrical productions.

In [1], for example, Vasilakos et al. describe a modern interactive theater based on mixed reality and ambient intelligence techniques designed to provide performers with creative tools that extend the grammar of traditional theater. Actors and dancers in different locations are filmed by multiple cameras and their images are rendered in 3D allowing them to be able to perform and dance in the same virtual location in real time.

In [2], similarly, Bernini et al. define the concept of augmented choreography in which virtual actors perform together with live performers to generate perceptions that influence the environment itself and the virtual actors.

In [3], Marner et al. present an augmented reality play in which the audience, interacting with the real and virtual actors, votes on how to move the stage action forward. The set is represented as a 3D virtual environment. Actors are tracked by sensors and cameras as they move around the stage, and projected content is controlled by their movements. Audience voting results are projected directly into the virtual environment.

A wide-ranging and up-to-date discussion of the relationship between digital technologies and theater can be found in [4], which understands live theater performance as an evolving art form in which augmented reality (AR) offers new ways to connect and engage the audience and expand stage potential. In addition, the addition of 3D tracking and interactive projections enables new opportunities for theatrical performance and storytelling.

In [5], Mavridis et al. explore the use of robotics and holograms to create interactive extended-reality theatrical performances. The proposal is still in its early stages but hints at the technological, creative, and artistic potential of this performance mode realized in the form of a theater space that includes multiple sensors, a screen, a pseudo-3D holographic transparency, an audience area, a humanoid robot, and human actors. The screen and hologram can display static and moving images, online virtual worlds populated by characters, or a mix of these elements. The robotic and virtual characters can be autonomous, partially autonomous, puppet, scripted, or controlled in real time by human actors (embodied telepresence). In addition, multiple modes of remote participation can be supported (e.g., through video conferencing, remote control of robots and/or virtual characters, etc.).

Finally, in [6], Papathanasiou et al. explore the use of combined artificial-intelligence, location-based narratives, and user participation as means of creating effective extended reality XR experiences of impact and increasing empathy and engagement.

3   XR Theater: Requirements

Thirteen months after the kickoff, as a preliminary outcome of the RASTA project, within the framework of an architecture agreed upon by all project partners, our research unit collected and formalized the requirements for the system for the realization of interactive theatrical events in extended reality. The requirements gathering was conducted through state-of-the-art analysis and interactive discussion sessions (focus groups) with Salento theater directors, actors, and producers.

A summary description of the use cases and sequence diagrams in standard UML follows.

3.1   Use Case: Augmented Reality Performative Action Composition

The use case (UC) represented in Fig. 1 defines activities for creating an enriched theatrical performance through augmented, immersive, group reality techniques. It starts with the composition or choice of a script, and then creates the actual virtual environment. The definition of the theatrical performance is done by successive refinements, and the system allows interaction between the three main figures involved in this phase: script writer, content creator, performer.

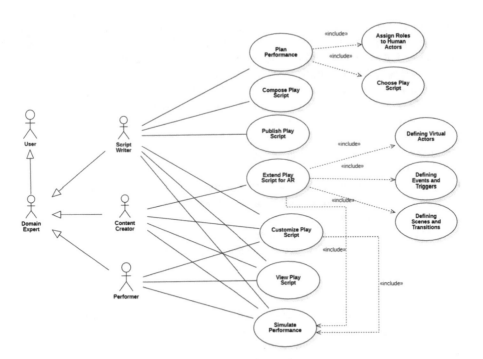

Fig. 1. Augmented Reality Performative Action Composition.

Script Writer UC:

- Plan Performance. Plan the theatrical performance by choosing what to stage and assigning each actor the characters to play.
- Compose Play Script. Compose the script (in text format).
- Publish play script. Once the script creation phase is completed, the script is made available to the actors who are to play it and to the content creator who will be responsible for creating the virtual environment.
- Customize Play Script. The script writer may receive requests to modify and customize the script from the performer and the content creator. He evaluates the suggestions received and, if he finds them valid, makes changes to the script to be staged. It may request changes in the sets and in general 3D contents and interactions made by the content creator, for example, after simulating the performance.
- View Play Script. The script writer can view/edit the script.
- Simulate Performance. Once the content creator has finished creating the virtual environment, the script writer can simulate the performance to check whether it is in line with what he had in mind at the time of creation and whether there are any changes to be made.

Content Creator UC:

- Extend Play Script for AR. Once the script writer has decided which script he wants to put on stage, the Content Creator comes into play, who takes care of extending the script with augmented reality content, i.e., creates the so-called virtual environment composed of sets and/or virtual actors. The Content Creator, following the directions in the script made available by the script writer, deals with the technical details related to the creation of a theatrical performance enriched with augmented reality content. Therefore, he/she deals with the definition of the Augmented Reality scenes, the ways of transition between one scene and another, the creation of the virtual actors, the definition of events and triggers, the ways in which, both the performers and the audience, will be able to interact with the system, and the part related to immersive video-mapping.
- Customize Play Script. Allows the content creator to receive modification requests made by performers and script writers. Evaluates the requests received and decides whether to make changes. Can request script writer for changes to the script.
- View Play Script. Can view/comment the script.
- Simulate Performance. Can perform technical rehearsals by simulating the performance.

Performer UC:

- Customize Play Script. Performers can send editing and customization requests to both the script writer and the content creator.
- View Play Script. Can view the script.
- Simulate Performance. Once the content creator has finished creating the virtual environment, performers can simulate the performance to see if there are any changes that need to be made or simply to test.

3.2 Use Case: Augmented Reality Performative Action

The use case defines the activities required to stage a theatrical performance enriched with augmented reality content and automated storytelling. The use case is shown in Fig. 2. There are two main actors:

- Performer who enacts the theatrical action.
- Spectator witnesses the theatrical action by taking part in it through interactions with the system guided by the performer.

An important entity in this use case is also the virtual performer (agent) i.e., a software component that is responsible for simulating an interaction with a virtual character. This interaction can take place in different ways, with voice, with gestures, with body language, with sensor-operated commands, etc.

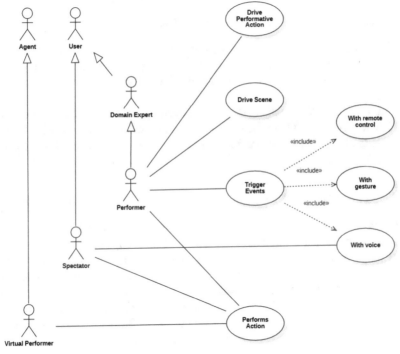

Fig. 2. Augmented Reality Performative Action.

Performer UC:

- Drive Performative Action. The performer is responsible for driving the narrative by acting at a high level i.e., deciding what to enact at a certain time based on the audience itself or its reactions.
- Drive Scene. He/she can decide to customize the narrative by choosing to move from one scene to another by varying the order to his/her liking.

- Trigger Events. The performer is provided with equipment consisting of sensors, remote controls, cameras, microphones, etc., through which he or she can guide the stage action by operating events in the narrative.
- Performs Action. Three entities actively participate in the performative action: the performer, the spectator, and one or more virtual performers (who are agents guided by system logic and user-triggered events). The performer can interact in different ways with the virtual performers.

  Spectator UC:

- Trigger Events with voice. The spectator can interact with the system through voice but only when the performer so decides.
- Performs Action. The spectator actively participates in the performative action by interacting through gestures with the virtual environment or virtual performers when the human performer so decides.

3.3   Sequence Diagram: Play Script Composition for Script Writer

The Play Script composition for Script Writer is shown in Fig. 3. The Script Writer, using the Story Composer interface, composes a new Play Script that is saved within the Stories DB. Once the Play Script is finished, the Script Writer publishes it in the Stories DB.

Next, the Script Writer plans the performance by choosing the script from those published and adapting it to the actor and the location where the performance will take place, via the Story Composer interface, then saving the information entered into the Stories DB.

3.4   Sequence Diagram: Virtual Environment Creation for Content Creator

The virtual environment creation for Content Creator is shown in Fig. 4. The Content Creator consults the list of planned performances by selecting one of them, thanks to the Story Composer interface that gets the information from the Stories DB. Referring to the information in the Play Script, he creates the virtual environment which includes one or more scene, made of zero or more virtual objects and virtual actors, also defining how to interact with them. To do this, it can download 3D models (props, reconstructions of buildings, characters, etc.) from the 3D-DB and use the Story Composer interface that dialogues with the 3D Scene Builder to recreate the various scenes in the script.

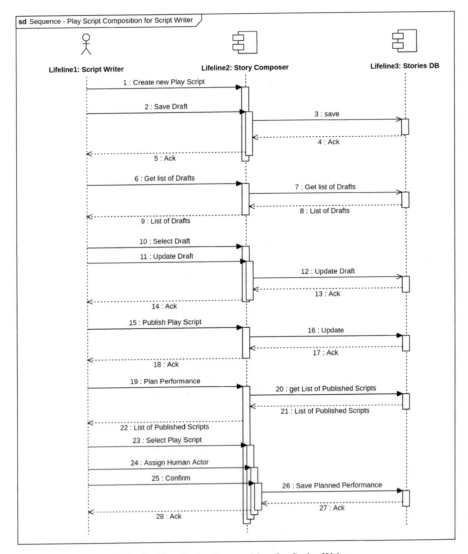

Fig. 3. Play Script Composition for Script Writer.

3.5 Sequence Diagram: Customize Play Script

The Customize Play Script is shown in Fig. 5. The domain expert (script writer, content creator and performer) can simulate the performance using the 3D Stories Visualizer. If they decide that changes are appropriate, they create one or more requests for the Script Writer and/or Content Creator. These requests are saved in the stories DB.

The Script Writer and/or Content Creator opens the list of requests and decides whether or not to make changes to respond to the Domain Expert who made the request.

Then, they make the required changes saving them in a new version of the Play Script.

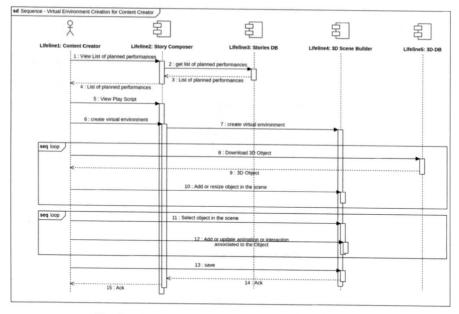

Fig. 4. Virtual Environment Creation for Content Creator.

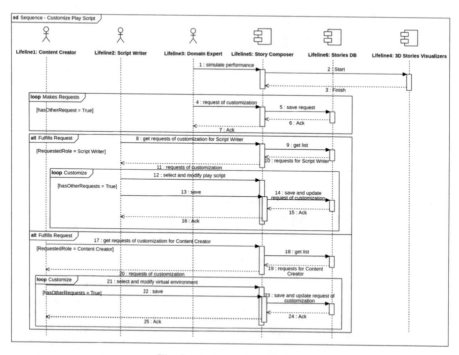

Fig. 5. Customize Play Script.

3.6 Sequence Diagram: Perform Augmented Reality Performative Action

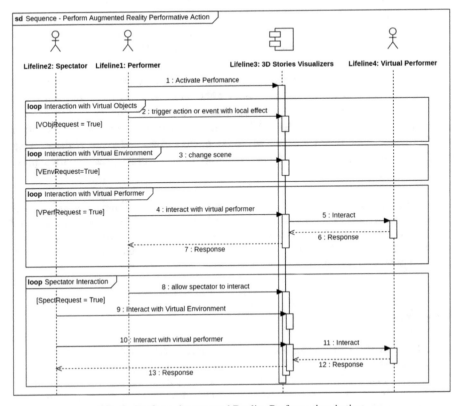

Fig. 6. Perform Augmented Reality Performative Action.

The Augmented Reality Performative Action is shown in Fig. 6. The performer uses the 3D Stories Visualizer to stage the performative action. He can guide the action by choosing what to stage based on the audience and their reactions.

He can guide the scenes by changing their order as he wishes. It can trigger events in the story using different input modes. Can interact with Virtual Performers in different ways.

At the same time the spectator can interact with the system when the performer decides.

4 Conclusions and Future Developments

The paper reports, in standard UML, the description of the main requirements for a system for preparing and staging interactive theater shows in extended reality, describing its main use cases and sequence diagrams.

In the following phases, the RASTA project involves the detailed definition of the hardware and software components of the described system, the implementation of the software components, the acquisition of the hardware components, the development/integration of the software system, the realization of three extended-reality theatrical performances in Apulia, at the Castle of Corigliano D'Otranto, the Fortified Citadel of Acaya, and the Castle of Carlo V in Lecce, and the evaluation of the audience and stakeholders' satisfaction.

Specifically, for the implementation aspects, the detailed structure of the software is being defined, and the technical tests of integration between the subsystems Text To Speech, Speech To Text, Large Multimodal Model, 3D Rendering and Interactions, Immersive Projections, Gesture Recognition, Retrieval Augmented Generation, and Digital Content Repository are being conducted. Also underway is the retrieval and digitization of iconographic, textual material, period artifacts, and theatrical scene settings.

Acknowledgments. The RASTA project is funded by the Ministry of-University and Research from the resources of PON "Research and Innovation" 2014–2020 and FSC Funds pursuant to and in accordance with Art. 13 of the Notice for the submission of Industrial Research and Experimental Development projects in the 12 areas of smart specialization identified by the PNR 2015–2020 and Art. 1 of DD No. 551 of April 27, 2020.

References

1. Vasilakos, A.V., et al.: Interactive theatre via mixed reality and Ambient Intelligence. Information Sciences **178**(3), 679–693 (2008). https://doi.org/10.1016/j.ins.2007.08.029
2. Bernini, D., De Michelis, G., Plumari, M., Tisato, F., Cremaschi, M.: Towards augmented choreography. In: Brooks, A.L. (ed.) Arts and Technology. LNICSSITE, vol. 101, pp. 10–17. Springer, Heidelberg (2012). https://doi.org/10.1007/978-3-642-33329-3_2
3. Marner, M.R., Haren, S., Gardiner, M., Thomas, B.H.: Exploring interactivity and augmented reality in theater: a case study of Half Real. In: 2012 IEEE International Symposium on Mixed and Augmented Reality-Arts, Media, and Humanities (ISMAR-AMH), pp. 81–86. IEEE, November 2012
4. Lisowski, D., Ponto, K., Fan, S., Probst, C., Sprecher, B.: Augmented Reality into Live Theatrical Performance. In: Nee, A.Y.C., Ong, S.K. (eds.) Springer Handbook of Augmented Reality, pp. 433–450. Springer International Publishing, Cham (2023). https://doi.org/10.1007/978-3-030-67822-7_18
5. Mavridis, N., Hanson, D.: The IbnSina center: an augmented reality theater with intelligent robotic and virtual characters. In: RO-MAN 2009 - The 18th IEEE International Symposium on Robot and Human Interactive Communication, Toyama, Japan, pp. 681–686 (2009). https://doi.org/10.1109/ROMAN.2009.5326148
6. Papathanasiou, A., Chavez, N.T., Bokovou, P., Chrysanthopoulou, C., Gerothanasi, A.D., Dimitriadi, N.M.: Artificial intelligence, augmented reality, location-based narratives, and user participation as means for creating engaging extended reality experiences. In: Proceedings of 5th International Conference Digital Culture & AudioVisual Challenges (DCAC-2023), 14p. (2022)

Author Index